Ecosystem Monitoring: Collective Species and Environmental Information

Ecosystem Monitoring: Collective Species and Environmental Information

Editors

Matteo Convertino
Jie Li

MDPI

Basel • Beijing • Wuhan • Barcelona • Belgrade • Novi Sad • Cluj • Manchester

Editors

Matteo Convertino
Tsinghua University
Shenzhen, China

Jie Li
University of Amsterdam
Amsterdam, The Netherlands

Editorial Office
MDPI
St. Alban-Anlage 66
4052 Basel, Switzerland

This is a reprint of articles from the Topic published online in the open access journals *Entropy* (ISSN 1099-4300), *Land* (ISSN 2073-445X), *Environments* (ISSN 2076-3298), and *Remote Sensing* (ISSN 2072-4292) (available at: https://www.mdpi.com/topics/Ecosystem_Monitoring).

For citation purposes, cite each article independently as indicated on the article page online and as indicated below:

Lastname, A.A.; Lastname, B.B. Article Title. *Journal Name* **Year**, *Volume Number*, Page Range.

ISBN 978-3-0365-9260-2 (Hbk)
ISBN 978-3-0365-9261-9 (PDF)
doi.org/10.3390/books978-3-0365-9261-9

Cover image courtesy of Matteo Convertino

Contents

About the Editors

Matteo Convertino

Matteo Convertino is an associate professor at Tsinghua University, Shenzhen International Graduate School, where he is the principal investigator of the fuTuRE EcoSystems Lab (TREES) and the Associate Director of the State Key Lab of Ecological Remediation and Carbon Sequestration. His expertise is in ecosystem patterns, networks and flows, ecosystem health, biocomplexity engineering, ecosystem data science and analytics, ecological forecasting, systemic risk and strategic portfolio management, biogeomorphology, and socio-eco-hydro-climatology. Convertino's work is focused on developing nature-based solutions to counter ecological imbalances, namely ecohydrological engineering for restoration and protection, and policies that enhance the water and carbon cycles and biodiversity. His work is based on the identification of optimal ecological connections and flows (dynamical trees) constituting the backbone of ecosystem health. The current applications focus on coastal and urban ecosystems from a holistic basin perspective where everything flows and can be positively engineered by trading off risks and decisions. Such research is performed via pattern-oriented theoretical and computational models based on information and network sciences which develop into digital ecosystem models for guiding scientific investigation and eco-engineering applications.

Dr. Convertino has produced 95+ publications and technical reports. Dr. Convertino, among others, was granted the Pengcheng Peacock Talents Award from the Shenzhen government (2022–2024), the Ministry of Science and Education Foreign Talents Award (2021), the SOUSEI top 20% performing scientist award from Hokkaido University, the 2016 Top 10 Team honorific mention for the Dengue and Influenza Forecasting Challenge sponsored by the Office of Science and Technology Policy (US White House), and the 2011 Young International Research Scientist Fellowship from the Chinese Academy of Sciences.

Jie Li

Jie Li is a postdoctoral researcher at the University of Amsterdam, where his research focus is on building information theoretical higher-order network models and applying the models to unravel the intricacies of multiplex disease networks within the scope of an EU-funded project. The major goal of the EU project is to understand the causative mechanisms underlying the comorbidity of cardiovascular diseases and depression and identify the significant biomarkers responsible for the development of the complex comorbidity. The core of his work is to infer higher-order interactions based on information theory and develop novel methods to analyze a comprehensive disease network combining high-order and pairwise interactions into one graph. Dr. Li's current research interests lie primarily in higher-order relationships, information theory, multiplex disease networks, and their applications in biomedical fields related to CVD and depression comorbidities. His expertise traverses a diverse spectrum, from bio-complexity engineering to ecological pattern forecasting, from digital signal processing, data analytics, and visualization to causal inference and network-based analyses. Dr. Li's research efforts have resulted in multiple impactful papers, technical reports within the project framework, and presentations at international conferences. Prior to being a postdoctoral researcher, he completed his Ph.D. at Hokkaido University, where he was involved in a research project on information dynamics for complex ecosystem prediction and design led by Dr. Matteo Convertino.

Preface

The collective behavior of species is the by-product of species interactions and habitat structural organization and flows, all shaped by evolution and systemic environmental pressure. This universal equation is hidden by data whose availability, uncertainty, and relevance may not allow us to fully predict the ecological patterns considered, a by-product of collective behavior. This is why this Topic aims to highlight the saliency of data and the nexus between data, patterns, and eco-environmental determinants, including methodological advancement to extract salient features from data. Specifically, ecological information is key to assessing ecosystem health, performing robust predictions of ecosystem function from water to carbon flow, and extracting indicators for precise ecosystem management, planning, and engineering.

Matteo Convertino and Jie Li
Editors

environments

MDPI

Editorial

Sensing Linked Cues for Ecosystem Risk and Decisions

Matteo Convertino [1,2]

[1] fuTuRE EcoSystems Lab (TREES), Institute of Environment and Ecology, Tsinghua Shenzhen International Graduate School, Tsinghua University, Shenzhen 518055, China; matteo@sz.tsinghua.edu.cn

[2] Shenzhen Key Laboratory of Ecological Remediation and Carbon Sequestration, Tsinghua Shenzhen International Graduate School, Shenzhen 518055, China

Ecological indicators of ecosystem anomalies are fundamentally important to sensing how close we are to slow or catastrophic ecosystem shifts and to targeting systemic controls for preservation, restoration and eco-based development. Ecosystem anomalies, I argue, are grounded in *ecohydrological determinants* and lead to alterations in socio-ecological functions and services, including the collapse of species or hydroclimatological disasters such as floods, droughts and heatwaves on land and in the ocean. Therefore, linked ecological cues in the form of multiscale data are salient for predicting the risk of ecological change.

The aim of this Special Issue was to gather advances in ecosystem monitoring and monitored data, including technology and ecological data (phenotypical, phylogenetic, eDNA, macroecological, etc.), data fusion, pattern reconstruction and analysis, and inference models for the extraction of predictive information aimed at guiding ecosystem engineering (integrated ecological and environmental engineering), considering both predictions and field restoration.

The centrality of data must be seen as connected data as follows.

(1) Ecological *data* address biodiversity and water as green-blue foundational elements beyond biogeochemical fluxes that are the byproducts of the baseline ecological configuration. Species sense the quality of the environment, and ecological data reflect the functioning of eco-environmental ties. There is no environment that is fully abiotic, and yet efforts to compile ecological data must be comprehensive of the flows of ecosystems over time;

(2) The spatial connections among habitats (natural and self-emergent habitats and those of human-made *design*, which are reflected in geomorphological and infrastructural data, respectively) are the basis of any ecological function with strong climate feedback; thus, "climate neutral" efforts must consider the engineering of salient hydrologic flows and eco-geomorphological connections (broadly defined as *ecological ties*) whose scale-free organization is the optimal configuration of our ecosystem;

(3) Networks of people's *decisions*, from the behavior of citizens to stakeholder development and management strategies, are critical for an ecosystem's function and intelligence, in which the latter is as much a conscious action as the reactions of species to information sensed in ecosystems. All these decisions are associated with ecological information (extracted by models as perceptrons) for which digitized information carries values and thresholds with respect to the functions of ecosystems to create forecasts, assess indicators and ecosystem states and define ecosystem services and controls (what is needed and/or desired, for which the definition of optimal trade-offs is essential).

Despite their tremendous importance for understanding the function, integrity, and future trajectories of biodiversity, *ecological networks* (or, more broadly, ecological ties) are traditionally restricted to the biological interactions of species. However, ecological networks represent the structures of food webs, hydro-bio-geochemical/energy flows, and the many and diverse types of interactions between all species in ecosystems the

Citation: Convertino, M. Sensing Linked Cues for Ecosystem Risk and Decisions. *Environments* **2023**, *10*, 169. https://doi.org/10.3390/environments10100169

Received: 11 September 2023
Accepted: 19 September 2023
Published: 29 September 2023

underpinning ecosystemic function that defines fitness and risks. Multilayer networks, sensu lato, are connecting people, habitats, and climate with feedback that affects our conscious and unconscious behaviors, health, evolution and existence in the long term. In general, any tie, or set of knots, is ecological information about biotic components in "abiotic" environments that we need to sense, map and frame.

Can we infer visible and invisible collective networks from ecosystem patterns? More importantly, can we intelligently engineer salient eco-hydro-geomorphological networks to adaptively optimize our collective (biodiverse) beliefs and decisions, enhancing climate/human-impacted ecosystem services? Can we design key indicators, controls, plans, portfolio investments and policies for our desired future ecosystems? Indeed, we can, and we must.

Various initiatives are targeting global information gathering of ecological communities and their restoration, such as GeoBON (https://geobon.org/ebvs/indicators/) (accessed on 10 September 2023), Restor (https://restor.eco/), Allen Coral Atlas (https://allencoralatlas.org/), the UN Biodiversity Lab (https://unbiodiversitylab.org/en/), Global Forest Watch (https://www.globalforestwatch.org/), NEON (https://www.neonscience.org/), BioTIME (https://biotime.st-andrews.ac.uk/), the Living Planet (https://www.livingplanetindex.org/), PREDICTS (https://www.nhm.ac.uk/our-science/our-work/biodiversity/predicts.html) and GBIF (https://www.gbif.org/). Global environmental databases such as BioClim (https://www.worldclim.org/data/bioclim.html), WorldClim (https://www.worldclim.org/), Copernicus Climate Data Store (https://cds.climate.copernicus.eu/#!/home), and NOAA Climate Data (https://www.ncei.noaa.gov/cdo-web/) address the "abiotic" spheres of ecosystems. It is desirable that these eco-environmental databases are used together to pinpoint risks and solutions to global challenges, considering local–global "butterfly effects" in space–time (i.e., ecological ties).

In this Special Issue, many papers highlighted data and methods used to infer patterns across multiple scales and ecosystems, as well as to provide solutions, including predictive capabilities. For marine ecosystems, the delicate nature of the phytoplankton–environmental nexus was highlighted is in determining the extent and persistence of algal blooms [1], and the ways in which the phenology of coastal vegetation in a cold temperate intertidal system impacts remote sensing (and the subsequent classification of coastal habitats) was addressed [2]. Both studies actually emphasize how ecological conditions affect the information that can be gathered and yet add intrinsically uncontrollable (but measurable) uncertainty into monitoring technology; this is rather important and unappreciated since a large number of scientists and policy makers assume that all data are the undisputable, golden truth. This far from reality, and data fusion and selection should be dynamical processes based on the value of information constrained via predictive patterns predict.

Other papers showed the potential of extracting vegetation information from tree attributes [3] to study gross ecosystem production [4] and plant seasonal phenomena like flowering [5]. More importantly, several studies highlighted the critical role of hydrogeomorphology in shaping vegetation patterns [6] by also introducing new methods such as the use of a "geodetector" [7] which includes spatial and risk dependencies. Species have been shown to be bioindicators of ecosystem structures, such as geese for basin vegetation [8] and fish in rivers, which are also affected by climate and other anthropogenic factors [9].

Hydrological dynamics was also studied in its complexity, considering river runoff [10] and its consequences when poorly managed, i.e., floods [11]. Hydrological dynamics which also experience variability due to changes in temperature extremes can trigger wildfires [12] in water-depleted landscapes where vegetation is largely combustible.

The roles of human decisions, such as land management practices, which are largely affecting woody invasive species [13] as undesired species, and human disturbances like mines, which alter vegetation [14], are critical in positive and negative human–ecological feedback respectively. Capturing this feedback is necessary, including in important natural world heritage sites, such as through remote sensing [15]. The advancement and refinement of methods in treating ecological data, for example, for tracking salient changes in species distributions [16], is constantly important due to the availability of new technology such as satellite imagery [17] and small-scale biological data [1].

In conclusion, ecological data are the sine qua non condition for making optimal *ecosystem decisions* in which the *collective design and engineering* of ecological components (changing an ecological structure by taking advantage of species' collective behaviors and human enhancements) optimizes systemic function. We argue that we must transition from a reductionist way of thinking to consequentialist thinking in which data-informed, nature-based patterns are the ultimate objective achieved via optimal strategic decisions. Top-down ecosystem inputs (natural flows and infrastructure) coupled with well-placed bottom-up ecological components and enhancers create self-organized habitats and ecosystems: this is *Pareto optimal dynamics*, leading to scale-free ecological patterns.

This is particularly important when thinking about the future climate and the co-existence of natural and future human habitats which support each other in risks and needs. The collectivity of data, design (natural and human-made) and decisions is necessary for all ecosystems in which we are the primary *ecosystem engineers*.

Acknowledgments: We wish to thank the authors for their contributions and their willingness to share innovative ideas and methods in this Special Issue. In addition, we would like to express our appreciation to the reviewers for the considerable amount of time they invested in providing accurate and fair manuscript evaluations. Finally, we would like to express our pleasure in working with staff of the *Entropy* Editorial Office Entropy for this fruitful and excellent cooperation. M.C. acknowledges the SZ Pencheng Peacock Talents funding, B class.

Conflicts of Interest: The author declares no conflict of interest.

References

1. Wang, H.; Galbraith, E.; Convertino, M. Algal Bloom Ties: Spreading Network Inference and Extreme Eco-Environmental Feedback. *Entropy* **2023**, *25*, 636. [CrossRef] [PubMed]
2. Légaré, B.; Bélanger, S.; Singh, R.K.; Bernatchez, P.; Cusson, M. Remote Sensing of Coastal Vegetation Phenology in a Cold Temperate Intertidal System: Implications for Classification of Coastal Habitats. *Remote Sens.* **2022**, *14*, 3000. [CrossRef]
3. Vivar-Vivar, E.D.; Pompa-García, M.; Martínez-Rivas, J.A.; Mora-Tembre, L.A. UAV-Based Characterization of Tree-Attributes and Multispectral Indices in an Uneven-Aged Mixed Conifer-Broadleaf Forest. *Remote Sens.* **2022**, *14*, 2775. [CrossRef]
4. Zhao, N.; Wang, H.; Zhong, J.; Bai, Y.; Yi, S. Evaluation of the Gross Ecosystem Product and Analysis of the Transformation Path of "Two Mountains" in Hulunbuir City, China. *Land* **2023**, *12*, 63. [CrossRef]
5. Jolly, B.; Dymond, J.R.; Shepherd, J.D.; Greene, T.; Schindler, J. Detection of Southern Beech Heavy Flowering Using Sentinel-2 Imagery. *Remote Sens.* **2022**, *14*, 1573. [CrossRef]
6. Sun, X.; Xiao, Y. Vegetation Growth Trends of Grasslands and Impact Factors in the Three Rivers Headwater Region. *Land* **2022**, *11*, 2201. [CrossRef]
7. Yao, B.; Ma, L.; Si, H.; Li, S.; Gong, X.; Wang, X. Spatial Pattern of Changing Vegetation Dynamics and Its Driving Factors across the Yangtze River Basin in Chongqing: A Geodetector-Based Study. *Land* **2023**, *12*, 269. [CrossRef]
8. He, K.; Lei, J.; Jia, Y.; Wu, E.; Sun, G.; Lu, C.; Zeng, Q.; Lei, G. Temporal Dynamics of the Goose Habitat in the Middle and Lower Reaches of the Yangtze River. *Remote Sens.* **2022**, *14*, 1883. [CrossRef]
9. Arciszewski, T.J.; Ussery, E.J.; McMaster, M.E. Incorporating Industrial and Climatic Covariates into Analyses of Fish Health Indicators Measured in a Stream in Canada's Oil Sands Region. *Environments* **2022**, *9*, 73. [CrossRef]
10. Liu, X.; Zhao, H. Multiscale Analysis of Runoff Complexity in the Yanhe Watershed. *Entropy* **2022**, *24*, 1088. [CrossRef] [PubMed]
11. Zhang, W.; Liu, X.; Yu, W.; Cui, C.; Zheng, A. Spatial-Temporal Sensitivity Analysis of Flood Control Capability in China Based on MADM-GIS Model. *Entropy* **2022**, *24*, 772. [CrossRef] [PubMed]
12. Petrie, M.D.; Savage, N.P.; Stephen, H. High and Low Air Temperatures and Natural Wildfire Ignitions in the Sierra Nevada Region. *Environments* **2022**, *9*, 96. [CrossRef]
13. Adoyo, B.; Schaffner, U.; Mukhovi, S.; Kiteme, B.; Mbaabu, P.R.; Eckert, S.; Choge, S.; Ehrensperger, A. Pathways towards the Sustainable Management of Woody Invasive Species: Understanding What Drives Land Users' Decisions to Adopt and Use Land Management Practices. *Land* **2022**, *11*, 550. [CrossRef]
14. Ali, K.; Khan, N.; Ullah, R.; Shah, M.; Khan, M.E.H.; Jones, D.A.; Dewidar, M. Spatial Pattern and Key Environmental Determinants of Vegetation in Sand Mining and Non-Mining Sites along the Panjkora River Basin. *Land* **2022**, *11*, 1801. [CrossRef]
15. Zhang, N.; Xiong, K.; Xiao, H.; Zhang, J.; Shen, C. Ecological Environment Dynamic Monitoring and Driving Force Analysis of Karst World Heritage Sites Based on Remote-Sensing: A Case Study of Shibing Karst. *Land* **2023**, *12*, 184. [CrossRef]

16. Muhammed, K.; Anandhi, A.; Chen, G. Comparing Methods for Estimating Habitat Suitability. *Land* **2022**, *11*, 1754. [CrossRef]
17. Venter, Z.S.; Roos, R.E.; Nowell, M.S.; Rusch, G.M.; Kvifte, G.M.; Sydenham, M.A.K. Comparing Global Sentinel-2 Land Cover Maps for Regional Species Distribution Modeling. *Remote Sens.* **2023**, *15*, 1749. [CrossRef]

entropy

Article

Algal Bloom Ties: Spreading Network Inference and Extreme Eco-Environmental Feedback

Haojiong Wang [1], Elroy Galbraith [1] and Matteo Convertino [2,3,*]

[1] Laboratory of Information Communication Networks, Graduate School of Information Science and Technology, Hokkaido University, Sapporo 060-0814, Japan
[2] fuTuRE EcoSystems Lab (TREES), Institute of Environment and Ecology, Tsinghua Shenzhen International Graduate School, Tsinghua University, Shenzhen 518055, China
[3] Shenzhen Key Laboratory of Ecological Remediation and Carbon Sequestration, Tsinghua Shenzhen International Graduate School, Shenzhen 518055, China
* Correspondence: matteo@sz.tsinghua.edu.cn

Abstract: Coastal marine ecosystems worldwide are increasingly affected by tide alterations and anthropogenic disturbances affecting the water quality and leading to frequent algal blooms. Increased bloom persistence is a serious threat due to the long-lasting impacts on ecological processes and services, such as carbon cycling and sequestration. The exploration of eco-environmental feedback and algal bloom patterns remains challenging and poorly investigated, mostly due to the paucity of data and lack of model-free approaches to infer universal bloom dynamics. Florida Bay, taken as an epitome for biodiversity and blooms, has long experienced algal blooms in its central and western regions, and, in 2006, an unprecedented bloom occurred in the eastern habitats rich in corals and vulnerable habitats. With global aims, we analyze the occurrence of blooms in Florida Bay from three perspectives: (1) the spatial spreading networks of chlorophyll-a (CHLa) that pinpoint the source and unbalanced habitats; (2) the fluctuations of water quality factors pre- and post-bloom outbreaks to assess the environmental impacts of ecological imbalances and target the prevention and control of algal blooms; and (3) the topological co-evolution of biogeochemical and spreading networks to quantify ecosystem stability and the likelihood of ecological shifts toward endemic blooms in the long term. Here, we propose the transfer entropy (TE) difference to infer salient dynamical inter actions between the spatial areas and biogeochemical factors (ecosystem connectome) underpinning bloom emergence and spread as well as environmental effects. A Pareto principle, defining the top 20% of areal interactions, is found to identify bloom spreading and the salient eco-environmental interactions of CHLa associated with endemic and epidemic regimes. We quantify the spatial dynamics of algal blooms and, thus, obtain areas in critical need for ecological monitoring and potential bloom control. The results show that algal blooms are increasingly persistent over space with long-term negative effects on water quality factors, in particular, about how blooms affect temperature locally. A dichotomy is reported between spatial ecological corridors of spreading and biogeochemical networks as well as divergence from the optimal eco-organization: randomization of the former due to nutrient overload and temperature increase leads to scale-free CHLa spreading and extreme outbreaks a posteriori. Subsequently, the occurrence of blooms increases bloom persistence, turbidity and salinity with potentially strong ecological effects on highly biodiverse and vulnerable habitats, such as tidal flats, salt-marshes and mangroves. The probabilistic distribution of CHLa is found to be indicative of endemic and epidemic regimes, where the former sets the system to higher energy dissipation, larger instability and lower predictability. Algal blooms are important ecosystem regulators of nutrient cycles; however, chlorophyll-a outbreaks cause vast ecosystem impacts, such as aquatic species mortality and carbon flux alteration due to their effects on water turbidity, nutrient cycling (nitrogen and phosphorus in particular), salinity and temperature. Beyond compromising the local water quality, other socio-ecological services are also compromised at large scales, including carbon sequestration, which affects climate regulation from local to global environments. Yet, ecological assessment models, such as the one presented, inferring bloom regions and their stability to pinpoint risks, are in need of application in aquatic ecosystems, such as subtropical and tropical bays, to assess optimal preventive controls.

Citation: Wang, H.; Galbraith, E.; Convertino, M. Algal Bloom Ties: Spreading Network Inference and Extreme Eco-Environmental Feedback. *Entropy* **2023**, *25*, 636. https://doi.org/10.3390/e25040636

Academic Editor: Yoh Iwasa

Received: 8 February 2023
Revised: 4 April 2023
Accepted: 7 April 2023
Published: 10 April 2023

Keywords: spatial network inference; biogeochemical networks; predictive causality; bloom prediction; Florida Bay

1. Introduction

1.1. *Algal Blooms as the Epitome of Marine Ecosystem Health*

Algal blooms are a manifestation of abnormal changes in phytoplankton communities in aquatic ecosystems, such as estuaries and lakes [1,2]. Despite discussions on the perceived global increase in algal blooms attributable to intensified monitoring and emerging bloom impacts, these blooms are increasing worldwide as highlighted from satellite images by Dai et al. [3], and thus they are posing various concerns for the local ecology and global climate. Blooms are highly destructive and persistent [4,5], causing various ecological catastrophes, such as the eduction of vegetated communities, widespread sponge mortality and loss of marine habitat geomorphological structure [6] due to, for instance, habitat calcification [7].

Despite this tremendous damage to aquatic ecosystems, the mutual influence between blooms and the environment has received little attention from scientists and policy makers. Algal blooms are, in fact, the byproduct of nitrogen (N) and phosphorus (P) change but can alter the N/P balance [8] and temperature [9] with implications on carbon sequestration of vegetation in blue carbon ecosystems affected by blooms [10], such as for seagrass. All these elements can be exacerbated by local and global climate change [11].

Despite the limited literature of the effects of blooms into the environment, some studies have explored the relationship between phytoplankton and water quality in bloom conditions [12], climatic and regional variations in phytoplankton as characteristic features of blooms [13,14] and habitat-specific effects that vary by local planktonic biogeochemical stress [15]. Fewer studies have inferred the spatial spreading of blooms characterized as complex networks and predicted blooms based on spatially explicit biogeochemical factors.

This type of biocomplexity study, such as the one we propose, would be necessary to define micro–macro feedback useful for risk assessment, management and policies aimed to minimize eco-environmental imbalances, leading to a decrease in ecosystem health, such as due to blooms. Blooms are the epitome of marine ecosystem health because their emergence is largely related to altered ecohydrological factors at the basin-scale, from land and ocean, leading to quick and persistent increases in phytoplankton with short-term impacts [16] and long-lasting systemic effects on the ecological function and the environment. This is beyond one species or humans only and is related to the progressive degeneration of ecosystem function from its optimality or baseline in relation to the initial or desired conditions.

1.2. *Complex Marine Ecosystems*

Marine microbial food webs consist of heterotrophic protists, phytoplankton, prokaryotes and viruses (i.e., the ocean microbiome). Together, they are responsible for a large part of the production, respiration and nutrient transfer in oceans; they affect, for instance, the carbon cycle both in blue carbon habitats and in the ocean via the carbon pump. As marine ecosystems are increasingly affected by anthropogenic disturbances both from land and ocean, predicting ecosystem responses above critical environmental pressure relies on a better understanding community dynamics, including their composition, spatial/temporal distribution and interactions.

Long-term observations are especially useful for this, and both Galbraith and Convertino [15] and Galbraith et al. [17] provided clear ecological patterns to use as indicators of ecosystem health in relation to ocean microbiome variability intended as a complex network. Chlorophyll-a (CHLa) seems to be the best indicator of community health; however, currently there is the need to quantify how much CHLa variability implies changes in ecological effects (e.g., blooms) and long-term effects, such as on the environment and ecosystem function (e.g., carbon cycle).

Coastal and marine ecosystems that experienced marine heatwaves, which were particularly significant in 2014–2015 worldwide, provide a unique opportunity to study how warming affects community dynamics (namely, microbiome interactions) and how imbalance of the latter affects the environment back in the long term. The presented tool for ecosystemic risk assessment and the results from FL Bay are the main innovations of this paper.

The topological network structure is an effective and intuitive way to describe the dynamical dependencies among diverse of analogous units of an ecosystem, or ecological communities composed of hundreds or thousands of populations of species [18,19]. This is particularly important for marine ecosystems where both structural networks (such as coastal and marine habitat connections and flows) and functional networks (such as prokaryotic and eukaryotic interactions) are not directly visible or known.

Yet, causal network discovery and inference models (e.g., see Li and Convertino [20] for an articulated discussion about ecosystems) are particularly important for mapping the ecological baseline on which current ecosystem assessment and future predictions of ecosystem patterns (tangibly liked to ecosystem services) can be made. Complex networks have great potential to help in solving contemporary real-world problems in a wide range of fields [21–26].

Complex networks have been used to analyze the dynamics of pseudo-periodic time series [27] and the functional dynamics of complex systems [28–31]. Furthermore, networks have become an excellent method for the study of functional and structural dependencies among very complex units with different temporal dynamics [32–35].

However, most of the considered networks in the literature and particularly those inferred in ecosystems, typically represent relationships based on known or assumed affiliations [36,37] or fixed connections [38]. This makes it difficult to represent the independent local properties of each node and, more importantly, the unique dependencies between different nodes.

This issue is particularly relevant for algal blooms where the biogeochemical networks are hypothesized to vary dramatically over time and space. This has been verified by recent studies on prokaryotic networks whose topological variability was strongly related to systemic ecological stress [15,17]. Nonetheless, no analyses have been made so far on bloom spreading networks, and this research presents a novel template for characterizing and predicting algal blooms based on chlorophyll-a and associated water quality factors.

1.3. *Ecological Patterns as Chlorophyll-a Spreading Networks*

Species, including eukaryotes at the microscale, operate in dynamical ecosystems where the ability to respond to changing environmental flows is paramount. An effective collective response, affecting the re-balancing of optimal ecosystem states requires suitable information transfer among species; thus, ecosystems critically depend on eco-environmental interaction networks. This underpins the process of ecosystem evolution toward low entropy states (characterized by scale-free distribution of CHLa as investigated in this paper) [39] as well as adaptation to new environmental stress states [40], some of which can be undesired, such as those with persistent and large blooms. In this paper, we highlight the central role of information transfer as a salient feature for collective eco-environmental dynamics leading to algal blooms.

Connectomics is broadly defined as the study of structural and functional networks (the connectome), which are maps of a system (such as the nervous system), mainly in the brain; however, this concept has been extended to ecosystems (see Convertino and Valverde Jr. [41]) to characterize both functional species interaction networks, their stimuli with the environment or the envirome itself as set of interdependent environmental processes [15] and habitat networks [20]. The connectome enables understanding of how spreading information is processed (coded, stored, transmitted and decoded in an information sense, which can be any ecological information) at and among different scales of the system (e.g., one node and the whole system, while also considering cross-scale dependencies).

As the connectome is the salient information of ecosystems, its knowledge allows one to improve predictive skills in the short and long term to represent ecosystem patterns. For the aforementioned needs, i.e., to detect the trajectories of spreading blooms and their potential environmental impacts, we demonstrate the capability of an information-theoretic approach to infer bloom networks and biogeochemical feedback. The optimal information flow model was developed initially for inferring species interaction networks in any ecosystem from abundance data [20] and was later applied to predict fish biodiversity patterns Li and Convertino [42] and eco-environmental interactions of the ocean microbiome [17].

The ecological time series underpinning ecological dynamics are particularly important for assessing ecological states and early warning signals of shifts [43] before the inference of ecological networks. The proposed model applies transfer entropy (TE) differences (to target the salient directed interactions) to infer a spatial network strategy that can identify the sources and sinks of bloom outbreak as well as foretell changes probabilistically in water quality factors (in average and fluctuations) when blooms happen.

Through the model, we specifically infer and analyze the spatial ecological corridors determining bloom spread and direct interactions between CHLa and environmental factors to quantify the environmental effects of ecological dysbiosis; previous efforts (see Wang and Convertino [44]) focused on the whole set of biogeochemical interactions useful for forecasting outbreaks, except for bloom spreading networks.

Previously, CHLa has often been used as an indicator of blooms given its sensitivity to environmental changes, ease of monitoring and ability to reflect phytoplankton biomass effectively [45] but has not yet been verified as a systemic indicator of ecosystem health related to ecosystem function. We discuss the results of applying this model to algal blooms observed in Florida Bay (Florida, USA) in the Florida Everglades National Park between 2005 and 2006 when a recurrence of large blooms was observed.

Due to its unique lagoon configuration and climate, Florida Bay (Figure 1) regularly experiences algal blooms [46] as frequently as many other aquatic ecosystems in subtropical and tropical climates. Thus, for algal blooms, there is the need for a powerful dynamic prediction model to support decision-making and bloom prevention.

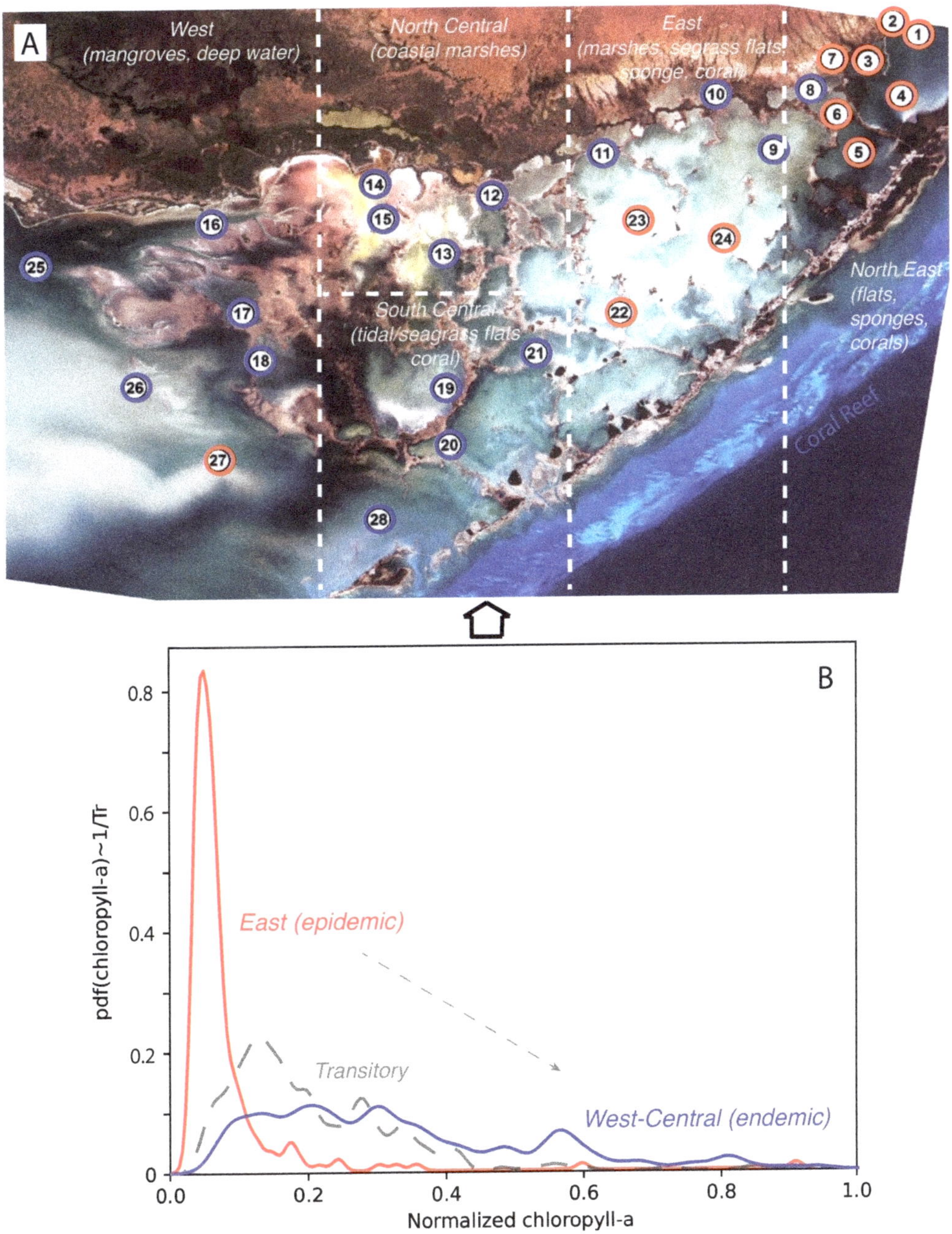

Figure 1. Florida Bay and area classification based on CHLa dynamics. The red–blue classification in plot (**A**) is related to the probabilistic structure of CHLa as highlighted in plot (**B**). Plot A also highlights the main habitats and species present in FL Bay.

2. Materials and Methods

The proposed TE network inference model that can be used for variable interaction discovery and prediction at multiple scales is explained. Its structure is graphically shown in Figure 2. The model is a further refinement on the one proposed by Li and Convertino [20] for the use of TE differences to prune the network and define salient predictors of ecological patterns, which are algal blooms in this case.

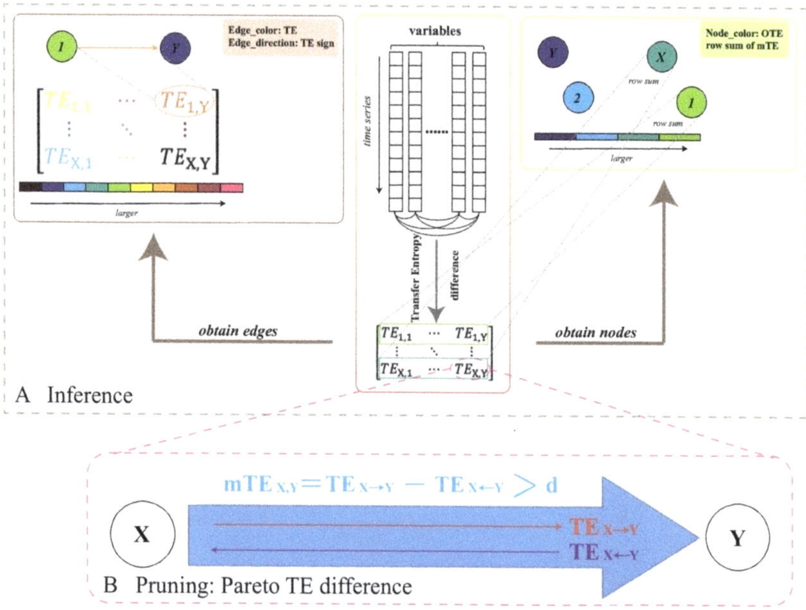

Figure 2. Ecological corridor inference model. The structure of the TE inference model. Here, variables are annotated as X and Y generically. X can be thought of as $CHLa$ and Y as all other environmental variables. The first step of the proposed model is to infer variable pairwise interaction as TE and node collective influence (OTE), determined via Equations (2) and (5), respectively. The second step is to prune the network considering only salient Pareto interactions via thresholding TE differences with a threshold d of causal significance, which is set to consider the top 20% of TEs (Equation (4)) that are necessary and sufficient to predict bloom spread.

2.1. Datasets

The Florida International University Southeast Environmental Research Center (FIU SERC) established a water quality monitoring system of 28 spatially distributed stations in Florida Bay (Figure 1), where each station (considered as a node in a network perspective) collects monthly data on chlorophyll-a (CHLa), total organic carbon, inorganic and organic nitrogen and phosphorus (TN and TP), turbidity (TURB), pH, salinity (SAL), water temperature (TEMP) and dissolved oxygen (see Boyer and Briceño [47] and Nelson et al. [14] for a description on how data are measured).

We used a threshold-based quantile regression method (analogous to Nelson et al. [14]) to establish an average threshold of $\geq 2\ \mu gL^{-1}$ on CHLa, universally applied to all stations, to distinguish bloom from non-bloom states across all stations. Initially, the dataset for this study spanned 2004 to 2006, corresponding to before, during and after a severe bloom outbreak in Florida Bay in 2005 [48] in terms of a CHLa extreme. In 1999, several blooms were observed in the same area but with lower CHLa extremes [14].

Then, the dataset (comprising all 2004, 2005 and 2006 CHLa monthly data) was filtered to include only those months and stations with CHLa values exceeding the critical blooming threshold, i.e., those months and stations indicating sustained bloom conditions. As a result, the final dataset contained 18, 63 and 136 rows of measurements (i.e., months) for the 2004, 2005 and 2006 bloom periods (pre-, peri- and post-bloom), respectively. More generically, 2005 can be considered as epitomic of bloom outbreaks, while 2004 and 2006 are representative of early and post-bloom periods.

2.2. Ecosystem Organization and Connectome

The entropy of the ecosystem, manifesting ecological disorganization in relation to CHLa variability, is dependent on the probability distribution functions (pdfs) that affect TE calculated on the pdf divergence and asynchronicity. The TE variability of an area, or the whole system, can be decomposed into eco-environmental interactions (considering CHLa and environmental factors acting as determinants or effect of ecological imbalance) and the ecological areal interactions underpinning bloom spread. This variability affects the organization propagation of CHLa (i.e., how randomly distributed CHLa is) and, in an information-balance equation, can be written as the spatio-temporal convolution of the aforementioned components composing the ecosystem connectome, i.e.,

$$\overbrace{H(\text{CHLa})}^{\text{eco-function}} = \sum_{m,n} \int_0^t \overbrace{\underbrace{(1 - TE(X_m, \text{CHLa}_m))}_{\text{eco-env feedback}} * \underbrace{(1 - TE(\text{CHLa}_{m,n}))}_{\text{eco-corridors}}}^{\text{eco-connectome}} \, d\tau \,, \tag{1}$$

where X stands for all other environmental variables except for CHLa, and m, n stands for the location of each area being monitored over the period t. The specific TE chosen in Equation (1) is related to TE analytics and the posed objectives, to be later specified. It should be noted that the time delay τ between eco-env factors in Equation (1) has been set to one due to the sub-monthly variability of CHLa and the resolution of the data.

Equation (1) is focused on CHLa patterns where networks are the backbone determinants of the ecological "weave" (CHLa interconnected patterns) that can be potentially controlled. Space and time are the dimensions along which CHLa is considered, plus other dimensions along gradients of environmental features on which stress–response patterns and related features (e.g., early warning signals and risk thresholds) can be derived. The networks define sources, sinks, pathways and determinants to guide monitoring and environmental control for bloom prevention.

In this paper, we specifically analyze the spatial ecological corridors determining bloom spread and direct interactions between CHLa and environmental factors (second and first term in Equation (1), where, for the latter, only $\text{CHLa}_m \rightarrow X_m$ interactions are considered) to quantify environmental effects of ecological dysbiosis; Wang and Convertino [44] focused instead on the whole set of biogeochemical interactions useful for forecasting except for bloom spreading networks.

2.3. Eco-Environmental Network Inference

Transfer entropy (TE), constructed from information entropy [49], measures the causal relationship between two asynchronous and divergent variables (expressed as a time series) X and Y (in the bivariate form, yet not accounting for second-order indirect interactions) by quantifying the predictive information flow between them [50]. Previously, the TE-based model, called the optimal information flow model (in relation to the maximization of ecosystemic entropy to gather the largest information), was used to discover causal relationships in human and aquatic ecosystems, e.g., for bacteria [15,17,25] and fish interactions [20] and to assess ecosystem health.

The information flow, and thus the predictive relationship between variables, is bidirectional. In this paper, we took the form of bivariate TE (while skipping interactions higher than the third-order, which was our first modeling assumption considering the weakly third-order interactions between environmental factors [51]) and calculated the difference between the pairwise information flows to identify the strongest causal factor, i.e., $TE_{X \to Y}$ and $TE_{X \leftarrow Y}$ (where X and Y can be either ecological, such as CHLa, or environmental variables) as follows:

$$\begin{aligned} TE_{X,Y} &= TE_{X \to Y} - TE_{X \leftarrow Y} \\ &= \sum p(y_{t+1}, y_m, x_n) log \frac{p(y_{t+1}|y_m, x_n)}{p(y_{t+1}|y_m)} \\ &- \sum p(x_{t+1}, x_n, y_m) log \frac{p(x_{t+1}|x_n, y_m)}{p(x_{t+1}|x_n)}, \end{aligned} \tag{2}$$

where x_{t+1} and y_{t+1} are the values of variables X and Y at time $t+1$ (yet, $\Delta t = 1$ month, which is our second modeling assumption considering the fact that CHLa values are very sensitive to past changes in the immediate past reflecting Markovian dynamics [51], while long-term increasing trends can lead to extreme CHLa shifts). x_n and y_m denote the histories of time-varying variables X and Y up to $t-n+1$ and $t-m+1$, respectively. $TE_{X \to Y}$ is the transfer entropy of time series variable X to Y, whereas $TE_{X \leftarrow Y}$ indicates the transfer entropy of Y to X.

In this study, we considered only positive $TE_{X,Y}$ where $X =$ CHLa and Y are all other environmental factors for eco-environmental feedback in Equation (1) and considered all $TE_{X,Y}$ where X and Y are both CHLa in two different nodes. Additionally, in the TE calculation, we did not investigate the optimal time delay between X and Y nor the optimal set of factors that are predictive of CHL, as in [20], due to: (i) the fact that bloom eco-env feedback occurs at scale lower than one month (at which data are available) and (ii) our interest into the entire systemic dynamics. This first part of all TE inference is considering all pairs of variables (Figure 2A).

The unbounded causality matrix, or more precisely the predictive causality matrix **TE** unconstrained to any prediction of biodiversity patterns as in [42], based on calculated TEs without the optimization of Δt and predictive environmental factors of ecological patterns in an optimal information flow perspective, can be constructed as follows:

$$\mathbf{TE} = \begin{bmatrix} TE_{1,1} & \cdots & TE_{1,Y} \\ \vdots & \ddots & \vdots \\ TE_{X,1} & \cdots & TE_{X,Y} \end{bmatrix}. \tag{3}$$

where $TE_{X,Y}$ is indeed a difference of transfer entropies as in the transfer entropy graph neural network model (TEGNN) (originally developed by Duan et al. [52] and applied to algal blooms by [51]) in contrast to the optimal information flow model (OIF) originally developed by Li and Convertino [20]. For each year, two networks were constructed with each defined by an underlying matrix of transfer entropy differences **TE**. One inferred matrix was a spatial network in which the 28 stations were nodes, and the causal influences among them were the edges.

The time series used to calculate the transfer entropy differences (Equation (2)) in this network were the time series of CHLa measurements at each station. The second inferred network was a water quality network, in which the nodes were the water quality factors (CHLa, TN, TP, SAL, TEMP and TURB), and the edges were the causal influence among them. In this study, however, the causality matrix underlying the water quality network was further filtered to focus only on the effect of CHLa on other water quality factors in relation to the objective to quantify this eco-environmental feedback; the reverse effect of water quality factors on CHLa and the interactions among water quality factors were not considered in this task but were in Convertino and Wang [51] and Wang and Convertino [44].

Following the Pareto principle [53], the largest 20% of values (and not 20% of events) identify the most influential variables (stations or factors), which are Pareto elements with the largest portfolio effect (à la Anderson et al. [54], which is about variance-mean scaling related to CHLa distributions), yet defining the risk of blooms. Therefore, we only retained the largest 20% of values in the mTE in order to focus on the most influential variables.

$$mTE_{X,Y} = \begin{cases} TE_{X,Y}, & TE_{X,Y} > d \\ 0, & \text{otherwise} \end{cases}, \tag{4}$$

where d is the threshold value, or Pareto critical value, of significant causal relationships as necessary and sufficient to predict bloom dynamics. If $TE_{X,Y} > d$, then X is a significant cause of Y; otherwise X is a weak cause of Y, or X and Y are mutually causal without any preferential direction in causality (Equation (4) and Figure 2). Additionally, the values in the matrices for each of the three years were normalized to enable better comparison of the inferred interactions. This second part of salient TE selection is about the network pruning (Figure 2B).

2.4. Eco-Environmental Factor Predictive Causality

The total outgoing transfer entropy (OTE as in Galbraith et al. [17] reflecting the total direct influence of one variable for all other influenced variables) of a node can be used to measure its influence on the collective dynamics (Equation (5)) as follows:

$$OTE = \sum_{Y} mTE_{X,Y}, \tag{5}$$

where OTE is the sum of the Xth row in the mTE matrix (after the application of the threshold as in Equation (4). The larger the OTE, the stronger the influence of node X on all other directly connected nodes in the network. In an information-balance perspective, OTE is the cumulative effect of all environmental or ecological variables for a node.

3. Results and Discussion

3.1. Spatio-Temporal Spreading and Fluctuations

To infer and characterize the spreading networks of blooms, while underpinning the ecological risk, we considered Florida Bay blooms between 2004 and 2006. We inferred a novel spatial influence network underpinning bloom spread among a set of spatially distributed water monitoring stations. This was achieved by deriving a TE matrix from spatio-temporal patterns of CHLa derived from monitored stations (see Section 2.3). The TE matrix for 2004 suggests that the study site was free of severe blooms, except for a few stations in the northwest: specifically, stations 16, 14, 25 and 26 (Figure 1) at least in 2004 where the resurgence of blooms was observed after the large bloom in 1999 [14,51].

Ecological spreading corridors are defined by the most divergent and asynchronous CHLa among nodes, while defining the most likely interdependent area, at least in a predictive causality sense (causality considering all other feasible connections, which are all other nodes in this case). Divergence and asynchronicity, as highlighted by Li and Convertino [20], are related to the difference in pdfs of CHLa (in two nodes) at different or equivalent time periods, respectively.

Spreading can be related to marine currents; however, in this study, the purpose is not to define the precise mechanisms underpinning the ecological patterns but rather to define the patterns' backbone networks, which are the salient spreading networks. This also identifies the potential coastal areas of influence of biogeochemical loads in FL Bay and the maximum extent of blooms—something that is poorly quantified but necessary for bloom prevention.

In analogy, runoff in terrestrial basins are predicted equivalently to CHLa, where the amount (and distribution) of water in different locations changes in an asynchronous way and is dependent on river network spreading to define the timing and divergent volume. True causality, leaving aside the feasibility of its assessment, must be included considering all areas where CHLa can spread, which, in a bay, is virtually everywhere; however, this is challenged by the data limitation that is constrained only to the used stations in this case.

The properties of network edges, representing eco-environmental interactions, depend on mTE (Equation (4)). The edge directions of the spreading network (Figure 3) suggest that an algal bloom would have initiated around station 16 and then moved west with a preferential direction toward the northwest. The edge colors (proportional to mTE) suggest that the bloom was moderately strong but localized in 2004 (Figure 3A) with a high probability to continue growing in the bay (due to mTE directions). The spatial spreading mTE matrix for 2005 revealed large and widespread bloom outbreaks that were concentrated in the western and central areas of the bay (Figure 3B).

In that year, the spatial influence was the strongest near stations 25, 16, 14 and 12 in the northwest and station 28 in the south region. The edge colors indicate that the bloom at all stations was moderately strong and also very likely to continue in the NE direction. Station 28 seems largely affected by many other stations in the bloom spread and yet is likely a sink node with potentially strong ecological effects also considering its proximity to the FL coral reef.

The matrix for 2006 (Figure 3C) shows the most extreme area interactions as well as a reversal in the spreading of blooms, i.e., moving from NE to central areas. The edge colors imply that bloom activity was extremely high, covering a wide area of Florida Bay. Nonetheless, the resulting graph suggests that, after the largest outbreak, the bloom moved from the easternmost into the north-central area, while the bloom in the west region dissipated.

We show how, by analyzing the information flow among spatially distributed nodes, it is possible to model the spatial spread of a phenomenon, such as algal blooms. In addition, this approach is able to detect sources, sinks, directions and salient pathways of bloom spreading. Due to various unaccounted factors, such as wind intensity and direction, current direction and bathymetry, there is a certain dynamic spatial change of blooms that is not attributed to the aforementioned factors. However, the model can take into account any environmental factor if available and can attribute the degree of variability of CHLa.

In a complex network sense, the bloom spatial network in 2004 is small in scale and regular in topology but has an obvious active station (station 16) that is an actively connected hub for bloom spreading. Therefore, it is much easier to take measures against blooms at this time (whether possible) or to prevent triggers by controlling environmental determinants. This area is well-know to be heavily influenced by nutrient efflux from the Everglades [14].

Particularly with the outbreaks of blooms in 2005 and later in 2006, the network has many more areas that are very active and affected, yet bloom management becomes more difficult. Over time, a spreading network transition is observed from a regular/small-world in 2004, to scale-free in 2005 and regular (or uniform) topology in 2006 with long-range connections.

3.2. Water Quality Trends and Bloom Impacts

The investigation of the impact of CHLa extremes (the magnitude, duration and frequency) on ecosystem health is a poorly covered topic in science. To explore how algal blooms impacted the water quality in Florida Bay, we analyzed how CHLa impacted other water quality variables using TE (see Section 2.3). We focused our analysis on how CHLa implicated potential changes in water quality—in terms of predictive causality—for stations where extreme blooms were most likely.

At the most active station in 2004 (i.e., station 16, characterized by coastal marshes, which is likely the source of blooms; see Figure 3A), blooms did not affect TN, TP, SAL and TURB, except for a slight effect on the water temperature (see Figure 4A). Rather, TN, TP, SAL and TURB, likely driven by a riverine efflux in the bay, triggered CHLa changes leading to blooms as highlighted in Wang and Convertino [44] and Convertino and Wang [51]. In 2005, the impact of blooms on other water quality factors was mostly evident at station 25, which is a deep-water mangrove habitat, where the blooms were the most intense (see Figure 3B).

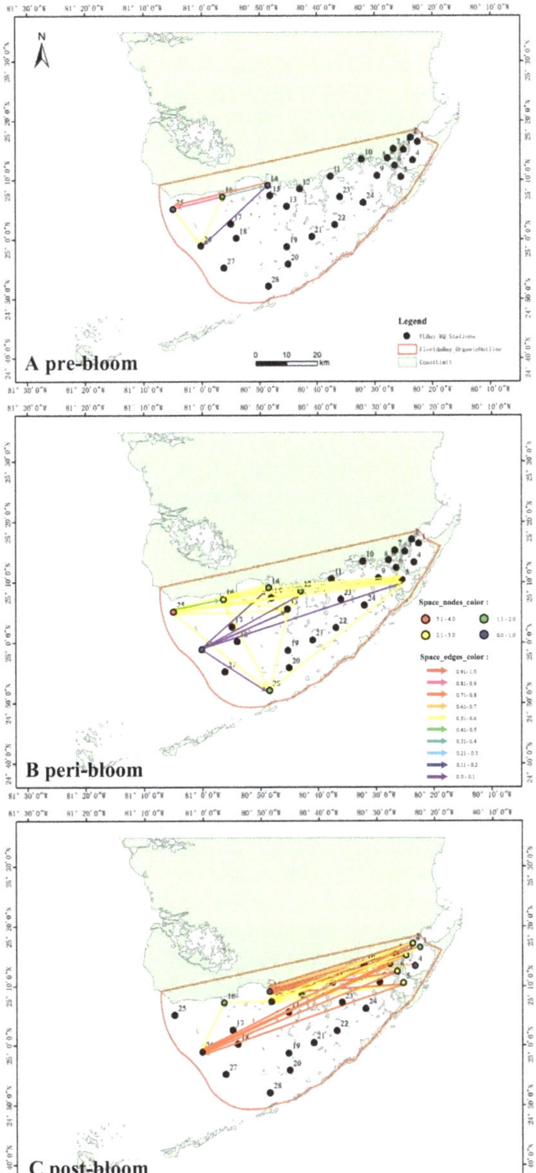

Figure 3. Inferred spatial CHLa for the 2004, 2005 and 2006 pre-, peri- and post-bloom periods in Florida Bay. Link and node color (from blue to red) is proportional to mTE based on CHLa interdependence between node pairs and *OTE* considering only $TE_{\mathrm{CHLa}->Env}$ where Env stands for all other environmental factors. East to west node and link dynamic increases are observed from 2004 to 2006 as well as a spreading network transition from regular/small-world to scale-free and regular (or uniform) with long-range connections for 2004, 2005 and 2006 (**A–C**). Each year corresponds to a different bloom precursor area and environmental factors (the central and northwest areas more affected by nutrients), widespread and extremely localized outbreak (the northeast more affected by temperature and turbidity and sequential effects of spreading).

CHLa induced not only water temperature changes but also variations in the total nitrogen and salinity (TN and SAL) with a higher impact on the latter (Figure 4B). In 2006 (see Figure 3C), where blooms were the most extreme but localized (the NW area), the effect of blooms on water quality peaked at station 3, followed by stations 5 and 2 and then station 6 in terms of magnitude (Figure 4C). Stations 2 and 3 experienced blooms throughout the year, while station 6 had a relatively short bloom (7 months as reported in the data).

At stations 3 and 6 (characterized more by tidal flats), blooms induced changes in the water temperature, salinity, total phosphorus and turbidity, while, at stations 2 and 5 (characterized more by submerged marshes), blooms led to substantial fluctuations in the total nitrogen, total phosphorus, salinity and turbidity. Information flow patterns (TE patterns) suggest that blooms first strongly caused water temperature alterations, then enhanced the salinity and nitrogen and later impacted other nutrients (phosphorous) and the turbidity. This is aligned with an understanding of the underlying microbiological processes [51].

In the vicinity of station 2, blooms caused a large change in salinity, while the effects on TN, TP and TURB were less significant. As blooms are a manifestation of eutrophication in water bodies, large amounts of phytoplankton cause dramatic changes in the total phosphorus and turbidity, such as near station 3, with a minor influence on the temperature and salinity. Around station 5, the bloom had a strong influence on the turbidity and salinity, with a minor impact on the TN and TP due to the deeper water in this area.

Despite the bloom near station 6 being relatively short, it still caused elevated changes in both the salinity and turbidity and, in a minor way, in the water temperature and total phosphorus. In general, the occurrence of blooms had serious effects on the total phosphorus, salinity and turbidity in the eastern zone of Florida Bay; a worrisome condition because of the highly valuable biodiversity in that area comprising a wide set of sponge, fish and coral species.

Our results reveal that algal bloom severity also caused environmental degradation a posteriori beyond the direct causal effects of environmental change (particularly from temperature in the ocean and nutrients from estuarine efflux) in triggering blooms a priori. Certainly, the primary causal pathway is about temperature leading to CHLa changes; however, the inferred networks also manifests the feedback of CHLa change on temperature. While this can be minor with respect to the first mechanism, it is also possible in relation to algal overgrowth and local temperature increase.

This substantiates environmental changes due to ecological imbalances [9], such as the oceanic positive feedback mechanism, which can lead to further increases in phytoplankton growth, chlorophyll-a concentration and temperature. Blooms are ecological processes that consume energy and yet increase the local temperature—precisely, algal blooms absorb light from the sun and carbon from the atmosphere, which increases the temperature of surface water. Whether this can be captured by our data or other data is an open question, but what is certainly true is that the bidirectional CHLa-temperature feedback is inferred as well as the CHLa-salinity.

Rising temperature, also related to local eutrophication, implies more evaporation from waterbodies and yet higher salinity if the hydrology is not changed. Of course, if the algae grow overly much (in term of biomass), a large amount of oxygen is depleted when they die, and this creates hypoxia and cascading risks, such as the death of species and the emergence of toxins. This can also lead to an exceeded capacity of zooplankton to sink carbon to the bottom of the ocean and, thus, an increase in the size and frequency in blooms, which is not good for the generated temperature, which is a co-occurring risk factor.

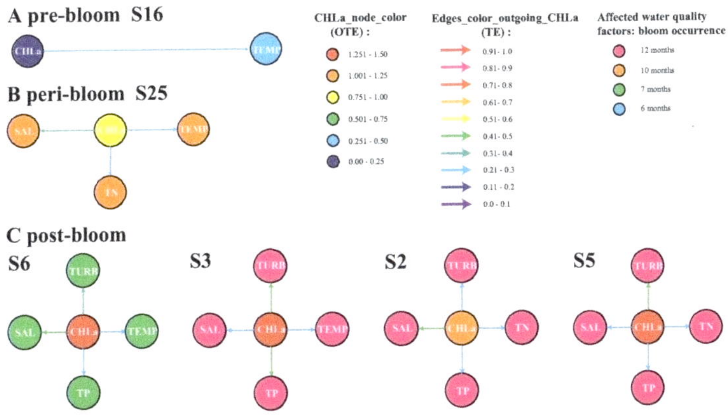

Figure 4. Inferred biogeochemical networks for the 2004, 2005 and 2006 pre-, peri- and post-bloom periods in Florida Bay. The purpose was to quantify local eco-environmental impacts for bloom sources. Only four nodes in 2006 and one node for 2004 and 2005 were considered because those are the most active in terms of the CHLa *OTE*. However, blooms are spreading phenomena, and other nodes are involved. Stations 16 and 25 are characterized by mangrove habitats in the west region, while stations 2, 3, 5 and 6 (displayed proportionally to a gradient of potential impact of CHLa on the environment) are characterized by coastal marshes and marine flat habitats in the east region of Florida Bay. The color of the directed edges is proportional to ranges of mTE for $TE_{\mathrm{CHLa}->Env}$ only. The node color for CHLa is proportional to *OTE* and, for other water quality factors, depends on the frequency of the local blooms during that year (manifesting the potential impact of CHLa on the environment): specifically, blue, green, orange and pink are for 6, 7, 10 and 12 months of bloom occurrence.

3.3. Bloom Intensity and Area Dependency

We explored the interaction dynamics of blooms by analyzing the annual probability distribution, or pdf, of the outgoing transfer entropy (OTE; see Figure 5) pre-, peri- and post-bloom. *OTE* quantified the extent to which blooms around one area can predict CHLa dynamics (in terms of the value and distribution) in other areas: higher *OTE* values indicate higher area interactions with higher spreading and predictability. In 2004, the *OTE* ranged between 0 and 1.7; most values with a non-zero probability were between 0 and 0.6.

It can be seen that most of the stations have no bloom, resulting in a low probability of large values of *OTE* but a high probability of low *OTE*. The pdf is bimodal with a leptokurtic character. In an ecological sense, the dynamics are characterized by highly localized blooms and few traces of bloom emergence in other areas. Thus, the bloom spatial network system was relatively contained in 2004 and corresponds to a regular/small-world topology (Figure 3). This also corresponds to simple low-TE dynamics of eco-environmental interactions (Figure 4).

In 2005, the *OTE* range increased to a maximum of 4.0, with most OTEs having a higher probability than in 2004. In 2006, the range of OTEs increased even further to a maximum of 13, with all *OTE* values having a higher probability. This also corresponds to a shift in the pdf to being more platykurtic while highlighting more widespread and common bloom dynamics.

From the perspective of complex networks, the number of nodes with large *OTE* values increased over time. This indicates an explosive spread of blooms across FL Bay. Therefore, the initial energy dissipation became higher over time. In 2005, the system was in an active and complex state, which makes the management of blooms extremely challenging. The 2006 pdf has higher entropy because it is a distribution closer to a Poisson pdf than previous years.

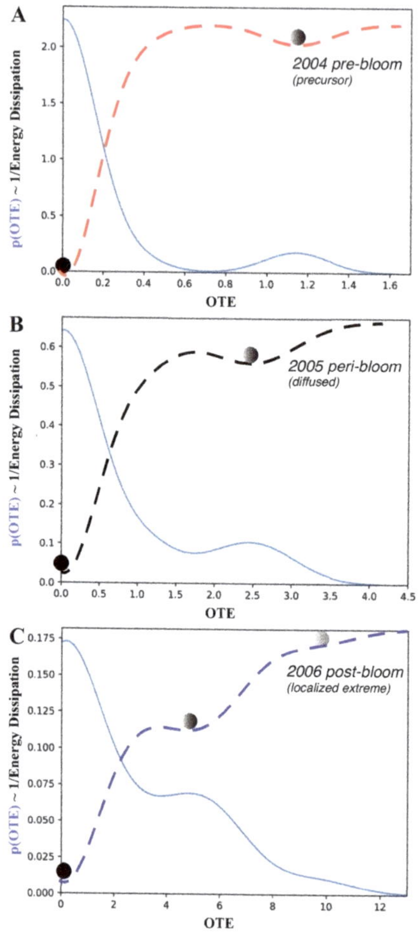

Figure 5. Probability distribution of the CHLa collective influence and ecosystem potential. (**A–C**) are for the 2004, 2005 and 2006 pre-, peri- and post-bloom periods in Florida Bay. CHLa's collective influence was assessed based on *OTE* range and distribution, where the latter defines the energy potential (in dashed red, black and blue for 2004, 2005 and 2006 aligned with the distinct epidemic, transitory and endemic dynamics as in Figure 1B), stability of ecosystem states and transition probabilities from one to another.

The pdf of the OTE proves that OTE reflects the probabilistic state of ecosystems with particular reference to algal blooms in this case. The higher the entropy, the larger the effect of blooms and the higher the ecological effects; interestingly, for FL Bay, we notice that, the higher the entropy, the more scale-free the bloom spreading network is, although a time delay may exist between ecological effects (CHLa, which is more random, such as in 2006) and the largest spreading network (which was in 2005) signifying potential long-term effects.

By flipping the pdf, it is possible to obtain information on the ecosystem potential landscape about the energy dissipation, likelihood of shifts and relative stability of bloom conditions (Figure 5). The energy dissipation of the system, which is the potential amount of energy consumed by ecological processes, is visualized, where $\propto max[p(OTE)] - p(OTE)$, which scales with $\sim 1/p(OTE)$; therefore, the higher the leptokurtic character of the pdf is, the lower the energy dissipation (such as in 2004).

The energy potential also gives the number of ecological states (metastable states are identified by the point where the pre- and post-curvature of the energy landscape diverges in sign; these are represented by the balls in Figure 5), the probability of a configuration to be stable (the lower the energy potential with respect to all other states is, the higher the stability) and the likely shifts among states (proportional to the slope of the energy potential), all of which define the "resilience" of the ecosystem, which is the rapidity to bounce back to initial states.

Higher entropy corresponds to higher energy dissipation in relation to larger and more random OTEs. This implies a lower probability of CHLa stable states, which are much closer to each other and increase in number, implying a higher likelihood of shifts with larger ecological impacts. For FL Bay, the energy dissipation also increases in average value for the pre-, peri- and post-bloom periods indicating a diminishing resilience and loss of complexity of the system; this also highlights the persistent effects of blooms despite their relatively short duration.

4. Conclusions

This study uniquely proposed a model based on optimal transfer entropy (TE) with TE differences to infer bloom spatial dependencies, which were used to pinpoint risk areas and pathways to target monitoring and controls. Blooms showed non-trivial spreading patterns manifested by network transitions with different stability results that determined their persistence and potential ecological effects. For FL Bay, we predominantly highlighted the spatial trends and the neglected impacts of algal blooms on the water quality. The following specific results are worth highlighting.

- We showed how CHLa patterns carry information regarding the underpinning eco-hydrological networks (and associated spreading determinants, such as nutrients) that support ecosystem function and services. Salient Pareto interactions were defined via thresholding TE differences with a threshold of causal significance that was set to consider the top 20% of TEs (related to the tail of scale-free CHLa probability distribution function), i.e., necessary and sufficient interactions to predict the risk of bloom spreading.
  More generally, the discovery and inference of the "ecosystem connectome" (as biogeochemical determinant and spreading networks) allows for the assessment of ecosystem health (quantified by the proximity to an optimal condition, such as the non-bloom state) as well as the investigation of causal determinants and their sources, proximity to ecosystem shifts and targeted ecohydrological controls.
- Through spatial analysis of bloom spreading networks, we showed how regions not previously involved in blooms (i.e., the highly biodiverse NE tidal-flat habitats with corals and sponges) were caused by large imbalances of CHLa in the western and central blooms, which were causally involved. The latter regions were characterized by CHLa that was more randomly distributed and a higher probability of CHLa extremes. This probabilistic structure, reflecting the spatial distribution of CHLa, is likely tipping eastern regions to similar bloom endemics. From the perspective of complex networks, this bloom event (2004–2006) evolved from a spatial network with a localized trigger area and a small-world topology to a random topology with long-range spatial diffusion.
  In 2005, when most stations were blooming, the spatial spreading network was scale-free (theoretically optimal in a purely topological and predictive sense [55,56]) with a random biogeochemical network, including CHLa (topologically suboptimal), which underpins the dichotomy between structural and functional networks for ecological risks.
- In terms of temporal dynamics, subsequent to the first bloom outbreak, persistent and recurring blooms were observed for several NE areas with long-lasting environmental

impacts on turbidity and salinity aggravated by temperature increases. Bloom sources were related to central coastal marshes and, to a lower extent, mangrove habitats. We further showed that blooms were a recurring and persistent phenomenon over a long period of time with continuous outbreaks in interdependent regions. This led to higher energy dissipation and larger instability dictated by the more random distribution of CHLa, which was associated with a more uniform network with long-range connectivity regardless of habitats, likely leading to the loss of ecological heterogeneity.

- The analysis of biogeochemical factors affecting water quality showed that the occurrence of blooms could only affect small fluctuations of temperature at the beginning of the blooms; however, repeated bloom outbreaks largely affected other biogeochemical factors (such as salinity, turbidity and CHLa triggering hysteresis or memory effects) that are poorly systemically controllable due to the loss of vegetation and other keynote species.
  The concentration of CHLa can be influenced by temperature and salinity, and changes in the CHLa concentration can, in turn, have indirect effects on water temperature through various ecological processes. In some regions, facilitated by shallow-water habitats, a water temperature increase can stimulate phytoplankton growth and increase the concentration of CHLa. The increased CHLa can, in turn, absorb more sunlight, which can lead to local warming of the water.
  In the long term, the persistence of blooms, i.e., high CHLa, may also alter nutrient cycling as highlighted by other studies with the term "oceanic positive feedback mechanism" [11], and our model was able to infer this secondary causal pathway together with the primary one, where temperature change led to CHLa change and blooms. This underscores that bloom management should start from the source, otherwise blooms' environmental impacts will gradually expand and become uncontrollable, thus, also affecting the ecosystem stability and resilience and settling into undesired ecological states.

Although the intensity, duration and spatial distribution of blooms are governed by a multiplicity of factors, CHLa variability (independently of any trigger) still has a wide degree of predictability and control in an ecosystem perspective considering both predictive and ecological engineering models. We proposed a data-based inferential model to be used for ecological intelligence to look into patterns of risk (source and pathways), trajectories and determinants.

Our proposed spatial and biogeochemical network inference model provides valuable information for the forecasting and management of blooms—for instance, by pinpointing monitoring and nature-based solutions in source areas, such as coastal blue-carbon habitats to inhibit progressive eco-environmental imbalances and the related impacts. Further work will look into the precise quantification of critical thresholds (habitat- and climate-specific or universal) as early warning signals of environmental factors (including controls) that lead to persistent blooms and accounting for systemic stress, that is reflected by the condition of habitats as their ecological history.

Author Contributions: M.C. designed and guided the study and wrote and finalized the manuscript. H.W. performed the calculations and created the first figures and draft. E.G. helped in the writing of the paper. All authors have read and agreed to the published version of the manuscript.

Funding: M.C. acknowledges the Shenzhen Pengcheng Peacock Talents funding (B class) and the Tsinghua University SIGS start-up funding.

Data Availability Statement: Data of Florida Bay blooms are available from Nelson et al. (2017), Mar. Ecol. Prog. Ser., that originally published the data.

Conflicts of Interest: The authors declare no conflict of interest.

References

1. Smayda, T.J. What is a bloom? A commentary. *Limnol. Oceanogr.* **1997**, *42*, 1132–1136. [CrossRef]
2. Paerl, H.W.; Fulton, R.S.; Moisander, P.H.; Dyble, J. Harmful freshwater algal blooms, with an emphasis on cyanobacteria. *Sci. World J.* **2001**, *1*, 76–113. [CrossRef] [PubMed]
3. Dai, Y.; Yang, S.; Zhao, D.; Hu, C.; Xu, W.; Anderson, D.M.; Li, Y.; Song, X.P.; Boyce, D.G.; Gibson, L.; et al. Coastal phytoplankton blooms expand and intensify in the 21st century. *Nature* **2023**, *615*, 1–5. [CrossRef] [PubMed]
4. Lotze, H.K.; Worm, B.; Sommer, U. Propagule banks, herbivory and nutrient supply control population development and dominance patterns in macroalgal blooms. *Oikos* **2000**, *89*, 46–58. [CrossRef]
5. Cosper, E.M.; Dennison, W.C.; Carpenter, E.J.; Bricelj, V.M.; Mitchell, J.G.; Kuenstner, S.H.; Colflesh, D.; Dewey, M. Recurrent and persistent brown tide blooms perturb coastal marine ecosystem. *Estuaries* **1987**, *10*, 284–290. [CrossRef]
6. Butler, M.J., IV; Hunt, J.H.; Herrnkind, W.F.; Childress, M.J.; Bertelsen, R.; Sharp, W.; Matthews, T.; Field, J.M.; Marshall, H.G. Cascading disturbances in Florida Bay, USA: Cyanobacteria blooms, sponge mortality, and implications for juvenile spiny lobsters Panulirus argus. *Mar. Ecol. Prog. Ser.* **1995**, *129*, 119–125. [CrossRef]
7. Frieder, C.A.; Kessouri, F.; Ho, M.; Sutula, M.; Bianchi, D.; McWilliams, J.C.; Deutsch, C.; Howard, E. Effects of urban eutrophication on pelagic habitat capacity in the Southern California Bight. *ESS Open Arch.* **2023**. Available online: https://d197for5662m48.cloudfront.net/documents/publicationstatus/130200/preprint_pdf/cca98c7aa83c14dc943466ce32711731.pdf (accessed on 15 March 2023).
8. Inomura, K.; Deutsch, C.; Jahn, O.; Dutkiewicz, S.; Follows, M.J. Global patterns in marine organic matter stoichiometry driven by phytoplankton ecophysiology. *Nat. Geosci.* **2022**, *15*, 1034–1040. [CrossRef] [PubMed]
9. Dunstan, P.K.; Foster, S.D.; King, E.; Risbey, J.; O'Kane, T.J.; Monselesan, D.; Hobday, A.J.; Hartog, J.R.; Thompson, P.A. Global patterns of change and variation in sea surface temperature and chlorophyll a. *Sci. Rep.* **2018**, *8*, 14624. [CrossRef] [PubMed]
10. Spivak, A.C.; Sanderman, J.; Bowen, J.L.; Canuel, E.A.; Hopkinson, C.S. Global-change controls on soil-carbon accumulation and loss in coastal vegetated ecosystems. *Nat. Geosci.* **2019**, *12*, 685–692. [CrossRef]
11. Gobler, C.J. Climate change and harmful algal blooms: Insights and perspective. *Harmful Algae* **2020**, *91*, 101731. [CrossRef] [PubMed]
12. Burd, A.B.; Jackson, G.A. An analysis of water column distributions in Florida Bay. *Estuaries* **2002**, *25*, 570–585. [CrossRef]
13. Briceño, H.O.; Boyer, J.N. Climatic controls on phytoplankton biomass in a sub-tropical estuary, Florida Bay, USA. *Estuaries Coasts* **2010**, *33*, 541–553. [CrossRef]
14. Nelson, N.G.; Munoz-Carpena, R.; Phlips, E.J. A novel quantile method reveals spatiotemporal shifts in phytoplankton biomass descriptors between bloom and non-bloom conditions in a subtropical estuary. *Mar. Ecol. Prog. Ser.* **2017**, *567*, 57–78. [CrossRef]
15. Galbraith, E.; Convertino, M. The Eco-Evo Mandala: Simplifying Bacterioplankton Complexity into Ecohealth Signatures. *Entropy* **2021**, *23*, 1471. [CrossRef]
16. Malone, T.C.; Newton, A. The globalization of cultural eutrophication in the coastal ocean: Causes and consequences. *Front. Mar. Sci.* **2020**, *7*, 670. [CrossRef]
17. Galbraith, E.; Frade, P.; Convertino, M. Metabolic shifts of oceans: Summoning bacterial interactions. *Ecol. Ind.* **2022**, *138*, 108871. [CrossRef]
18. Wang, X.F. Complex networks: Topology, dynamics and synchronization. *Int. J. Bifurc. Chaos* **2002**, *12*, 885–916. [CrossRef]
19. Ilany, A.; Booms, A.S.; Holekamp, K.E. Topological effects of network structure on long-term social network dynamics in a wild mammal. *Ecol. Lett.* **2015**, *18*, 687–695. [CrossRef]
20. Li, J.; Convertino, M. Inferring ecosystem networks as information flows. *Sci. Rep.* **2021**, *11*, 1–22. [CrossRef]
21. Reijneveld, J.C.; Ponten, S.C.; Berendse, H.W.; Stam, C.J. The application of graph theoretical analysis to complex networks in the brain. *Clin. Neurophysiol.* **2007**, *118*, 2317–2331. [CrossRef] [PubMed]
22. Donges, J.F.; Zou, Y.; Marwan, N.; Kurths, J. Complex networks in climate dynamics. *Eur. Phys. J. Spec. Top.* **2009**, *174*, 157–179. [CrossRef]
23. Coscia, M.; Giannotti, F.; Pedreschi, D. A classification for community discovery methods in complex networks. *Stat. Anal. Data Min. ASA Data Sci. J.* **2011**, *4*, 512–546. [CrossRef]
24. Feldhoff, J.H.; Lange, S.; Volkholz, J.; Donges, J.F.; Kurths, J.; Gerstengarbe, F.W. Complex networks for climate model evaluation with application to statistical versus dynamical modeling of South American climate. *Clim. Dyn.* **2015**, *44*, 1567–1581. [CrossRef]
25. Li, J.; Convertino, M. Optimal microbiome networks: Macroecology and criticality. *Entropy* **2019**, *21*, 506. [CrossRef]
26. Zou, Y.; Donner, R.V.; Marwan, N.; Donges, J.F.; Kurths, J. Complex network approaches to nonlinear time series analysis. *Phys. Rep.* **2019**, *787*, 1–97. [CrossRef]
27. Zhang, J.; Small, M. Complex network from pseudoperiodic time series: Topology versus dynamics. *Phys. Rev. Lett.* **2006**, *96*, 238701. [CrossRef]
28. Gfeller, D.; De Los Rios, P.; Caflisch, A.; Rao, F. Complex network analysis of free-energy landscapes. *Proc. Natl. Acad. Sci. USA* **2007**, *104*, 1817–1822. [CrossRef] [PubMed]
29. Lin, J.; Ban, Y. Complex network topology of transportation systems. *Transp. Rev.* **2013**, *33*, 658–685. [CrossRef]
30. Zhou, C.; Zemanová, L.; Zamora, G.; Hilgetag, C.C.; Kurths, J. Hierarchical organization unveiled by functional connectivity in complex brain networks. *Phys. Rev. Lett.* **2006**, *97*, 238103. [CrossRef]

31. Zhou, C.; Zemanová, L.; Zamora-Lopez, G.; Hilgetag, C.C.; Kurths, J. Structure–function relationship in complex brain networks expressed by hierarchical synchronization. *N. J. Phys.* **2007**, *9*, 178. [CrossRef]
32. Albert, R.; Barabási, A.L. Statistical mechanics of complex networks. *Rev. Modern Phys.* **2002**, *74*, 47. [CrossRef]
33. Boccaletti, S.; Latora, V.; Moreno, Y.; Chavez, M.; Hwang, D.U. Complex networks: Structure and dynamics. *Phys. Rep.* **2006**, *424*, 175–308. [CrossRef]
34. Costa, L.d.F.; Rodrigues, F.A.; Travieso, G.; Villas Boas, P.R. Characterization of complex networks: A survey of measurements. *Adv. Phys.* **2007**, *56*, 167–242. [CrossRef]
35. Newman, M.E. The structure and function of complex networks. *SIAM Rev.* **2003**, *45*, 167–256. [CrossRef]
36. Andrade Jr, J.S.; Herrmann, H.J.; Andrade, R.F.; Da Silva, L.R. Apollonian networks: Simultaneously scale-free, small world, Euclidean, space filling, and with matching graphs. *Phys. Rev. Lett.* **2005**, *94*, 018702. [CrossRef]
37. Dorogovtsev, S.N.; Goltsev, A.V.; Mendes, J.F.F. Pseudofractal scale-free web. *Phys. Rev. E* **2002**, *65*, 066122. [CrossRef]
38. Kumar, R.; Raghavan, P.; Rajagopalan, S.; Sivakumar, D.; Tomkins, A.; Upfal, E. Stochastic models for the web graph. In Proceedings of the 41st Annual Symposium on Foundations of Computer Science, Redondo Beach, CA, USA, 12–14 November 2000; pp. 57–65.
39. Hidalgo, J.; Grilli, J.; Suweis, S.; Munoz, M.A.; Banavar, J.R.; Maritan, A. Information-based fitness and the emergence of criticality in living systems. *Proc. Natl. Acad. Sci. USA* **2014**, *111*, 10095–10100. [CrossRef] [PubMed]
40. Ramírez-Carrillo, E.; López-Corona, O.; Toledo-Roy, J.C.; Lovett, J.C.; de León-González, F.; Osorio-Olvera, L.; Equihua, J.; Robredo, E.; Frank, A.; Dirzo, R.; et al. Assessing sustainability in North America's ecosystems using criticality and information theory. *PLoS ONE* **2018**, *13*, e0200382. [CrossRef] [PubMed]
41. Convertino, M.; Valverde Jr, L.J. Toward a pluralistic conception of resilience. *Ecol. Indic.* **2019**, *107*, 105510. [CrossRef]
42. Li, J.; Convertino, M. Temperature increase drives critical slowing down of fish ecosystems. *PLoS ONE* **2021**, *16*, e0246222. [CrossRef] [PubMed]
43. Convertino, M.; Reddy, A.; Liu, Y.; Munoz-Zanzi, C. Eco-epidemiological scaling of leptospirosis: Vulnerability mapping and early warning forecasts. *Sci. Total Environ.* **2021**, *799*, 149102. [CrossRef] [PubMed]
44. Wang, H.; Convertino, M. Algal Bloom Ties: Systemic Biogeochemical Stress and Chlorophyll-a Shift Forecasting. *Preprints* **2023**, *submitted*.
45. Boyer, J.N.; Kelble, C.R.; Ortner, P.B.; Rudnick, D.T. Phytoplankton bloom status: Chlorophyll a biomass as an indicator of water quality condition in the southern estuaries of Florida, USA. *Ecol. Ind.* **2009**, *9*, S56–S67. [CrossRef]
46. Phlips, E.J.; Badylak, S.; Lynch, T.C. Blooms of the picoplanktonic cyanobacterium Synechococcus in Florida Bay, a subtropical inner-shelf lagoon. *Limnol. Oceanogr.* **1999**, *44*, 1166–1175. [CrossRef]
47. Boyer, J.N.; Briceño, H.O. South Florida Coastal Water Quality Monitoring Network. FY2006 Cumulative Report South Florida Water Management District, Southeast Environmental Research Center, Florida International University. 2007. Available online: http://serc.fiu.edu/wqmnetwork/ (accessed on 15 January 2023).
48. Cloern, J.E.; Jassby, A.D. Patterns and scales of phytoplankton variability in estuarine–coastal ecosystems. *Estuaries Coasts* **2010**, *33*, 230–241. [CrossRef]
49. Shannon, C.E. A mathematical theory of communication. *Bell Syst. Tech. J.* **1948**, *27*, 379–423. [CrossRef]
50. Schreiber, T. Measuring information transfer. *Phys. Rev. Lett.* **2000**, *85*, 461. [CrossRef] [PubMed]
51. Convertino, M.; Wang, H. Envirome Disorganization and Ecological Riskscapes: The Algal Bloom Epitome. In *Risk Assessment for Environmental Health*; CRC Press: Boca Raton, FL, USA, 2022; pp. 327–346.
52. Duan, Z.; Xu, H.; Huang, Y.; Feng, J.; Wang, Y. Multivariate time series forecasting with transfer entropy graph. *Tsinghua Sci. Technol.* **2022**, *28*, 141–149. [CrossRef]
53. Sanders, R. The Pareto principle: Its use and abuse. *J. Serv. Mark.* **1987**, *1*, 37–40. [CrossRef]
54. Anderson, S.C.; Cooper, A.B.; Dulvy, N.K. Ecological prophets: Quantifying metapopulation portfolio effects. *Methods Ecol. Evol.* **2013**, *4*, 971–981. [CrossRef]
55. Martinello, M.; Hidalgo, J.; Maritan, A.; Di Santo, S.; Plenz, D.; Muñoz, M.A. Neutral theory and scale-free neural dynamics. *Phys. Rev. X* **2017**, *7*, 041071. [CrossRef]
56. Xu, L.; Patterson, D.; Levin, S.A.; Wang, J. Non-equilibrium early-warning signals for critical transitions in ecological systems. *Proc. Natl. Acad. Sci. USA* **2023**, *120*, e2218663120. [CrossRef] [PubMed]

remote sensing MDPI

Article

Remote Sensing of Coastal Vegetation Phenology in a Cold Temperate Intertidal System: Implications for Classification of Coastal Habitats

Brigitte Légaré [1,*,†], Simon Bélanger [1,†], Rakesh Kumar Singh [1], Pascal Bernatchez [1] and Mathieu Cusson [2]

1 Département de Biologie, Université du Québec à Rimouski, Chimie et Géographie, Québec-Océan et BORÉAS, 300 Allée des Ursulines, Rimouski, QC G5L 3A1, Canada; simon_belanger@uqar.ca (S.B.); rakeshkumar_singh@uqar.ca (R.K.S.); pascal_bernatchez@uqar.ca (P.B.)
2 Département des Sciences Fondamentales, Université du Québec à Chicoutimi, Québec-Océan 555, Boulevard de l'Université, Chicoutimi, QC G7H 2B1, Canada; mathieu_cusson@uqac.ca
* Correspondence: brigitte.legare@uqar.ca
† These authors contributed equally to this work.

Citation: Légaré, B.; Bélanger, S.; Singh, R.K.; Bernatchez, P.; Cusson, M. Remote Sensing of Coastal Vegetation Phenology in a Cold Temperate Intertidal System: Implications for Classification of Coastal Habitats. *Remote Sens.* **2022**, *14*, 3000. https://doi.org/10.3390/rs14133000

Academic Editors: Matteo Convertino and Jie Li

Received: 30 April 2022
Accepted: 19 June 2022
Published: 23 June 2022

Abstract: Intertidal vegetation provides important ecological functions, such as food and shelter for wildlife and ecological services with increased coastline protection from erosion. In cold temperate and subarctic environments, the short growing season has a significant impact on the phenological response of the different vegetation types, which must be considered for their mapping using satellite remote sensing technologies. This study focuses on the effect of the phenology of vegetation in the intertidal ecosystems on remote sensing outputs. The studied sites were dominated by eelgrass (*Zostera marina* L.), saltmarsh cordgrass (*Spartina alterniflora*), creeping saltbush (*Atriplex prostrata*), macroalgae (*Ascophyllum nodosum*, and *Fucus vesiculosus*) attached to scattered boulders. In situ data were collected on ten occasions from May through October 2019 and included biophysical properties (e.g., leaf area index) and hyperspectral reflectance spectra ($R_{rs}(\lambda)$). The results indicate that even when substantial vegetation growth is observed, the variation in $R_{rs}(\lambda)$ is not significant at the beginning of the growing season, limiting the spectral separability using multispectral imagery. The spectral separability between vegetation types was maximum at the beginning of the season (early June) when the vegetation had not reached its maximum growth. Seasonal time series of the normalized difference vegetation index (NDVI) values were derived from multispectral sensors (Sentinel-2 multispectral instrument (MSI) and PlanetScope) and were validated using in situ-derived NDVI. The results indicate that the phenology of intertidal vegetation can be monitored by satellite if the number of observations obtained at a low tide is sufficient, which helps to discriminate plant species and, therefore, the mapping of vegetation. The optimal period for vegetation mapping was September for the study area.

Keywords: vegetation phenology; spectral signature; intertidal coastal ecosystem; remote sensing; eelgrass (*Zostera marina* L.); saltmarsh; classification

1. Introduction

Intertidal ecosystems are productive and dynamic environments located between low and high tide levels. These environments play various essential ecological and biogeochemical roles that benefit both terrestrial and marine ecosystems by sequestering carbon and nitrogen and regulating biogeochemical cycles [1,2]. With the increase in population, industrial development, and over-exploitation of resources over the past decades, coastal and littoral environments are among the most vulnerable ecosystems in the world [1,3–6]. The anthropogenic pressures associated with coastal ecosystems and the increase in sea levels linked to climate change contribute to the loss of 20 to 70% of these ecosystems [7,8]. In addition, global warming causes a rise in the number and intensity of winter storms and reduces the coastal ice cover in cold temperate regions and higher latitudes, leading to rapid changes in coastal environment functions and structures [8]. Management activity relies on the provision of spatial and explicit information on coastal vegetation distribution, making it more important than ever [9].

Satellite remote sensing technology is increasingly used to map vegetation cover of habitats, varying from forests to coasts and their temporal evolution [10,11]. However, more in situ observations are needed to obtain accurate mapping and to orient the development of satellite-based classification algorithms [12,13]. Optical remote sensing images provide essential information on vegetation cover, such as the spatial extent presence and the dominant vegetation type, the biomass estimate, the percentage cover, and the leaf area index (LAI), [14–16]. Recent advances in satellite sensors and algorithms now provide the ability to monitor coastal vegetation at relatively high spatial (<10 m pixel size) and temporal (weekly) scales [17]. Spaceborne sensors commonly used to map coastal vegetation ecosystems or habitats include Landsat-8 [18], Sentinel-2 [17,19], RapidEye [19], and PlanetScope [20].

Mapping coastal vegetation using satellite remote sensing imagery relies on the assumption that the dominant vegetation type presents unique spectral signatures to distinguish from others [9,21]. However, reflectance spectra of coastal vegetation frequently overlap, change rapidly over the season, and are not always distinguishable by using multispectral sensors alone due to the low number of broad spectral bands. Understanding vegetation spectral reflectance variability is essential and can be used to map different vegetation species using remote sensing images [22,23]. In addition, understanding the seasonal evolution, i.e., the phenology [24], of the reflectance spectra of key vegetation types can provide additional clues for distinguishing them from space [25,26]. It can help, for example, to identify the season when the difference between the band is at its greatest.

The first objective of this study was to document with in situ measurements the seasonal evolution of dominant vegetation reflectance spectra, along with biophysical properties, in a cold temperate intertidal system. Analysis of spectral measurements is crucial to determine appropriate spectral resolutions and a classification scheme. The four prevalent vegetation types included eelgrass (*Zostera Marina*), macroalgae (*Ascophyllum nodosum, Fucus vesiculosus*), saltmarsh cordgrass (*Spartina alterniflora*), and creeping saltbush (*Atriplex prostrata*). The second objective was to assess the potential of combining multispectral sensors onboard satellite constellations offering high spatial and temporal resolution (i.e., Sentinel-2 Multispectral Instrument (MSI), Landsat-8 operational land imager (OLI), RapidEye (RE), and PlanetScope (PS)) to quantify the phenological change of intertidal vegetation. Finally, we identified the best seasonal window to classify and map coastal ecosystems during the relatively short growing season (<6 months) in the studied system.

2. Materials and Methods

2.1. Study Area

The study area is an intertidal saltmarsh and seagrass meadow near the Baie de l'Isle-Verte National Wildlife Area located along the south shore of the St. Lawrence Estuary, Québec, Canada (Figure 1). The Baie de l'Isle-Verte (BIV), located east of the sampling site, is a protected area of 322 hectares; it was created in 1980 by Environment Canada to protect the intertidal cordgrass marsh of l'Isle-Verte and the adjacent coastal habitats, which are used by waterfowl and numerous other fauna species.

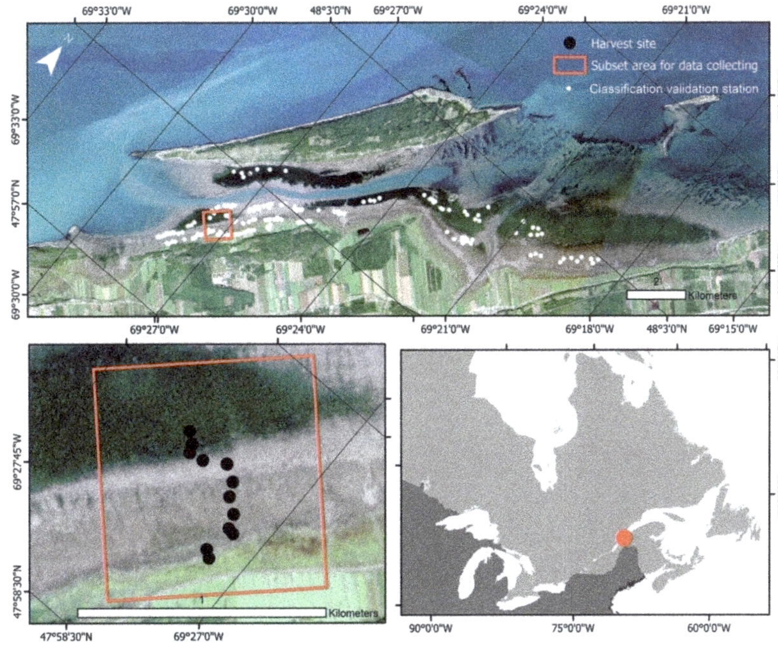

Figure 1. True-color image of l'Isle-Verte from PlanetScope captured on 3 September 2019. The outer red box delimits the subset used for the study area, and the black dots identify the 13 harvest sites. White dots on the upper panel are the locations selected for the classification validation.

The sampling site is protected from wave action by the Isle-Verte island (municipality of Notre-Dame-des-Sept-Douleurs) (69°27′00″W; 47°58′30″N). This area is home to one of the largest eelgrass meadows of the St. Lawrence Estuary, a rich and productive environment that has been monitored by the Department of Fisheries and Oceans since the 1980s [27,28]. The climate of the study area is cold (mean annual temperature of 3.5 °C), humid, and highly influenced by estuarine and marine conditions. In the winter, temperatures are well below the freezing point and most of the bay is covered by landfast ice and ice formed along and attached to the shore, also known as ice foot [29] with a mean winter temperature of −10.8 °C. In the summer, the temperatures on the shore remain colder than inland due to the marine influence. The tides are semi-diurnal, with an average tidal range of 3.5 m [30]. The spring tides can reach 4.7 m and are only 1.5 m during neap tides. The estuarine waters are brackish (about 20–25 PSU) and relatively turbid in the study area, limiting the detection of submerged vegetation by remote sensing [30–32].

2.2. In Situ Measurement

The sampling area was divided into four zones based on the frequency of immersion, which is a function of elevation, the ice in the winter, the water temperature, and salinity (Figure 2). The zones are: (1) the eelgrass bed of *Zostera marina* (EG); (2) the mud flat with microphytobenthos (SD: sediment) and the macroalgae zone composed of Fucus and Ascophyllum, mainly attached to scattered boulders (MA); (3) the low marsh cordgrass zone dominated by *Spartina alterniflora* (SC); and (4) the high marsh zone mainly dominated by *Atriplex prostrata* (CS, creeping saltbush) [30,31,33]. During the winter, ice scouring creates ice-made tidal pools, which increase the spatial heterogeneity within the shore habitats [34–36].

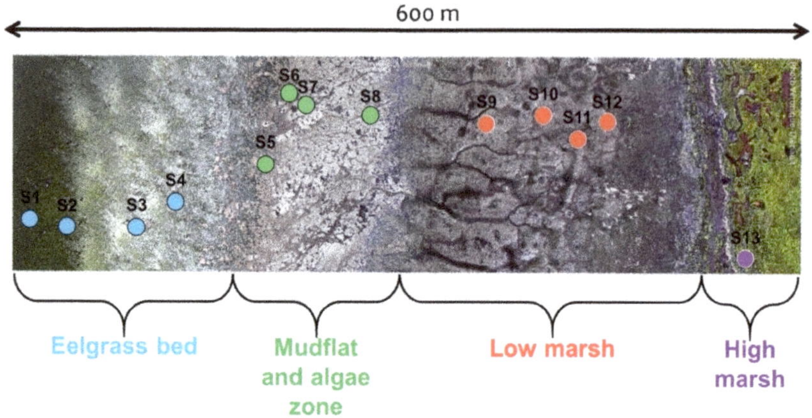

Figure 2. Drone image collected on 1 September 2019 presenting the delimitation of the four zones of the study site and the transect of 13 harvest points.

A 600 m transect perpendicular to the shoreline was set up to cover the four zones identified above (Figure 2). Thirteen stations, or plots, were placed along the transect, covering each zone of the coast, in relatively homogeneous patches of vegetation (>100 m^2), except for SD and MA where scattered boulders create high spatial heterogeneity from a remote sensing perspective. A complete species composition list, including the percentage cover estimates, was established for every plot in 50 × 50 cm (2500 cm^2) quadrats (Table 1).

Table 1. Zone type, percentage coverage at the beginning and the end of the season, and the vegetation biomass harvested at the end of the season in g/m^2.

Station	Zone Type	Coverage Beginning of the Season (%)	Coverage End of the Season (%)	Total Biomass (g/m^2)	Latitude (N)	Longitude (W)
1	EG	100	100	122.4	47°58′59.52″	−69°27′24.84″
2	EG	100	100	45.0	47°58′58.08″	−69°27′23.04″
3	EG	50	100	48.3	47°58′57.00″	−69°27′21.96″
4	EG	20	50	16.4	47°58′57.72″	−69°27′20.16″
5	SD	100	100	NA	47°58′57.36″	−69°27′19.44″
6	MA	100	100	476.7	47°58′59.16″	−69°27′15.12″
7	MA	100	100	1018.7	47°58′58.80″	−69°27′15.12″
8	MA	100	100	1009.0	47°58′57.72″	−69°27′12.14″
9	SC	75	100	72.4	47°58′56.28″	−69°27′10.80″
10	SC	90	100	200.6	47°58′54.84″	−69°27′08.28″
11	SC	100	100	146.8	47°58′53.04″	−69°27′06.84″
12	SC	100	100	219.7	47°58′52.68″	−69°27′06.12″
13	CS	100	100	40.4	47°58′48.36″	−69°27′06.48″

Note: only one station was sampled for CS due to time constraints (i.e., low tide sampling) and difficulty in identifying CS at the beginning of the season.

Each station was geolocated using a GPS, marked with a wooden post, and visited up to 10 times from 17 May–25 October 2019 (Table 2). All stations were sampled during the low tide, approximately every two weeks during the maximum growth period (defined as mid-May to early August for this study) and then every three to four weeks for the rest of the season until late October due to a decrease of change in the vegetation in the summer, or for practical reasons (instrument availability and other academic activities).

Table 2. Dates, sky, tidal levels and number of stations of the in situ data collection.

Date of Sampling	Weather Conditions	Tide Level (m)	Number of Stations Visited
2019-05-17	Cloudy	0.5	10
2019-05-23	Partly cloudy	1.1	13
2019-06-05	Partly cloudy	0.5	13
2019-06-18	Clear sky	0.6	13
2019-07-04	Clear sky	0.3	13
2019-07-18	Partly cloudy	0.8	13
2019-08-13	Clear sky	0.9	13
2019-09-03	Clear sky	0.3	13
2019-10-01	Cloudy	0.2	13
2019-10-25	Cloudy	0.8	12

Plant allometry was determined for each station along with radiometric measurements and vertical photographs for the vegetation percent cover (every time). More specifically, plant allometry included the length and width of leaves, and it was determined for eelgrass (EG), the creeping saltbush (CS), and the saltmarsh cordgrass (SC) for every sampling date. For each harvested site, the number of plants (or shoots) were calculated inside the quadrat of ~2500 cm^2 Furthermore, the number of leaves by plant and their lengths and widths for EG and SC were measured on five plants chosen randomly inside the station. For the CS, the lengths of the stems were measured using five plants, chosen randomly. The lengths of the leaves and stems were measured using tape directly on the field. Solely at the end of the season, the vegetation inside the quadrat was harvested for biomass determination. The leaf area index (LAI) was calculated using a similar method by [37] and applied for the EG and SC. Three shoots were collected in the field and photographed with their leaves (flattened out on a white board next to a ruler). The leaf surface area (S_{Leaf}) was determined using the wand tool of ImageJ software [38]. S_{Leaf} was measured twice for the photosynthetic leaf surface (using only the green parts of the leaves) and for the total leaf surface (including the necrotic tissue). The LAI was then calculated based on the following, by Watson [37]:

$$LAI = \frac{S_{\text{Leaf}} \times D_{\text{shoot}}}{1000} \quad (1)$$

where S_{Leaf} is the mean surface area (cm^2) measured for the three shoots of a given quadrat and shoot density (D_{shoot}) is the density (N shoots m^{-2}) for the original value for the same quadrat.

2.2.1. In Situ Radiometry

Field radiometry was performed using a VIS-NIR analytical spectral device (ASD) spectroradiometer (Handheld2-pro model) recording wavelengths from 325 to 1075 nm. The upwelling radiance (L, Wm^{-2}sr^{-1}nm^{-1}) spectra were measured at nadir. To avoid any disturbance of the light field, the radiance measurements were collected with a 5 m bare optical fiber fitted to the ASD, having a field-of-view (FOV) of 25°. The fiber was attached at the end of a rod fixed to a tripod pointing the surface at nadir from 113 cm above the surface to obtain a ~2500 cm^2 surface area (Figure 3), similar to the quadrat area. For each spectrum recorded, the ASD collects five spectra and computes the average and standard deviation of the spectra. Before each surface radiance measurement (L_{surf}), a calibrated white reference (Spectralon plate, L_{ref}) with a known reflectance value was measured (99.57%) at nadir (R_{panel}). At least two sets of measurements (replicates) were made at each site. More measurements were performed under changing illumination due to the presence of thin clouds or cloudy conditions.

Figure 3. Acquisition of the emerged surface reflectance with the ASD fitted to a tripod (**A**) and typical vegetation target photographed at nadir: CS (**B**), EG (**C**), SC (**D**), and MA (**E**).

The surface reflectance, R_{surf} (unit less), was calculated following Equation (2).

$$R_{\text{surf}} = \frac{L_{\text{surf}} \times R_{\text{panel}}}{L_{\text{ref}}} \quad (2)$$

Because the R_{surf} were not collected on the same day and time of the day, under different sky conditions, sun-view geometry, and instrument settings (i.e., integration time) (Table 2), variations in magnitude due to illumination differences were present [39]. Similar issues were noticed in the literature [17], and a correction was then applied to R_{surf} following [40]. Briefly, the multiplicative scatter correction method (MSC; [40,41]) minimizes the uncertainty in the reflectance spectra due to different measurement conditions while keeping the main spectral features. MSC applies a simple correction to the raw reflectance value at each wavelength of the sample spectra based on the reference spectra. It uses the offset and the slope values estimated by linear least-squares regression of all the raw spectra collected in the field against a reference spectrum [41]. In MSC, the linear regressions are fit to a local wavelength region using a moving window of a specified length. The reference spectra applied in the MSC to the sample spectra were the mean spectra of each station's spectra collected during the growing season.

The spectral similarity between the R_{surf} spectra was estimated using the spectral angle mapper (SAM; rad), used [42] as:

$$SAM = \cos^{-1}\left(\frac{\sum_{k=1}^{B} x_k y_k}{\sqrt{\sum_{k=1}^{B} x_k^2} \times \sqrt{\sum_{k=1}^{B} y_k^2}}\right) \quad (3)$$

where B is the number of spectral bands, k the ith waveband, x and y are two R_{surf} spectra to compare, respectively. The SAM was applied to both raw and MSC-corrected spectra, but only the latter were used for further analysis. The SAM was used to examine the spectra similarities/differences among different species, and for a given species during the growing season.

2.2.2. Spectral Resampling and Vegetation Indices

For comparison with satellite data, the measured hyperspectral R_{surf} were resampled to the multispectral bands of OLI, MSI, RapidEye, and PlanetScope using the sensor spectral response function (SRF) provided by the space agencies. For all sensors, only the visible and the near-infrared bands (400–900 nm) were used for the analysis.

Many different vegetation indices (VIs) have previously been applied to multispectral remote sensing data [43]. In the present study, VIs suitable for the satellite sensors were tested (Table 3) to identify which indices respond best to seasonal variability and vegetation density metrics (i.e., LAI).

Table 3. Vegetation indices applied to the reflectance spectra and the satellite images.

Name	Equation	Source
Green–Red Vegetation Index (GRVI)	$\frac{\rho_{green} - \rho_{red}}{\rho_{green} + \rho_{red}}$	Motohka et al., 2010 [44]
Normalize Difference Aquatic Vegetation Index (NDAVI)	$\frac{\rho_{nir} - \rho_{blue}}{\rho_{nir} + \rho_{blue}}$	Villa et al., 2014 [45]
Normalize Difference Vegetation Index (NDVI)	$\frac{\rho_{nir} - \rho_{red}}{\rho_{nir} + \rho_{red}}$	Tucker, 1979 [46]
Soil Adjusted Vegetation Index (SAVI)	$\frac{\rho_{nir} - \rho_{red}}{\rho_{nir} + \rho_{red} + 1} \times (1 + L)$	Huete, 1988 [47]
Water Adjusted Vegetation Index (WAVI)	$\frac{\rho_{nir} - \rho_{blue}}{\rho_{nir} + \rho_{blue} + 1} \times (1 + L)$	Villa et al., 2014 [45]

Note: where L is the magnitude of green vegetation cover.

2.3. Satellite Data

Cloud-free satellite images from OLI, MSI, RapidEye, and PlanetScope captured between May and October 2019 were downloaded for the data providers (USGS EarthExplorer, earthexplorer.usgs.gov; Copernicus Open Access Hub, scihub.copernicus.eu; and PlanetExplorer planet.com/explorer/ all accessed on 16 June 2022).

A total of 16 images (Sentinel-2 (6), Landsat-8 (2), RapidEye (2), and PlanetScope (6)) were selected based on the date, the state of the tide, and the cloud cover, i.e., a low tide under a clear sky (less than 15% of cloud cover) (Table 4).

Level 1 images (top-of-atmosphere radiance) were atmospherically corrected to retrieve the surface reflectance using ACOLITE (version Python 20190326.0), which was adapted for all sensors. To compute the surface reflectance, the dark spectrum fitting method was adopted using the standard parameters proposed by ACOLITE [48]. Briefly, after cloud masking and gas transmission correction of top-of-atmosphere reflectance, ACOLITE found multiple dark pixels in sub-scenes, to define a "dark spectrum" based on the lowest reflectance values in any spectral bands. The dark spectrum was then used to estimate the atmospheric path reflectance according to the best fitting aerosol model. Sky and sun glint components were also estimated over water surfaces, but not for land surfaces. The images were then projected, and the VIs were calculated. The homogeneous area of each vegetation species and sediment were delimited based on GPS collected on the field and from drone images (collected in September) to compare VIs from the images to the in situ. The VI values were extracted from those areas covering the stations visited in the field. For MA, due to the small patch size, subpixel fractions were calculated based on the sensor's pixel size using a very high spatial resolution drone image. For each area, the mean values of the VIs were calculated.

Table 4. List of the images acquired for the study and the water level when the sensor collected the images.

Sensors	OLI	MSI	RapidEye	PlanetScope
Acquisition dates/Water level		2019-05-05 (B)/0.6 m		2019-05-18 (DR)/0.4 m
		2019-06-04 (B)/0.5 m		2019-06-08 (DC)/0.8 m
				2019-06-18 (DC)/0.6 m
		2019-07-04 (B)/0.6 m	2019-07-04/0.6 m	2019-07-07 (DC)/0.5 m
				2019-08-05 (DC)/0.6 m
	2019-08-14/0.9 m			2019-08-15 (DR)/0.8 m
		2019-09-12 (B)/1.2 m	2019-09-03/0.3 m	2019-09-03 (DR)/0.3 m
		2019-09-30 (A)/0.1 m		
	2019-10-10/1.5 m	2019-10-20 (A)/1.6 m		2019-10-03 (DC)/0.8 m
Source	USGS	Copernicus	Planet	Planet

Note: A and B in the MSI column depict the satellite Sentinel-2A and Sentinel-2B, respectively. For PlanetScope—DC and DR correspond to Dove Classic and Dove-R sensors, respectively.

2.4. Classification

Image classification was performed using a machine learning algorithm named the extreme gradient boosted decision tree algorithm (hereafter referred to as XGBoost) [49]. To classify the vegetation, the XGBoost creates decision trees based on the input features and assigns specific weights to these features based on their significance in identifying a vegetation species, thereby defining the feature importance for the model. To train the XGBoost model, the in situ R_{surf} spectra associated with each vegetation type and sediment were used [50]. The in situ dataset was divided into training and testing datasets with a 4:1 ratio to train the model effectively. We first trained the model using the R_{surf} of each vegetation type and sediment collected during the growing season. This model was applied to the Sentinel-2 image time series (hereafter referred to as XGBoost^{noSeason}). Six images were classified from June to October (one per month). To cover all months of the growing season, an image from August 2020 was used as no clear sky images were available in August 2019. This process allowed us to identify the most favorable month to distinguish intertidal vegetation. Furthermore, to identify the best period in the season to classify the intertidal coastal habitats, XGBoost models were trained using the in situ R_{surf} collected closest to the date of the satellite image acquisition (hereafter referred to as XGBoost^{Season}).

The classification accuracy was evaluated using 170 different data points for which the vegetation classes were known based on the information collected from the study area and photo-interpretation of the drone or aerial photographs (white dots on Figure 1). A confusion matrix was generated and the overall accuracy, Kappa coefficient (κ), and per-class F1-score were calculated to assess the performance of the classification. The κ values ranged between 0 and 1; they are measures of accuracy that account for the random chance of correct classification [51]. The F1-score (ranging between 0–1) is the harmonic average of precision and recall for every class to deal with unbalanced validation data [52].

2.5. Satellite Vegetation Index Validation

The VI values from the reflectance spectra (in situ) and the multispectral satellite images were statistically analyzed for the different vegetation types. It was then possible to determine whether the VI values between vegetation types differed between in situ and satellite data. The accuracies of the satellite VI retrievals were evaluated using the root mean square error (RMSE), the mean relative difference (Bias), and Pearson's correlation coefficient (r), expressed respectively as:

$$RMSE = \sqrt{\frac{\sum_{i=1}^{N}(y_i - x_i)^2}{N}} \tag{4}$$

$$Bias = \frac{1}{N}\sum_{i=1}^{N}(y_i - x_i) \tag{5}$$

$$r = \frac{\sum_{i=1}^{N}(x_i - \overline{x})(y_i - \overline{y})}{\sqrt{\sum_{i=1}^{N}(x_i - \overline{x})^2}\sqrt{\sum_{i=1}^{N}(y_i - \overline{y})^2}} \tag{6}$$

where N is the number of samples in the data set, x_i the in situ value, and y_i the satellite estimated value of the parameter of interest; $\overline{x}$ and $\overline{y}$ are the corresponding mean values of the samples. We used a significance level of $\alpha = 0.05$ for all statistical analyses.

3. Results

3.1. Field Observations

3.1.1. Biophysical Parameters

Over the season, the number of leaves per shoot, the leaves, and the stem maximum length for *Zostera marina* (EG), *Spartina alterniflora* (SC), and *Atriplex prostrata* (CS) were measured (Figure 4A). For the SC and the CS, the maximum length increased from May to August, with lengths reaching about 35 cm (except for station S9 that remained shorter than 20 cm), which was 30 times longer than at the beginning of the season in mid-May. A sharp decrease in the absolute length of the stem and leaves occurred after the peak reach in August, with leaf length values at ~10 cm in October. In contrast, the EG leaf length at the beginning of the season was much longer (10 to 25 cm) than the other species. It must be noted that the length of the leaf was inversely proportional to the elevation relative to the sea level, with longer leaves in deeper waters (S1–S3). More than twice shorter leaves were observed at station S4 located at the fringe of the meadow in sightly shallower waters (Figure 2). The maximum lengths of the eelgrass leaves increased until September, reaching lengths >20 cm with a maximum of 80 cm and the absolute lengths decreased at all stations thereafter.

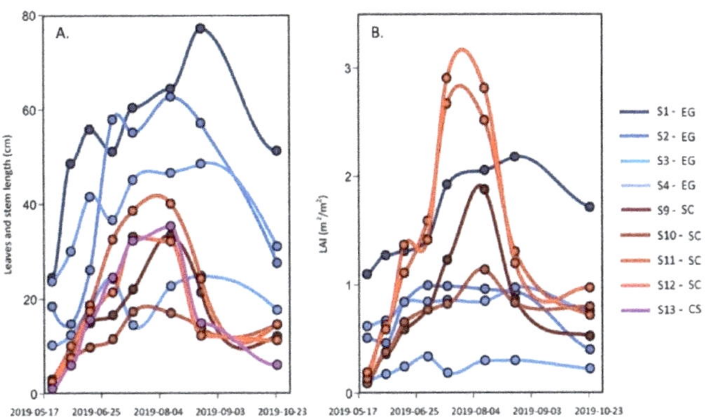

Figure 4. Seasonal evolution of (**A**) the leaves and stem maximal length and (**B**) the leaf area index (LAI) of the EG and SC, and the steam length of the CS from 23 May–25 October 2019.

The leaf area index (LAI), a commonly used proxy to quantify the density of vegetation canopy, allows the estimation of the plant canopy surface area per m^2 of the substrate (Figure 4B). For the EG, the LAI was somewhat more stable throughout the season and lower than the SC in the middle of the summer. The seasonal evolution of EG LAI is explained with tissue growth until August, and thereafter—a decrease (due to the appearance of necrosis on the leaves, which reduced the LAI based on the green part of the leaves). As for the SC, the LAI increased from May to July (S11 ad S12) or August (S9 and S10) reaching values > 1 m^2 m^{-2} (maximum close to 3 m^2 m^{-2}), and then decreased to values of 0.5 to 1 m^2 m^{-2} in October due to the pigment degradation in the tissue caused by the temperature change.

3.1.2. Spectral Signatures of Vegetation Types and Sediments

Figure 5 presents the average and standard deviation (shaded area) of each vegetation type calculated on all available raw and MSC-corrected R_{surf}, respectively. As expected, much larger variability was observed in raw spectra compared to the MSC-corrected, which was due to the seasonal evolution of the spectra, but also, to some extent, due to the illumination conditions that varied from date-to-date (e.g., a clear sky versus cloudy or partly cloudy; (Table 2). Despite a marked seasonality in terms of leaf length and LAI (Figure 4), the standard deviation of raw R_{surf} was much lower for *Spartina alterniflora* (SC) compared to the other vegetation species. The largest variability was observed in the NIR for creeping saltbush, followed by macroalgae and eelgrass. A homogeneous spectral variability was also observed for bare sediments, except a notable trough around 676 nm corresponding to the red peak of absorption because of chlorophyll-a. The MSC normalization removed most of the variability, including part of the seasonal variability. Interestingly, the spectral shape of the MSC-corrected spectra for MA in the visible bands showed virtually no variability, suggesting a very stable pigment composition across the growing season for this type of vegetation. For comparison, EG and SC showed the largest variability in the visible bands after the application of the MSC, followed by SD.

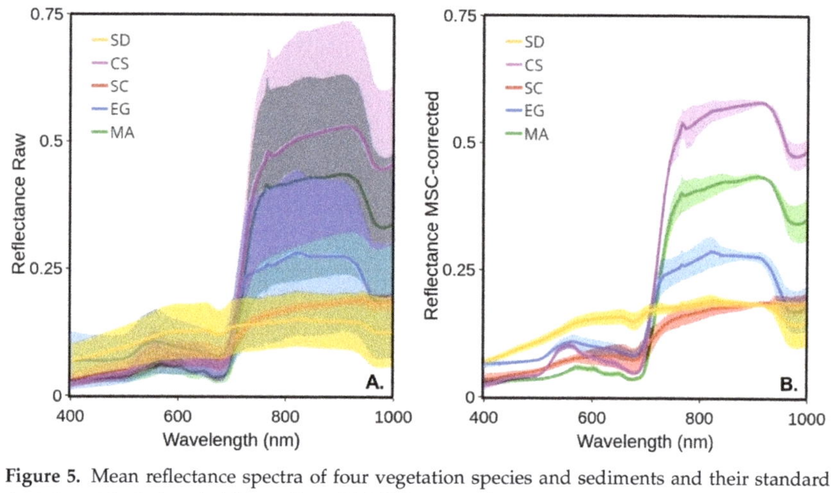

Figure 5. Mean reflectance spectra of four vegetation species and sediments and their standard deviations (shaded area): (**A**) raw R_{surf}; (**B**) MSC-corrected R_{surf}.

The SAM quantifies the spectral similarity and difference between the vegetation species (Figure 6). Sediment (SD) spectra differed markedly from all vegetation types (SAM > 0.29 rad), with the largest difference with MA. The spectral differences between the EG and MA, EG and CS, and MA and SC showed intermediate values (0.23 < SAM < 0.27 rad). Between the different vegetation types, MA and CS showed the smallest differences (0.1 rad), followed by SC versus CS (0.17 rad).

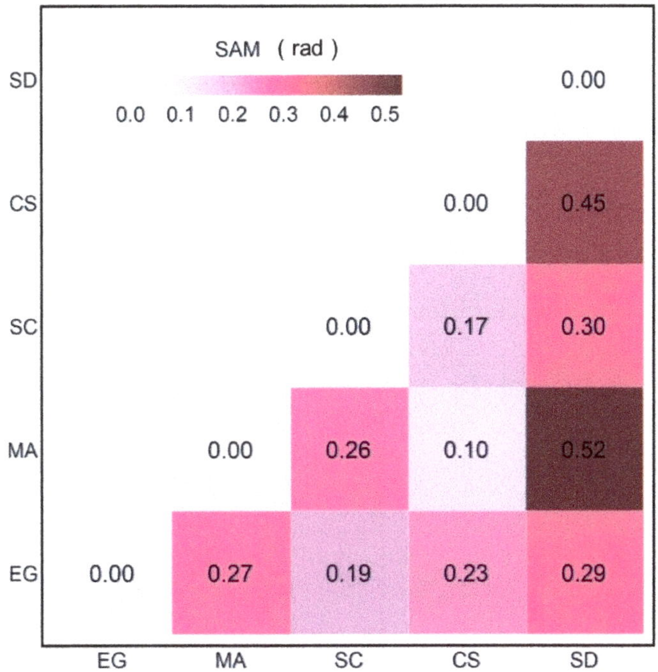

Figure 6. SAM analyses were applied to the MSC-corrected R_{surf} of four vegetation species and sediments.

3.1.3. Spectral Phenology

In general, the spectral shape of vegetation species remained almost the same throughout the season. However, there can be spectral crossovers between species due to changes in the absorption strengths and pigment concentrations. Therefore, it is necessary to consider intra- and inter-species spectral variability throughout the season.

In the visible spectrum, the reflectance peak of the healthy EG was stable over the season and was located between 555 and 564 nm (Figure 7C). The EG showed a very low reflectance peak at the beginning of the season (early May) before jumping up to the highest reflectance for this season at the end of May. Afterward, the spectra remained, with mid- to low peaks and values observed in the middle and at the end of the season. Two major absorption bands were visible and located in the blue (489–496 nm) and red (671–677 nm), respectively. Those absorption bands are typical of the absorption features of chlorophyll and carotenoids. Note that the spectrum observed on 23 May differed from the other. The exact reason for this anomaly could not be confirmed due to the unavailability of supporting data. At any other moment through the season, the EG spectra stayed similar without a clear seasonal evolution.

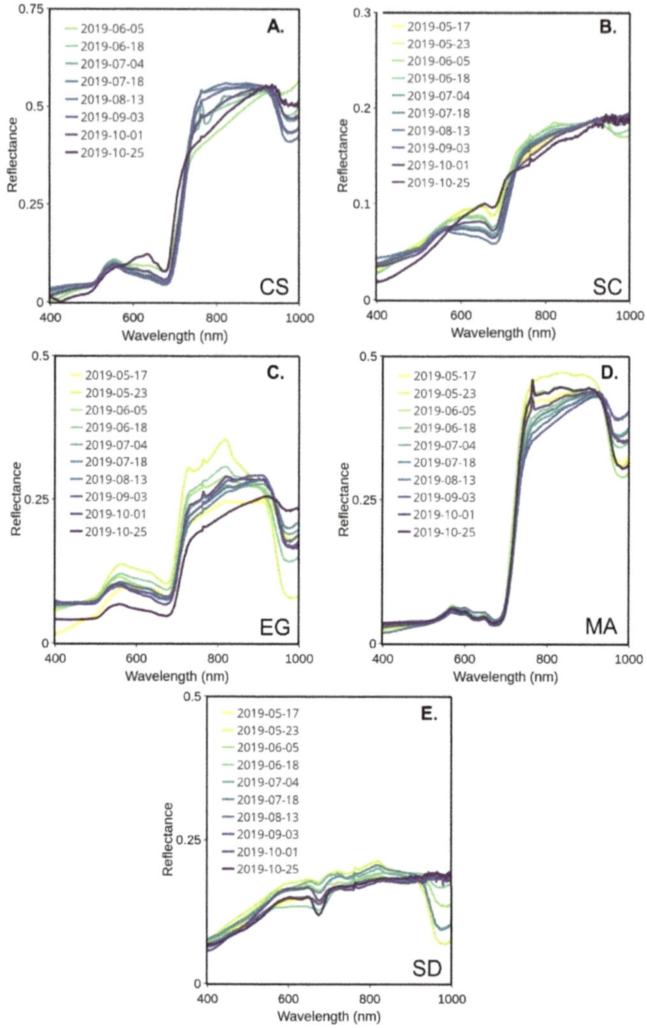

Figure 7. MSC-corrected reflectance spectra collected through the summer for four vegetation species and sediments collected from 17 May to 25 October 2019 for CS (**A**), SC (**B**), EG (**C**), MA (**D**), and SD (**E**).

The macroalgae spectra remained stable over the season (Figure 7D), with almost no change in the visible range compared to the NIR. The location of the green reflectance peak was regular, around 568 to 574 nm. Still, two absorption bands could be located, one in the green (504–520 nm) and the other in the red (671–678 nm). Secondary reflectance peaks are found in the red near 650 and 670 nm, respectively (c.f. Figure 7).

The peaks of reflectance of CS and SC were shifted to the longer wavelengths in the beginning (May to end of June) and at the end (mid-September to November) of the season (Figure 7A,B), but were almost the same as the EG in the middle of the summer (July to the middle of September). For the SC, the peak reflectance location shifted from 556 to 648 nm from the summer to the fall (Figure 7B). As for CS, the reflectance peak was located between 554 nm in the summer and shifted to 648 nm in October (Figure 7A). Both species presented similar absorption bands in blue (488–499 nm) and in red (672–680 nm) portions of the visible. The later corresponds to the red absorption peak of chlorophyll-a. The seasonal shift in the reflectance peak can be explained by the composition of the dominant pigment changing at the beginning and end of the season.

SAM was used to assess the seasonal variation in the spectral reflectance for each vegetation type and bare sediment (Figure 8). For MA and SC, the spectra did not change over the season with SAM values constantly below 0.18 and 0.14 rad, respectively (Figure 8B,D). Even though the SAM values were low for SC, we observed a difference in the spectral shape between the beginning/end and the middle of the season, but the differences remained small in terms of SAM. For example, the August spectrum was clearly different relative to the spectra measured in May, June, and the end of October (0.11 < SAM < 0.14 rad; c.f. Figure 8B). Among the four vegetation types, the CS showed the clearest seasonal evolution in terms of reflectance (Figure 8A). The values obtained by the SAM ranged from 0.02 to 0.27 rad. Spectra at the beginning (May and June) and at the end (September and October) of the season were similar, and different from those from the middle of the season (July to August). As for EG, the SAM values ranged from 0.04 to 0.38 rad (Figure 8C), with an anomalous value for the spectra collected on 23 May. The high SAM values for those spectra may be attributed to a particular moment corresponding to a high EG growth rate, or to the unstable sky conditions while collecting in situ radiometric data. As mentioned previously, the exact reason for this anomaly could not be confirmed due to the unavailability of supporting data. At any moment, the EG spectra stayed remarkably similar without a clear seasonal evolution.

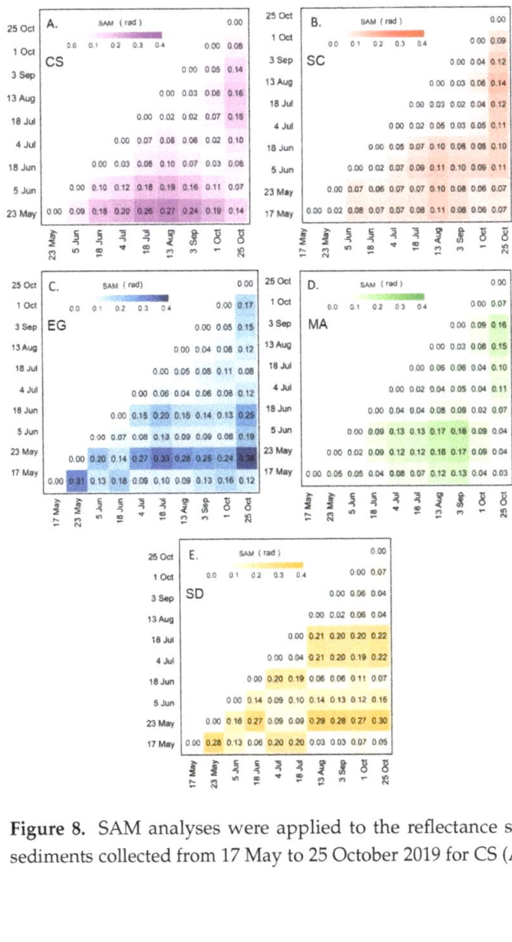

Figure 8. SAM analyses were applied to the reflectance spectra of four vegetation species and sediments collected from 17 May to 25 October 2019 for CS (**A**), SC (**B**), EG (**C**), MA (**D**), and SD (**E**).

Figure 9 shows the SAM calculated for a vegetation type using the spectra at a given date, against another date. This analysis helps to understand the variations of spectral changes of a species relative to another during the summer season. The most significant difference between all vegetation species spectra occurred at the beginning (May and June) and the season's end (September and October). For example, Figure 9A compares creeping saltbush (CS) with the other vegetation types (EG, MA, SC) and sediments (SD). The largest difference between CS and the other was in May, except for SC, where the SAM remained stable during the growing season (0.2 rad). In June and July, CS and MA were very similar and difficult to distinguish spectrally (SAM < 0.1 rad) Figure 9A,D. A similar remark can be made for EG, where the lowest SAM were generally found in July. It also suggests that EG differs more from the other vegetation types in the spring (end of May) and in early October. Saltmarsh cordgrass (SC) is more easily distinguished from EG and MA at the beginning and end of the season (Figure 9B), but more different from sediment in July. The sediment spectra are different from all ofthe vegetation spectra (Figure 9E). Overall, this analysis indicates that it can be difficult to distinguish many vegetation species based on a single date in the middle of the growing season (June to August).

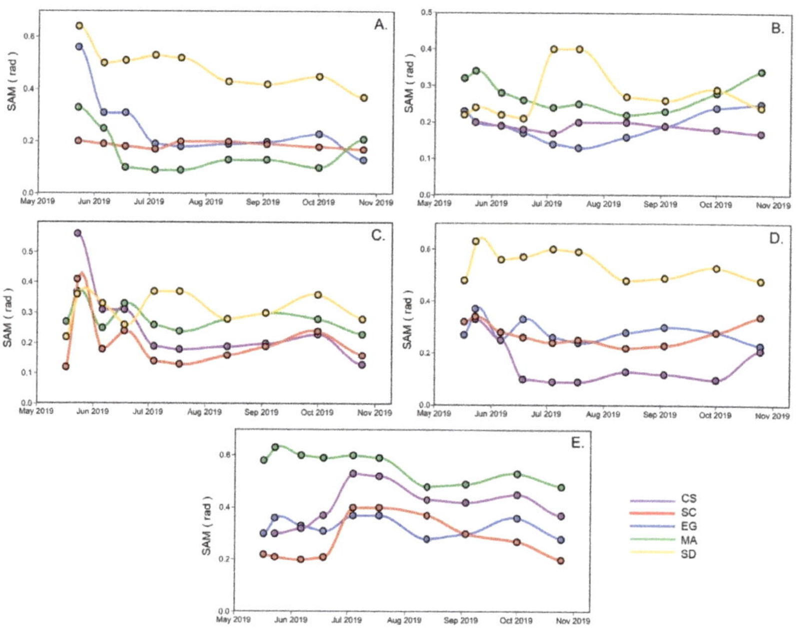

Figure 9. SAM analyses applied to the reflectance spectra according to the vegetation species and sediment over time, collected from May to the end of October for CS (**A**), SC (**B**), EG (**C**), MA (**D**), and SD (**E**).

3.2. *Classification of Vegetation Type*

To examine the implication of spectral and seasonal variability of the vegetation type for mapping the coastal intertidal zone using multispectral imagery, a classification algorithm was applied to the Sentinel-2 MSI time-series covering the growing season (Table 4). This sensor was selected because (i) it provides the best response to seasonal variability and availability over the critical dates for the vegetation phenology; and (ii) it has 13 spectral bands, including key bands in the red-edge portion of the spectrum that is suitable for vegetation mapping.

The XGBoost algorithm was first applied to the image using the in situ spectra associated with the date closest to the acquired image (XGBoost^{Season}, Figure 10). By changing the input spectra, the algorithm demonstrated a successful identification of the vegetation species, which confirmed the occurrences and spatial distributions by each species (Table 5). For the first and last images (May and October), most of the classification mistakes were due to confusion between saltmarsh cordgrass (*Spartina alterniflora*, SC) and the sediment (not shown) due to their similar spectral signatures at this time of the year (Figure 9B,E). On every image, the algorithm seems to have a problem identifying the macroalgae (MA). This might have been caused by the spatial resolutions of the images and the sizes of the MA patches attached to the boulders that were generally smaller than the pixel sizes. Note that none of the classified images present MA on the rocky shores of the island (west tip and north east coast of the island). Based on the classification evaluation metrics (overall accuracy and κ indices; (Table 5), early June and the end of September images were best suited for the coastal vegetation cover mapping. In other dates, the overall accuracy reached about 70% with the worst performance obtained in October (accuracy of 65% and κ of 0.56).

Table 5. Model validation statistics (overall accuracy (κ), and mean F1-score) of XGBoost^{Season}.

Date	Overall Accuracy (%)	κ	Mean F1-Score
2019-05-05	74	0.68	0.25
2019-06-04	71	0.63	0.20
2019-07-04	72	0.65	0.20
2020-08-05	69	0.61	0.24
2019-09-30	87	0.84	0.29
2019-10-20	65	0.56	0.23

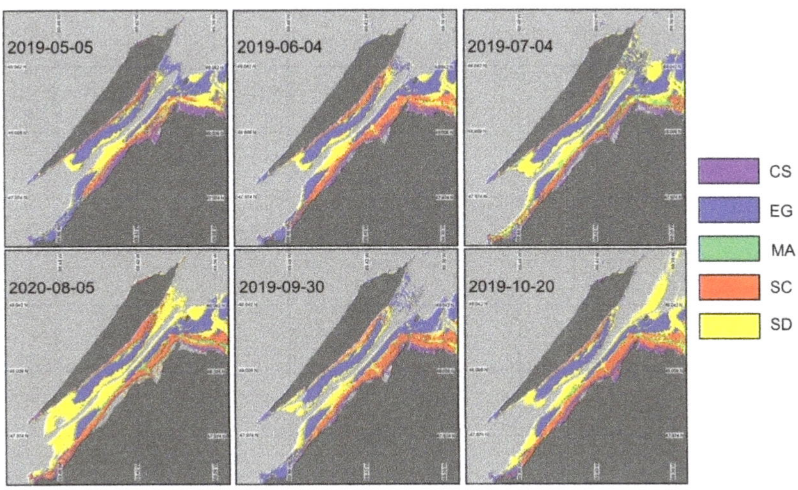

Figure 10. Classification of coastal and intertidal vegetation using XGBoost^{Season} applied to the Sentinel-2 time series.

Even though XGBoost^{Season} can classify the intertidal zone with significantly good accuracy, it is not practical to train separate models for every month of the growing season. Therefore, we trained a single XGBoost model (XGBoost^{noSeason}) (Table 6) using the same set of in situ spectra, without accounting for the seasonality. The goal was to evaluate the sensitivity of the classification to seasonal variability in the vegetation spectra.

XGBoost^{noSeason} yielded marked differences in vegetation mapping (Figure 11) throughout the season when spectral evolution was not taken into account. More sediment was identified in May, reflecting the early stage in the vegetation development and/or the low LAI. Macroalgae occupied a relatively large surface area, which was not consistent with the field observations. As expected, the classified image from May has the lowest accuracy and κ index (57% and 0.44, respectively). Between May and June, we see an expansion of the EG and SC surface areas. The images from July and August have a problem identifying other vegetation types than EG. In the summertime, the vegetation spectra from all types were too similar and could not be distinguished by the model trained without accounting for the seasonality in R_{surf}. The best performance of XGBoost^{noSeason} was obtained in September, with accuracy and κ values of 87% and 0.85, respectively. At this time, the vegetation was fully grown and the senescence had just started, yielding spectrally distinguishable surfaces. The October image yielded the second-best classification metric despite some senescence within the vegetation. It is interesting to note that the October image did not perform as well with the in situ spectra associated with the same date. Although the results of the two classifications for the October image are quite different, the validation results are almost similar. This can be explained by confusion from the algorithms between different species. For example, in the XGBoost^{Season}, the majority of the confusion is with the SD and the MA, while in the XGBoost^{noSeason}, the confusion is between SC and SD, resulting in similar validation statistics.

Table 6. Model validation statistics (overall accuracy (κ), and mean F1-score) of XGBoost^{noSeason}.

Date	Overall Accuracy (%)	κ	Mean F1-Score
2019-05-05	57	0.44	0.18
2019-06-04	59	0.47	0.17
2019-07-04	63	0.52	0.18
2020-08-05	57	0.45	0.19
2019-09-30	87	0.85	0.29
2019-10-20	65	0.55	0.23

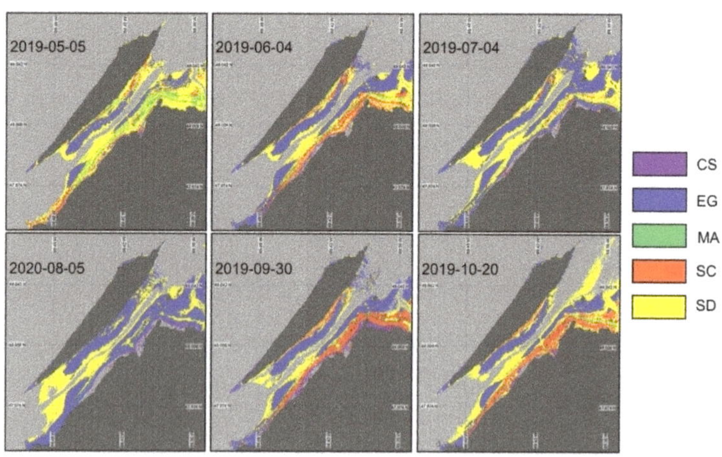

Figure 11. Classification of coastal and intertidal vegetation using XGBoost^{noSeason} applied to the Sentinel-2 time series.

Overall, only the eelgrass meadows were well classified (>85% accuracy) in all images, although they were over-represented in the middle of the summer. In contrast, confusion between MA and EG was observed for most images, as observed on the rocky coast of the island at the western tip and along the northeast coast.

3.3. *Remote Sensing of Coastal Vegetation Phenology*

We further assessed the potential of recent satellite constellations to monitor the phenology of intertidal vegetation found in the study area. The VI time series were computed on both in situ and satellite-derived R_{surf} from OLI, MSI, RapidEye, and PlanetScope. We first evaluated the robustness of each VI in terms of its capacity to predict the leaf area index (Section 3.3.1), a classical proxy for vegetation biomass, as well as its response to the seasonal variability in the vegetation. It turns out that the NDVI provided the highest correlation with LAI, which was selected for further analysis. After a comparison of in situ NDVI ($NDVI_{in\ situ}$) with satellite-derived NDVI ($NDVI_{sat}$) (Section 3.3.2), we examined the potential of this spectral index to monitor coastal vegetation phenology in our study area (Section 3.3.3).

3.3.1. Vegetation Indices as Predictor of LAI

Among the VIs listed in Table 3, the NDVI presented the best predictor of the LAI (highest r and RMSE). Figure 12 shows the linear relationship between LAI and the NDVI for EG and SC, separately. For both species, we found an important variability in terms of LAI at some stations at a given NDVI value. This is particularly evident at stations S1 and S2, where a dense EG cover (100% coverage throughout the growing season) produced almost constant NDVI value, while the LAI varied by a factor of 2. At other stations where the initial % coverage was lower (Table 1), seasonal variability was more evident, especially for the SC. Note that the MSC correction of R_{surf} tends to reduce the variability in NDVI, but increases the Pearson correlation with the LAI (Table 7), confirming the relevance for that correction.

Table 7. Pearson's correlation coefficient (r) of the LAI versus VI relationships obtained for EG and SC. The highest (**bold**) and lowest (*italic*) values are highlighted.

VIs	Raw Spectra		MSC-Corrected Spectra	
	EG	SC	EG	SC
NDVI	**0.59**	**0.50**	**0.65**	**0.51**
SAVI	0.38	0.36	0.53	0.41
WAVI	0.31	*0.04*	0.54	*0.01*
NDAVI	*0.20*	0.13	*0.21*	0.14
GRVI	0.28	0.44	0.33	0.49

3.3.2. Satellite-Derived NDVI Validation

Within a given time frame of two days, 15 match-ups, i.e., co-incident in situ and satellite observations, were obtained considering the four satellite sensors evaluated (Figure 13). The relatively strong linear relationship (r = 0.76) between the MSC-corrected $NDVI_{in\ situ}$ and $NDVI_{sat}$, and the low bias (−0.054) and RMSD (0.16) indicate good agreement between the in situ and satellite NDVI. The $NDVI_{in\ situ}$ values followed were clustered by vegetation type showing that the MA had the lowest values, followed by SC, EG, and CS. For all sensors, a much larger range of $NDVI_{sat}$ (−0.24 to 0.95) was obtained compared to $NDVI_{in\ situ}$ (0.13 to 0.75). This was also true for individual vegetation types (e.g., MA and EG), where the $NDVI_{in\ situ}$ varied in a narrow range compared to the satellite retrievals. One explanation for the low range in $NDVI_{in\ situ}$ was due to the MSC normalization of the in situ spectra (see below). Sentinel-2 MSI and Landsat-8 OLI presented the best relations between the

$NDVI_{in\ situ}$ and satellite values. The RapidEye and PlanetScope sensor provided generally good results for most of the vegetation species, except for the MA. These sensors had lower spectral resolutions (i.e., broad spectral bands), yielding outliers, as shown in Figure 13.

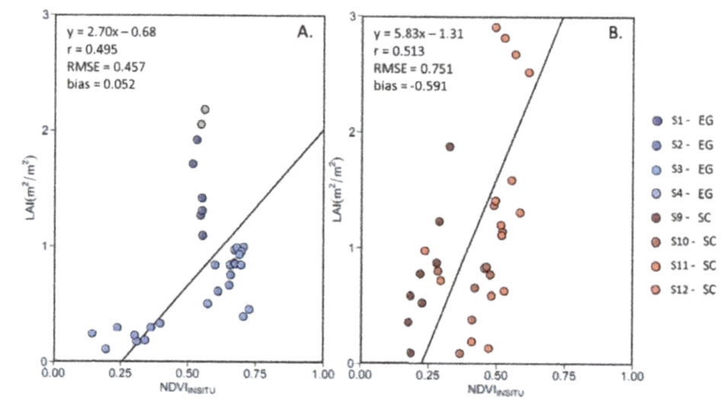

Figure 12. Linear relationship between the in situ MSC-corrected NDVI and LAI for (**A**) EG and (**B**) SC. Grey data points represent outliers.

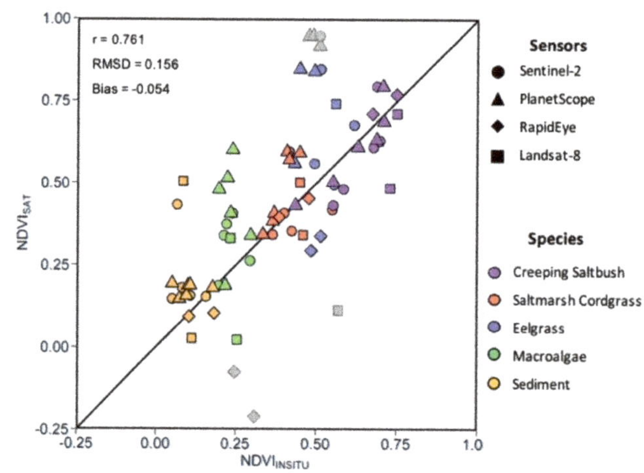

Figure 13. Scatter plot of in situ and satellite-derived NDVI. Different symbols are for sensors and colors for vegetation types (see legend). Grey data points represent outliers.

3.3.3. NDVI-Based Phenology

Figure 14 shows the seasonal evolution of NDVI for each vegetation type calculated from the in situ MSC-corrected (solid lines) and raw (dashed lines) reflectance spectra. Interestingly, there is not much crossover with the NDVI value among species; each has its specific range (as in Figure 13). The shape and range is similar for both normalized and raw spectra for SD and MA, but the normalization reduces the range for CS, SC, and EG. As expected, bare sediment (SD) or mud had the lowest NDVI values (<0.15), followed by SC, EG, CS, and finally MA. For SD and MA, the NDVI was relatively constant over the season with values of 0.04 ± 0.11 and 0.82 ± 0.07, respectively. Note that MA NDVI was for "pure" spectra while satellite-derived MA NDVI values considered the subpixel coverage (sediment mixed with MA). We can see a clear seasonal evolution for the CS with NDVI as obtained from MSC-corrected reflectance ranging between 0.44 and 0.76. For SC, the NDVI peaked in August at 0.49 and was minimum in May with a value of 0.33. For these two vegetation types, in particular, the MSC-correction substantially reduced the range of variability Figure 14. EG NDVI constantly increased from May to early October when computed on MSC-corrected spectra but showed two peaks in early July and early September when computed on raw spectra. Note that the decrease in NDVI in mid-summer (July–August) was likely due to the presence of necroses in the leaves. EG seasonal evolution also varied among stations with more marked evolution at S3 and S4 where the initial coverages were 50% and 20%, respectively (Table 1; Figure 4).

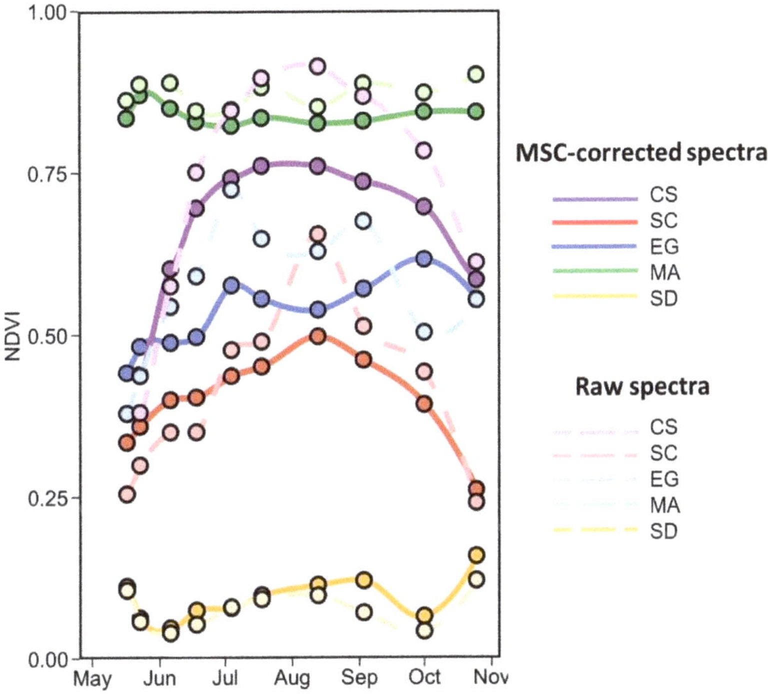

Figure 14. Temporal evolution of the NDVI values from the in situ spectra collected from May to the end of October of four vegetation species and sediments. Dashed lines are NDVI calculated on raw reflectance spectra.

Since only two images were available for Landsat-8 OLI and RapidEye sensors, respectively,we assessed the space-based phenology using Sentinel-2 and PlanetScope constellations. Six images were available for the Sentinel-2A and 2B MSI sensor from May until the end of October 2019, covering most of the critical moments for the coastal and saltmarsh growing season (Figure 15), but with a more than 2-month gap between 4 July and 12 September.

Eight images from mid-May to early October 2019 were available for the PlanetScope sensor (Figure 16). PS images filled the gap of MSI images, but no images beyond October 3 were available for the senescence period.

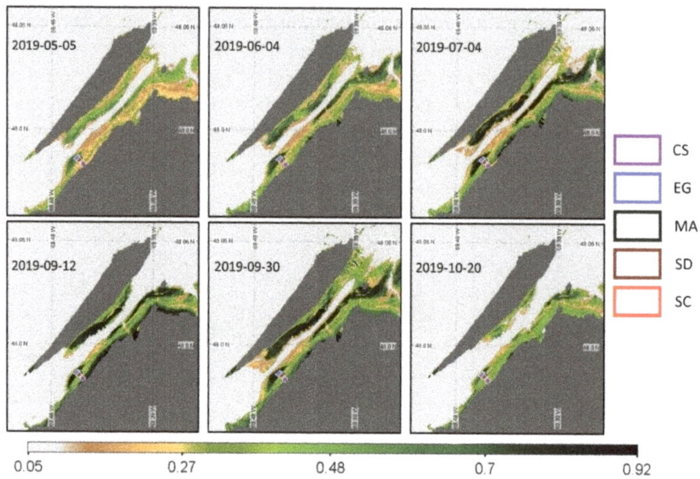

Figure 15. Time series of NDVI from the Sentinel-2 MSI acquisition obtained at a low tide. Homogeneous regions of interest (ROIs) for the four vegetation types and sediments where the NDVI values were extracted are also shown.

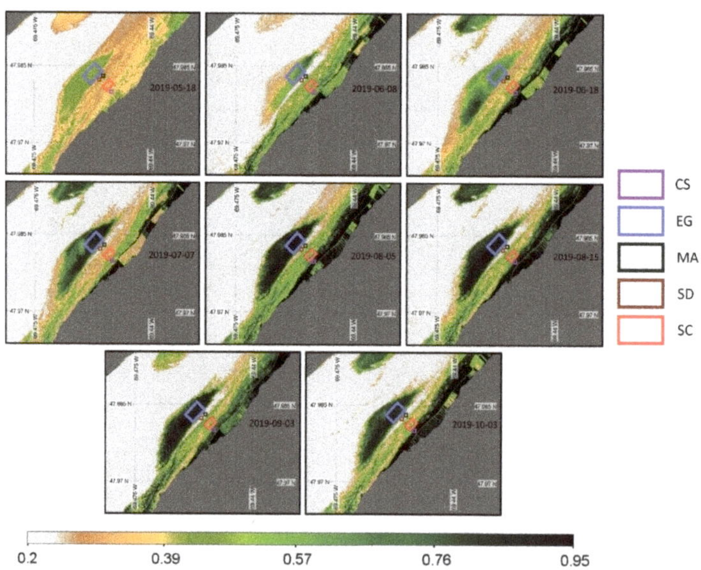

Figure 16. Time series of NDVI from the PlanetScope sensor acquisition obtained at a low tide and zoomed in on the region of interest. Homogeneous regions of interest (ROIs) for the four vegetation types and sediments where the NDVI values were extracted are also shown.

Figure 17 depicts the phenology of NDVI by combining the two satellite sensors for the pixels extracted from five homogeneous regions of interest (ROI) located in the sampling area (Figure 16). Steady growth in NDVI for all vegetation types was identifiable from May to August, while the sediment stayed relatively constant. The agreement between MSI and PS was very good for CS and SC throughout the season, except in September/October for SC. The phenology of these two vegetation types agree well with in situ observations (Figure 14) with peaks in August reaching 0.75 and 0.55 for CS and SC, respectively.

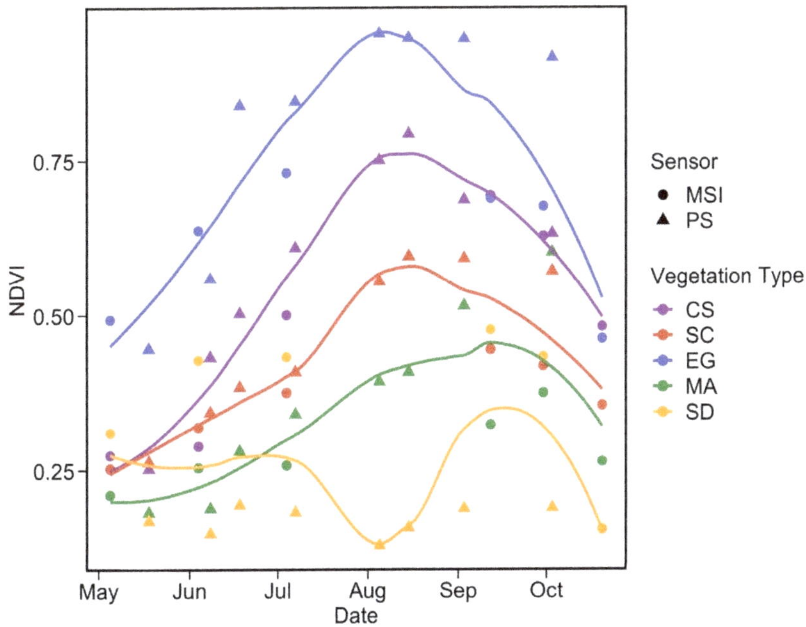

Figure 17. Satellite-based phenology of coastal vegetation from MSI and PS sensors for the four vegetation types and sediments. NDVI pixels values were extracted from homogeneous surfaces depicted in Figure 16. The LOESS fit was applied for each vegetation type to better visualize the phenology.

A more noisy signal was obtained for EG, which showed higher NDVI relative to other vegetation types. This result contrasts with in situ observations (Figure 14), but not with Figures 15 and 16 where dense EG stations (S1 and S2) constantly showed higher $NDVI_{sat}$ compared to $NDVI_{in\ situ}$. EG mapping was more sensitive to tidal differences among satellite acquisitions as it stood in deeper water (e.g., <0.2 m above sea level) relative to CS and SC. The phenology of this vegetation type was more evident from space than for field observations probably due to the percentage coverage of our selected stations.

Macroalgae phenology extracted from satellite images was more pronounced than expected from the in situ observations. Again, the large difference in MA phenology between in situ and the satellite was due to the spectral resolution versus the size of the vegetation patches. The lack of a continuous cover of MA limited our interpretation for this type of vegetation. The SD showed the lowest values and remained relatively constant throughout the season for a given sensor, although a small peak was observed in the late summer. A relatively important discrepancy was obtained between MSI and PS. The later yielded NDVI within the range of the in situ data with non-seasonal variability. In contrast, MSI values for the sediments were relatively high (0.25 to 0.4).

In summary, the images from PS and MSI sensors, allowed us to detect the seasonal evolution of the vegetation. By combining the time series of the two sensors, we can better see the complementary data, offering a better understanding and monitoring of the phenology of the vegetation.

4. Discussion

In this study, we showed that spectral seasonal evolution in coastal and intertidal vegetation types can be detected by in situ and spaceborne sensors. This evolution is related to the biophysical metric of the vegetation, such as the LAI, but it depends on the vegetation type and the pigment composition. A better understanding of the vegetation phenology provides key insights for the coastal mapping based on spectral classification, with strengths and limitations discussed in the following sections.

We observed an important evolution of the vegetation biophysical metrics, with a growth phase from May to September, which was particularly evident for saltmarsh cordgrass (*Spartina alterniflora*) and creeping saltbush. This observation was consistent with other studies [53–55]. For eelgrass, the growth continued until the beginning of October, while the LAI showed an increase between May and June, remained constant until September, and was followed by a small decrease at the end of the season. Similar results were shown in other studies [56–58].

Our vegetation R_{surf} spectra measurement is consistent with other studies [17,59–61]. The in situ spectra analyses suggest that the general shapes of R_{surf} are similar for all vegetation types (EG, MA, SC, and CS), but markedly different from the SD. Even if the shape is similar, they are different from each other and evolve over the season, as demonstrated using the SAM and the NDVI. This seasonal evolution is crucial for ecosystem remote sensing and is rarely mentioned and documented in other studies [62]. The spectra from mid-June to the beginning of September often overlapped. At that time of the growing season, the distinction among vegetation types was less obvious. Furthermore, the evaluation of the pure vegetation spectra revealed that MA and EG were stable throughout the season under 100% cover. In contrast, CS and SC presented more variability during the growing season with similar spectra at the start (May and June) and at the end (September and October), but different in the middle of the season (July to August).

Sentinel-2 images were chosen to classify the intertidal and coastal vegetation types in the study area. Using the identical spectra for all images, the XGBoost algorithm (XGBoost^{noSeason}) was not able to distinguish the vegetation types in July and August. Nevertheless, good results were obtained in June and at the end of the growing season (September and October). The highest accuracy (κ of 0.85) was obtained with the September 30th image at the onset of the senescence phase, suggesting that fully-grown vegetation is spectrally distinguishable from each other. We also demonstrate that accounting for vegetation phenology in the classification training dataset (XGBoost^{Season}) improved the classification accuracy by about 15%.

The NDVI was the most suitable proxy for the LAI among different VIs, and it was selected to evaluate the phenology from remote sensing using in situ spectra and multispectral images from four sensors (Landsat-8 OLI, Sentinel-2 MSI, PlanetScope, and RapidEye). The NDVI has been widely demonstrated to be a good descriptor of vegetation dynamics for many types of ecosystems, including wetlands [17,63–66]. It is also widely used in satellite-based phenology monitoring, as it could be applied to almost any multispectral sensors on a wide range of platforms (in situ, drone, plane, satellite) [67,68]. Based on in situ NDVI measurements, SC and CS clearly showed a seasonal evolution, while MA, EG, and SD were more stable throughout the season. Frequent observations are needed to quantify the seasonal variability and evolution of the vegetation in a cold temperate environment due to a relatively short growing season (<6 months). Changes in coastal vegetation, such as growth, flowering, senescence, and shedding of leaves, take weeks and even months [68–70]. For this reason, we recommend visiting the field as soon as the ice starts to melt (i.e., late March or beginning of April) [69,71]. In our study, we started the fieldwork in mid-May when the EG was already growing, with some stations showing 100% areal coverage.

By combining MSI and PS images, we could also track the vegetation phenology with a clearer signal for EG compared to the in situ. In general, satellite-derived phenology was consistent with the in situ NDVI values shown in this study and as presented by [17,60,72,73]. For example, Zoffoli et al. [17] also used MSI images to document the phenology of intertidal seagrass (*Zostera noltii*) in France. For comparison, these authors obtained 22 MSI images obtained at a low tide under cloud-free conditions for a ten-month period (March to December). With only six MSI images in our region, the combination with the PlanetScope constellation was necessary to fully capture the growing seasons and the following senescence. However, if our result demonstrates the feasibility, differences in spectral and spatial resolutions between sensors may complicate the combination of multisensor data for phenology monitoring.

As for Landsat-8 and Sentinel-2—both EO missions are dedicated to land cover monitoring and are widely used for phenology assessment. They both acquire images at lower temporal and spatial resolutions compared to PlanetScope. Even if the images are not available at a rapid rate due to tidal height and cloud cover restrictions, we can expect monthly images for the Sentinel-2 sensor, but much less for Landsat (one or two per year). Sentinel-2 images are easily usable and provide much better spectral resolution compared to the very-frequency acquisition of PlanetScope. Therefore, Sentinel-2 images allow further mapping capability by applying a classification algorithm using a high number of bands (10 bands). Still, Landsat is a very useful sensor, offering long-term time series for detecting and relatively good spectral capability for coastal mapping [74–76]. In addition, the temporal resolution will improve with the recent launch of Landsat-9 in September 2021, opening the door for data fusion from MSI and OLI [73,77,78]. The fusion of Landsat and Sentinel-2 images has tremendous potential to improve the ability to detect vegetation change and to cover all key periods of the vegetation phenology. However, in our case, combining the data was not necessary because the Sentinel-2 time series was already covering all of the vegetation phenology.

The PlanetScope sensor constellation is relatively new and offers many advantages, including high temporal resolution obtained at high spatial resolution compared to Landsat and Sentinel-2. Those images appear to be promising for extracting natural resources information, including intertidal ecosystem estimates and potentially even vegetation diversity estimates [20,79,80]. As seen in our results, the sensor provides many images covering most of the vegetation key. The spatial resolution of PlanetScope images provides many details, but it is not without issues. Indeed, the variable radiometric quality, inconsistent radiometric calibration across multiple platforms, and low spectral resolution are central challenges for marine and coastal applications, such as vegetation classification and ecosystem monitoring [20,81–83]. The low spectral resolution of the first Dove generation limits its use for the vegetation type classification. Furthermore, the noise level is high, and the radiometric quality and inconsistency are low on clusters of pixels, especially in homogeneous pixels. This indicates the low signal-to-noise ratio of PlanetScope images [20,84]. This issue was encountered during the atmospheric correction, and even after the correction, noise can still be detected in the NDVI images. The next generation of PlanetScope sensors launched in January 2022 with a greater number of bands (eight bands) and a radiometric signal that may have helped resolve most of the limitations of the early sensor fleets.

The presence of water overlying the vegetation at the time of in situ data acquisition or in the images highly affected the spectral reflectance due to its high absorption in the red and the infrared. With high water levels within the vegetation, the species identification can be difficult, and further processes will be needed to use the images and spectra [85–88]. For this reason, the state of the tidal level could significantly influence the spatiotemporal distribution of remotely-sensed parameters, such as vegetation NDVI, which uses red and infrared wavelengths. For example, the bathymetric map combined with water level measurements during the data acquisition could be used to correct the water column interference to retrieve the bottom reflectance [88]. However, estuarine water masses of the St. Lawrence are characterized by high concentrations of colored dissolved organic matter (CDOM) and suspended sediments that severely limit the light penetration, even in the visible bands [89], and impair water column correction. Due to the loss of spectral information in the NIR, submerged vegetation indices need to be based on visible bands only [40,88], or located in the red-edge portion of the spectra [59]. In conclusion, we recommend selecting images at the lowest tides possible to maximize the area to be mapped and to minimize the effect of water on the vegetation. The maximum tidal height that allows the vegetation to be mapped using multispectral data requires prior knowledge of the area. Furthermore, the mapped area will mainly depend on the area characteristics, such as the bathymetry/elevation and location of the vegetation.

Many atmospheric correction (AC) algorithms can be applied to multispectral images; this is a crucial step in the processing of remote sensing data for aquatic and coastal applications. Ideally, AC aims to separate the top-of-atmosphere observation by the satellite sensor into the signal from the atmosphere and the signal from the surface to retrieve surface reflectance [90]. By using PlanetScope and RapidEye sensors, we are limited by the atmospheric correction algorithm due to a lack of spectral bands (as, for example, in the SWIR). Even though we can apply different atmospheric corrections, the algorithm applied to the image needs to be the same to compare the sensors together. Here, we adopted ACOLITE, as it allows the application of the dark spectrum fitting (DSF) atmospheric correction method to all imagery evaluated in this study. Furthermore, ACOLITE has been developed for the coastal environment and is currently widely used for aquatic-based applications, such as coastal water monitoring [91–94]. The sensitivity of satellite-derived NDVI phenology to AC could have been quantified, which was out of the scope of this study.

The area covered by our sampling sites was minimal (472,800 m^2) as we focused our effort on one EG meadow and intertidal vegetation section to maximize our frequent visits to the sites. This area was selected for its diversity of vegetation cover and easy access but was nevertheless quite representative of the entire coast of the region. The small size of the studied area helped us to develop excellent knowledge of the vegetation dynamics and the ecosystem structure. However, the sites were not optimal for the detection of small macroalgae patches (e.g., <1–2 m^2) with the limited satellite spatial resolution.

5. Conclusions

In this work, we assessed the seasonal dynamics of four typical intertidal vegetation types encountered in cold temperate coastal littoral, including macroalgae (*Ascophyllum nodosum* and *Fucus vesiculosus*), eelgrass (*Z. marina*), saltmarsh cordgrass (*S. alterniflora*), and creeping saltbush (*A. prostrata*). The seasonal evolution was determined based on biophysical characteristics (leaf area index), in situ reflectance spectra, vegetation indices, and classification of multispectral images. We identified a significant seasonal change in phenology of saltmarsh cordgrass and creeping saltbush. Even though some seasonal change could be observed for some vegetation types, no significant changes were observed in the in situ reflectance spectra for eelgrass and macroalgae. Moreover, we evaluated the potential of the NDVI to quantify the vegetation phenology from space. We demonstrated that the NDVI was the best vegetation index proxy to track the phenology that could be applied to multispectral cameras (including drones). Satellite-based NDVI, which strongly correlates with in situ values for saltmarsh cordgrass and creeping saltbush, were used to assess the potential of multispectral instruments to assess the phenology. By combining Sentinel-2 and planet imagery, we showed that the seasonal evolution of eelgrass NDVI was more evident than with in situ measures, likely because of the initial coverage of the quadrats (2500 cm^2). The extreme gradient boosted decision tree (XGBoost) algorithm was applied to a monthly time series of Sentinel-2 using in situ spectra as input spectral classes. The results indicate September as the best month of the year to classify coastal vegetation in our cold temperate environment, i.e., when the vegetation is fully grown and spectrally distinguishable.

Further work is required to monitor the vegetation species from this complex ecosystem located in a cold temperate climate with a relatively short growing season. We intend to extend the vegetation species mapping, especially for marshes that have a high plant diversity (*Salicornia maritima*, *Spartina pectinata*, *Spartina patens*, etc.). Satellite remote sensing provides access to spatial scales, enabling the environment to be documented over vast areas. Widening the study area to cover all the coasts of the St. Lawrence maritime estuary and Gulf system would be interesting to know their conditions and interannual evolution. In addition, it will extend our knowledge of vegetated coastal ecosystems and their overall importance to the environment. Furthermore, it would be interesting to evaluate the carbon stock sequestration rates in coastal habitats (seagrass and marshes). Remote sensing tools are nowadays developed to quantify the extent of seagrasses and marshes, the species composition of these environments, and the above-ground biomass. In addition, some authors have demonstrated the possibility of estimating carbon stocks using empirical algorithms [95–97]. With these types of data, it will be possible to document and monitor changes in carbon stocks and estimate emissions as functions of ecosystem degradation, conservation, and restoration. Finally, historical data could be used to assess the history of carbon stocks and emissions and the spatial distribution and changes of vegetation species.

Author Contributions: B.L. contributed to the data collection and processing and wrote the original draft of this manuscript. S.B. supervised the work, came up with the original idea, contributed to the revision, wrote the manuscript versions, and provided funding. R.K.S. contributed to the atmospheric correction of satellite imagery, data processing (XGBoost classification), and revised the manuscript. P.B. co-supervised B.L., contributed to the fieldwork planning, and revised the manuscript. M.C. led the R.Q.M. project (see below) and provided funding. He contributed to the fieldwork planning and revised the latter version of the manuscript. All authors have read and agreed to the published version of the manuscript.

Funding: This research was mainly funded by the program Odyssée Saint-Laurent from the Réseau Québec Maritime (grant number 2018-38-04) attributed to MC (PI), PB, and SB (co-PIs). SB was provided funding from an NSERC discovery grant (RGPIN-2019-06070) and from the WISE-Man Project funded by the Canadian Space Agency through the FAST program (FARIMA18).

Data Availability Statement: The data that support the findings of this study are available from the author upon reasonable request.

Acknowledgments: The authors acknowledge Véronique Thériault and Colin Surprenant for sharing the drone images. The students who contributed to the field work are also acknowledged: Loïc Théberge-Dallaire, Jimmy Mayrand, François Pierre Danhiez, Soham Mukherjee, Romy Léger-Daigle, and Christel Blot. Furthermore, the authors are thankful to the three reviewers for their comments and suggestions, which improved the quality and readability of the article.

Conflicts of Interest: The authors declare no conflict of interest.

Abbreviations

The following abbreviations are used in this manuscript:

CS	creeping saltbush
EG	eelgrass
GRVI	green-red vegetation index
MA	macroalgae
MSI	multispectral instrument
NDAVI	normalize difference aquatic vegetation index
NDVI	normalize difference vegetation index
OLI	operational land imagery
PS	PlanetScope
RE	RapidEye
R_{surf}	surface reflectance
SAVI	soil-adjusted vegetation index
SC	saltmarsh cordgrass
SD	sediment
WAVI	water adjusted vegetation index
XGBoost	extreme gradient boosted decision tree algorithm

References

1. Boesch, D.F.; Josselyn, M.N.; Mehta, A.J.; Morris, J.T.; Nuttle, K.; Simenstad, C.a.; Swift, D.J.P.; Nuttle, W.K. Scientific Assessment of Coastal Wetland Loss, Restoration and Management in Louisiana. *J. Coast. Res.* **1994**, 1–103.
2. Zhang, X.; Wang, L.; Jiang, X.; Zhu, C. *Modeling with Digital Ocean and Digital Coast*; Springer: Berlin/Heidelberg, Germany, 2017.
3. Cahoon, D.R.; Hensel, P.F.; Spencer, T.; Reed, D.; McKee, K.L.; Saintilan, N. *Coastal Wetland Vulnerability to Relative Sea-Level Rise: Wetland Elevation Trends and Process Controls*; Springer: Berlin/Heidelberg, Heidelberg, 2006; pp. 53–65.
4. Gedan, K.B.; Kirwan, M.L.; Wolanski, E.; Barbier, E.B.; Silliman, B.R. The present and future role of coastal wetland vegetation in protecting shorelines: Answering recent challenges to the paradigm. *Clim. Chang.* **2011**, *106*, 7–29. [CrossRef]
5. Nicholls, R.J. Coastal flooding and wetland loss in the 21st century: Changes under the SRES climate and socio-economic scenarios. *Glob. Environ. Chang.* **2004**, *14*, 69–86. [CrossRef]
6. Airoldi, L.; Beck, M.W. Loss, Status and Trends for Coastal Marine Habitats of Europe. *Oceanogr. Mar. Biol.* **2007**, *25*, 345–405.
7. GIEC. *Climate Change 2007: Impacts, Adaptation and Vulnerability*; GIEC: Genebra, Switzerland, 2009; pp. 214–219.
8. Harley, C.D.; Hughes, A.R.; Hultgren, K.M.; Miner, B.G.; Sorte, C.J.; Thornber, C.S.; Rodriguez, L.F.; Tomanek, L.; Williams, S.L. The impacts of climate change in coastal marine systems. *Ecol. Lett.* **2006**, *9*, 228–241. [CrossRef]
9. Dekker, A.; Brando, V.; Anstee, J.; Suzanne, F.; Malthus, T.; Karpouzli, E. Remote Sensing of Seagrass Ecosystems: Use of Spaceborne and Airborne Sensors. In *Seagrasses: Biology, Ecology and Conservation*; Larkum, A.W.D., Orth, R., Duarte, C.M., Eds.; Springer: Dordrecht, The Netherlands, 2006; Chapter 15; pp. 347–359.
10. Hossain, M.S.; Bujang, J.S.; Zakaria, M.H.; Hashim, M. The application of remote sensing to seagrass ecosystems: An overview and future research prospects. *Int. J. Remote Sens.* **2015**, *36*, 61–114. [CrossRef]
11. Xie, Y.; Sha, Z.; Yu, M. Remote sensing imagery in vegetation mapping: A review. *J. Plant Ecol.* **2008**, *1*, 9–23. [CrossRef]
12. Carlotto, M.J. Effect of errors in ground truth on classification accuracy. *Int. J. Remote Sens.* **2009**, *30*, 4831–4849. [CrossRef]
13. Satyanarayana, B.; Mohamad, K.A.; Idris, I.F.; Husain, M.L.; Dahdouh-Guebas, F. Assessment of mangrove vegetation based on remote sensing and ground-truth measurements at Tumpat, Kelantan Delta, East Coast of Peninsular Malaysia. *Int. J. Remote Sens.* **2011**, *32*, 1635–1650. [CrossRef]
14. Hedley, J.D.; Russell, B.J.; Randolph, K.; Pérez-Castro, M.; Vásquez-Elizondo, R.M.; Enríquez, S.; Dierssen, H.M. Remote sensing of seagrass leaf area index and species: The capability of a model inversion method assessed by sensitivity analysis and hyperspectral data of Florida Bay. *Front. Mar. Sci.* **2017**, *4*, 1–20. [CrossRef]
15. Knyazikhin, Y.; Martonchik, J.V.; Diner, D.J.; Myneni, R.B.; Verstraete, M.; Pinty, B.; Gobron, N. Estimation of vegetation canopy leaf area index and fraction of absorbed photosynthetically active radiation from atmosphere-corrected MISR data. *J. Geophys. Res.* **1998**, *103*, 32239–32256. [CrossRef]
16. Phinn, S.; Roelfsema, C.; Dekker, A.; Brando, V.; Anstee, J. Mapping seagrass species, cover and biomass in shallow waters: An assessment of satellite multi-spectral and airborne hyper-spectral imaging systems in Moreton Bay (Australia). *Remote Sens. Environ.* **2008**, *112*, 3413–3425. [CrossRef]
17. Zoffoli, M.L.; Gernez, P.; Rosa, P.; Le Bris, A.; Brando, V.E.; Barillé, A.L.; Harin, N.; Peters, S.; Poser, K.; Spaias, L.; et al. Sentinel-2 remote sensing of Zostera noltei-dominated intertidal seagrass meadows. *Remote Sens. Environ.* **2020**, *251*, 112020. [CrossRef]
18. Wang, D.; Ma, R.; Xue, K.; Loiselle, S.A. The assessment of landsat-8 OLI atmospheric correction algorithms for inland waters. *Remote Sens.* **2019**, *11*, 169. [CrossRef]
19. Traganos, D.; Reinartz, P. Interannual change detection of mediterranean seagrasses using RapidEye image time series. *Front. Plant Sci.* **2018**, *9*, 1–15. [CrossRef]
20. Wicaksono, P.; Lazuardi, W. Assessment of PlanetScope images for benthic habitat and seagrass species mapping in a complex optically shallow water environment. *Int. J. Remote Sens.* **2018**, *39*, 5739–5765. [CrossRef]
21. Xue, J.; Su, B. Significant remote sensing vegetation indices: A review of developments and applications. *J. Sens.* **2017**, *2017*. [CrossRef]
22. Dekker, A.G.; Brando, V.E.; Anstee, J.M. Retrospective seagrass change detection in a shallow coastal tidal Australian lake. *Remote Sens. Environ.* **2005**, *97*, 415–433. [CrossRef]
23. Proença, B.; Frappart, F.; Lubac, B.; Marieu, V.; Ygorra, B.; Bombrun, L.; Michalet, R.; Sottolichio, A. Potential of High-Resolution Pléiades Imagery to Monitor Salt Marsh Evolution After Spartina Invasion. *Remote Sens.* **2019**, *11*, 968. [CrossRef]
24. Bolton, D.K.; Gray, J.M.; Melaas, E.K.; Moon, M.; Eklundh, L.; Friedl, M.A. Continental-scale land surface phenology from harmonized Landsat 8 and Sentinel-2 imagery. *Remote Sens. Environ.* **2020**, *240*, 111685. [CrossRef]
25. Drake, L.a.; Dobbs, F.C.; Zimmerman, R.C. Effects of epiphyte load on optical properties and photosynthetic potential of the seagrasses Thalassia testudinum Banks ex König and *Zostera marina* L. *Limnol. Oceanogr.* **2003**, *48*, 456–463. [CrossRef]
26. Marzialetti, F.; Giulio, S.; Malavasi, M.; Sperandii, M.G.; Acosta, A.T.R.; Carranza, M.L. Capturing coastal dune natural vegetation types using a phenology-based mapping approach: The potential of Sentinel-2. *Remote Sens.* **2019**, *11*, 1506. [CrossRef]
27. Babin-Roussel, V.; Didier, D.; Houde-Poirier, M.; Jean-Gagnon, F.; Lacombe, D.; Provencher- Nolet, L.; Morissette, A. *L'Île Verte: Portrait du territoire*; Technical Report; Département de biologie, chimie et géograpgie, Université du Québec à Rimouski: Rimouski, QC, Canada, 2011.
28. Martel, M.c.; Provencher, L.; Grant, C.; Ellefsen, H.F.; Pereira, S. *Distribution and Description of Eelgrass Beds in Québec*; Technical Report; Maurice Lamontagne Institute, Fisheries and Oceans Canada: Mont-Joli, QC, Canada, 2009.
29. US. Navy. *A Functional Glossary of Ice Terminology*; US Navy Hydrographic Office: Washington, DC, USA, 1952.

30. Environment and Climate Change Canada. *Baie-de-L'Isle-Verte National Wildlife Area Management Plan 2018*; Environment and Climate Change Canada, Canadian Wildlife Service, Quebec Region: Gatineau, QC, Canada, 2018; pp. 1–76.
31. Gauthier, J.; Rosa, J.; Lehoux, D. *Les Marécages Intertidaux Dans L'Estuaire du Saint-Laurent*; Environnement Canada, Service Canadien de la Faune: Gatineau, QC, Canada, 1980.
32. La Société De Conservation De La Baie de l'Isle-Verte. *Mise en valeur de l'habitat de poisson de la Réserve Nationale de Faune de L'Isle-Verte*; Rapport conjoint Société de conservation de la baie de l'Isle-Verte et Groupe Environnement Shooner pour la Direction de la gestion de l'habitat du poisson (DGHP); ministère des Pêches et des Océans Canada: Ottawa, ON, Canada, 1995.
33. Quintin, C.; Bernatchez, P.; Bu, T. Géomorphologie et diversité végétale des marais du Cap Marteau et de l'Isle-Verte, estuaire du Saint-Laurent, Québec. *Géographie Phys. Et Quat.* **2006**, *60*, 149–164. . [CrossRef]
34. Dionne, J.C. Schorre Morphology on the south shore of the St.Lawrence esturary. *Am. J. Sci.* **1968**, *266*, 380–388. [CrossRef]
35. Garneau, M. Cartographie et phyto-écologie du territoire côtier Cacouna-Isle-Verte. Master's Thesis, Université Laval, Québec, QC, Candada, 1984.
36. Dionne, J.C. An Estimate of Shore Ice Action in a Spartina Tidal Marsh, St . Lawrence Estuary, An Estimate of Shore Ice Action in a Spartina Tidal Marsh, St . Lawrence Estuary, Quebec, Canada. *J. Coast. Res.* **1989**, *5*, 281–293.
37. Watson, D.J. Comparative Physiological Studies on the Growth of Field Crops: II The Effect of Varying Mutrient Supply on Net Assimilation Rate and Leaf Area. *Ann. Bot.* **1947**, *11*, 375–407. [CrossRef]
38. Rasband, W.S. ImageJ: Image Processing and Analysis in Java. Astrophysics Source Code Library. 2012. ascl–1206. Available online: https://ui.adsabs.harvard.edu/abs/2012ascl.soft06013R/abstract (accessed on 16 June 2022).
39. Qu, Y. Sea Surface Albedo. In *Comprehensive Remote Sensing*; Liang, S., Ed.; Elsevier: Amsterdam, The Netherlands, 2018; pp. 163–185.
40. Fyfe, S.K. Spatial and temporal variation in spectral reflectance: Are seagrass species spectrally distinct? *Limnol. Oceanogr.* **2003**, *48*, 464–479. [CrossRef]
41. Isaksson, T.; Kowalski, B. Piece-wise multiplicative scatter correction applied to near-infrared diffuse transmittance data from meat products. *Appl. Spectrosc.* **1993**, *47*, 702–709. [CrossRef]
42. Kruse, F.A.; Lefkoff, A.B.; Boardman, J.W.; Heidebrecht, K.B.; Shapiro, A.T.; Barloon, P.J.; Goetz, A.F. The spectral image processing system (SIPS)-interactive visualization and analysis of imaging spectrometer data. *Remote Sens. Environ.* **1993**, *44*, 145–163. [CrossRef]
43. Bargain, A.; Robin, M.; Méléder, V.; Rosa, P.; Le Menn, E.; Harin, N.; Barillé, L. Seasonal spectral variation of Zostera noltii and its influence on pigment-based Vegetation Indices. *J. Exp. Mar. Biol. Ecol.* **2013**, *446*, 86–94. [CrossRef]
44. Motohka, T.; Nasahara, K.N.; Oguma, H.; Tsuchida, S. Applicability of Green-Red Vegetation Index for remote sensing of vegetation phenology. *Remote Sens.* **2010**, *2*, 2369–2387. [CrossRef]
45. Villa, P.; Mousivand, A.; Bresciani, M. Aquatic vegetation indices assessment through radiative transfer modeling and linear mixture simulation. *Int. J. Appl. Earth Obs. Geoinf.* **2014**, *30*, 113–127. [CrossRef]
46. Tucker, C.J. Red and photographic infrared linear combinations for monitoring vegetation. *Remote Sens. Environ.* **1979**, *8*, 127–150. [CrossRef]
47. Huete, A. A Soil-Adjusted Vegetation Index (SAVI). *Remote Sens. Environ.* **1988**, *25*, 295–309. [CrossRef]
48. Vanhellemont, Q. Adaptation of the dark spectrum fitting atmospheric correction for aquatic applications of the Landsat and Sentinel-2 archives. *Remote Sens. Environ.* **2019**, *225*, 175–192. [CrossRef]
49. Chen, T.; Guestrin, C. XGBoost: A Scalable Tree Boosting System. In Proceedings of the 22nd ACM SIGKDD International Conference on Knowledge Discovery and Data Mining (KDD'16), San Francisco, CA, USA, 13–17 August 2016; ACM Press: New York, NY, USA, 2016; pp. 785–794.
50. Ghatkar, J.G.; Singh, R.K.; Shanmugam, P. Classification of algal bloom species from remote sensing data using an extreme gradient boosted decision tree model. *Int. J. Remote Sens.* **2019**, *40*, 9412–9438. [CrossRef]
51. Foody, G. On the compensation for chance agreement in image classification accuracy assessment. *Photogramm. Eng. Remote Sens.* **1992**, *58*, 1459–1460.
52. Lipton, Z.C.; Elkan, C.; Narayanaswamy, B. Thresholding Classifiers to Maximize F1 Score. *arXiv* **2014**, arXiv:1402.1892.
53. Anderson, C.M.; Treshow, M. A review of environmental and genetic factors that affect height in Spartina alterniflora loisel. (Salt marsh cord grass). *Estuaries* **1980**, *3*, 168–176. [CrossRef]
54. Egan, T.P. The Ecology, Physiology, and Molecular Biology of the Halophyte Atriplex Prostrata Boucher (Chenopodiaceae). Ph.D. Thesis, Ohio University, Athens, OH, USA, 1999.
55. Madakadze, I.C.; Coulman, B.E.; Mcelroy, A.R.; Stewart, K.A.; Smith, D.L. Evaluation of selected warm-season grasses for biomass production in areas with a short growing season. *Bioresour. Technol.* **1998**, *65*, 1–12. [CrossRef]
56. Cimon, S.; Deslauriers, A.; Cusson, M. Multiple stressors and disturbance effects on eelgrass and epifaunal macroinvertebrate assemblage structure. *Mar. Ecol. Prog. Ser.* **2021**, *657*, 93–108. [CrossRef]
57. Nellis, P.; Dorion, D.; Pereira, S.; Ellefsen, H.f. Monitoring of vegetation and fish in six eelgrass beds in Quebec (2005–2010). In *Canadian Technical Report of Fisheries and Aquatic Sciences 2985*; Pêches et Océans Canada, Institut Maurice-Lamontagne, 850, route de la Mer: Mont-Joli, QC, Canada, 2012; pp. 1–96.
58. Postlethwaite, V.R.; McGowan, A.E.; Kohfeld, K.E.; Robinson, C.L.; Pellatt, M.G. Low blue carbon storage in eelgrass (Zostera marina) meadows on the Pacific Coast of Canada. *PLoS ONE* **2018**, *13*, e0198348. [CrossRef]

59. Pu, R.; Bell, S.; Baggett, L.; Meyer, C.; Zhao, Y. Discrimination of seagrass species and cover classes with in situ hyperspectral data. *J. Coast. Res.* **2012**, *28*, 1330–1344. [CrossRef]
60. Tian, Y.Q.; Yu, Q.; Zimmerman, M.J.; Flint, S.; Waldron, M.C. Differentiating aquatic plant communities in a eutrophic river using hyperspectral and multispectral remote sensing. *Freshw. Biol.* **2010**, *55*, 1658–1673. [CrossRef]
61. Wicaksono, P.; Fauzan, M.A.; Kumara, I.S.W.; Yogyantoro, R.N.; Lazuardi, W.; Zhafarina, Z. Analysis of reflectance spectra of tropical seagrass species and their value for mapping using multispectral satellite images. *Int. J. Remote Sens.* **2019**, *40*, 8955–8978. [CrossRef]
62. Klemas, V. Remote Sensing Techniques for Studying Coastal Ecosystems: An Overview. *J. Coast. Res.* **2011**, *27*, 2–17. [CrossRef]
63. Dong, Z.; Wang, Z.; Liu, D.; Song, K.; Li, L.; Jia, M.; Ding, Z. Mapping Wetland Areas Using Landsat-Derived NDVI and LSWI: A Case Study of West Songnen Plain, Northeast China. *J. Indian Soc. Remote Sens.* **2014**, *42*, 569–576. [CrossRef]
64. Lv, J.; Jiang, W.; Wang, W.; Wu, Z.; Liu, Y.; Wang, X.; Li, Z. Wetland loss identification and evaluation based on landscape and remote sensing indices in Xiong'an new area. *Remote Sens.* **2019**, *11*, 2834. [CrossRef]
65. Suir, G.M.; Sasser, C.E. Use of NDVI and Landscape Metrics to Assess Effects of Riverine Inputs on Wetland Productivity and Stability. *Wetlands* **2019**, *39*, 815–830. [CrossRef]
66. White, D.C.; Lewis, M.M.; Green, G.; Gotch, T.B. A generalizable NDVI-based wetland delineation indicator for remote monitoring of groundwater flows in the Australian Great Artesian Basin. *Ecol. Indic.* **2016**, *60*, 1309–1320. [CrossRef]
67. Kang, X.; Hao, Y.; Cui, X.; Chen, H.; Huang, S.; Du, Y.; Li, W.; Kardol, P.; Xiao, X.; Cui, L. Variability and changes in climate, phenology, and gross primary production of an alpine wetland ecosystem. *Remote Sens.* **2016**, *8*, 391. [CrossRef]
68. Potter, C.; Alexander, O. Changes in vegetation phenology and productivity in Alaska over the past two decades. *Remote Sens.* **2020**, *12*, 1546. [CrossRef]
69. Blok, S.E.; Olesen, B.; Krause-Jensen, D. Life history events of eelgrass Zostera marina L. populations across gradients of latitude and temperature. *Mar. Ecol. Prog. Ser.* **2018**, *590*, 79–93. [CrossRef]
70. Smith, G.M.; Spencer, T.; Murray, A.L.; French, J.R. Assessing seasonal vegetation change in coastal wetlands with airborne remote sensing: An outline methodology. *Mangroves Salt Marshes* **1998**, *2*, 15–28. [CrossRef]
71. Robertson, A.I.; Mann, K.H. Disturbance by ice and life-history adaptations of the seagrass Zostera marina. *Mar. Biol.* **1984**, *80*, 131–141. [CrossRef]
72. Hmimina, G.; Dufrêne, E.; Pontailler, J.Y.; Delpierre, N.; Aubinet, M.; Caquet, B.; de Grandcourt, A.; Burban, B.; Flechard, C.; Granier, A.; et al. Evaluation of the potential of MODIS satellite data to predict vegetation phenology in different biomes: An investigation using ground-based NDVI measurements. *Remote Sens. Environ.* **2013**, *132*, 145–158. [CrossRef]
73. Wang, Q.; Blackburn, G.A.; Onojeghuo, A.O.; Dash, J.; Zhou, L.; Zhang, Y.; Atkinson, P.M. Fusion of Landsat 8 OLI and Sentinel-2 MSI Data. *IEEE Trans. Geosci. Remote Sens.* **2017**, *55*, 3885–3899. [CrossRef]
74. Alam, S.M.R.; Hossain, M.S. A Rule-Based Classification Method for Mapping Saltmarsh Land-Cover in South-Eastern Bangladesh from Landsat-8 OLI. *Can. J. Remote Sens.* **2020**, *47*, 1–25. [CrossRef]
75. Borfecchia, F.; Consalvi, N.; Micheli, C.; Carli, F.M.; Cognetti De Martiis, S.; Gnisci, V.; Piermattei, V.; Belmonte, A.; De Cecco, L.; Bonamano, S.; et al. Landsat 8 OLI satellite data for mapping of the Posidonia oceanica and benthic habitats of coastal ecosystems. *Int. J. Remote Sens.* **2019**, *40*, 1548–1575. [CrossRef]
76. Rapinel, S.; Bouzillé, J.B.; Oszwald, J.; Bonis, A. Use of bi-Seasonal Landsat-8 Imagery for Mapping Marshland Plant Community Combinations at the Regional Scale. *Wetlands* **2015**, *35*, 1043–1054. [CrossRef]
77. Li, P.; Ke, Y.; Wang, D.; Ji, H.; Chen, S.; Chen, M.; Lyu, M.; Zhou, D. Human impact on suspended particulate matter in the Yellow River Estuary, China: Evidence from remote sensing data fusion using an improved spatiotemporal fusion method. *Sci. Total Environ.* **2021**, *750*, 141612. [CrossRef] [PubMed]
78. Shao, Z.; Cai, J.; Fu, P.; Hu, L.; Liu, T. Deep learning-based fusion of Landsat-8 and Sentinel-2 images for a harmonized surface reflectance product. *Remote Sens. Environ.* **2019**, *235*, 111425. [CrossRef]
79. Hoa, N.H. Comparison of Various Spectral Indices for Estimating Mangrove Covers Using Planetscope Data: A Case Study in Xuan Thuy National Park, Nam Dinh Province. *J. For. Sci. Technol.* **2017**, *5*, 74–83.
80. Nazeer, M.; Waqas, M.; Shahzad, M.I.; Zia, I.; Wu, W. Coastline vulnerability assessment through landsat and cubesats in a coastal mega city. *Remote Sens.* **2020**, *12*, 749. [CrossRef]
81. Latte, N.; Lejeune, P. PlanetScope radiometric normalization and sentinel-2 super-resolution (2.5 m): A straightforward spectral-spatial fusion of multi-satellite multi-sensor images using residual convolutional neural networks. *Remote Sens.* **2020**, *12*, 2366. [CrossRef]
82. Mudereri, B.T.; Dube, T.; Adel-Rahman, E.M.; Niassy, S.; Kimathi, E.; Khan, Z.; Landmann, T. A comparative analysis of planetscope and sentinel sentinel-2 space-borne sensors in mapping striga weed using guided regularised random forest classification ensemble. *Int. Arch. Photogramm. Remote Sens. Spat. Inf. Sci. ISPRS Arch.* **2019**, *42*, 701–708. [CrossRef]
83. Traganos, D.; Cerra, D.; Reinartz, P. Cubesat-derived detection of seagrasses using planet imagery following unmixing-based denoising: Is small the next big? *Int. Arch. Photogramm. Remote Sens. Spat. Inf. Sci.* **2017**, *42*, 283–287. [CrossRef]
84. Dobrinic, D. Horizontal Accuracy Assessment of Planetscope, Rapideye and Worldview-2 Satellite Imagery. *Int. Multidiscip. Sci. Geoconference SGEM* **2018**, *18*, 129–136. [CrossRef]
85. Cho, H.J.; Lu, D. A water-depth correction algorithm for submerged vegetation spectra. *Remote Sens. Lett.* **2010**, *1*, 29–35. [CrossRef]

86. Hu, C. A novel ocean color index to detect floating algae in the global oceans. *Remote Sens. Environ.* **2009**, *113*, 2118–2129. [CrossRef]
87. Hyun, J.C.; Kirui, P.; Natarajan, H. Test of multi-spectral vegetation index for floating and canopy-forming submerged vegetation. *Int. J. Environ. Res. Public Health* **2008**, *5*, 477–483. [CrossRef]
88. Pe'Eri, S.; Morrison, J.R.; Short, F.; Mathieson, A.; Lippmann, T. Eelgrass and macroalgal mapping to develop nutrient criteria in new hampshire's estuaries using hyperspectral imagery. *J. Coast. Res.* **2016**, *76*, 209–218. [CrossRef]
89. Mohammadpour, G.; Gagné, J.P.; Larouche, P.; Montes-Hugo, M.A. Optical properties of size fractions of suspended particulate matter in littoral waters of Québec. *Biogeosciences* **2017**, *14*, 5297–5312. [CrossRef]
90. Vanhellemont, Q.; Ruddick, K. Atmospheric correction of metre-scale optical satellite data for inland and coastal water applications. *Remote Sens. Environ.* **2018**, *216*, 586–597. [CrossRef]
91. Fan, Y.; Li, W.; Gatebe, C.K.; Jamet, C.; Zibordi, G.; Schroeder, T.; Stamnes, K. Atmospheric correction over coastal waters using multilayer neural networks. *Remote Sens. Environ.* **2017**, *199*, 218–240. [CrossRef]
92. Ilori, C.O.; Pahlevan, N.; Knudby, A. Analyzing performances of different atmospheric correction techniques for Landsat 8: Application for coastal remote sensing. *Remote Sens.* **2019**, *11*, 469. [CrossRef]
93. Kuhn, C.; de Matos Valerio, A.; Ward, N.; Loken, L.; Sawakuchi, H.O.; Kampel, M.; Richey, J.; Stadler, P.; Crawford, J.; Striegl, R.; et al. Performance of Landsat-8 and Sentinel-2 surface reflectance products for river remote sensing retrievals of chlorophyll-a and turbidity. *Remote Sens. Environ.* **2019**, *224*, 104–118. [CrossRef]
94. Vanhellemont, Q.; Ruddick, K. Acolite for Sentinel-2: Aquatic applications of MSI imagery. *Eur. Space Agency* **2016**, *SP-740*, 9–13.
95. Angelopoulou, T.; Tziolas, N.; Balafoutis, A.; Zalidis, G.; Bochtis, D. Remote sensing techniques for soil organic carbon estimation: A review. *Remote Sens.* **2019**, *11*, 676. [CrossRef]
96. DeFries, R.; Achard, F.; Brown, S.; Herold, M.; Murdiyarso, D.; Schlamadinger, B.; de Souza, C., Jr. Earth observations for estimating greenhouse gas emissions from deforestation in developing countries. *Environ. Sci. Policy* **2007**, *10*, 385–394. [CrossRef]
97. Patenaude, G.; Milne, R.; Dawson, T.P. Synthesis of remote sensing approaches for forest carbon estimation: Reporting to the Kyoto Protocol. *Environ. Sci. Policy* **2005**, *8*, 161–178. [CrossRef]

land

MDPI

Article

Evaluation of the Gross Ecosystem Product and Analysis of the Transformation Path of "Two Mountains" in Hulunbuir City, China

Na Zhao [1,2], Hui Wang [1,*], Jingqiu Zhong [1,3,4], Yun Bai [2] and Sang Yi [2]

1 School of Geography, Liaoning Normal University, Dalian 116029, China
2 Department of Business and Tourism, Hulunbuir Vocational Technical College, Hulunbuir 021000, China
3 Center for Studies of Marine Economy and Sustainable Development, Liaoning Normal University, Dalian 116029, China
4 State Key Laboratory of Resources and Environmental Information System, Institute of Geographic Sciences and Natural Resources Research, CAS, Beijing 100101, China
* Correspondence: wanghui@lnnu.edu.cn; Tel.: +86-13840841336

Abstract: The objective assessment of ecological systems forms the basis of solving ecological environmental problems. Evaluating the ecosystem status of each county through the gross ecosystem product (GEP) can reveal the value of each ecosystem. In this study, we used the eco-economic method to calculate the GEP and the green gold index (GGI) of 13 counties in Hulunbuir City between 2015 and 2020. The results show that: (1) The GEP of Hulunbuir City in 2020 was 980.025 billion yuan. The GGI was 8.36, which was much higher than the national average. (2) Forestry and pastoral regions were the main contributors to the regulation service. (3) Hulunbuir City had the largest forest value, while the farmland value was the lowest. The most important sources of forest, grassland, wetland, water, and farmland value were Oroqen, Xin Right Banner, Xin Left Banner, Xin Right Banner, and Morin Banner, respectively. Based on our analysis, we found significant results through the transformation of the "Two Mountains" in Erguna, Genhe, and Zhalantun. The other counties in our study must optimize ecological research with respect to the traditional economic model. Our results provide a scientific reference for the application of the "Two Mountains" base in each county.

Keywords: gross ecosystem product (GEP); green gold index (GGI); "Two Mountains" transformation; county territory

Citation: Zhao, N.; Wang, H.; Zhong, J.; Bai, Y.; Yi, S. Evaluation of the Gross Ecosystem Product and Analysis of the Transformation Path of "Two Mountains" in Hulunbuir City, China. *Land* **2023**, *12*, 63. https://doi.org/10.3390/land12010063

Academic Editors: Matteo Convertino and Jie Li

Received: 19 November 2022
Revised: 16 December 2022
Accepted: 19 December 2022
Published: 26 December 2022

1. Introduction

The rapid development of the global economy has negatively affected the ecological environment. Ecological and environmental crises, such as ecosystem function degradation, excessive natural resource consumption, biodiversity reduction, and pollution, have adversely affected the well-being of humans [1]. These global concerns have led to an increased focus on ecological and environmental management and sustainability [2]. Ecological resources not only provide a variety of tangible products but also provide ecosystem services [3]. Assessment of ecosystem status will aid in the sustainable utilization of ecological resources and help in the development of solutions [4]. Since Constanza [5] and Daily [6] published their articles on the value of ecosystem services, numerous scholars have conducted a series of similar studies worldwide [7–11]. As a result of growing research, scholars have proposed the incorporation of ecological benefits into the evaluation systems of economic and social development [12]. Consequently, countries worldwide have started accounting and auditing the value of natural resources. Concepts such as green gross domestic product, system of environmental-economic accounting, and gross economic-ecological product have emerged [13–15]. Ouyang (2013) proposed the concept of the gross ecosystem product (GEP), corresponding to the gross domestic product

(GDP) [16]. GEP is the sum of the final material products and service values provided by ecosystems for human well-being and sustainable economic and social development, including the ecological product value, ecological regulation service value, and ecological culture service value [17]. Thus, the GEP forms an independent ecosystem value accounting system. The evaluation of GEP directly reveals the substantial value of an ecosystem in the study area [18]. Ma [19], Bai [20], and Dong [21] evaluated GEP in China, Yunnan Province, and Ordos City, respectively. County region is the basic administrative unit in China and the micro-subject and key link of ecological civilization construction [22]. Scholars realized the importance of assessing the effectiveness of ecological protection at the county scale, and GEP research is gradually focusing on that scale. You [22], Yu [23], and Pema [24,25] evaluated the GEP of China's Eshan County, the Chenggong District, and Garzê Tibetan Autonomous Prefecture and Xishui County, respectively. However, few studies have compared GEP among counties. Not only can county-scale GEP be used to understand the status of these local ecosystems, but more importantly, the joint effect of comparing GEP among counties plus environmental protection can help improve the GEP at a larger scale. Therefore, this study calculated, compared, and analyzed the GEP of 13 counties in Hulunbuir City and obtained the GEP of Hulunbuir City by summing the GEP of the counties. Compared with a direct accounting of Hulunbuir City GEP, the results of this study have more application value and social significance. By focusing on individual counties, targeted ecological protection measures can be implemented to ultimately improve the ecosystems of the greater Hulunbuir City area, which is one of the innovations of this study.

GEP accounting methods mainly include energy-value and quality-value evaluation methods [26], with the quality-value method being used globally. The GEP accounting process comprises two parts: (1) ecosystem service quality and (2) ecosystem service value [16]. The output of ecological products and the quantity of the ecosystem service function, namely the ecosystem service quality, are calculated through various ecological models. The units of ecosystem service quality are not the same. Accounting for the ecosystem services' value requires transforming the ecosystem service quality into a uniform and directly linear unit [27]. Then, the values can be summed up to obtain the GEP. In previous studies, no uniform standard was developed for this calculation. GEP values are different when different evaluation methods, evaluation indexes, and price parameters are selected for the same region. Jin [28] and Fan [29] both calculated GEP in Guizhou Province, but the methods and indicators of evaluation were different, leading to varied results. These inconsistencies skew our understanding of GEP in a study area. To improve the science, standardization, and operability of GEP accounting, the National Development and Reform Commission of China issued the "GEP Accounting Specification (Trial)" in 2022.

While exploring the replicable, extensible, and demonstrable accounting models of GEP, scholars have promoted the transformation of GEP from an "accounting value" to a "policy point" [30]. Jin [28] incorporated GEP into the evaluation and analysis of ecological compensation performance assessment in Guizhou Province. To better understand how much ecosystem value there is, Dong [21] applied GEP to a comparative analysis of ecological stock and flow in Ordos City, China. Chen [31] studied the coupling relationship between GEP and ecological carrying capacity in Changting County, China, to provide a basis for evaluating the effectiveness of ecological protection. Similarly, Lin [32] applied GEP to identify ecological protection space in the Yangtze River Delta region, which is critical for maintaining regional ecological security. In practice, GEP is primarily studied in combination with GDP, which can provide a theoretical basis for government performance appraisal. The "clear waters and green mountains are as good as mountains of gold and silver" theory (referred to as the "Two Mountains" theory) links GEP with GDP [33]. The "clear waters and green mountains" represent the competitive natural resources and good ecological environment that provide ecological products and services for people's lives and survival. The "gold and silver mountains" represent regional economic conditions and people's livelihood related to income level [34]. The "Two Mountains" theory points out that the natural ecosystem not only has considerable ecological benefits, but the ecological value can also be converted into economic benefits and contribute to human well-being [35]. Quantifying the value of "clear waters and green mountains" is the basis for an efficient transformation from ecological resources to ecological assets and capital, and the GEP can be used to evaluate this value [36]. Ma [19] first used the green gold index (GGI) to link "clear waters and green mountains" with "gold and silver mountains," which is relatively mature in practical application. Cheng [37] and Chen [38] measured the transformation relationship between "clear waters and green mountains" and "gold and silver mountains" in Quzhou City and Ninghai County of Zhejiang Province, respectively, through the GGI, and discussed the transformation path of these "Two Mountains." As the basic unit of the "Two Mountains" theory, the county has the comparative advantages of small size, rapid transformation, and rapid effect [34]. Since 2017, the Ministry of Ecology and Environment has named 136 "Two Mountains" bases, of which the county-scale accounts for up to 75%.

Hulunbuir City is an important ecological barrier in China and a world-class eco-city with unique ecological advantages and capitalization. In recent years, Hulunbuir City has paid more attention to ecological protection, including controlling pollution and restoring ecosystems. However, there has been no quantitative analysis of the number of "ecological properties," the degree of environmental improvements, or the ecological differences among each county in Hulunbuir City. Therefore, referencing the 2022 "GEP Accounting Specification (Trial)," we screened evaluation indicators that best aligned with the ecosystem characteristics of Hulunbuir City so that the GEP would more truly reflect the region's environment and protections. Then, we conducted quantitative GEP analysis of the 13 counties in Hulunbuir City at the end of the 12th Five-Year Plan (2015) and 13th Five-Year Plan (2020) in accordance with the accounting method in the 2022 document. Then, the GGI of each county was calculated according to the GEP results. Finally, based on the GEP and GGI, the transformation path of "Two Mountains" in each county was explored, and the application of the "Two Mountains" base was further promoted in Hulunbuir City and surrounding counties. This study fills the gap of GEP comparative research at the county scale and applies the indicators and methods in the 2022 specification to measure the GEP of 13 counties. This study is an early implementer of the specification and has set an example for similar accounting in other counties. Furthermore, our study of the transformation path of "Two Mountains" at the county level can be replicated in the theoretical and practical research of similar counties in China or even around the world.

2. Materials and Methods

2.1. Study Area

Hulunbuir City (115°31′–126°04′ E, 47°05′–53°20′ N) is located in the northeastern Inner Mongolia Autonomous Region, with a total area of 262,000 km^2, accounting for 21.4% of the total area of Inner Mongolia. It has rich and diverse land resources and types and is characterized by abundant resources, such as grasslands, forests, minerals, water, and biology. Hulunbuir City is famous for "Great Grasslands, Great Forests, Great Wetlands, Great Lakes, and Great Snow." It has been dominated by agriculture and animal husbandry (single-industry areas) and has developed into a regional economy based on the planting, animal husbandry, coal power, coal chemicals, and processing and manufacturing industries, with logistics trade and tourism as important components. The city now has jurisdiction over 14 counties, forming four ecological economic zones according to the landform, land type, and industrial types [39]. Oroqen Autonomous Banner (Oroqen), Yakeshi City (Yakeshi), Genhe City (Genhe), and Erguna City (Erguna) belong to the forestry region. Xin Barag Right Banner (Xin Right Banner), Ewenki Autonomous Banner (Ewenki), Xin Barag Left Banner (Xin Left Banner), and Prairie Chenbarhu Banner (Chen Banner) belong to the pastoral region. Zhalantun City (Zhalantun), Arun Banner, and Daur Autonomous Banner of Morin Dawa (Morin Banner) belong to the agricultural region. Manzhouli City (Manzhouli) and Hailar Area (Hailar) are collectively referred to as the central urban region.

It can be seen in Figure 1 that there are several types of ecosystems in Hulunbuir City, including forests, shrubland, grassland, wetlands, water bodies (including lakes and rivers), and farmland. Among these, the forest ecosystem area is the largest, accounting for 54.77% of Hulunbuir City, mainly distributed in Oroqen, Yakeshi, Erguna, Genhe, Zhalantun, and Ewenki. The grassland ecosystem accounted for 34.43%, which was concentrated in Xin Right Banner, Xin Left Banner, Chen Banner, Ewenki, Hailar, and Manzhouli. The farmland ecosystem is distributed in Morin Banner, Arun Banner, Zhalantun, Oroqen, Yakeshi, Erguna, and other regions in a pattern of large dispersion and small aggregation. The water body ecosystem is mainly distributed in Xin Right Banner, Xin Left Banner, and Morin Banner. The wetland ecosystem is distributed in New Right Banner, Xin Left Banner, and E Banner.

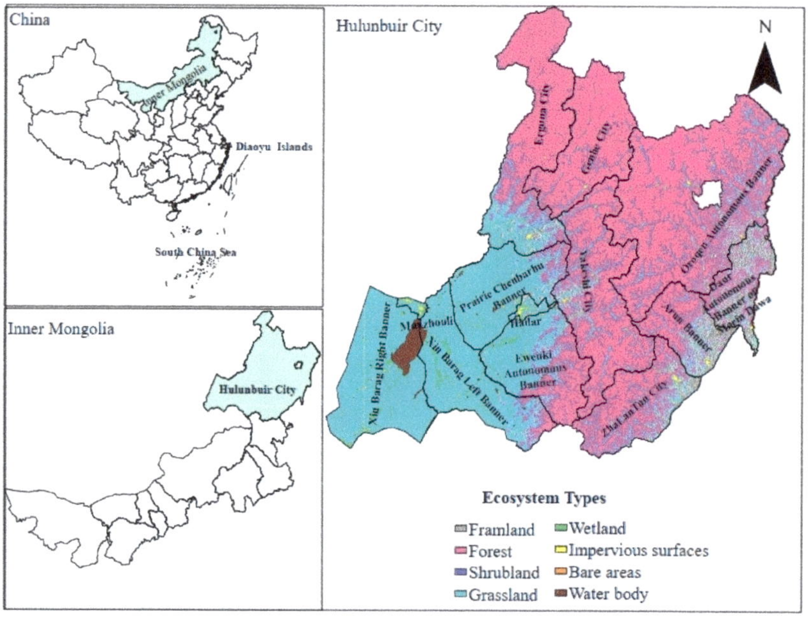

Figure 1. The ecosystem types of the study area (Hulunbuir City, China).

2.2. *Methodology*

2.2.1. Data Preprocessing

The Land-Use and Land-Cover Change (LUCC) of this study adopted 30 m land-use data released by the Remote Sensing Institute of the Chinese Academy of Sciences. Digital elevation models were obtained from the Resource and Environmental Science and Data Center of the Chinese Academy of Sciences. Meteorological data, such as rainfall, surface runoff, and evapotranspiration, were obtained from the Global Land Data Assimilation System and Hulunbuir Meteorological Bureau. Soil data were obtained from the National Tibetan Plateau/Third Pole Environment Data Center. The Net Primary Productivity and Normalized Difference Vegetation Index data were obtained from MOD17 and MOD13 products from the National Aeronautics and Space Administration. Data for the reservoir storage capacity and water resource consumption were obtained from the Hulunbuir Water Resources Bulletin. Social and economic data were all obtained from the statistical yearbook for Hulunbuir City and relevant data from the Statistics Bureau. The prices of all types of products and alternative products were derived from the Forest Ecosystem Service Function Assessment Specifications [40]. The above raster data were formed by stitching and cutting. The data year was the end of the 12th Five-Year Plan (2015) and the end of the 13th Five-Year Plan (2020) of Hulunbuir City.

2.2.2. Assessment Method for Ecosystem Service Quality

Water conservation is the interception and retention of precipitation by the ecosystem, thus increasing the available water, improving water quality, and regulating runoff. The water balance model was used to calculate water conservation:

$$Q_w = \sum_{i=1}^{n} (P - R - ET) \times A_i \times 10^{-3}, \tag{1}$$

where Q_w is the water conservation (m^3), P is rainfall (mm), R is surface runoff (mm), ET is evapotranspiration (mm), A_i is the area of the ecosystem i (m^2), and i and n are the ecosystem category and quantity, respectively.

Soil conservation is a function of reducing both the erosion capacity of rainwater and soil loss due to ecosystem action. The revised universal soil loss equation (RUSLE) was used to estimate soil conservation:

$$Q_s = \sum_{i=1}^{n} [R \times K \times L \times S \times (1 - C)] \times A_i \times 10^{-4}, \tag{2}$$

where Q_s is soil conservation (t), R is the rainfall erosivity factor ($MJ \cdot mm \cdot hm^{-2} \cdot h^{-1} \cdot a^{-1}$), K is the soil erosivity factor ($t \cdot hm^2 \cdot h \cdot hm^{-2} \cdot MJ^{-1} \cdot mm^{-1}$), L is the slope length factor (dimensionless), S is the slope factor (dimensionless), C is the vegetation cover factor (dimensionless), and A_i is the area of the ecosystem i (m^2).

Carbon fixation and oxygen release occur when ecosystems immobilize carbon in plants and soil to reduce the concentration of carbon dioxide in the air while releasing oxygen through photosynthesis. Carbon fixation and oxygen release were calculated via the carbon sequestration mechanism model:

$$Q_c = \sum_{i=1}^{n} 1.62NPP \times A_i \times 10^{-6} \text{ and} \tag{3}$$

$$Q_o = \sum_{i=1}^{n} 1.2NPP \times A_i, \tag{4}$$

where Q_c is carbon fixation (t), Q_o is oxygen release (t), and NPP is the net primary productivity of vegetation ($gC \cdot m^{-2}$). According to the photosynthesis equation, plants require 1.62 g of CO_2 to produce 1 g of dry matter and release 1.20 g of O_2. Finally, A_i is the area of the ecosystem i (m^2).

Climatic regulation is the effect of ecosystem cooling and humidification through vegetation transpiration and water surface evaporation. We calculated the climate regulation function using the evapotranspiration model:

$$E_{pt} = \sum_{i=1}^{n} EPP_i \times S_i \times D \times \frac{1}{3600r} \text{ and} \tag{5}$$

$$E_{we} = W_a \times E_p \times \gamma \times 10^{-3}, \tag{6}$$

where E_{pt} is the energy consumed by vegetation transpiration ($kW \cdot h$), E_{we} is the energy consumed by the evaporation of water ($kW \cdot h$), EPP_i is the transpiration consumption per unit area of the ecosystem i ($kJ \cdot m^{-2} \cdot d^{-1}$), S_i is the area of the ecosystem i (m^2), D is the number of days with air conditioning use (d), r is the energy efficiency ratio of the air conditioner (value of 3), W_a is the area of wetlands and water (m^2), E_p is the annual evaporation (mm), and γ is the power consumption of the humidifier to convert 1 m^3 of water into steam ($kW \cdot h$, $\gamma = 120$).

Air purification is the absorption and filtration of pollutants in the atmosphere by vegetation, such as SO_2, NO_x, and particulate matter, to reduce the concentration of air pollutants and improve air quality. We used the pollutant purification model to evaluate the air purification function:

$$Q_{ap} = \sum_{i=1}^{n} \sum_{j=1}^{m} Q_{ij} \times A_j \times 10^{-6}, \tag{7}$$

where Q_{ap} is the purification amount of air pollutants (t), Q_{ij} is the unit area purification of class i air pollutants in the class j ecosystem (t·km^{-2}), A_j is the area of the ecosystem j (m^2), i and n are the categories and quantities of the air pollutants, respectively, and j and m are the categories and quantities of the ecosystems, respectively.

Water purification is the adsorption and degradation of water pollutants by the ecosystem. The main pollutants in the water bodies are the chemical oxygen demand (COD), total nitrogen (TN), and total phosphorus (TP). The ecosystem reduces the concentration of COD, TN, and TP, and purifies the water environment. We used the pollutant purification model to evaluate the air purification function:

$$Q_{wp} = \sum_{i=1}^{n} \sum_{j=1}^{m} P_{ij} \times A_j \times 10^{-6}, \tag{8}$$

where Q_{wp} is the purification amount of water pollutants (t) and P_{ij} is the unit area purification of class i water pollutants in a class j ecosystem (t·km^{-2}).

Windbreak and sand fixation reduce wind erosion and sand damage by increasing the soil wind resistance. The revised wind erosion equation (RWEQ) was used to quantify the amount of windbreak and sand fixation:

$$Q_{sf} = \sum_{i=1}^{n} \left[0.1699 \times (WF \times EF \times SCF \times K')^{1.3711} \times \left(1 - C^{1.3711}\right) \times A_i \times 10^{-3} \right], \tag{9}$$

where Q_{sf} is the amount of windbreak and sand fixation (t), WF is the climate erosion factor (kg·m^{-1}), EF is the soil erosion factor (dimensionless), SCF is the soil crust factor (dimensionless), K' is the surface roughness factor (dimensionless), and C is the vegetation cover factor (dimensionless).

Flood storage is the ability of an ecosystem to reduce flood damage by regulating storm runoff and reducing flood peaks. Models for reservoir flood control capacity, lake adjustable storage volume, surface water lag, and vegetation regulation and storage were adopted to quantify the storage capacity of reservoirs, lakes, swamps, and vegetation, respectively.

$$C_r = 0.16C_t, \tag{10}$$

$$C_l = e^{5.653} \times A^{0.680} \times T \times 10^4, \tag{11}$$

$$C_m = 0.3S \times 10^6, \text{ and} \tag{12}$$

$$C_v = \sum_{i=1}^{n} \left(P - R_{fi}\right) \times A_i \times 10^{-3}, \tag{13}$$

where $C_r, C_l, C_m,$ and C_v are the flood control capacities of the reservoir, lake, swamp, and vegetation, respectively (m^3), C_t is the total reservoir storage capacity (m^3), A is the lake area (km^2), S is the swamp area (km^2), P is the rainstorm rainfall (mm), and R_{fi} is the rainstorm runoff of ecosystem i (mm).

For functional quantity accounting, the water balance model, RUSLE, carbon sequestration mechanism model, and RWEQ were suitable for the calculation of various ecosystem functions. The functional coefficients of the evapotranspiration model, pollutant purification model, and flood and storage model are different in the application of different ecosystem types.

2.2.3. Evaluation Method for Ecosystem Service Values

After measuring the ecosystem service functions, the market value, shadow project, replacement costs, and other methods were used to calculate the various ecosystem service values (Table 1).

Table 1. Evaluation method for ecosystem service value.

Category	Index	Methodology	Description
Ecological product value	Value of agricultural, forestry, animal husbandry, and fishery products	Market value method	Market price of agriculture, forestry, animal husbandry, and fishery, respectively
	Water resources value	Market value method	Market price of water for different purposes
Ecological regulation service value	Water conservation value	Shadow project method	Cost of building the reservoir
	Soil conservation value	Replacement cost method	Cost of dredging the reservoir, and cost of non-point source pollution treatment
	Value of carbon fixation and oxygen release	Reforestation cost method	Cost of afforestation
	Value of climate regulating function	Replacement cost method	Price of domestic electricity
	Value of air purification function	Replacement cost method	Cost of air pollutant control
	Value of water purification function	Replacement cost method	Cost of water pollutant control
	Wind-breaking and sand-fixing value	Replacement cost method	Cost of sand control projects
	Value of flood storage function	Shadow project method	Cost of building the reservoir
Ecological culture service value	Landscape recreation value	Travel cost method	Sum of direct tourist costs and consumer surplus

3. Results

3.1. Accounting for the GEP

The GEP of Hulunbuir City in 2020 was 980.025 billion yuan, showing an increase of 31.94% compared with that in 2015. In the county, the GEP of the forestry and central urban region accounted for 53.90% and 3.13%, respectively, and that of the pastoral and agricultural regions were 257.107 and 163.951 billion yuan, respectively. Compared with that in 2015, the GEP of the forestry, pastoral, agricultural, and central urban regions in 2020 all increased by different amplitudes. The value of the ecological product in Hulunbuir City increased by 8.30% from 46.774 billion yuan in 2015 to 50.657 billion yuan in 2020. Among them, the value of the pastoral and agricultural regions increased by 72.05% and 11.09%, respectively, while the value of the forestry and central urban regions decreased by different amplitudes. Agricultural regions are the main supply areas of ecosystem products, with an output value of 26.476 billion yuan in 2020. The ecological regulation service is the main source of GEP in Hulunbuir City. The value of the ecological regulation service was 852.2 billion yuan in 2020, showing an increase of 219.135 billion yuan compared with that in 2015. The forestry region was the main contributing area for the regulation service. The value of regulation services in the forestry region increased from 332.428 billion yuan in 2015 to 484.054 billion yuan in 2020. The contribution of ecological regulation services in the pastoral region was only inferior to that in the forestry region. The value of regulation services in the agricultural regions in 2020 was 126.874 billion yuan, with an increase of 34.96% compared with the value in 2015. The contribution of regulation services in the central urban regions was the lowest, with a value of 3.851 billion yuan in 2015 and 4.1 billion yuan in 2020. The value of cultural ecosystem services in Hulunbuir City was 77.169 billion yuan in 2020, showing an increase of 22.64% compared with the value in 2015. The value of culture ecosystem services in the central urban region was only lower than that of the forestry region, yielding a value of 24.722 billion yuan in 2020 (Table 2).

Table 2. Accounting results of GEP on each region in 2015 and 2020 (100 million·a^{-1}).

Region	Year	Ecological Product Service Value	Ecological Regulation Service Value	Ecological Culture Service Value	GEP
Forestry region	2015	146.68	3324.28	230.38	3701.34
	2020	122.48	4840.54	319.66	5282.68
Pastoral region	2015	58.44	2027.74	72.18	2158.36
	2020	100.55	2371.72	98.79	2571.07
Agricultural region	2015	238.32	940.12	90.70	1269.13
	2020	264.76	1268.74	106.02	1639.51
Central urban region	2015	24.29	38.51	235.93	298.74
	2020	18.77	41.00	247.22	306.99

Among the three ecological service values, the main service function of the forestry, pastoral, and agricultural regions was the regulation service, and the main function of the central urban region was the culture ecosystem service. The cultural service function was the second largest function in the forestry region, with an increasing proportion. Recently, the forestry region has explored the cultural attributes of ecological products and continued to create high-grade ecological tourism. Compared with 2015, the number of tourists in pastoral regions increased in 2020, resulting in an increase in the proportion of the cultural service value. However, this also yields problems and hidden dangers to the ecological environment of pastoral regions. For example, the destruction of grasslands is common. The proportion of the ecological regulation service value in pastoral regions slightly decreased. After returning farmland to grassland, the proportion of the product value provided by pastoral regions has considerably increased according to the balance of grass storage. Services provided by ecological products are the second largest function in agricultural regions, but the proportion of ecological products decreased in 2020. The proportion of regulation services in agricultural regions increased, indicating that the return of farmland to forest and grassland has been effective. The second major function of the central urban region is to regulate services. Government offices in the central urban region attach importance to ecological environmental protection, increases in afforestation efforts, and improvements in vegetation coverage. The proportion of the ecological regulation service value in the central urban region has increased compared with that in 2015 (Table 2).

3.2. Calculation of Ecological Regulation Services Value

According to the above analysis, the ecological regulation service is the main value source of GEP in Hulunbuir City. The value of the eight types of regulation services in the 13 counties and districts of Hulunbuir City was classified by the natural break point method, and distribution maps of the eight types of regulation services in 2015 and 2020 were obtained (Figure 2). Oroqen is the main supply area of water conservation. The level of the water conservation supply area in Erguna and Genhe increased from third in 2015 to fourth in 2020, mainly due to the significant increase in the shrubby land area. As the ecosystem area in Morin Banner decreased by 15.01% in 2020 compared with that in 2015, the water conservation supply grade of Morin Banner decreased to the first level (Figure 2a,b). The rainfall erosivity of Oroqen in 2020 was weakening, but the vegetation coverage of Oroqen decreased, resulting in a reduction in the soil conservation grade in 2020. The rainfall erosivity of Erguna, Genhe, Yakeshi, Zhalantun, Ewenki, Arun Banner, and Morin Banner increased in 2020, but the vegetation coverage and ecosystem area increased, and there was an increase in the overall soil conservation grade. Higher vegetation coverage can hinder rainfall, with a reduction in the actual soil erosion (Figure 2c,d). Carbon fixation and oxygen release are the main functions of ecological regulation services in Hulunbuir City, and their value accounted for more than 30% of the regulation services. Oroqen is the highest-grade supply area for carbon fixation and oxygen release. The functional levels of carbon fixation and oxygen release in Genhe, Chen Banner, Xin Left Banner, Xin Right

Banner, and Morin Banner decreased in 2020, mainly due to the decrease in the ecosystem area (Figure 2e,f). The value of climate regulation accounted for more than 40% of the value of ecological regulation services in Hulunbuir City. The level of the supply area for the climate regulation function changed negligibly between 2015 and 2020. The service function of climate regulation is composed of two parts: vegetation transpiration and water surface evapotranspiration. Areas with high vegetation coverage and a large water area are the main functional areas for climate regulation. The forest area of Oroqen and Yakeshi ranked first and second in Hulunbuir City, respectively; therefore, the service level of climate regulation in these two regions was high. Most of Hulun Lake is located in the Xin Right Banner, which is also in the top tier of the climate regulation value (Figure 2g,h). As forests have the strongest ability to remove SO_2, NOx, and particulate matter from air pollutants, Oroqen, with the largest forest area, has the strongest ability to remove air pollutants. Yakeshi, Erguna, and Genhe are in a Level 4 supply zone. Grasslands can also purify SO_2, NOx, and particulate matter. The air purification of Chen Banner, Xin Left Banner, Xin Right Banner, Ewenki, and Zhalantun was in the three-level supply area. The air purification level in Morin Banner was downgraded from Level 2 to 1, mainly due to a 15.20% reduction in the ecosystem area (woodland, shrubland, and grassland) in 2020 compared with that in 2015 (Figure 2i,j). Wetlands and water bodies are the main ecological areas for water quality purification. Xin Left Banner and Xin Right Banner have large areas of wetlands and water bodies, such that they are the main functional areas for water quality purification. In 2020, the wetland area in Hailar and Manzhouli decreased by 17.97% and 35.35%, respectively, compared with that in 2015, such that the grade of the water purification supply area decreased from Level 2 to 1 (Figure 2k,l). Changes in the windbreak and sand fixation functions were mainly related to regional meteorological and vegetation factors. Xin Left Banner and Xin Right Banner were the main functional areas for windbreak and sand fixation. The windbreak and sand fixation capacity of Oroqen and Genhe increased by 61.65% and 46.43%, respectively, in 2020, with additional improvements to the functional grade (Figure 2m,n). The grade of the flood storage function supply area in Hulunbuir City changed negligibly between 2015 and 2020. The flood storage function is mainly related to the storage capacity of reservoirs, lakes, marshes, and vegetation. Erguna, Genhe, Xin Left Banner, Ewenki, and Xin Right Banner have high comprehensive flood storage capacities and are the main supply areas for flood storage.

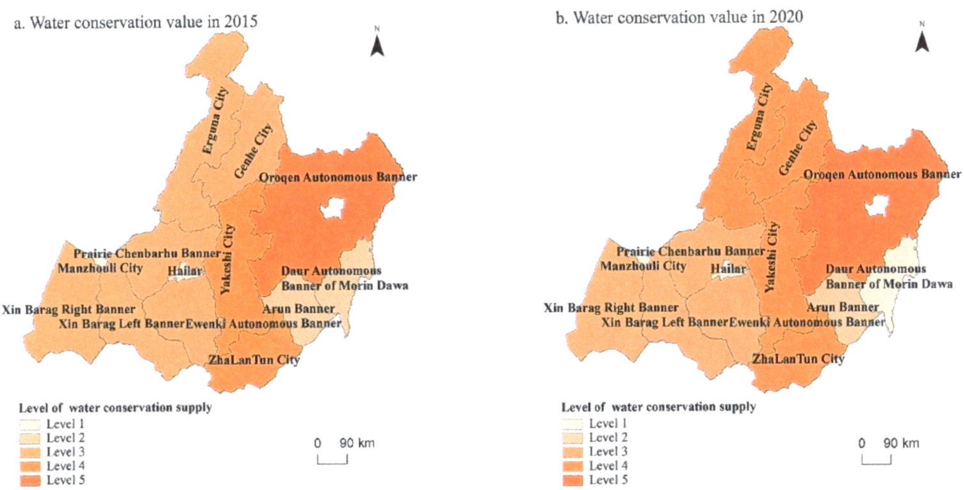

Figure 2. *Cont.*

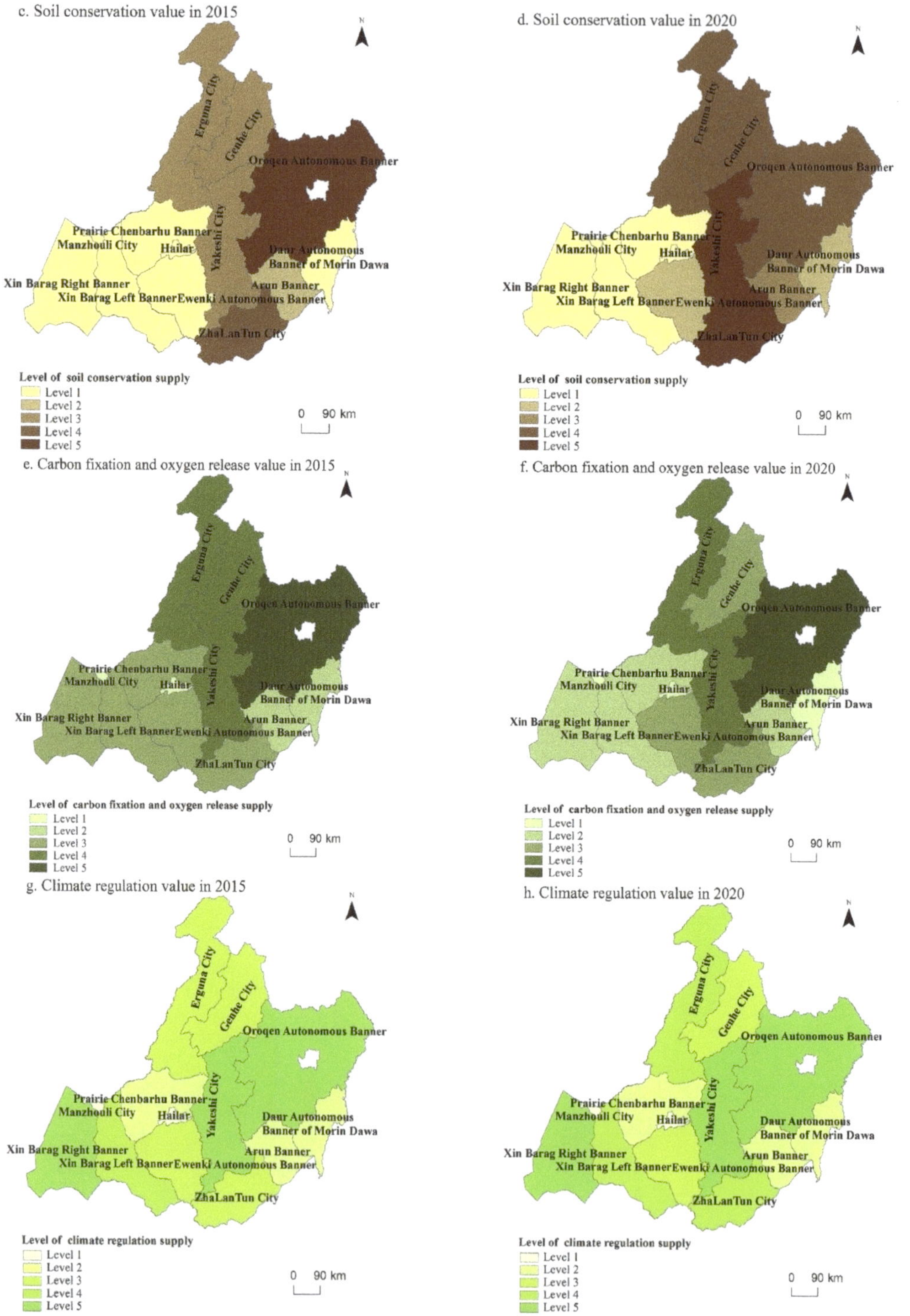

Figure 2. *Cont.*

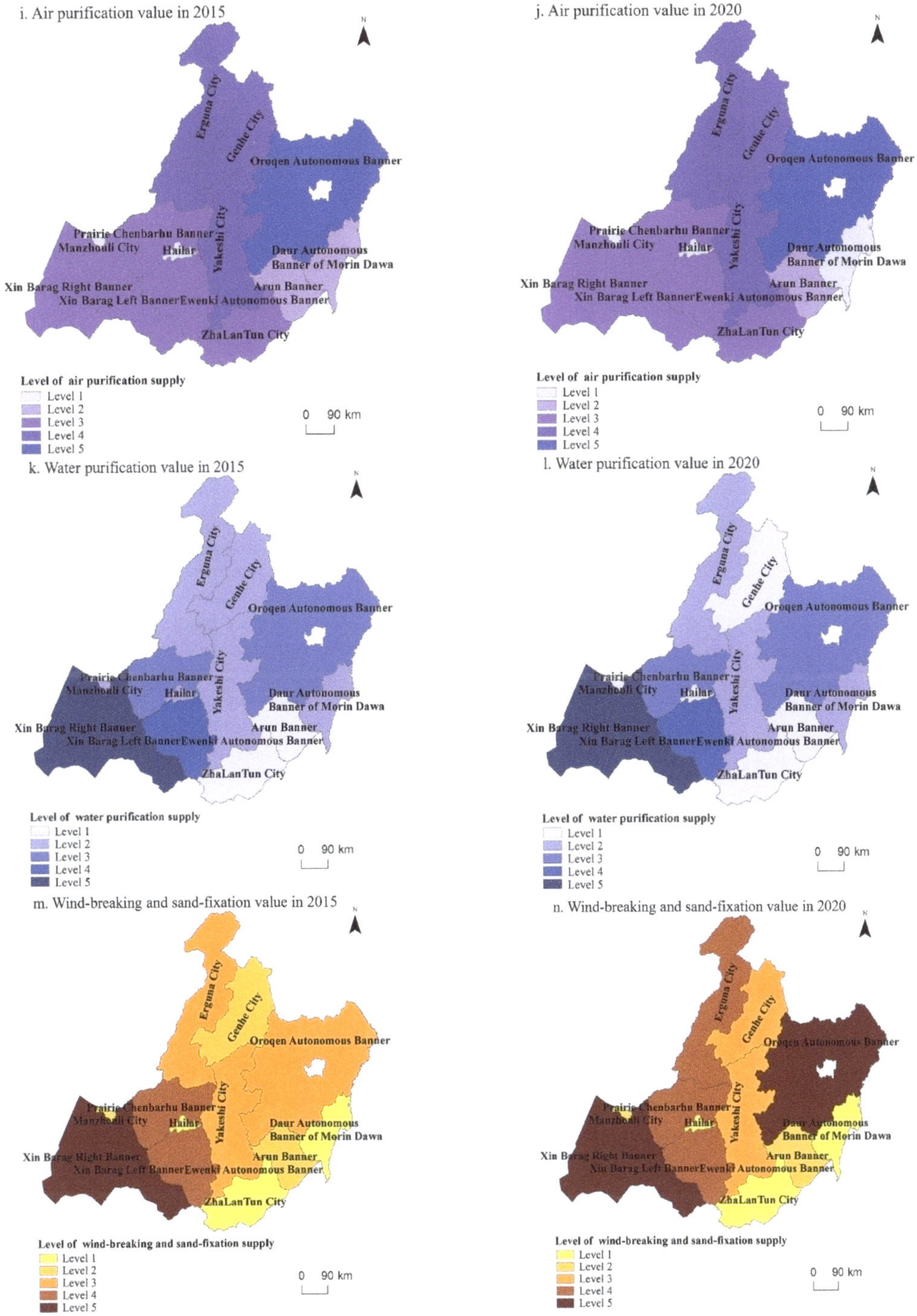

Figure 2. *Cont.*

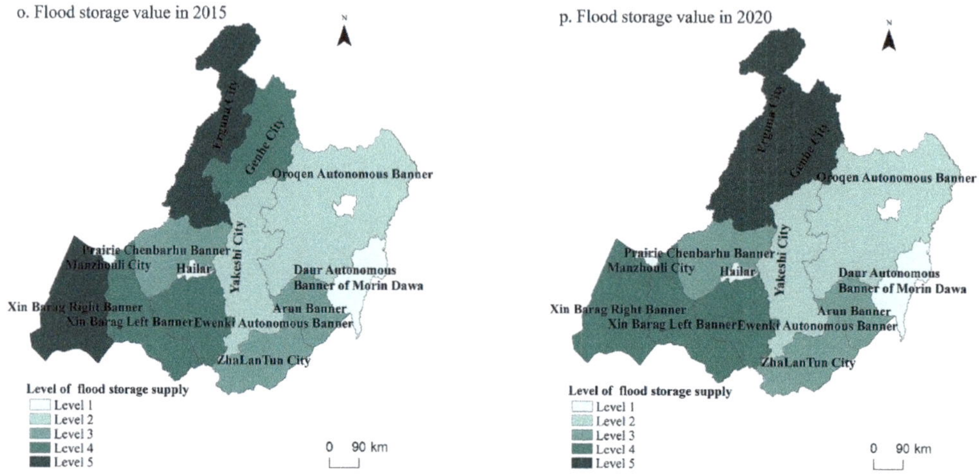

Figure 2. Spatial distribution map of eight regulatory function grades.

3.3. GEP Assessment in Major Ecological Lands

Overall, the forest ecosystem of Hulunbuir City created the highest value. The GEP of the forest ecosystem accounted for 62.32% in 2020, showing an increase of 180.535 billion yuan compared with that in 2015, owing to construction results from key forestry projects, such as the implementation of the natural forest protection project, returning farmland to forest, forest tending, and closing mountains for forest cultivation. During the implementation of the natural forest protection project, the forest resources management and protection system was implemented, which reduced the consumption of forest resources and realized the growth of both forest area and forest stock. Forest ecosystem services have been enhanced, and the capacity of carbon sinks has been improved. Through the implementation of returning farmland to forest, the forest coverage rate of Hulunbuir City has increased by 0.5%, and the ecological environment has been significantly improved. Forest-tending projects improve the growth environment of trees and increase the survival rate of seedlings. Forest closure is a traditional forest cultivation method in China, which clearly affects soil and water conservation, increases species diversity, and reduces forest diseases and pests. The vast grassland ecosystem area played an important role in Hulunbuir City in 2015, accounting for 25.72% of the GEP. Unfortunately, the role of grassland ecosystems gradually declined in 2020, mainly due to the inappropriate reclamation of grassland via human activity, predatory exploitation, overgrazing, and other behaviors, resulting in an increasing reduction in the grassland area, serious degradation, and even desertification. Hulunbuir City has diverse wetland types and is an important ecological region in the cold and arid regions of northern China. The ecological value of wetlands accounted for 4.77% in 2020, which was 0.69% lower than that in 2015. Natural factors, such as insufficient water sources, and human factors, such as agricultural reclamation, grazing, and mowing of wetland vegetation, contributed to wetland ecosystem disturbances. Hulunbuir City has more than 3000 rivers and more than 500 lakes. Water ecosystem services are equally important. The GEP of the water ecosystem accounted for 7.69% and 6.73% in 2015 and 2020, respectively. The most important sources of forest, grassland, wetland, water, and farmland value were found to be Oroqen, Xin Right Banner, Xin Left Banner, Xin Right Banner, and Morin Banner, respectively. The value of the forest ecosystem was mainly reflected in five aspects: climate regulation, carbon sequestration and oxygen release, water conservation, flood storage, and forest tourism. The grassland ecosystem value was mainly concentrated in five categories: carbon sequestration and oxygen release, climate regulation, water conservation, livestock product supply, and grassland tourism. Categories with a high contribution to the wetland ecosystem value were flood storage, wetland tourism, climate regulation, carbon sequestration, and oxygen release. The value of the water ecosystem was mainly derived from five categories: climate regulation, water tourism, flood storage, water resources supply, and fishery product supply (Table 3).

Table 3. Contribution rate of ecosystem functional services value of LUCC.

Ecosystem	Products Value	Water Resources	Water Conservation	Soil Conservation	Carbon Fixation and Oxygen Release	Climate Regulating	Air Purification	Water Purification	Wind-Breaking and Sand-Fixing	Flood Storage	Landscape Recreation
Forest	0.44%		14.41%	0.62%	32.67%	38.38%	0.19%		0.01%	7.40%	5.89%
Grassland	8.95%		15.91%	0.27%	37.64%	29.36%	0.21%		0.04%	2.42%	5.20%
Wetland			0.80%	0.01%	1.87%	28.46%		0.21%		39.23%	29.43%
Water	1.52%	2.91%				61.92%				10.37%	23.28%

Figure 3 shows the proportion of the ecosystem value composition in each region. The forestry region not only has a large area of forest, but also Heishantou grassland, Erguna wetland, Genhe wetland, and other important ecosystems. Therefore, the values of grassland and wetland in the forestry region were 8.97% and 2.86%, respectively. The pastoral region not only has a vast grassland, but also contains Honghuerji forest, Hulun Lake, Bier Lake, Huihe wetland, and other important functional areas of the ecosystem. Therefore, the proportions of the forest and water values in the pastoral region were 15.94% and 12.86%, respectively. The Chaihe National Forest Park, Yaru River, and Nierji Reservoir were observed to be important ecological functional areas as agricultural regions. The values of forests, grasslands, farmland, water bodies, and wetlands in the agricultural region were 58.41%, 15.90%, 12.26%, 9.61%, and 3.83%, respectively. The proportion of forest value in Hailar district was the highest, followed by the grassland, water, and farmland values. The water ecosystem of Manzhouli played the greatest role, mainly because part of Hulun Lake is located in Manzhouli. The wetland value of Manzhouli was relatively high where the Erka wetland is located.

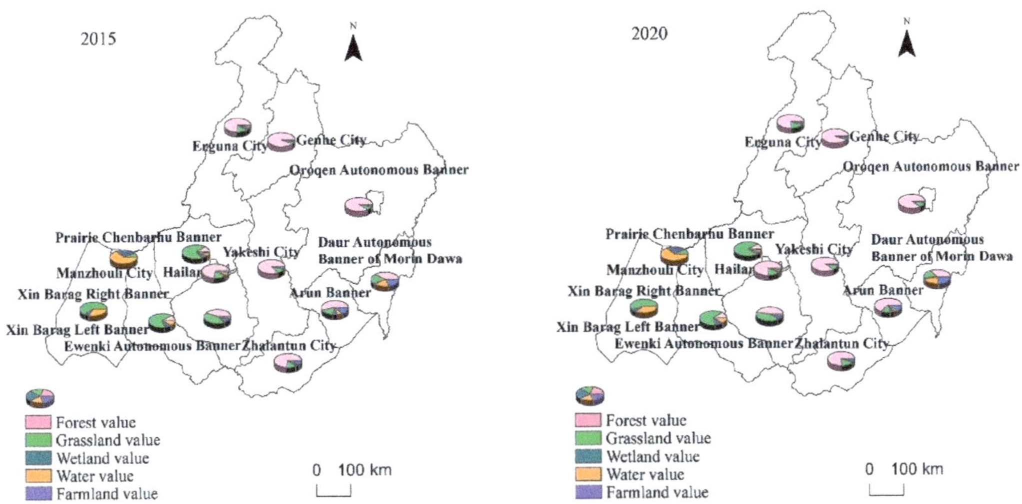

Figure 3. Proportion composition of ecological-type values of each region in 2015 and 2020.

3.4. GEP Application

The GEP per unit area can reflect the ecosystem supply capacity. A higher GEP per unit area indicates a stronger ecosystem supply capacity in the region. The GEP per unit area of Hulunbuir City in 2020 was 3.7736 million yuan, showing an increase of 31.94% compared with 2015, indicating that the overall ecosystem supply capacity of Hulunbuir City has been greatly improved. Compared with Hulunbuir City, Manzhouli, Hailar,

Zhalantun, Genhe, Erguna, and Yakeshi have greater ecosystem supply potentials, while other regions have lower supply capacities. The GEP per capita refers to the GEP enjoyed by each person. The GEP per capita could reflect people's ecological welfare level, with a higher value indicating more ecological welfare everyone enjoys. The GEP per capita of Hulunbuir City in 2020 was 438,200 yuan, with an increase of 40.79% compared with 2015, indicating that the per capita ecological well-being level of Hulunbuir City has significantly increased. Compared with Hulunbuir City, people in Xin Right Banner, Erguna, Xin Left Banner, Genhe, Oroqen, Chen Banner, and Arun Banner enjoy more ecological welfare per capita, while the level of ecological well-being per capita in other regions is low. The GGI can reflect the quantitative relationship between the "Two Mountains." The GGI of Hulunbuir City in 2020 was 8.36, which was considerably higher than the national average of 1.01, indicating that the resource value of "clear waters and green mountains" in Hulunbuir City was considerably greater than that of "gold and silver mountains." There was significant potential for ecological factors to be transformed into production factors, and ecological wealth to be transformed into material wealth. Compared with those in 2015, the growth rates of the GEP, GDP, and GGI of Hulunbuir City in 2020 were 31.94%, 20.84%, and 9.18%, respectively, showing that the ecological protection of Hulunbuir City has achieved success. The regional economy has sustained development, and the ability to transform "clear waters and green mountains" into "gold and silver mountains" is also improving. Additionally, ecological protection achievements have far exceeded the level of economic growth, laying a solid foundation for the ecological status of Hulunbuir City, but we must also continue to explore the green-economic development model. Comparing the GGI of each county in Hulunbuir City, we found that in 2015 and 2020, the GGI values of Erguna, Genhe, Oroqen, Xin Left Banner, Xin Right Banner, and Yakeshi were higher than the overall level of Hulunbuir City, indicating that these six regions effectively use ecological advantages to develop the regional economy, while those of the other regions were lower than the overall level of Hulunbuir City over the five years. GGI levels at Hailar and Manzhouli were lower than the national average. By comparing the GGI growth rate of each county in Hulunbuir City, we can conclude that the GGI growth rate of Erguna, Genhe, and Zhalantun was higher than the overall GGI growth rate of Hulunbuir City, indicating that the transformation of the "Two Mountains" in these three regions has achieved remarkable results. The GGI growth rate of Yakeshi, Arun Banner, Chen Banner, Oroqen, and Morin Banner was lower than the overall GGI growth rate of Hulunbuir City. The "Two Mountains" in these regions have a large transformation space; therefore, they must fully rely on advantageous ecological resources to transform them into economic development power. The transformation capacity of the "Two Mountains" in Hailar, Manzhouli, Xin Right Banner, Ewenki, and Xin Left Banner was decreasing, indicating that we must further promote the ecological article based on the traditional economic model (Figure 4).

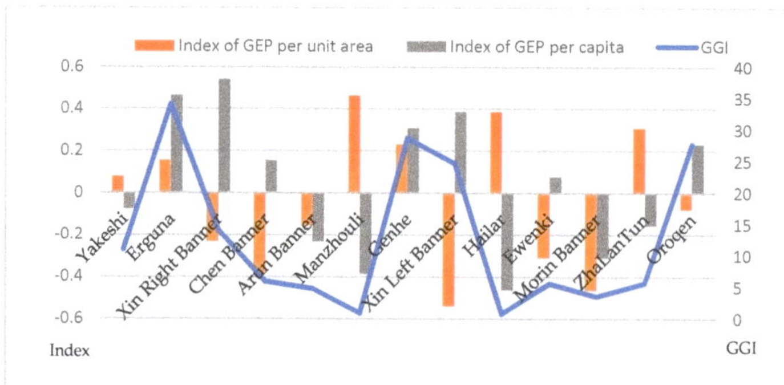

Figure 4. The index of GEP per unit area, index of GEP per capita, and GGI of each region in 2020.

3.5. *Transformation Path of "Two Mountains"*

According to the above analysis, we explored the transformation path of the "Two Mountains" in each county of Hulunbuir City. Considering the ecosystems and industrial types of counties included in the four ecological economic zones, and that the transformation paths of the "Two Mountains" are replicable and transferable, this study selected the four ecological economies of forestry, pastoral, agricultural, and central urban regions as the entry point of analysis.

3.5.1. Transformation Path of "Two Mountains" in the Forestry Region

The forestry region is the main contribution area of regulation service, with a regulation service value accounting for 91.63% of the GEP and a growth rate of 45.62%, ranking first among the four ecological economic zones. Therefore, the regulation service is the key point of "Two Mountains" transformation in the forestry region. The eight main types of regulation services of the forestry region were water conservation, soil conservation, carbon fixation, oxygen release, climate regulation, air purification, windbreak, and sand fixation. Among them, carbon fixation and oxygen release function were the most prominent, with a value of up to 180 billion yuan. The forestry region can directly realize the transformation of the carbon sequestration and oxygen release values for the "Two Mountains" through carbon sink trading. The transformation of other regulatory values can also be indirectly completed through the ecological compensation mechanism, i.e., the principle of "who uses, who pays." There was a significant correlation between the increase in the ecosystem regulation service value and the increase in vegetation coverage. The vegetation coverage of the forestry region was 84.41% in 2020. Therefore, the forestry region should continue to implement various ecological protection construction projects and transform the ecological value of the regulation service into economic value. The second major function of forestry regions is to provide cultural services, which accounted for 6.05%, and the growth rate is the first among the four ecological economic zones, indicating that the ecological cultural function of the forestry region has great potential. The main natural resources in the forestry region are forests and wetlands. The forestry region continues to explore the cultural elements of forest and wetland tourism while building a well-known forest and wetland tourism destination and transforming forest and wetland ecosystem value into tourism economic value. The ecological products value in the forestry region accounted for 2.32% in 2020, which decreased by 16.49% compared with 2015, indicating that the ecological products value needs to be further explored. Forestry regions should focus on brand agriculture and realize the circular transformation of the "Two Mountains." The GGI of the forestry region is 25.37, which is much higher than the average of Hulunbuir City (8.36), indicating that the ecological level of the forestry region is much higher than the economic level. The focus of the "Two Mountains" transformation in the forestry region is the rapid transformation from GEP to GDP.

3.5.2. Transformation Path of "Two Mountains" in the Pastoral Region

The ecological regulation services value in the pastoral region accounts for 92.25% of GEP, with a growth rate of 16.96%. Therefore, regulation services are also the focus of "Two Mountains" transformation in the pastoral region. The area of grassland, wetland, and water bodies in the pastoral region is large. Among the eight regulation services, climate regulation, water purification, windbreak, sand fixation, and flood storage are the main ecological functions in the pastoral region. The ecological regulation value should be transformed into economic value via ecological compensation. Grassland is an ecosystem with both ecological and production functions. The grassland ecosystem not only has important ecological functions, but also provides pasture resources for livestock to maintain metabolism. The ecological products supply value in the pastoral region accounted for 3.91%. The growth rate was 72.05%, which was much higher than that of other ecological economic zones, indicating that the supply function of ecological products in the pastoral region had great potential. We must, therefore, extend the industrial chain of pastoral products, enhance the brand value of livestock products, and transform the ecological products value of the pastoral region into economic value. The cultural ecosystem services value in the pastoral region accounted for 3.84% of the GEP, with a growth rate of 36.87%, which was slightly lower than that of the forestry region, indicating that the cultural ecosystem services value of the pastoral region had greater potential. Pastoral regions should integrate the development of grassland tourism with cultural understudies, actively promote "driving + experience" grassland tourism, conduct a series of grassland cultural activities, and create high-end grassland tourism destinations, to improve the ecotourism value of pastoral regions. The GGI of the pastoral region was 12.55, which was slightly higher than the average of Hulunbuir City, indicating that the ecological surplus of the pastoral region was slight. The focus of the transformation of "Two Mountains" in the pastoral region was to promote the transformation from GEP to GDP while maintaining ecological balance.

3.5.3. Transformation Path of "Two Mountains" in the Agricultural Region

The ecological regulation service value in the agricultural region accounts for 77.38% of GEP. Therefore, the regulation service is also the focus of "Two Mountains" transformation in the agricultural region. The growth rate was 34.96%, indicating that the ecosystem regulation function in the agricultural region has been greatly improved through ecological construction projects, such as returning farmland to forest and grassland. The ecological products supply service is the second largest function of the agricultural region, accounting for 16.15%, which is the highest in the four ecological economic zones. The growth rate was 11.09%, indicating that the product supply function in the agricultural region continued to strengthen. It is necessary to continue to develop and innovate green and low-pollution ecological agricultural products and enhance the brand value of geographical indication agricultural products. The value of agricultural products can be directly transformed into industrial economic value. The cultural ecosystem services value in the agricultural region accounted for 6.47% of GEP, with a growth rate of 16.89%, indicating that the cultural ecosystem services in the agricultural region need to be explored further. We will deepen the integration of culture and tourism in the agricultural region. The ecological cultural value of the agricultural area is transformed into tourism economic value by conducting agricultural expo activities. The GGI of the agricultural region is 4.70, which is lower than the average of Hulunbuir City, indicating that the ecological level of the agricultural region is lower than the economic level. The focus of "Two Mountains" transformation is the two-way transformation of GEP and GDP.

3.5.4. Transformation Path of "Two Mountains" in the Central Urban Region

The main function of the central urban region was the culture ecosystem service. The culture ecosystem service value in the central urban region accounts for 80.53% of GEP, with a growth rate of 4.79%. The transformation of the ecotourism value is the focus of the transformation of the "Two Mountains" in the central urban regions. As the main collection and distribution center of tourists in Hulunbuir City, the central urban regions have many tourists, but negligible ecotourism resources. By adding ecological resources and cultural elements and performing research and other measures of ecological cultural activities, tourists will be attracted to the ecotourism products of the region. The proportion of the central urban regulation service was 13.36%. The regulation function of Hailar mainly comes from urban greening and Xishan National Park, while the regulation service of Manzhouli mainly focuses on urban greening and Hulun Lake. Therefore, attention should be paid to the ecological protection of Xishan National Park and Hulun Lake. The proportion of the ecological products supply service in the central urban region is 6.11%, which is 22.75% lower than that in 2015, indicating that the value of ecological products in the central urban region needs to be further explored. The GGI of the central urban region is 0.84, which is far lower than the average of Hulunbuir City, indicating that the ecological level of the central urban region is far lower than the economic level. The focus of the transformation of "Two Mountains" in the central urban region is the transformation of GDP into GEP, and then GEP into GDP.

4. Discussion

4.1. Differences in GEP Results

Evaluation methods, selection of evaluation indicators, and price parameters will all impact GEP results. If different accounting methods and indexing systems are selected for the same region regarding different price parameters, the calculated amount of ecosystem functions and values will vary. Regarding similar ecological niche areas of relevant research results combined with the practical situation of the ecological resources of Hulunbuir City, this study determined the index for 11 types of accounting. Accounting methods focus on references to the 2022 "GEP Accounting Specification (Trial)" issued by the National Development and Reform Commission of China. The price is widely used in the practical life parameters and some references. After comparing the calculated research results with the results of a previously published study on calculating the GEP for the Inner Mongolia Autonomous Region [41], a report on the first comprehensive evaluation of forest ecosystem service value in Hulunbuir City [42], and the National County/Municipal District Gross Ecosystem Product (GEP) Research Report 2021 [43], we can conclude that the evaluation results of this study were within a reasonable range.

4.2. Research Limitations and Prospects

Owing to the impact of the COVID-19 pandemic, Hulunbuir City tourism data in 2020 cannot reflect the real values of regional cultural services. This study selected tourism data from 2019 to calculate the values for recreational cultural services in 2020. Due to difficulties in data acquisition, the functional value of pest and disease control was not accounted for. There are inevitable errors in the acquisition and processing of remote sensing data, meteorological data, soil data, and other data. Although there were some errors in the research results, this study accurately reflects the state and changes in the ecological environment of each county in Hulunbuir City during the 12th and 13th Five-Year plans. It also provides a scientific basis for future policy formulation with respect to ecological protection and construction in each region. In a future study, the GEP accounting of Hulunbuir City will be calculated every five years as a normalization work. The next study is to calculate the GEP of 12 cities in Inner Mongolia and verify the importance of Hulunbuir City's ecological status through the calculation results. It is important to promote the further development of ecological civilization construction in Inner Mongolia.

5. Conclusions

We selected 2015 and 2020 as the research nodes to compare and analyze the characteristics of GEP in 13 counties. We verified the value status of "clear waters and green mountains" in the study area to explore the effects of ecological protection in the 12th and 13th Five-Year plans of Hulunbuir City. Based on the results of the two stages, we explored the transformation path of the "Two Mountains" value in 13 counties. The main conclusions were as follows.

The GEP of Hulunbuir City in 2020 was 980.025 billion yuan, of which the product value was 50.657 billion yuan, the regulation service value was 852.2 billion yuan, and the cultural service value was 77.169 billion yuan. The GEP ranking for the counties in Hulunbuir City was as follows: forestry region > pastoral region > agricultural region > central urban region. The agricultural region was the main supply area of ecosystem products, the forestry and pastoral regions were the main contribution areas of regulating services, and the value of cultural ecosystem services in the central urban region was second only to the forestry region. Oroqen is the main supply area for water conservation. The soil conservation capacity improved in Erguna, Genhe, Yakeshi, Zhalantun, Ewenki, Arun Banner, and Morin Banner. Carbon fixation and oxygen release were the main functions of the ecological regulation service in Hulunbuir City. Oroqen had the highest carbon fixation and oxygen release capacity. Climate regulation was the most important ecological regulation function of Hulunbuir City. Oroqen, Yakeshi, and Xin Right Banner were the main source areas for the climate regulation value. Oroqen had the strongest air purification capacity. Xin Left Banner and Xin Right Banner were the main areas of the water purification function, as well as of wind-breaking and sand-fixing. The flood regulation and storage capacity of Erguna, Genhe, Xin Left Banner, Ewenki, and Xin Right Banner were high. The GEP of major ecological lands was in the order of forest > grassland > water body > wetland > cropland. By region, the order of the ecosystem value is forest, grassland, wetland, farmland, and water bodies in forestry regions. The ecosystem value in pastoral regions was ranked as follows: grassland, forest, water body, wetland, and farmland. The order of the ecosystem value in agricultural regions is forest, grassland, farmland, water body, and wetland. The order of the ecosystem value in Hailar is forest, grassland, water body, and farmland. The order of the ecosystem value in Manzhouli is water body, wetland, grassland, forest, and farmland. The GGI of Hulunbuir City in 2020 was 8.36. The GEP per unit area and GEP per capita in the forestry region were high, and the GGI was much higher than 8.36. The GEP per unit area of the pastoral region was low, while the GEP per capita and the GGI were high. The GEP per unit area and GEP per capita in the agricultural region were low, and the GGI was slightly lower than 8.36. The GEP per unit area in the central urban region was high, while the GEP per capita was low, and the GGI was much lower than 8.36. The growth rate of GGI was high in Erguna, Genhe, and Zhalantun, somewhat low in Yakeshi, Arun Banner, Chen Banner, Oroqen, and Morin Banner, and exhibited a downward trend in Hailar, Manzhouli, Xin Right Banner, Ewenki, and Xin Left Banner.

In this study, the GEP was calculated on the county scale and applied to the transformation practice of the "Two Mountains." These research ideas and methods can be extended to similar regions in China and around the world, with considerable application value and social significance. Firstly, through comparative evaluation of GEP among counties, more targeted ecological protection measures can be implemented, and ultimately the overall ecological protection of cities and counties can be improved. Secondly, GEP was calculated in accordance with the accounting methods in the specification issued by the national authority in 2022, so that the accounting results can be traced, verified, and comparable. For the 13 counties included in this study, the accounting results were determined to be comparable, which is an innovation of this study. Thirdly, we promoted the transformation of GEP from an "accounting value" to a "policy point." The transformation path of "Two Mountains" in the forestry and pastoral regions can be extended to the counties with good ecological environments and poor economies. The transformation path of "Two Mountains" in the agricultural region can be extended to the counties with poor ecological environments and economies. The transformation path of "Two Mountains" in the central urban region is suitable for areas with developed economies and poor ecological environments. In future research, the GEP accounting scope should be expanded to calculate the GEP of all counties in Inner Mongolia and throughout China. The accounting results are applied to the transformation practice of "Two Mountains" to actively promote the construction of ecological civilization at the county scale.

Author Contributions: Conceptualization: N.Z. and H.W.; methodology: N.Z. and J.Z.; software: N.Z. and Y.B.; validation: N.Z., H.W., J.Z., Y.B. and S.Y.; formal analysis: N.Z. and H.W.; investigation: N.Z. and S.Y.; resources: N.Z. and Y.B.; data curation: N.Z. and S.Y.; writing—original draft preparation: N.Z., H.W., J.Z., Y.B. and S.Y.; writing—review and editing: N.Z., H.W., J.Z., Y.B. and S.Y.; visualization: N.Z. and J.Z.; supervision: H.W.; project administration: N.Z. All authors have read and agreed to the published version of the manuscript.

Funding: This research was funded by the Hulunbuir Federation of Social Sciences, China (HSK202211) and the Inner Mongolia University Science and Technology Research Project, China (NJSY22298).

Data Availability Statement: The data used in this study are from the Remote Sensing Institute of the Chinese Academy of Sciences (https://data.casearth.cn/sdo/detail/5fbc7904819aec1ea2dd7061

(accessed on 2 June 2022)), the Resource and Environmental Science and Data Center of the Chinese Academy of Sciences (https://www.resdc.cn/ (accessed on 7 June 2022)), the Global Land Data Assimilation System (http://disc.sci.gsfc.nasa.gov/hydrology/data-holdings (accessed on 17 June 2022)), the National Tibetan Plateau/Third Pole Environment Data Center (https://disc.gsfc.nasa.gov/datasets/GLDAS_CLSM025_DA1_D_2.2/summary?keywords=runoff (accessed on 22 June 2022)), and the National Aeronautics and Space Administration (http://www.nasa.gov/ (accessed on 2 July 2022)).

Acknowledgments: The authors thank the Hulunbuir Bureau of Statistics in China for technical and data support.

Conflicts of Interest: The authors declare no conflict of interest. The funders had no role in the design of the study, the collection, analyses, or interpretation of data, the writing of the manuscript, or the decision to publish the results.

References

1. Bai, Y.; Zhuang, C.; Ouyang, Z. Spatial characteristics between biodiversity and ecosystem services in a human-dominated watershed. *Ecol. Complex.* **2011**, *8*, 177–183. [CrossRef]
2. Bai, Y.; Jiang, B.; Wang, M. New ecological redline policy (ERP) to secure ecosystem services in China. *Land Use Policy* **2016**, *55*, 348–351. [CrossRef]
3. Gao, J.; Fan, X.; Li, H. Research on constituent elements, operation modes and political demands for capitalizing ecological assets. *Res. Environ. Sci.* **2016**, *29*, 315–322. [CrossRef]
4. Zhao, Y.; Wen, Q.; Ai, J. Ecosystem Service Value of Forests in Yunnan Province. *For. Res.* **2010**, *23*, 184–190. [CrossRef]
5. Costanza, R.; d'Arge, R.; de Groot, R.; Farber, S.; Grasso, M.; Hannon, B.; Limburg, K.; Naeem, S.; O'Neill, R.; Paruelo, J. The value of the world's ecosystem services and natural capital. *Nature* **1997**, *387*, 253–260. [CrossRef]
6. Daily, G. *Nature's Services: Societal Dependence on Natural Ecosystems*; Island Press: Washington, DC, USA, 1997.
7. Chiabai, A.; Travisi, C.; Markandya, A. Economic assessment of forest ecosystem services losses: Cost of policy inaction. *Environ. Resour. Econ.* **2011**, *50*, 405–445. [CrossRef]
8. Lamarque, P.; Tappeiner, U.; Turner, C. Stakeholder perceptions of grassland ecosystem services in relation to knowledge on soil fertility and biodiversity. *Reg. Environ. Chang.* **2011**, *11*, 791–804. [CrossRef]
9. Engle, V. Estimating the provision of ecosystem services by Gulf of Mexico coastal wetlands. *Wetlands* **2011**, *31*, 179–193. [CrossRef]
10. Zhao, J.; Xu, Y.; Xiao, H. Ecosystem services evaluation based on comprehensive national power for sustainable development: The evaluation on 13 countries. *Wetlands* **2003**, *23*, 121–127. [CrossRef]
11. Xie, G.; Zhang, Y.; Lu, C. Study on valuation of rangeland ecosystem services of China. *J. Natl. Res.* **2001**, *16*, 47–53. [CrossRef]
12. UN; European Commission. *System of National Accounts*; China Statistics Press: Beijing, China, 2012.
13. European Commission; Organisation for Economic Co-Operation and Development; United Nations; World Bank. System of Environmental-Economic Accounting, 2012: Experimental Ecosystem Accounting. Available online: http://unstats.un.org/unsd/envaccounting/eea_white_cover.pdf (accessed on 20 July 2022).
14. Gary, S.; Andrew, K.; Mark, E. Creating physical environmental asset accounts from markets for ecosystem conservation. *Ecol. Econ.* **2012**, *82*, 114–122.
15. Wang, J.; Ma, G.; Yu, F. Study on China's Gross Economic-ecological Product Accounting in 2015. *China Popul. Resour. Environ.* **2018**, *28*, 1–7.
16. Ouyang, Z.; Zhu, C.; Yang, G. Gross ecosystem product: Concept, accounting framework and case study. *Acta Ecol. Sin.* **2013**, *33*, 6747–6761. [CrossRef]
17. Ouyang, Z.; Zheng, H.; Xiao, Y.; Polasky, S. Improvements in ecosystem services from investments in natural capital. *Science* **2016**, *352*, 1455–1459. [CrossRef] [PubMed]
18. Liu, T.; He, L.; Zhao, H. Discussion on the generalization path of regional ecological product value realization. *Ecol. Environ. Sci.* **2022**, *31*, 1059–1070.
19. Ma, G.; Yu, F.; Wang, J. Measuring gross ecosystem product (GEP) of 2015 for terrestrial ecosystem in China. *China Environ. Sci.* **2017**, *37*, 1474–1482.
20. Bai, Y.; Li, H.; Wang, X. Evaluating natural resource assets and gross ecosystem products using ecological accounting system: A case study in Yunnan Province. *J. Nat. Resour.* **2017**, *32*, 1100–1112.
21. Dong, T.; Zhang, L.; Xiao, Y. Assessment of ecological assets and gross ecosystem product value in Ordos City. *Acta Ecol. Sin.* **2019**, *39*, 3062–3074.
22. You, X.; He, D.; Xiao, Y. Assessment of ecological protection effectiveness in a county area: Using Eshan County as an example. *Acta Ecol. Sin.* **2019**, *39*, 3051–3061.
23. Yu, M.; Jin, H.; Li, Q. Gross Ecosystem Product (GEP) Accounting for Chenggong District. *J. West China For. Sci.* **2020**, *49*, 41–55.
24. Pema, D.; Xiao, Y.; Ouyang, Z. Gross ecosystem product accounting for the Garzê Tibetan Autonomous Prefecture. *Acta Ecol. Sin.* **2017**, *37*, 6302–6312.

25. Pema, D.; Xiao, Y.; Ouyang, Z. Assessment of ecological conservation effect in Xishui county based on gross ecosystem product. *Acta Ecol. Sin.* **2020**, *40*, 499–509.
26. Ma, G.; Zhao, X.; Wu, Q. Concept definition and system construction of gross ecosystem production. *Resour. Sci.* **2015**, *37*, 1709–1715.
27. Gao, X.; Li, J.; Xu, J. *National Economic Accounting Theory and Practice in China*; Renmin University of China Press: Beijing, China, 2012.
28. Jin, L.; Li, J.; Kong, D. Evaluation of the incorporation of gross ecosystem product into performance appraisals for ecological compensation. *Acta Ecol. Sin.* **2019**, *39*, 24–36.
29. Fan, Z.; Li, W. Research on the Realization Mechanism of Ecological Product Value—A Case Study of Guizhou Province. *J. Hebei GEO Univ.* **2020**, *43*, 82–90. [CrossRef]
30. Yin, Y.; Xi, F.; Wang, J. Application of ecosystem value in policy system design: A case study of Fuzhou City, China. *Chin. J. Appl. Ecol.* **2021**, *32*, 3815–3823. [CrossRef]
31. Chen, H.; Wang, Y.; Huang, Y. Evaluation of Regional Ecosystem Services Grade Coupling Ecological Carrying Capacity and Gross Ecosystem Product-A Case Study of Changting County, Fujian Province. *J. Soil Water Conserv.* **2021**, *35*, 150–160. [CrossRef]
32. Lin, Y.; Xu, X. Spatiotemporal variations of gross ecosystem product and identification of important ecological protection spaces in the Yangtze River Delta. *Resour. Sci.* **2022**, *44*, 847–859. [CrossRef]
33. Wang, J.; Su, J.; Wan, J. The theoretical connotation of "green water and green mountains are golden mountains and silver mountains" and its realization mechanism innovation. *Environ. Prot.* **2017**, *45*, 13–17. [CrossRef]
34. Rong, B.; Yang, S.; Chu, C. Typical Modes of the Transformation from "Lucid Waters and Lush Mountains" to "Gold and Silver Mountains" at County Level. *Chin. J. Environ. Manag.* **2021**, *13*, 20–26. [CrossRef]
35. Ouyang, Z.; Lin, Y.; Song, C. Research on Gross Ecosystem Product (GEP): Case study of Lishui City, Zhejiang Province. *Environ. Sustain. Dev.* **2020**, *45*, 80–85. [CrossRef]
36. Wei, H.; Fan, W.; Wang, X. Integrating supply and social demand in ecosystem services assessment: A review. *Ecosyst. Serv.* **2017**, *25*, 15–27. [CrossRef]
37. Cheng, C.; Ge, C.; Du, Y. Green Gold Index accounting for Quzhou City in Zhejiang Province. *Acta Ecol. Sin.* **2019**, *39*, 37–44.
38. Chen, M.; Ji, R.; Liu, X. Gross ecosystem product accounting for 'Two Mountains'Bases and transformation analysis: The case study of Ninghai County. *Acta Ecol. Sin.* **2021**, *41*, 5899–5907.
39. Qu, X.; Dou, H.; Gao, S. Spatial and Temporal Variation of NPP and Its Response to Climate in Hulun Buir. *Desert Oasis Meteorol.* **2019**, *13*, 100–105. [CrossRef]
40. National Forestry and Grassland Administration. *Specifications for Assessment of Forest Ecosystem Services*; Standards Press of China: Beijing, China, 2008.
41. Jin, C.; Lu, Y. Review and Prospect of Research on Value Realization of Ecological Products in China. *Econ. Geogr.* **2021**, *41*, 207–213. [CrossRef]
42. Available online: https://www.maxlaw.cn/hulunbeier/news/956142773098.shtml (accessed on 27 July 2022).
43. Available online: https://yrd.huanqiu.com/article/45ytDLdbvVL (accessed on 27 July 2022).

remote sensing MDPI

Article

UAV-Based Characterization of Tree-Attributes and Multispectral Indices in an Uneven-Aged Mixed Conifer-Broadleaf Forest

Eduardo D. Vivar-Vivar [1], Marín Pompa-García [2,*], José A. Martínez-Rivas [2] and Luis A. Mora-Tembre [3]

1 Maestría en Geomática Aplicada a Recursos Forestales y Ambientales, FCFyA, Universidad Juárez del Estado de Durango, Río Papaloapan y Blvd, Durango Valle del Sur s/n, Durango 34120, Mexico; unam.vivar@gmail.com
2 Laboratorio de Dendroecología, FCFyA, Universidad Juárez del Estado de Durango, Río Papaloapan y Blvd, Durango Valle del Sur s/n, Durango 34120, Mexico; mtz.alexis05@gmail.com
3 Secretaría de Ecología y Medio Ambiente del Estado de Quintana Roo, Efraín Aguilar Núm. 418, Col. Campestre de la Ciudad de Chetumal, Chetumal 77030, Mexico; krotalo25@gmail.com
* Correspondence: mpgarcia@ujed.mx; Tel.: +52-61-81-301096

Abstract: Unmanned aerial vehicles (UAVs) have contributed considerably to forest monitoring. However, gaps in the knowledge still remain, particularly for natural forests. Species diversity, stand heterogeneity, and the irregular spatial arrangement of trees provide unique opportunities to improve our perspective of forest stands and the ecological processes that occur therein. In this study, we calculated individual tree metrics, including several multispectral indices, in order to discern the spectral reflectance of a natural stand as a pioneer area in Mexican forests. Using data obtained by UAV DJI 4, and in the free software environments OpenDroneMap and QGIS, we calculated tree height, crown area, number of trees and multispectral indices. Digital photogrammetric procedures, such as the ForestTools, Structure from Motion and Multi-View Stereo algorithms, yielded results that improved stand mapping and the estimation of stand attributes. Automated tree detection and quantification were limited by the presence of overlapping crowns but compensated by the novel stand density mapping and estimates of crown attributes. Height estimation was in line with expectations (R^2 = 0.91, RMSE = 0.36) and is therefore a useful parameter with which to complement forest inventories. The diverse spectral indices applied yielded differential results regarding the potential vegetation activity present and were found to be complementary to each other. However, seasonal monitoring and careful estimation of photosynthetic activity are recommended in order to determine the seasonality of plant response. This research contributes to the monitoring of natural forest stands and, coupled with accurate in situ measurements, could refine forest productivity parameters as a strategy for the validity of results. The metrics are reliable and rapid and could serve as model inputs in modern inventories. Nevertheless, increased efforts in the configuration of new technologies and algorithms are required, including full consideration of the costs implied by their adoption.

Keywords: index vegetation; UAV; natural forest; estimation attributes; forest productivity; crown delineation; automated tree detection

Citation: Vivar-Vivar, E.D.; Pompa-García, M.; Martínez-Rivas, J.A.; Mora-Tembre, L.A. UAV-Based Characterization of Tree-Attributes and Multispectral Indices in an Uneven-Aged Mixed Conifer-Broadleaf Forest. *Remote Sens.* **2022**, *14*, 2775. https://doi.org/10.3390/rs14122775

Academic Editor: Nikolay Strigul

Received: 21 April 2022
Accepted: 7 June 2022
Published: 9 June 2022

1. Introduction

Accurate forest ecosystem monitoring has become a constant among the requirements of large-scale environmental projects [1]. For example, the scientific community seeks improved inputs for models of monitoring greenhouse gases [2] and estimation of carbon and biomass storage [3], among others, including reducing emissions from deforestation and forest degradation (REDD) [4].

Remote sensing technology has become more flexible in recent years and offers a promising perspective [5]. At present, it is not only possible to obtain a greater spatial resolution of attributes that are visible to the naked eye, but also to detect multispectral attributes beyond the visible spectrum [6–8]. For example, experimentation with radiation emitted by the multispectral indices is crucial to further our understanding of ecological mechanisms that remain unclear, particularly in heterogeneous forests [9]. To this end, the application of drone technology has been rapidly extended at small and medium scales [10], including in the acquisition of highly accurate estimates at the individual tree level that serves to improve forest management [11,12]. As a result, direct and destructive estimates, which are normally so time-consuming, have been substituted by indirect estimates, such as those obtained by using unmanned aerial vehicles (UAV), with the opportunities offered by near-real-time monitoring with multiple sensors [13].

The Mexican Sierra Madre Occidental is known to occupy a special place in forest diversity [14], where different conifer and broadleaf species coexist within a small site. These areas constitute natural laboratories for gathering scientific knowledge regarding dendroecological attributes, including the use of multispectral indices since these can reflect different ecological inter-relationships [15]. For example, the estimation of structural attributes to individual tree levels, such as height and crown area, is of great utility for forest inventories and the determination of biomass and carbon values [3]. For their part, the multispectral indices are indicators of crown vigor, phenology, structural characteristics, defoliation risk and photosynthesis rates, among other variables of forest productivity [16,17]. Monitoring of these variables in complex, unevenly aged and mixed forests can therefore offer new elements for decision-making on the part of forest managers [18], considering that these systems are the greatest reservoirs of carbon and are widely representative of the Mexican forests.

Although different studies have estimated the structural attributes of heterogeneous forests [19], few have integrated the evaluation of spectral attributes [20]. Moreover, estimation of aerial structures in natural forests remains a challenge [11].

For this reason, a complete census of the trees in a pilot site was conducted in order to evaluate the capacity of a UAV to (1) calculate the metrics of tree height, crown area and the number of trees, and (2) examine the applicability of multispectral indices for monitoring in a mixed and heterogeneous stand of conifer and broadleaf species in northern Mexico. We hypothesized that the UAV will provide reliable information pertaining to the attributes of individual trees, thus serving to improve our future perspective of the vegetation properties.

2. Materials and Methods

The study site is found in the area known as "El Cordoncito" in Mesa de Pawiranachi, in the municipality of Guachochi, in the Sierra Madre Occidental mountain range of northern Mexico (27°80′5700N, 107°60′4100W; 2400 masl) (Figure 1).

Located in the region of the Holarctic and Neotropical transition, the zone presents a great complexity of ecosystems predominated by pine and oak forests, as a consequence of the variables of physiography and climatic units. This region supplies more than 25% of the timber production in Mexico, and is one of the most important timber reserves in the country, provides a wide variety of environmental services and has a predominantly indigenous population [21]. The vegetation includes forests of pine-oak species, such as *Pinus engelmannii* Carr., *P. arizonica Engelm.*, *P. leiophylla* Schiede ex Schltdl. &; Cham., *Quercus arizonica* Sarg., *Q. crassifolia* Humb. &; Bonpl. and *Q. durifolia* Seemen ex Loes., as well other broadleaf species including *Arbutus arizonica* (A. Gray) Sarg.; *A. bicolor* S. González, M. González &; P.D. Sørensen; *Juniperus deppeana* Steud. There are also patches of tropical montane cloud forest and communities such as chaparral (primary and secondary) and forest clearing vegetation [14]. The dominant soils are Regosols and Leptosols of alluvial origins. The predominant climate is semi-cold and semi-humid, with long and cold summers and monsoon rains accompanied by winter precipitation with an annual mean value of 779 mm and mean annual temperatures of 5 to 12 °C.

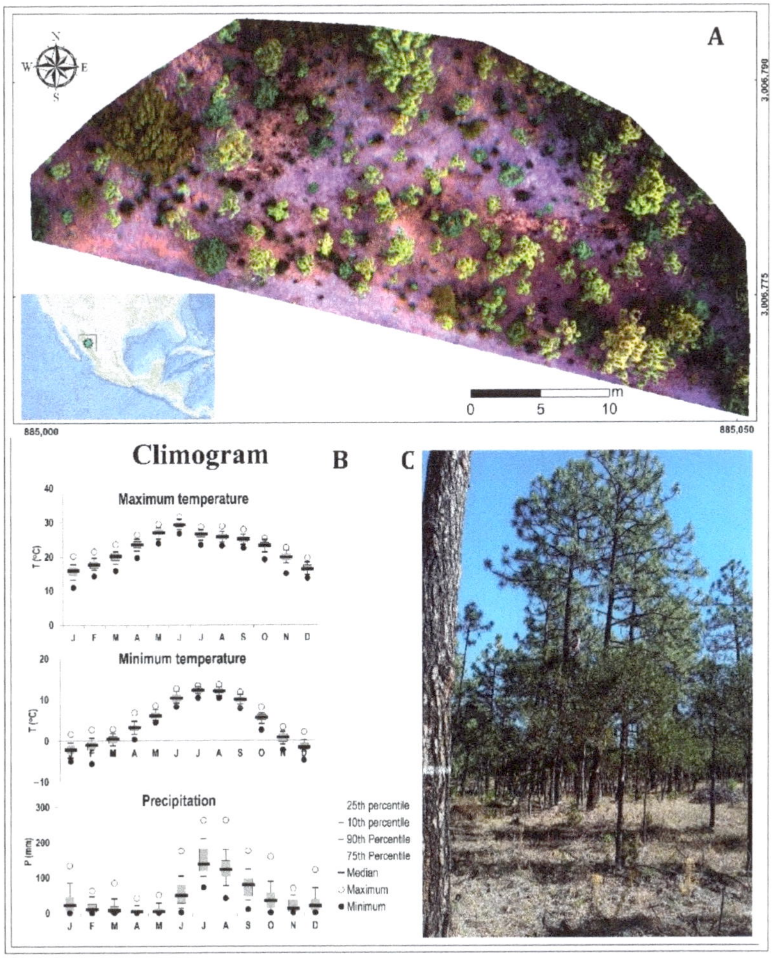

Figure 1. Location and view of Mesa del Pawiranachi (**A**) and climogram of the Papajichi ejido in Chihuahua, Mexico (**B**) and the perspective of site (**C**).

Workflow

To obtain field data during the month of October; each individual tree within the study area was labeled by fixing an aluminum plate to the base. A complete census of all individuals was conducted, recording the following dasometric variables at individual tree level: diameter at breast height (DBH, cm) and basal diameter (BD, cm), using a diametric tape; and commercial height (CH, m) and total height (TH, m), which were measured directly by climbing the trees and using a length meter (Figure 2).

The study area was overflown using a DJI Phantom multispectral (P4M) quadcopter (Figure 2). The P4M camera has a total of six imaging sensors, five of which are multispectral (bands: blue = 450 ± 16 nm, green = 560 ± 16 nm, red = 650 ± 16 nm, RedEdge = 730 ± 16 nm, near-infrared = 840 ± 26 nm) and one RGB sensor, all with a global 2 MP shutter. The UAV was flown in order to obtain and subsequently process 400 aerial photographs of the study area, taken from an altitude of 50 m, with overlaps between the images and lines of 80 and 75%, respectively. A subsequent flight was conducted from east to west in order to capture RGB and spectral images. Both flights were conducted on 16 October 2021, which was a sunny day, with suitable wind conditions (<25 kph) and a mean temperature of 19 °C.

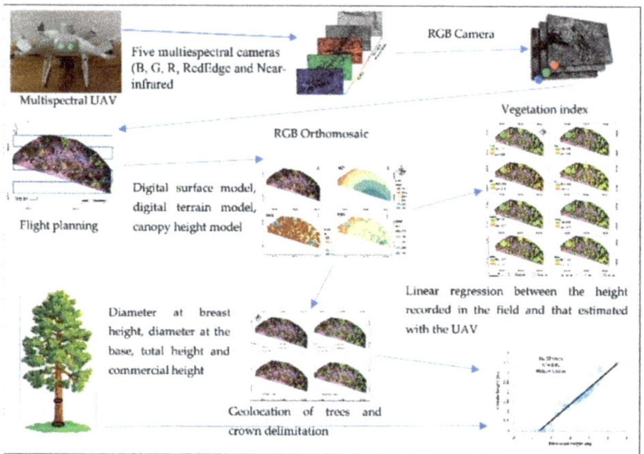

Figure 2. Workflow of the processing stages of images taken by an unmanned aerial vehicle (DJI Phantom 4 Multispectral UAV).

The UAV used had a georeferencing system on board during the flight and it was not necessary to use the real-time-kinematic (RTK) system since the georeferencing system could attain vertical and horizontal location pressures of ± 0.1 and ± 0.3 m, respectively [22]. For processing, we used a computer with an AMD Ryzen 3900x processor with 24 cores at 3.8 GHz, with an integrated Nvidia Quadro p620 quad-core 2 GHz video card and 32 GB of RAM; this was used in a Linux operating system environment based on the ubuntu distribution Pop!_OS version 22.04 LTS.

The images were processed and analyzed with photogrammetric procedures using the open-source software OpenDroneMap (ODM version: 2.8.4; Cleveland Metroparks, Ohio, USA [23]). This software implements the algorithms Structure from Motion and Multi-View Stereo (SfM and MVS), producing 3D point clouds of 1000–20,000 points m^{-2}. We used VisualSfM to achieve the 3D reconstruction [12], due to its versatility in terms of reduced processing time. We then generated the RGB orthomosaic and multispectral orthomosaic. The digital surface (DSM) was generated considering the maximum elevation values from the trees in a point cloud. Where two points occurred on top of each other, only the tallest point was used. Gaps in the point cloud were filled using the dem-gap fill-steps process with the local gridding method. The digital terrain model (DTM) was obtained by classifying the point cloud using a simple morphological filter (SMRF).

Using the raster calculator tool of the open-source software QGIS, the canopy height model (CHM; Equation (1)) was generated in order to predict the potential height of each tree:

$$CHM = DSM - DTM \tag{1}$$

where CHM = canopy height model, DSM = digital surface model and DTM = digital terrain model.

Analysis of the canopy consisted of detecting and geolocating trees in the study area, estimating their heights and delimiting their crowns to obtain the values of crown diameter and area. The package ForestTools [24] of the statistical software R [25] was used as a tool to geolocate the individual trees and delimit their crowns through the variable window filter (VWF) algorithm and the algorithm of segmentation controlled by markers. This package automatically detects the crowns of the trees, obtains the tree height (TH, m), generates polygons and calculates the area of the crown (Ac, m^2).

Using the multispectral orthophoto, the reflectance level was calculated according to the wavelength of each band. In the QGIS software, the Semi-Automatic Classification

Plugin was executed; this tool allows us to calculate the reflectance of a chosen area according to the values of each band.

Given there are studies that have exhaustively verified their adequacy, and some may be redundant [26,27], we consider it appropriate to analyze them individually in order to obtain a more profound interpretation. Moreover, it has been documented that each is affected by sensor type and atmospheric effects, for which reason multi-analysis provides a rigor that guarantees their replicability and cooperation [28]. The calculation was performed using the raster calculator of the program QGIS through the following expressions (Equations (2)–(9)).

$$\mathrm{NDVI} = \frac{\mathrm{NIR} - \mathrm{RED}}{\mathrm{NIR} + \mathrm{RED}} \tag{2}$$

$$\mathrm{LCI} = \frac{\mathrm{NIR} - \mathrm{RedEdge}}{\mathrm{NIR} + \mathrm{RED}} \tag{3}$$

$$\mathrm{RVI} = \frac{\mathrm{NIR}}{\mathrm{RED}} \tag{4}$$

$$\mathrm{GNDVI} = \frac{\mathrm{NIR} - \mathrm{GREEN}}{\mathrm{NIR} + \mathrm{GREEN}} \tag{5}$$

$$\mathrm{NDRE} = \frac{\mathrm{NIR} - \mathrm{RedEdge}}{\mathrm{NIR} + \mathrm{RedEdge}} \tag{6}$$

$$\mathrm{NDGI} = \frac{\mathrm{GREEN} - \mathrm{RED}}{\mathrm{GREEN} + \mathrm{RED}} \tag{7}$$

$$\mathrm{TVI} = \sqrt{\frac{\mathrm{NIR} - \mathrm{RED}}{\mathrm{NIR} + \mathrm{RED}} + 0.05} \tag{8}$$

$$\mathrm{OSAVI} = \frac{\mathrm{NIR} - \mathrm{RED}}{\mathrm{NIR} + \mathrm{RED} + 0.16} \tag{9}$$

where NDVI = normalized difference vegetation index, GNDVI = green NDVI, LCI = leaf chlorophyll index, NDRE = normalized difference red edge index, OSAVI = optimized soil adjusted vegetation index, RVI = ratio vegetation index, TVI = transformed vegetation index, NDGI = normalized difference greenness index, NIR = near infrared band, RED = red band, RedEdge = red edge band and GREEN = green band.

To evaluate photosynthetic activity within the community, the index values were extracted for the crown of each tree. In order to obtain the statistics per genus, the zone statistics tool of the QGIS program was implemented.

With the exception of the crown, which is a novel parameter not geometrically comparable in the field, estimates of the variables at the individual tree level obtained with the UAV were evaluated with respect to the field measurements. In the case of height, given the operational difficulty of its measurement, a subsample of n = 57 trees was used to evaluate the accuracy of the estimates (https://youtu.be/EIkZQX8qI98; accessed on 22 March 2022). As a validation strategy, we manually digitized 47 trees and compared these with those values derived from the algorithm (Ac).

3. Results

A total of 400 images were obtained with the RGB sensor on a single flight and used to generate the RGB orthomosaic (the processing was 4 min), the digital terrain, the surface models and the canopy height models derived from the photogrammetric process (the processing was 8 min) with the software OpenDroneMap (Figure 3).

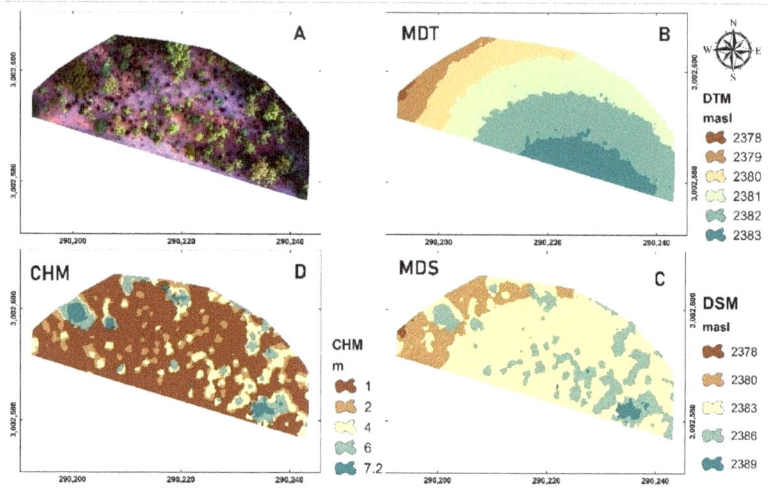

Figure 3. (**A**) = orthomosaic RGB, (**B**) = digital terrain model, (**C**) = digital surface model and (**D**) = canopy height model, derived from the UAV flight.

In the field study, 163 trees in "El Cordoncito" in the Mesa del Pawiranachi were recorded and measured. Table 1 presents the descriptive statistics of the metrics of the individuals taken in the field and calculated in the office. This area is a natural forest and subjected to timber forest management, the individual trees of which present a normal diameter of 1.8 to 62.3 cm with an average of 9.25 cm, and heights of 0.63 to 20.76 m with an average of 4.91 m. This indicates that the population present in "El Cordoncito" is in a state of growth since most of the individuals are juveniles with only a limited number of dominant trees present.

Table 1. Statistical description of the trees recorded in "El Cordoncito".

Variable	CH	TH	Hc	BD	ND
n	163	163	163	163	163
min	1.13	1.91	0.56	4	1.8
q1	1.86	3.495	1.555	9.25	6.2
average	2.211	4.911	2.7	13.064	9.247
median	2.16	4.45	2.18	11.5	8.1
q3	2.41	5.5	3.185	15.05	10.75
max	5.39	15.09	10.41	46.3	37.5
sd	0.664	2.262	1.782	6.446	5.231
se	0.052	0.177	0.14	0.505	0.41

where CH = commercial height (m), TH = total height (m), Hc = height of crown (m), BD = basal diameter (cm), ND = normal diameter (cm), n = number of trees, q1 = quartile 1, q3 = quartile 3, max = maximum, min = minimum, sd = standard deviation and se = standard error.

Statistics were also obtained at the genus level (Table 2). The greatest densities per genus were Pinus, Juniperus, Quercus and Arbutus (109, 43, 9 and 2, respectively), for which reason the forest in the community is Pinus-Juniperus.

Using the variable window filter algorithm of the package ForestTools, 132 trees were detected and their crowns delimited (Figure 4). The algorithm had a tree identification accuracy of 64.4% (163 digitalized–132 detected) with respect to the trees verified in the field. This can be attributed to the heterogeneity of the canopy structures, where some trees underlie the dominant individuals, as well as to the irregular spacing among the trees themselves.

Figure 5 shows the results of the linear regression applied to determine the relationship between the height and crown values recorded in the field and those estimated with the UAV. Furthermore, a graph of dispersion of the residuals of the field data against the predicted values is presented. Statistical evaluations of the characteristics recorded with the field data and those estimated with the UAV showed that 90% of the height and 91% of the crown delineation values are explained by the UAV.

Table 2. Statistical description of the trees recorded in "El Cordoncito".

Genus	Variable	n	min	q1	Average	Median	q3	max	sd	se
Arbutus	CH		1.54	1.585	1.63	1.63	1.675	1.72	0.127	0.09
	TH		4.05	4.318	4.585	4.585	4.852	5.12	0.757	0.535
	Hc	2	2.51	2.732	2.955	2.955	3.178	3.4	0.629	0.445
	BD		12.5	14.125	15.75	15.75	17.375	19	4.596	3.25
	ND		8	9.45	10.9	10.9	12.35	13.8	4.101	2.9
Juniperus	CH		1.13	1.43	1.745	1.71	2.03	2.41	0.365	0.056
	TH		1.91	2.58	3.549	3.45	4.47	6.34	1.122	0.171
	Hc	43	0.56	1.005	1.804	1.57	2.32	4.89	0.876	0.134
	BD		4	6.55	10.667	8.3	14.15	28.8	5.213	0.795
	ND		1.8	3.95	7.174	6.1	10.05	22.6	4.042	0.616
Pinus	CH		1.52	2.02	2.429	2.28	2.53	5.39	0.678	0.065
	TH		2.21	3.88	5.33	4.7	6.1	15.09	2.373	0.227
	Hc	109	0.6	1.75	2.901	2.47	3.32	10.41	1.856	0.178
	BD		6.3	10	13.504	11.6	14.7	46.3	6.025	0.577
	ND		3.9	6.9	9.742	8.2	10.3	37.5	5.141	0.492
Quercus	CH		1.66	1.72	1.927	1.92	2.02	2.31	0.208	0.069
	TH		3.3	4.91	6.411	5.86	8.81	10.11	2.446	0.815
	Hc	9	1.64	2.99	4.484	3.95	6.83	8.01	2.29	0.763
	BD		6.3	11.5	18.6	16.4	19.5	44.7	11.669	3.89
	ND		3.5	7.4	12.789	11.4	14.3	31.3	8.264	2.755

where CH = commercial height (m), Hc = height of crown (m), TH = total height (m), BD = basal diameter (cm), ND = normal diameter (cm), n = number of trees, q1 = quartile 1, q3 = quartile 3, max = maximum, min = minimum, sd = standard deviation and se = standard error.

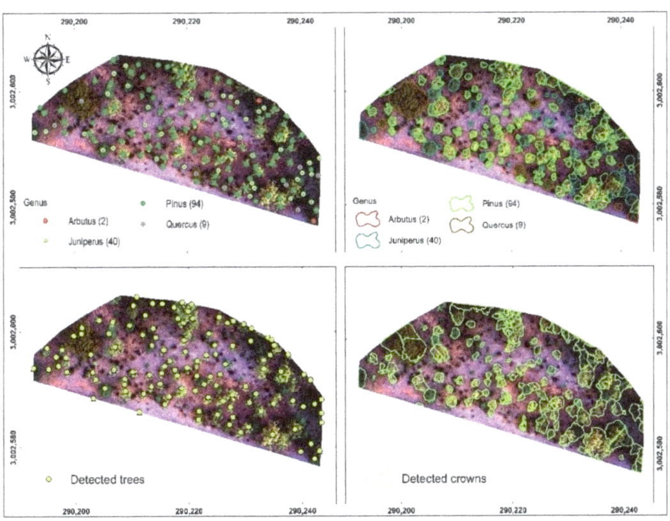

Figure 4. Digitalization of trees (**top left**), digitalization of crowns (**top right**), trees identified (**bottom left**) and crowns detected (**bottom right**) using the ForestTools algorithm.

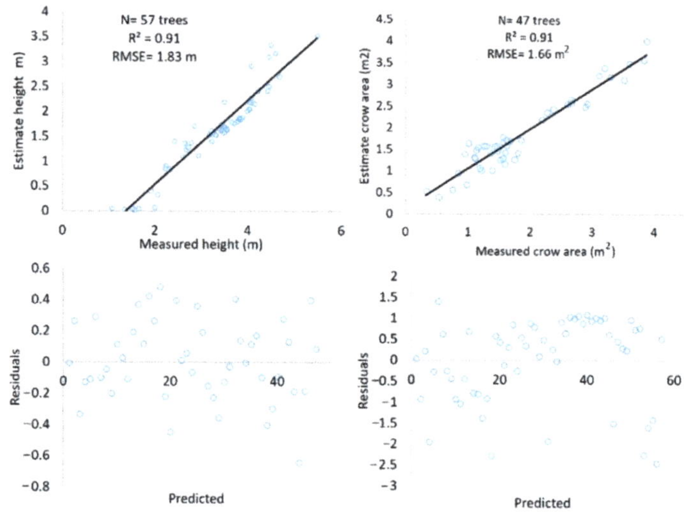

Figure 5. Regression analysis (**top**) and graph of residual vs. predicted values (**bottom**) for height (**left**) and crown delineation (**right**), at individual tree level with field data and estimates derived from the UAV, from 57 trees detected with the ForestTools algorithm.

Three crowns were randomly selected (one of each genus) in order to obtain the multispectral reflectance (Figure 6).

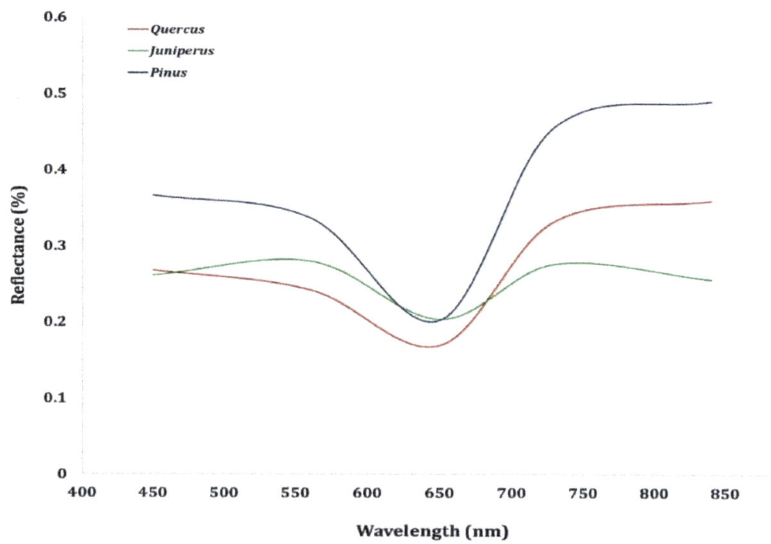

Figure 6. Spectral reflectance at different wavelengths in the crowns of three different genera.

With regard to estimations of the vegetation index, Figure 7 shows the different indices calculated for "El Cordoncito" in Mesa de Pawiranachi from the multispectral orthomosaic derived from the photogrammetric process with OpenDroneMap. Moreover, it was possible to determine the indices NDVI, NDGI, GNDVI, NDRE, OSAVI, LCI, TVI and RVI, obtaining values from −1 to 1, except for RVI, which presented values greater than 1 (Figure 7).

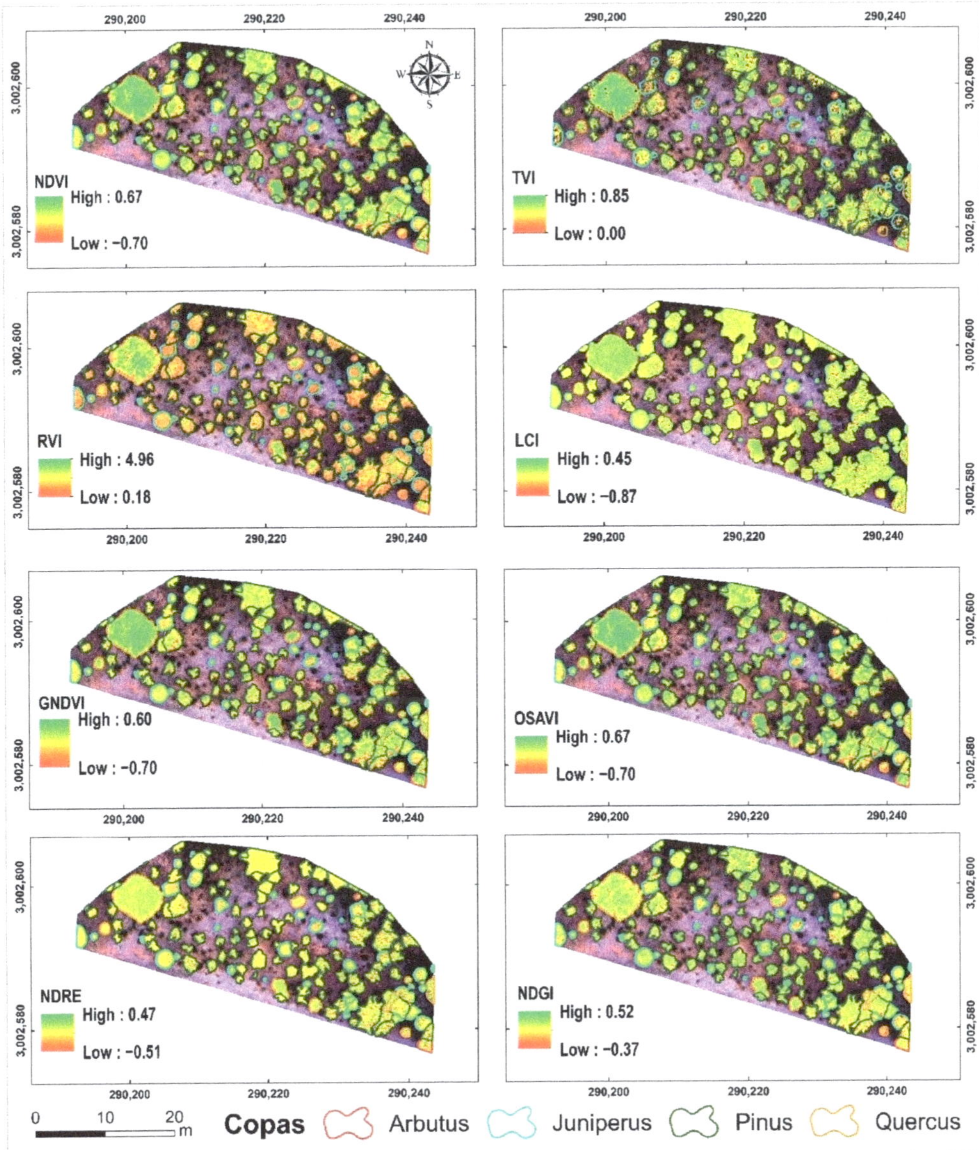

Figure 7. Vegetation indices of "El Cordoncito" in Mesa del Pawiranachi, NDVI = normalized difference vegetation index, GNDVI = green NDVI, LCI = leaf chlorophyll index, NDRE = normalized difference red edge index, OSAVI = optimized soil adjusted vegetation index, RVI = ratio vegetation index, TVI = transformed vegetation index, NDGI = normalized difference greenness index.

For evaluation of the indices according to the delimited trees, NDVI ranged from −0.701 to 0.665, TVI from 0.855 to 0, LCI from 0.48 to −0.95, GNDVI from 0.60 to −0.70, OSAVI from 0.67 to −0.70, NDRE from 0.47 to −0.55, NDGI from 0.47 to −0.55, and RVI from 0 to 4.13 (Figure 7).

The values of the vegetation indices of the crown were visibly higher (green and yellow colors) than the discriminated indices (herbaceous plant vegetation, bare soil, dead vegetal material—orange and red colors). However, both low and high values were found within the tree crowns. These are hypothetically attributable to different levels of photosynthetic activity, as well as open spaces. Figure 8 shows the distribution range of the maximum values of the vegetation indices per genus, which are commonly associated with seasonal extremes in variations of vegetation dynamics [29].

In terms of the distribution of the maximum values of the vegetation indices per genus, Quercus was highest in the indices TVI, NDVI, OSAVI, and LCI (Figures 8 and 9), while Arbutus was highest in NDRE and LCI, and Pinus in NDGI. Juniperus was the lowest across all of the indices (Table 3).

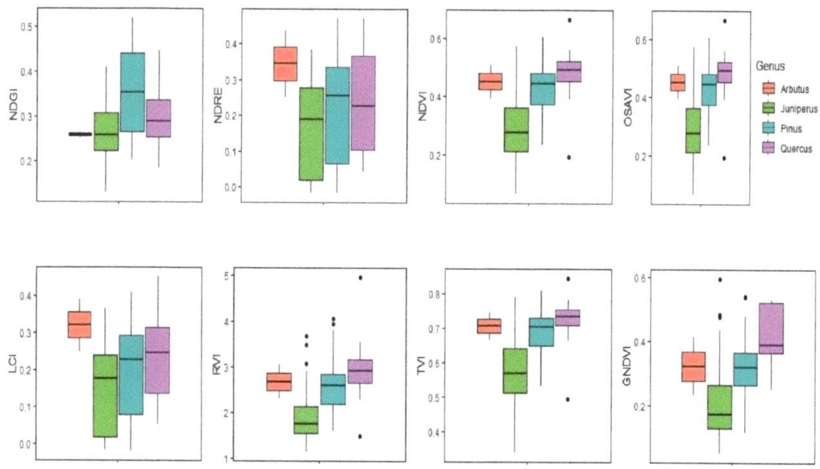

Figure 8. Distribution of the vegetation indices per genus in "El Cordoncito" in Mesa del Pawiranachi, NDVI = normalized difference vegetation index, GNDVI = green NDVI, LCI = leaf chlorophyll index, NDRE = normalized difference red edge index, OSAVI = optimized soil adjusted vegetation index, RVI = ratio vegetation index, TVI = transformed vegetation index, NDGI = normalized difference greenness index.

Table 3. Statistical description of the maximum values of the vegetation indices of the detected crowns.

Variable	GNDVI	LCI	NDGI	NDRE	NDVI	OSAVI	RVI	TVI
min	−0.024	0	−0.347	0	−0.206	−0.206	0.658	0.455
average	0.307	0.198	0.089	0.212	0.412	0.412	2.508	0.680
max	0.596	0.453	0.520	0.472	0.664	0.664	4.96	0. 845
sd	0.109	0.118	0.117	0.141	0.112	0.112	0.598	0.073

where min = minimum, max = maximum, sd = standard deviation, NDVI = normalized difference vegetation index, GNDVI = green NDVI, LCI = leaf chlorophyll index, NDRE = normalized difference red edge index, OSAVI = optimized soil adjusted vegetation index, RVI = ratio vegetation index, TVI = transformed vegetation index, NDGI = normalized difference greenness index.

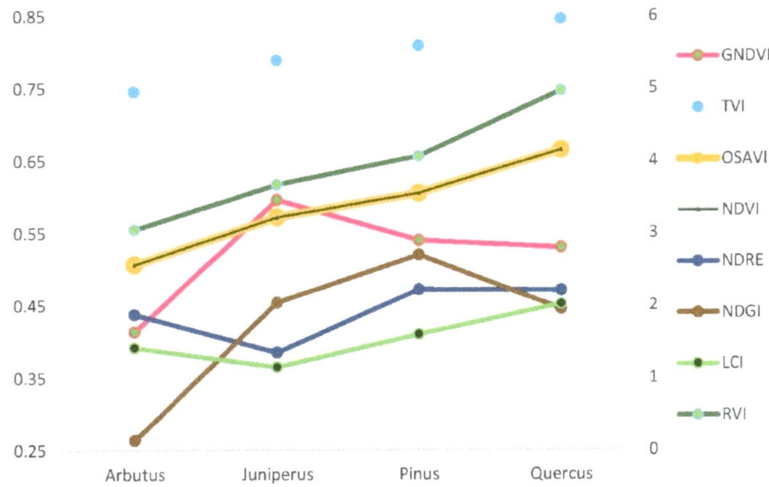

Figure 9. Comparison of the vegetation indices per genus in Mesa del Pawiranachi, NDVI = normalized difference vegetation index, GNDVI = green NDVI, LCI = leaf chlorophyll index, NDRE = normalized difference red edge index, OSAVI = optimized soil adjusted vegetation index, RVI = ratio vegetation index, TVI = transformed vegetation index, NDGI = normalized difference greenness index.

4. Discussion

Modern silviculture aims to optimize resources in the search for sustainable forest management [30]. This study developed procedures for the estimation of attributes at the level of the individual trees that are pioneers in Mexican natural forests. The number of trees, crown area and total height are the main measurements in the forest inventory and are critical to the support of appropriate decision-making. Similarly, multispectral indices were calculated as a strategy with which to improve knowledge of reflectance as an indicator of the ecological mechanisms faced by these ecosystems [31].

The quantification of such structural variables/traits has implications for the fauna in the habitat [32] and, in general, for the ecology of the forest stand [33]. For example, the geometry of the crown and the tree height is directly related to the capacities for carbon capture, while the photosynthetic rates of the trees can provide indications regarding vigor and climatic vulnerability [34]. These aerial parts of the tree represent variables of great ecological interest that merit further investigation in order to contribute to our knowledge.

4.1. Estimation of Attributes of Individual Trees

Regarding estimation of tree-level attributes using UAV technology, our study provides a perspective of such estimation in natural stands. From a practical point of view, this strategy enhances the use of individual tree-level attribute measurements, saves time and improves certainty. In spite of certain shortcomings (addressed below), we demonstrate that, when traditional measurement procedures are combined with UAV-derived geospatial information, knowledge of forest stands is enhanced and decision-making in forest management could be improved as a result. To scale up the potential for application to larger areas of forest, ideal experimentation sites must be identified. We consider that the study area was strategic in terms of the potential extrapolation of the application of the technique to other management scales. For example, the coexistence of four genera within a relatively limited area, together with the dendroecological structural variations present (see Table 2), confers advantages over traditional monospecific and contemporary studies and those of regular spatial arrangement [35].

In the first instance, quantification of the number of trees produced limited results, unlike those found in neighboring areas [12]. This could be explained by the difference in the structural complexity of the stand, as well as in the number of UAV flights performed. However, our findings can provide forest managers with a spatial perspective regarding the density of the forest that could be of utility for the purposes of exploration and planning, and that was previously unavailable or very difficult to achieve at ground level.

Delimitation of the tree crowns using UAV technology constitutes an important metric in biometry with implications for tree physiological development. In the field, the technician normally tries to indirectly estimate this parameter by assuming the area of a circle from the average of the greatest and smallest diameters of the crown. However, the bias inherent in this technique reduces the reliability of the results. For this reason, the estimates generated by the UAV present greater certainty since the algorithm acts to more closely delimit the irregular shape, as seen in [36]. The Structure from Motion (SfM) algorithm has been successfully used in conifer forests; however, its application in mixed forests with broadleaf species is still incipient [37]. Our results are favorable since tree density and distribution can be mapped along with the crown area. These metrics therefore offer an opportunity to influence estimates of biomass and/or carbon [38].

From an ecological perspective, the crown is a multipurpose ecological indicator, including estimates of the potential for carbon capture, aerial biomass storage, forest fire risk, requirements for cultivation work (pruning, thinning, etc.), density regimes, vegetation changes, regeneration strategies, classification of species and refuges for fauna [39,40]. Accurate estimation of the crown dimensions is therefore essential for precision silviculture [41].

One limitation of the detection algorithm is that the segmentation routine is strongly affected by stand characteristics such as density, species heterogeneity, and tree age [37]. Our strategy consisted of smoothing the canopy height model using filters, as seen in [39]. The result allowed better crown delimitation despite the differences in leaf and branch shapes among the studied genera. However, it is important to monitor the intensity of the filters and verify the field data, since small trees may be omitted, as well as those that were being suppressed by the adult trees. It is therefore advisable to complement the analysis with hyperspectral and LiDAR tools [42–44], although the financial implications should also be taken into account. In addition, segmentation techniques need to be refined to delimit the crowns from the tree tops. We therefore recommend conducting thorough initial tree top detection for use as an input to the canopy delimitation process

Although our study did not include an exhaustive analysis of the accuracy of auto-delimitation of the tree crown, we found the estimations to be good ($R^2 = 0.91$; RMSE = 1.83 m^2). Better results were clearly evident in the broadleaf species, supporting [37]. We attributed this to the fact that this particular leaf foliage gives rise to a homogeneous crown conformation, while the needle and branch arrangement in the conifers gives rise to greater inconsistencies in the segmentation process [44,45]. It is also advisable to differentially evaluate the algorithms according to species, given the variation that exists in the configuration of the irregular geometry that confers differing complexity according to genus.

Accurate height estimation is of crucial importance for both ecological and commercial reasons. It is a significant indicator of the productive capacity of the site and a fundamental requirement for subsequent estimates of stand structure [37]. Our methodology produced acceptable results ($R^2 = 0.91$, RMSE = 0.36) since, compared to previous studies such as those of [30,45], the statistical values are consistent. Any bias can be attributed to the difficulty in mapping the vegetation and leaf litter below the tree at the time of generating the digital surface model, and subsequent corrections are therefore advisable when field data on leaf litter thickness are available. However, the estimation capabilities of the UAV are significantly better than those of technician-led efforts, since the occlusion and overlapping of canopies make it difficult to accurately distinguish the canopy apex. Another limitation may be the seasonality of the estimations, since some species of the genera Arbutus and Quercus are devoid of foliage at certain times of the year, making it difficult to distinguish the uppermost tip of the tree. In normative terms, the error found is permissible under current regulations [46].

4.2. Multispectral Indices

Remote sensing is considered one of the most powerful technologies for vegetation assessment. The rapid and vertiginous development of this technology has proven to be an effective tool with which to further our understanding of vegetation dynamics [1,11]. In this study, we extracted the canopy reflectance spectrometry through multispectral indices using UAV at the individual-tree level.

It is recognized that the photosynthetic capacity of canopies is a crucial parameter with which to improve our understanding of the eco-physiological processes taking place between the forest and the atmosphere, i.e., the magnitude of photosynthetic variability in forest species remains a challenge that merits analysis beyond the leaf level, and its accurate estimation would reduce uncertainty in terrestrial biosphere models (e.g., carbon fluxes and others). With no intention of explaining the rates and thresholds of photosynthesis of the studied species, we discuss the spectral reflectance results calculated with multispectral vegetation indices (e.g., VI). This is a preliminary step to further exploring their use and linkage with ground-truth photosynthetic measurements (not measured in this study).

In addition, the spectral reflectance captured by a UAV sensor at canopy level is known to be associated, to a greater extent, with canopy geometry and dispersion of the foliage, etc., often producing spurious spectral variation that can be confused with the spectral signature associated with plant photosynthesis. As discussed below, each index presents differences in its spectral reflectance, but the best combination of these VI remains to be explored. Consequently, we discuss them as a potential source that could predict future photosynthetic productivity [47].

It is noteworthy that, although these VI are not direct measures of actual productivity rates, they are in line with previous studies that use reflectance as a viable tool with which to predict photosynthetic variables, or as an indicator of canopy "greenness" [48], although in this study we only used reflectance spectra and never used leaf-level data or phenological measurements.

In this sense, the NDVI represented the greenness of the individuals and is used by some authors as a proxy for relative biomass [49–51], although it is recommended to determine light use efficiency (LUE). By taking advantage of contrasts between electromagnetic bands and chlorophyll pigments, this index allowed us to differentially distinguish the trees in terms of vigor (Figure 8). The genus Quercus presented the highest values of this index, with advantageous implications for the conservation of this genus in the face of predicted climatic change [52]. Secondly, Pinus and Arbutus seem to share the same level of "greenness", although without very critical conditions. In contrast, the genus Juniperus seems to be the most heavily impacted by environmental or anthropogenic stressors (not studied here). Management strategies such as pruning, thinning, controlled burning and other regimes that could influence site productivity should therefore be considered. The reliability of NDVI in evergreen forests requires further assessment given that previous studies demonstrated uncoupling between NDVI magnitude and productivity due to a change in radiation-use efficiency [48,50,53]

The TVI is sensitive to crown structure [54] and is highly recommended for monitoring changes in the dynamics of the vegetation due to its property of contrasting the values of reflectance [55]. Its values confirmed Quercus as the genus with higher spectral reflectance than the other three genera.

A similar trend was followed by GNDVI, as an indicator of water from soil moisture and nitrogen consumption attributable to photosynthetic activity. Although our study did not contemplate chemical analysis of the soil, previous studies in neighboring sites have reported that nitrogen is not limiting [12]. Consequently, we hypothesized that water will be the limiting factor, as documented by [56,57]. These sites are of shallow soil depth and water from the monsoon rains tends to run off, leaving the trees under water stress and heavily reliant on winter rains to carry out their physiological processes [58].

The NDRE proved to be a useful complement with which to detect tree anomalies, since it optimizes the detailed data between the red bands and the NIR, complementing the benefits of the NDVI [59]. Thus, Arbutus differed markedly from the other three genera (Figure 8), meaning that this species undergoes marked changes in the properties of its spectral reflectance with respect to the other species. NDRE has been reported as a potential indicator of chlorophyll and changes in photosynthetic rate transitions (not measured here), making it very useful for programming in situ cultural activities that require these plants [59].

The LCI indicated the level of chlorophyll, where the genus Arbutus differed notably from the other genera. According to [13], this index is a good indicator of the leaf area index and is related to the maturity of the individual. For this reason, it merits continuous monitoring.

Regarding the OSAVI, this index evidenced that the highest chlorophyll values are presented by the genus Quercus, followed by Arbutus and the pines, whilst the Juniperus present a small scale. This index considers the soil as a factor of chlorophyll content, such that it is directly related to the amount of canopy foliage [60]. Coincidentally, [61] reports that its values are proportional to the robustness of the canopy. This makes sense since the oaks presented the largest crown dimensions (Table 2). As a consequence, this index could be appropriate for monitoring photosynthetic activity in closed, high-density canopies.

The RVI reflects the fact that the genus Quercus presents the highest chlorophyll values, which is attributed to its greater quantity of foliage and higher crown volume (Figure 8). These results agree with [62], who state that this index is an indicator of chlorophyll content as a factor that influences photosynthesis and nitrogen content.

The NDGI showed the best values for Juniperus and is an indicator [63] of changes in the status quo of the vegetation, e.g., post-disturbance. In other words, this greenness index reflects the reflective characteristics of plants, as determined by their momentary condition. Juniperus differs from other species in terms of cellular structure and moisture and chlorophyll content.

Our spectral multi-indices approach produced greater knowledge of the variation in spectral signatures, as a reflection of the potential photosynthetic activity in the studied trees. As shown in Figure 6, the wavelengths highlight a significant decrease in the percentage of genus-differentiated reflectance around 650 nm. These differences can be interpreted in different ways [63], but it is desirable in the future to take further considerations into account in order to make these data useful for interpretations of photosynthetic activity as such. For example, determination of the radiation use efficiency (RUE) from careful ground measurements would allow a better understanding of the rate of photosynthesis through NDVI. On the other hand, it is necessary to derive continuous spatial and temporal information regarding the tree physiology, complement with hyperspectral data, refine radiometric corrections and calibrate the sensors. This will act to reduce the uncertainty of monitoring the performance of photosynthesis at canopy level, and eventually at leaf level.

We also do not discount the possibility that our results could allow us to discriminate species and phenological processes [64–66]. In any case, it is crucial to determine the behavior of dendroecological variables collected in situ, since multispectral indices are complementary to the field information, which is essential in order to reach better conclusions.

As a limitation, the seasonality of image acquisition should be considered, since there are reports that multispectral indices are also multiseasonal [67]. The dynamics of tree phenology and physiological processes, therefore, merit further study with a temporal perspective. Particularly in the case of oaks, for example, there are seasons when the tree is devoid of foliage while the tree develops other physiological processes, such as root elongation, radial growth, bud development, etc.

Despite the promising results in terms of extracting phenotypic information of high spatial resolution and accurate spectral reflectance for tree-level applications, we highlight certain limiting factors that merit further research. For instance, a range of phenological stages, leaf optical properties, soil reflectance, canopy structure, sun directions, saturation phenomenon, sensor effects, etc. are all important factors governing spectral accuracy [68]. It is therefore advisable to exercise caution with the application of our approach, which is merely descriptive. Before linking multispectral imagery with vigor or tree productivity, fresh consideration of the use of UV-based VI is required. The lack of synchrony between canopy spectrometry and physiological processes means that these indices are not infallible indicators of instantaneous photosynthetic rates. The major problems include atmospheric effects, optical properties and canopy structural attributes, which must be measured in situ. Thus, the combination of VI with environmental and physiological variables requires the adoption of more innovative approaches. In addition, multispectral data acquisition and processing techniques imply high technical requirements and the need for specialized personnel.

5. Conclusions

This research was advantageous for forest monitoring in natural forests and could complement forest inventories and ecology studies. The long time periods required for individual field estimations are drastically reduced by the approach followed here. Our workflow proved to be an effective alternative for characterizing tree attributes. The information generated substantially facilitates applications using the knowledge of the studied species metrics. The accuracy of the metrics is reliable and the multispectral indices are useful indicators of potential photosynthetic capacity. They represent surrogate variables that can be used as input for models of forest ecosystem dynamics. The characterization of VI was a valuable indicator for distinguishing functional genus types. In particular, NDVI is a dominant and effective index for detecting photosynthetic activity, although this does not imply that it is infallible. As a result, one may consider the use of other vegetation indices as generated here.

The quality of the mapping allowed the proposal of new research paradigms, including the need to adjust the algorithms according to tree age, height and species group, since the mapping results were influenced by species composition. Likewise, stand density merits the exploration of additional technologies in order to reduce uncertainty, but the implications in terms of economic costs must be taken into account.

Author Contributions: E.D.V.-V. gathered field data and processing data; M.P.-G. conceived the experiment, lead the data analysis, wrote and edited the original manuscript; J.A.M.-R. processing data; L.A.M.-T. software. All authors have read and agreed to the published version of the manuscript.

Funding: This research was funded by CONACYT for funding provided through project A1-S-21471.

Data Availability Statement: Not applicable.

Acknowledgments: We thank to CONACYT for funding provided through project A1-S-21471, COCYTED and DendroRed, (http://dendrored.ujed.mx; accessed on 15 April 2022). Furthermore, we thank Ejido Papajichi, Andrés Cruz Cruz (El Kapy); Bersaín Acosta Barraza; Martín José Loya Barraza; Manuel de Jesús Espinoza Carrillo; José Pedro Lerma Chacarito; Alexis Arturo Chávez Cervantes; José Flores Hernández y Uriel Bustillos Espino for facilitating and supporting field data gathering.

Conflicts of Interest: The authors declare no conflict of interest.

References

1. Lechner, A.M.; Foody, G.M.; Boyd, D.S. Applications in Remote Sensing to Forest Ecology and Management. *One Earth* **2020**, *2*, 405–412. [CrossRef]
2. Wallerman, J.; Bohlin, J.; Nilsson, M.B.; Franssen, J.E. Drone-Based Forest Variables Mapping of ICOS Tower Surroundings. In *IGARSS 2018–2018 IEEE International Geoscience and Remote Sensing Symposium*; IEEE: Hoboken, NJ, USA, 2018; pp. 9003–9006. [CrossRef]

3. Jayathunga, S.; Owari, T.; Tsuyuki, S. The use of fixed–wing UAV photogrammetry with LiDAR DTM to estimate merchantable volume and carbon stock in living biomass over a mixed conifer–broadleaf forest. *Int. J. Appl. Earth Obs. Geoinf.* **2018**, *73*, 767–777. [CrossRef]
4. Overman, H.; Butt, N.; Cummings, A.R.; Luzar, J.B.; Fragoso, J. National REDD+ Implications for Tenured Indigenous Communities in Guyana, and Communities' Impact on Forest Carbon Stocks. *Forests* **2018**, *9*, 231. [CrossRef]
5. Grybas, H.; Congalton, R.G. A Comparison of Multi-Temporal RGB and Multispectral UAS Imagery for Tree Species Classification in Heterogeneous New Hampshire Forests. *Remote Sens.* **2021**, *13*, 2631. [CrossRef]
6. González, E.; Núñez, C.; Salinas, J.; Rodas, J.; Rodas, M.; Paiva, E.; Kali, Y.; Saad, M.; Lesme, F.; Lesme, J.; et al. Analysis and Application of Multispectral Image Processing Techniques Applied to Soybean Crops from Drones Vision System. In Proceedings of the 18th International Conference on Informatics in Control, Automation and Robotics (ICINCO), Online, 6–8 July 2021; pp. 707–715. [CrossRef]
7. Lausch, A.; Erasmi, S.; King, D.J.; Magdon, P.; Heurich, M. Understanding Forest Health with Remote Sensing-Part I—A Review of Spectral Traits, Processes and Remote-Sensing Characteristics. *Remote Sens.* **2016**, *8*, 1029. [CrossRef]
8. Dash, J.P.; Pearse, G.D.; Watt, M.S. UAV Multispectral Imagery Can Complement Satellite data for Monitoring Forest Health. *Remote Sens.* **2018**, *10*, 1216. [CrossRef]
9. Moncada, W.; Willems, B.; Pereda, A.; Aldana, C.; Gonzales, J. Tendencia anual, anomalías y predicción del comportamiento de cobertura de vegetación con imágenes Landsat y MOD13Q1, microcuenca Apacheta, Región Ayacucho. *Rev. Teledetección* **2022**, *59*, 73–86. [CrossRef]
10. Yu, R.; Luo, Y.; Zhou, Q.; Zhang, X.; Wu, D.; Ren, L. Early detection of pine wilt disease using deep learning algorithms and UAV-based multispectral imagery. *For. Ecol. Manag.* **2021**, *497*, 119493. [CrossRef]
11. Safonova, A.; Hamad, Y.; Dmitriev, E.; Georgiev, G.; Trenkin, V.; Georgieva, M.; Dimitrov, S.; Iliev, M. Individual Tree Crown Delineation for the Species Classification and Assessment of Vital Status of Forest Stands from UAV Images. *Drones* **2021**, *5*, 77. [CrossRef]
12. Gallardo-Salazar, J.L.; Pompa-García, M. Detecting Individual Tree Attributes and Multispectral Indices Using Unmanned Aerial Vehicles: Applications in a Pine Clonal Orchard. *Remote Sens.* **2020**, *12*, 4144. [CrossRef]
13. Liu, S.; Zeng, W.; Wu, L.; Lei, G.; Chen, H.; Gaiser, T.; Srivastava, A.K. Simulating the Leaf Area Index of Rice from Multispectral Images. *Remote Sens.* **2021**, *13*, 3663. [CrossRef]
14. González-Elizondo, M.S.; González-Elizondo, M.; Tena-Flores, J.A.; Ruacho-González, L.; López-Enríquez, I.L. Vegetación de la Sierra Madre Occidental, México: Una síntesis. *Acta Botánica Mex.* **2012**, *100*, 351–403. [CrossRef]
15. Fraser, B.T.; Congalton, R.G. Monitoring Fine-Scale Forest Health Using Unmanned Aerial Systems (UAS) Multispectral Models. *Remote Sens.* **2021**, *13*, 4873. [CrossRef]
16. Jiang, Q.; Fang, S.; Peng, Y.; Gong, Y.; Zhu, R.; Wu, X.; Ma, Y.; Duan, B.; Liu, J. UAV-Based Biomass Estimacion for Rice Combining Spectral, TIN-Based Structural and Meteorological Features. *Remote Sens.* **2019**, *11*, 890. [CrossRef]
17. Marques Ramos, A.P.; Prado Osco, L.; Garcia Furuya, D.E.; Nunes Gonçalves, W.; Cordeiro Santana, D.; Ribeiro Teodoro, L.P.; da Silva Junior, C.A.; Capristo-Silva, G.F.; Li, J.; Rojo Baio, F.H.; et al. A random forest ranking approach to predict yield in maize with uav-based vegetation spectral indices. *Comput. Electron. Agric.* **2020**, *178*, 105791. [CrossRef]
18. Reyes-Zurita, N.; Rodríguez-Ortiz, G.; Enríquez-del Valle, J.R.; Jiménez-Colmenares, C.L.; Rincón-Ramírez, J.A. Estimación de variables dasométricas en rodales bajo manejo forestal con vehículos aéreos no tripulados. *FIGEMPA: Investig. Desarro.* **2022**, *13*, 22–31. [CrossRef]
19. Erb, K.H.; Kastner, T.; Plutzar, C.; Bais, A.L.S.; Carvalhais, N.; Fetzel, T.; Gingrich, S.; Haberl, H.; Lauk, C.; Niedertscheider, M.; et al. Unexpectedly large impact of forest management and grazing on global vegetation biomass. *Nature* **2018**, *553*, 73–76. [CrossRef]
20. Otsu, K.; Pla, M.; Duane, A.; Cardil, A.; Brotons, L. Estimating the Threshold of Detection on Tree Crown Defoliation Using Vegetation Indices from UAS Multispectral Imagery. *Drones* **2019**, *3*, 80. [CrossRef]
21. Fulé, P.Z.; Ramos-Gómez, M.; Cortés-Montaño, C.; Miller, A.M. Fire regime in a Mexican forest under indigenous resource management. *Ecol. Appl.* **2011**, *21*, 764–775. [CrossRef]
22. DJI P4 Multispectral Specs. Available online: https://www.dji.com/p4-multispectral/specs (accessed on 21 March 2022).
23. OpenDroneMap/ODM. Available online: https://github.com/OpenDroneMap/ODM (accessed on 22 February 2022).
24. ForestTools: Analyzing Remotely Sensed Forest Data. Available online: https://CRAN.R-project.org/package=ForestTools (accessed on 22 January 2022).
25. The R Project for Statistical Computing. Available online: https://www.r-project.org/ (accessed on 22 January 2022).
26. Ganivet, E.; Bloomberg, M. Towards rapid assessments of tree species diversity and structure in fragmented tropical forests: A review of perspectives offered by remotely-sensed and field-based data. *For. Ecol. Manag.* **2019**, *432*, 40–53. [CrossRef]
27. Jurado, J.M.; Ortega, L.; Cubillas, J.J.; Feito, F.R. Multispectral Mapping on 3D Models and Multi-Temporal Monitoring for Individual Characterization of Olive Trees. *Remote Sens.* **2020**, *12*, 1106. [CrossRef]
28. Huang, S.; Tang, L.; Hupy, J.P.; Wang, Y.; Shao, G. A commentary review on the use of normalized difference vegetation index (NDVI) in the era of popular remote sensing. *J. For. Res.* **2021**, *32*, 1–6. [CrossRef]
29. Potter, C.S.; Brooks, V. Global analysis of empirical relations between annual climate and seasonality of NDVI. *Int. J. Remote Sens.* **1998**, *19*, 2921–2948. [CrossRef]

30. Hao, Z.; Lin, L.; Post, C.J.; Jiang, Y.; Li, M.; Wei, N.; Yu, K.; Liu, J. Assessing tree height and density of a young forest using a consumer unmanned aerial vehicle (UAV). *New For.* **2021**, *52*, 843–862. [CrossRef]
31. Kureel, N.; Sarup, J.; Matin, S.; Goswami, S.; Kureel, K. Modelling vegetation health and stress using hypersepctral remote sensing data. *Modeling Earth Syst. Environ.* **2022**, *8*, 733–748. [CrossRef]
32. Hardenbol, A.A.; Kuzmin, A.; Korhonen, L.; Korpelainen, P.; Kumpula, T.; Maltamo, M.; Kouki, J. Detection of aspen in conifer-dominated boreal forests with seasonal multispectral drone image point clouds. *Silva Fenn.* **2021**, *55*, 10515. [CrossRef]
33. Jayathunga, S.; Owari, T.; Tsuyuki, S. Digital Aerial Photogrammetry for Uneven-Aged Forest Management: Assessing the Potential to Reconstruct Canopy Structure and Estimate Living Biomass. *Remote Sens.* **2019**, *11*, 338. [CrossRef]
34. Su, J.; Liu, C.; Coombes, M.; Hu, X.; Wang, C.; Xu, X.; Li, Q.; Guo, L.; Chen, W.H. Wheat yellow rust monitoring by learning from multispectral UAV aerial imagery. *Comput. Electron. Agric.* **2018**, *155*, 157–166. [CrossRef]
35. Liu, K.; Shen, X.; Cao, L.; Wang, G.; Cao, F. Estimating forest structural attributes using UAV-LiDAR data in Ginkgo plantations. *ISPRS J. Photogramm. Remote Sens.* **2018**, *146*, 465–482. [CrossRef]
36. Weinstein, B.G.; Marconi, S.; Bohlman, S.A.; Zare, A.; White, E.P. Cross-site learning in deep learning RGB tree crown detection. *Ecol. Inform.* **2020**, *56*, 101061. [CrossRef]
37. Miraki, M.; Sohrabi, H.; Fatehi, P.; Kneubuehler, M. Individual tree crown delineation from high-resolution UAV images in broadleaf forest. *Ecol. Inform.* **2021**, *61*, 101207. [CrossRef]
38. Jones, A.R.; Raja Segaran, R.; Clarke, K.D.; Waycott, M.; Goh, W.S.H.; Gillanders, B.M. Estimating Mangrove Tree Biomass and Carbon Content: A cCmparison of Forest Inventory Techniques and Drone Imagery. *Front. Mar. Sci.* **2020**, *6*, 784. [CrossRef]
39. Panagiotidis, D.; Abdollahnejad, A.; Surový, P.; Chiteculo, V. Determining tree height and crown diameter from high-resolution UAV imagery. *Int. J. Remote Sens.* **2017**, *38*, 2392–2410. [CrossRef]
40. Mohan, M.; Silva, C.A.; Klauberg, C.; Jat, P.; Catts, G.; Cardil, A.; Hudak, A.T.; Dia, M. Individual Tree Detection from Unmanned Aerial Vehicle (UAV) Derived Canopy Height Model in an Open Canopy Mixed Conifer Forest. *Forests* **2017**, *8*, 340. [CrossRef]
41. Corona, P.; Chianucci, F.; Quatrini, V.; Civitarese, V.; Clementel, F.; Costa, C.; Floris, A.; Menesatti, P.; Puletti, N.; Sperandio, G.; et al. Precision forestry: Concepts, tools and perspectives in Italia. *For.-J. Silvic. For. Ecol.* **2017**, *14*, 1–21. [CrossRef]
42. Dalponte, M.; Bruzzone, L.; Gianelle, D. Tree species classification in the Southern Alps based on the fusion of very high geometrical resolution multispectral/hyperspectral images and LiDAR data. *Remote Sens. Environ.* **2012**, *123*, 258–270. [CrossRef]
43. Cunliffe, A.M.; Brazier, R.E.; Anderson, K. Ultra-fine grain landscape-scale quantification of dryland vegetation structure with drone-acquired structure-from-motion photogrammetry. *Remote Sens. Environ.* **2016**, *183*, 129–143. [CrossRef]
44. Zhen, Z.; Quackenbush, L.J.; Zhang, L. Trends in Automatic Individual Tree Crown Detection and Delineation—Evolution of LiDAR Data. *Remote Sens.* **2016**, *8*, 333. [CrossRef]
45. Shashkov, M.; Ivanova, N.; Shanin, V.; Grabarnik, P. Ground Surveys Versus UAV Photography: The Comparison of Two Tree Crown Mapping Techniques. In *Information Technologies in the Research of Biodiversity*; Springer: Cham, Switzerland, 2019; pp. 48–56. [CrossRef]
46. Norma Oficial Mexicana NOM-152-SEMARNAT-2006. Available online: http://www.diariooficial.gob.mx/nota_detalle.php?codigo=5064731&date=17/10/2008 (accessed on 20 March 2022).
47. Meneses, V.A.B.; Téllez, J.M.; Velasquez, D.F.A. Uso de drones para el análisis de imágenes multiespectrales en agricultura de precisión. *Limentech Cienc. y Tecnol. Aliment.* **2015**, *13*, 28–40. [CrossRef]
48. Gamon, J.A.; Field, C.B.; Goulden, M.L.; Griffin, K.L.; Hartley, A.E.; Joel, G.; Penuelas, J.; Valentini, R. Relationships between NDVI, Canopy Structure, and Photosynthesis in Three Californian Vegetation Types. *Ecol. Appl.* **1995**, *5*, 28–41. [CrossRef]
49. Riihimäki, H.; Heiskanen, J.; Luoto, M. The effect of topography on arctic-alpine aboveground biomass and NDVI patterns. *Int. J. Appl. Earth Obs. Geoinf.* **2017**, *56*, 44–53. [CrossRef]
50. Liu, S.; Cheng, F.; Dong, S.; Zhao, H.; Hou, X.; Wu, X. Spatiotemporal dynamics of grassland aboveground biomass on the Qinghai-Tibet Plateau based on validated MODIS NDVI. *Sci. Rep.* **2017**, *7*, 4182. [CrossRef]
51. Han, H.; Wan, R.; Li, B. Estimating Forest Aboveground Biomass Using Gaofen-1 Images, Sentinel-1 Images, and Machine Learning Algorithms: A Case Study of the Dabie Mountain Region, China. *Remote Sens.* **2022**, *14*, 176. [CrossRef]
52. Alla, A.Q.; Pasho, E.; Marku, V. Growth variability and contrasting climatic responses of two Quercus macrolepis stands from Southern Albania. *Trees* **2017**, *31*, 1491–1504. [CrossRef]
53. Vicente-Serrano, S.M.; Camarero, J.J.; Olano, J.M.; Martín-Hernández, N.; Peña-Gallardo, M.; Tomás-Burguera, M.; Gazol, A.; Azorin-Molina, C.; Bhuyan, U.; Kenawy, A.E. Diverse relationships between forest growth and the Normalized Difference Vegetation Index at a global scale. *Remote Sens. Environ.* **2016**, *187*, 14–29. [CrossRef]
54. Cui, B.; Zhao, Q.; Huang, W.; Song, X.; Ye, H.; Zhou, X. A New Integrated Vegetation Index for the Estimation of Winter Wheat Leaf Chlorophyll Content. *Remote Sens.* **2019**, *11*, 974. [CrossRef]
55. Σκιάνης, Γ.Α.; Βαιόπουλος, Δ.Α.; Νικολακόπουλος, Κ. Assessment of the TVI vegetation index with the aid of probability theory. *Geol. Soc. Am. Bull.* **2004**, *36*, 1338–1346. [CrossRef]
56. Girolimetto, D.; Venturini, V. Water Stress Estimation from NDVI-Ts Plot and the Wet Environment Evapotranspiration. *Adv. Remote Sens.* **2013**, *2*, 283–291. [CrossRef]
57. Thapa, S.; Rudd, J.C.; Xue, Q.; Bhandari, M.; Reddy, S.K.; Jessup, K.E.; Liu, S.; Devkota, R.N.; Baker, J.; Baker, S. Use of NDVI for characterizing winter wheat response to water stress in a semi-arid environment. *J. Crop Improv.* **2019**, *33*, 633–648. [CrossRef]

58. Pompa-García, M.; Camarero-Martínez, J.J. Potencial dendroclimático de la madera temprana y tardía de *Pinus cooperi* Blanco. *Agrociencia* **2015**, *49*, 177–187. Available online: http://www.scielo.org.mx/pdf/agro/v49n2/v49n2a6.pdf (accessed on 12 February 2022).
59. Jorge, J.; Vallbé, M.; Soler, J.A. Detection of irrigation inhomogeneities in an olive grove using the NDRE vegetation index obtained from UAV images. *Eur. J. Remote Sens.* **2019**, *52*, 169–177. [CrossRef]
60. Lei, S.; Luo, J.; Tao, X.; Qiu, Z. Remote Sensing Detecting of Yellow Leaf Disease of Arecanut Based on UAV Multisource Sensors. *Remote Sens.* **2021**, *13*, 4562. [CrossRef]
61. Steven, M.D. The Sensitivity of the OSAVI Vegetation Index to Observational Parameters. *Remote Sens. Environ.* **1998**, *63*, 49–60. [CrossRef]
62. Qi, H.; Wu, Z.; Zhang, L.; Li, J.; Zhou, J.; Jun, Z.; Zhu, B. Monitoring of peanut leaves chlorophyll content based on drone-based multispectral image feature extraction. *Comput. Electron. Agric.* **2021**, *187*, 106292. [CrossRef]
63. Nedkov, R. Normalized Differential Greenness Index for Vegetation Dynamics Assessment. *Comptes Rendus De L'academie Bulg. Des Sci.* **2017**, *70*, 1143–1146. Available online: https://www.researchgate.net/profile/R-Nedkov-2/publication/319141286_Normalized_differential_greenness_index_for_vegetation_dynamics_assessment/links/5994608d0f7e9b98953af1d6/Normalized-differential-greenness-index-for-vegetation-dynamics-assessment.pdf (accessed on 22 February 2022).
64. De Castro, A.I.; Torres-Sánchez, J.; Peña, J.M.; Jiménez-Brenes, F.M.; Csillik, O.; López-Granados, F. An Automatic Random Forest-OBIA Algorithm for Early Weed Mapping between and within Crop Rows Using UAV Imagery. *Remote Sens.* **2018**, *10*, 285. [CrossRef]
65. Liu, H. Classification of urban tree species using multi-features derived from four-season RedEdge-MX data. *Comput. Electron. Agric.* **2022**, *194*, 106794. [CrossRef]
66. Li, L.; Mu, X.; Chianucci, F.; Qi, J.; Jiang, J.; Zhou, J.; Chen, L.; Huang, H.; Yan, G.; Liu, S. Ultrahigh-resolution boreal forest canopy mapping: Combining UAV imagery and photogrammetric point clouds in a deep-learning-based approach. *Int. J. Appl. Earth Obs. Geoinf.* **2022**, *107*, 102686. [CrossRef]
67. Wan, L.; Cen, H.; Zhu, J.; Zhang, J.; Zhu, Y.; Sun, D.; Du, X.; Zhai, L.; Weng, H.; Li, Y.; et al. Grain yield prediction of rice using multi-temporal UAV-based RGB and multispectral images and model transfer—A case study of small farmlands in the South of China. *Agric. For. Meteorol.* **2020**, *291*, 108096. [CrossRef]
68. Asner, G.P. Biophysical and Biochemical Sources of Variability in Canopy Reflectance. *Remote Sens. Environ.* **1998**, *64*, 234–253. [CrossRef]

remote sensing

MDPI

Technical Note

Detection of Southern Beech Heavy Flowering Using Sentinel-2 Imagery

Ben Jolly [1,*], John R. Dymond [1], James D. Shepherd [1], Terry Greene [2] and Jan Schindler [1]

1 Manaaki Whenua—Landcare Research, Palmerston North 4472, New Zealand; dymondj@landcareresearch.co.nz (J.R.D.); shepherdj@landcareresearch.co.nz (J.D.S.); schindlerj@landcareresearch.co.nz (J.S.)
2 Department of Conservation, Christchurch 8140, New Zealand; tgreene@doc.govt.nz
* Correspondence: jollyb@landcareresearch.co.nz

Abstract: The southern beech (genus *Fuscospora* and *Lophozonia*) forest in New Zealand periodically has "mast" years, during which very large volumes of seeds are produced. This excessive seed production results in a population explosion of rodents and mustelids, which then puts pressure on native birds. To protect the birds, extra pest controls, costing in the order of NZD 20 million, are required in masting areas. To plan pest control and keep it cost-effective, it would be helpful to have a map of the masting areas. In this study, we developed a remote sensing method for the creation of a national beech flowering map. It used a temporal sequence of Sentinel-2 satellite imagery to determine areas in which a yellow index, which was based on red and green reflectance (red-green)/(red + green), was higher than normal in spring. The method was used to produce national maps of heavy beech flowering for the years 2017 to 2021. In 2018, which was a major beech masting year, of the 4.1 million ha of beech forest in New Zealand, 27.6% was observed to flower heavily. The overall classification accuracy of the map was 90.8%. The method is fully automated and could be used to help to identify areas of potentially excessive seed fall across the whole of New Zealand, several months in advance of when pest control would be required.

Keywords: southern beech; masting; beech flowering; seed fall; Fuscospora; Sentinel-2

Citation: Jolly, B.; Dymond, J.; Shepherd, J.; Greene, T.; Schindler, J. Detection of Southern Beech Heavy Flowering Using Sentinel-2 Imagery. *Remote Sens.* **2022**, *14*, 1573. https://doi.org/10.3390/rs14071573

Academic Editor: Brian Alan Johnson

Received: 17 February 2022
Accepted: 22 March 2022
Published: 24 March 2022

1. Introduction

New Zealand southern beech (genus *Fuscospora* and *Lophozonia*, formerly *Nothofagus* [1]) forest dominates over 2 million hectares (ha) of New Zealand forest and features in almost 2 million ha more [2]. It comprises five species: mountain beech (*Fuscospora cliffortioides*); red beech (*Fuscospora fusca*); silver beech (*Lophozonia menziesii*); black beech (*Fuscospora solandri*); and hard beech (*Fuscospora truncata*). The trees reproduce almost yearly, with periodic highly productive seasons known as "mast" years that produce large volumes of seeds [3]. The seeds are a significant food source for a number of birds and mammals. During mast years, the rodent population increases significantly [4,5], providing an abundant food source for mustelids (especially stoats—*Mustela erminea*) [6]. All rodent and mustelid species have been introduced to New Zealand and also prey on native bird species. When seeds begin to run out, increasing pressure is put on native biota as additional food sources are sought. The control of populations of introduced predators is essential for preserving populations of native and endemic birds, reptiles, and invertebrates in New Zealand, especially during beech mast years [6].

The New Zealand Department of Conservation (DOC) is responsible for managing the forests on public land, preserving native species, and coordinating pest control. The ability to predict the extent and intensity of significant mast events is a critical component for planning pest control efforts to limit the explosion of predator populations [4]. A number of approaches are currently used to plan management interventions: (i) modeling; (ii) field observations; and (iii) the in situ sampling of developing seed crops in tree canopies. (i) The "delta-T" (ΔT) model [7] uses the difference in mean temperature from the previous two summers ($T_{n-1} - T_{n-2}$) to predict likely seed fall for the following autumn at a national scale. However, it relies on temperature data, which are currently only available on a modeled 5 × 5 km grid, so it misses smaller scale micro-climate effects. While historical temperature is an important factor in synchronizing mast events [7], other factors, such as nutrient availability, also play a role [8]. New research suggests that rising temperatures caused by climate change may alter the mast cycle of beech trees [8,9], increasing the spatial and temporal complexity of mast patterns [8] and potentially de-synchronizing flowering and seeding, effectively reducing the impact and predictive power of temperature on the timing of the reproductive cycle [10]. (ii) Field staff from DOC are well placed to provide observations of beech flowering in certain areas as part of their normal duties. However, there are large areas of forest that remain unobserved and spatial extent is often difficult to define, especially at a regional or national scale. (iii) Extensive sampling campaigns are conducted during years in which a heavy mast is expected using helicopters so that staff can reach and clip the upper branches of trees in order to count the seeds. This task is expensive, labour intensive, and dangerous.

Remote sensing has proven to be an effective tool for monitoring vegetation phenology, particularly when a rich time series of imagery is available [11–16]. With sensors, such as Landsat 8 and Sentinel-2, it is possible to map phenology over millions of hectares and create national maps in great detail [11,17]. Phenological characteristics are usually derived by first fitting a curve to the time series of remote sensing data and then using either threshold-based methods, moving averages, inflection points or the time of maximum increase [18–22]. Seasonality is usually assumed [22]. As mast events are considered a deviation from the "median" annual cycle, it should be possible to use change detection techniques [23,24] to identify or even predict mast seasons when identifying features are visible from above [14,19,21].

Vegetation indices, such as the normalized difference vegetation index (NDVI) and the enhanced vegetation index (EVI), are often used to differentiate between flowers or seed pods/cones and to summarize data as one variable for analysis [14,19,22,25]. Often, multiple indices are combined in an effort to investigate multiple physical properties. For example, a study by Garcia et al. [21] investigating white spruce mast events found that vegetation indices targeting moisture were more effective than the traditional color-based indices, but they still struggled to reliably predict masting. Fernández-Martínez et al. [19] were more successful and used increasing EVI the winter before, along with weather data during spring, as an indicator of potential mast seasons for Mediterranean oaks. Neither study were able to find a usable signature for flowering or seed/cone production to map the mast events. Dixon et al. [14] and Chen et al. [25] both successfully used multi-scale imagery to map tree-scale flowering over landscapes: Dixon et al. [14] by training a random forest model with drone data from known events and Chen et al. [25] by developing an enhanced bloom index (EBI) from drone data and successfully translating that approach to CERES, PlanetScope, Sentinel-2, and Landsat data to increase coverage.

In this study, we investigated the use of freely available satellite remote sensing to identify large areas of significant southern beech flowering in New Zealand. We used imagery from the European Space Agency (ESA) Sentinel-2 "a" and "b" satellites to obtain a high rate of repeat passes in order to maximize the chances of multiple cloud-free observations and to produce detailed coverage at a national scale. Very little ground data on flowering are available. The irregular nature of the flowering events and the ruggedness and remoteness of much of the terrain made the planned collection of ground data for this study difficult. As significant flowering is an irregular phenological event, we modeled the phenology of beech forest per pixel and identified departures from this image-by-image during spring seasons to identify heavy beech flowering. Using this method, we produced national maps of heavy beech flowering for 5 years, 2017–2021, with a minimum mapping unit of 1 ha. We assessed the accuracy of the method by comparing it to a human operator at 1000 randomly selected sites. A national map of detectable heavy beech flowering could be produced at the end of every spring to assist DOC in identifying potential mast "hot spots" and thus, aid in the planning of pest control operations.

2. Materials and Methods

2.1. Area of Interest

This study covered the extent of known beech and mixed-beech forest in New Zealand, according to EcoSat Forests [2,26–28], and covered both North and South Islands from 36.1°S, 178.0°E to 46.4°S, 166.4°E. The terrain is largely mountainous with slopes of up to 45°, as any habitat deemed suitable for agriculture was cleared in the 1800s and early 1900s. Due to the large variations in both the altitude and latitude of the study area, the mean annual temperatures ranged from 3.8 °C to 16.6 °C and the mean cumulative annual rainfall ranged from 465 mm to 9305 mm.

2.2. Data

This study used the Copernicus Sentinel-2 Level-1C calibrated top of atmosphere (TOA) reflectance values, as downloaded from the ESA archives, with the edges masked to remove pixels that did not contain data for every band. The Sentinel-2 satellite mission consists of two satellites (Sentinel-2a and -2b) moving in sun-synchronous orbits that repeat every 10 days. These orbits are 180° out of phase with each other, which produces a 5-day revisit period. The swath width of each pass is 290 km, with five "passes" required to cover the mainland of New Zealand, each on a different day. The overlap between passes means that some areas of the country have a higher revisit rate. Each Sentinel-2 satellite is equipped with a multi-spectral imaging sensor that captures wavelengths from ultraviolet to short-wave infrared.

Our analysis used all available Sentinel-2 data over New Zealand from 1 September 2016 to 31 December 2021, with images every 10 days before Sentinel-2b came online 8 July 2017 and images every 5 days thereafter. A "mega-mast" occurred during spring 2018 and autumn 2019 [29,30]. Other years showed some small flowering events or none at all. Cloudy pixels were excluded from the temporal sequence using the methodology outlined in Shepherd et al. [31].

2.3. Methods

Southern beech flowers produce a reddening of the forest canopy that is visible from the ground, especially during a significant mast season. At the 10 m pixel nominal spatial resolution of Sentinel-2, the reddening of the canopy appears to the human eye as a subtle yellowing in the "natural color" (RGB) image as the red flowers increase the red component of a pixel but not to the point of dominating the green. Higher-wavelength image bands (and the associated indices) do not respond to this flowering, with the exception of Band 5 (red edge); however, this is also associated with the "flushes" of new foliage in late spring.

The yellowing effect of the canopy is subtle in natural color renders of Sentinel-2 imagery, as demonstrated in Figure 1a–d. Sub-figures (e–h) feature an exaggerated "red" (B4) band, which highlights the effect at the expense of non-forested areas. Figure 1 covers one month of imagery in an area of overlap between orbits 29 and 72 and thus, has approximately double the number of overpasses compared to other areas of the country. In that 30-day period, there were 13 overpasses that resulted in 5 usable cloud-free images, 4 of which are displayed. The flowering event was just starting on October 14th (Figure 1a,e), with the lower western and southern slopes in full flower 7 days later (b,f) and the upper and eastern slopes in full flower 15 days later (c,g) as the areas in (b,f) gave way to fresh green foliage. Eight days later, on 13 November (d,h), most of the flowering had vanished. Generally, there was a two- to three-week period, at most, in which a cloud-free image was required in order to observe flowering; however, there was no way of reliably knowing when this window would occur as it varied with season and location.

In order to detect the yellowing that is associated with southern beech flowering, we applied two approaches: the calculation of a normalized difference yellowing index (NDYI) to describe the red–green band relationship and the modeling of this index over the time series of the image archive to detect variations from the expected (non-flowering) state. The effect is subtle enough that the technique required was very sensitive to cloud and shadow contamination. In addition to the cloud masking performed above, "invalid" pixels from the cloud mask (cloud, shadow, snow or water) were buffered by 30 pixels (300 m) and patches of "valid" pixels smaller than 100 ha were re-classified as "invalid". Extra spectral filtering was then used to mask the remaining pixels that were too bright or dark to be forest (or useful): $B4_{red} < 650$, $B3_{green} < 900$, $B2_{blue} < 1000$, $B8_{NIR} > 1000$, $B4_{red} - B5_{red\ edge} < -1500$. The result was then further buffered by 3 pixels.

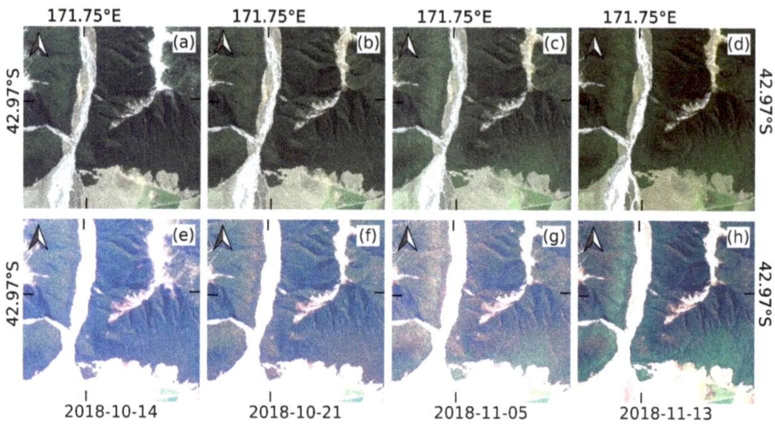

Figure 1. Sentinel-2 imagery from the Hawdon Valley, South Island, New Zealand, showing a flowering event during the spring of 2018: (**a–d**) are natural color; (**e–h**) are natural color with an exaggerated stretch to amplify the "red" band (Band 4). The images are organized by column, e.g., (**a**,**e**) are the natural color and stretched color for 14 October 2018, respectively.

The NDYI is similar to the well-known normalized difference vegetation index (NDVI) [32], using the "red" (Band 4665 ± 15 nm) and "green" (Band 3560 ± 18 nm) Sentinel-2 bands instead of the "near-infrared" and "red" bands. It is also very similar to the green-red vegetation index (GRVI) [33,34], with the order of the bands merely reversed:

$$\text{NDYI} = \frac{(B4_{red} - B3_{green})}{(B4_{red} + B3_{green})} \tag{1}$$

The NDYI was calculated using the Level-1C TOA reflectance product re-projected to the New Zealand Transverse Mercator (NZTM) coordinate reference system (EPSG:2193) and with any invalid pixels masked. As the NDYI represents the ratio of red–green and both are affected similarly by transient atmospheric conditions, it was negative over most areas of forest most of the time (i.e., when a pixel was "green"), increasing to near or slightly above 0 during heavy flowering events.

Substantial annual variation was present in the TOA reflectance observations over forest due to climatic conditions, vegetation phenology (e.g., new leaves), and sun angle. The NDYI signal visually observed during flowering was subtle enough within the context of a year of data that setting simple thresholds was inadequate to produce a reliable result. Additionally, flowering can occur at different times during the spring season, depending on latitude and altitude [3,8]. The temporal sequence of NDYI for a pixel in the Hawdon Valley (South Island, New Zealand) is shown in Figure 2 as an example. The orange line is a modeled NDYI time series that is unique to that pixel, using an approach similar to that used in the TMASK methodology developed for cloud detection in Landsat 8 imagery [35]. The model used robust regression to calculate unique per pixel (i, j) coefficients for the sine and cosine terms for intra- ($a_{1,i,j}$, $a_{2,i,j}$) and inter-annual ($a_{3,i,j}$, $a_{4,i,j}$) variability, as well as a constant term ($c_{i,j}$). Where x is the number of days since the start of the temporal sequence, T_{yr} is the number of days per year, and T_{all} is the number of days in the sequence:

$$\mathrm{NDYI}_{mod}(i,j,x) = c_{i,j} + a_{1,i,j}sin(2\pi\frac{x}{T_{yr}}) + a_{2,i,j}cos(2\pi\frac{x}{T_{yr}}) + a_{3,i,j}sin(2\pi\frac{x}{T_{all}}) + a_{4,i,j}cos(2\pi\frac{x}{T_{all}}) \quad (2)$$

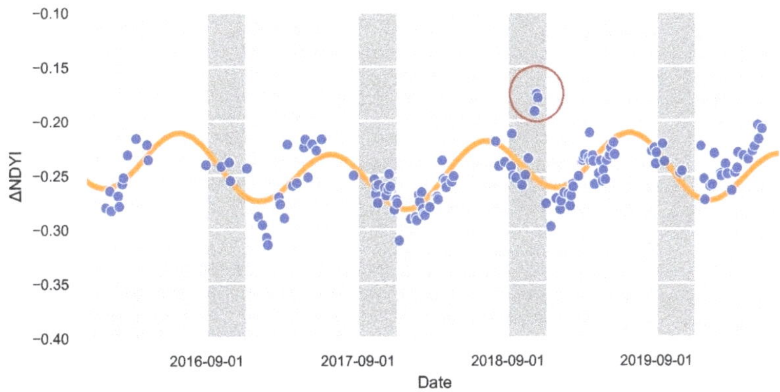

Figure 2. The observed normalized difference yellowing index (NDYI) time series for an example pixel from Figure 1 in the Hawdon Valley (blue) with superimposed modeled values (orange). The gray areas indicate austral spring seasons (September–November) where the NDYI is expected to peak during a mast. The red circle shows the NDYI that was higher than expected during the spring of 2018 (the "mega-mast" season).

The observed NDYI for each pixel and date were then subtracted from the modeled value for that pixel and date to produce ΔNDYI and the maximum value was found for each flowering season (1 September to 10 December) using:

$$\Delta\mathrm{NDYI} = \mathrm{NDYI} - \mathrm{NDYI}_{mod} \quad (3)$$

Finally, ΔNDYI values were converted into a map of "heavy flowering detected" vs. "heavy flowering not detected" by following a method similar to Shepherd et al. [36]. First, a high ΔNDYI threshold of 0.08 was chosen by assessing ΔNDYI against seed trap data [4] during the 2018 "mega-mast". This threshold was used to create "seed" areas, which were grown outward by progressively lowering the threshold to 0.04. The resulting "flowering" pixels were then buffered by 2 pixels, followed by a 5 × 5 majority filter that was then eroded by 2 pixels. "Heavy flowering not detected" patches that were smaller than 1 ha (minimum mapping unit) were removed by being re-coded as "heavy flowering detected" or "no data" (majority of surrounding pixels), then the "heavy flowering detected" patches that were smaller than 1 ha were re-coded as "heavy flowering not detected" to reduce small-scale noise.

As no reliable spatial dataset of beech flowering exists beyond occasional field reports from DOC staff, the national-scale map for the 2018 mast year was accuracy assessed by a human operator. At 1000 randomly selected sites, 500 in "heavy flowering detected" and 500 in "heavy flowering not detected", the operator determined whether heavy flowering was observed in the temporal sequence of 2018 cloud-free imagery in comparison to a median spring image (excluding 2018). Heavy flowering was easily observed in the temporal progression of spectral reflectance relative to the median image, especially when the spatial extent of the flowering moved upward in elevation as the season progressed (using the exaggerated red band stretch shown in Figure 1). A confusion matrix of proportions was estimated using the method of Card [37], from which precision and recall were calculated [38].

3. Results

The austral spring (September–November) of 2017 was a light flowering season for southern beech in New Zealand. This was followed by a "mega-mast" in 2018, with heavy flowering observed from the ground during spring and a corresponding heavy seed fall the following autumn. Maps of the maximum spring ΔNDYI were produced for each year of data, with the results for the 2018 season shown in Figure 3. The spatial patterns corresponded with anecdotal reports from DOC staff who were based at field offices around New Zealand. There was heavy flowering throughout most of the northwestern corner of the South Island and sporadic heavy flowering in eastern Fiordland. The inset of Figure 3 highlights the level of detail available and shows the heavy flowering on the lower slopes of the Hawdon and Poulter Valleys, dissipating as altitude increases up the valley walls (the black areas are not beech forest; they are either alpine or riverbed in this location).

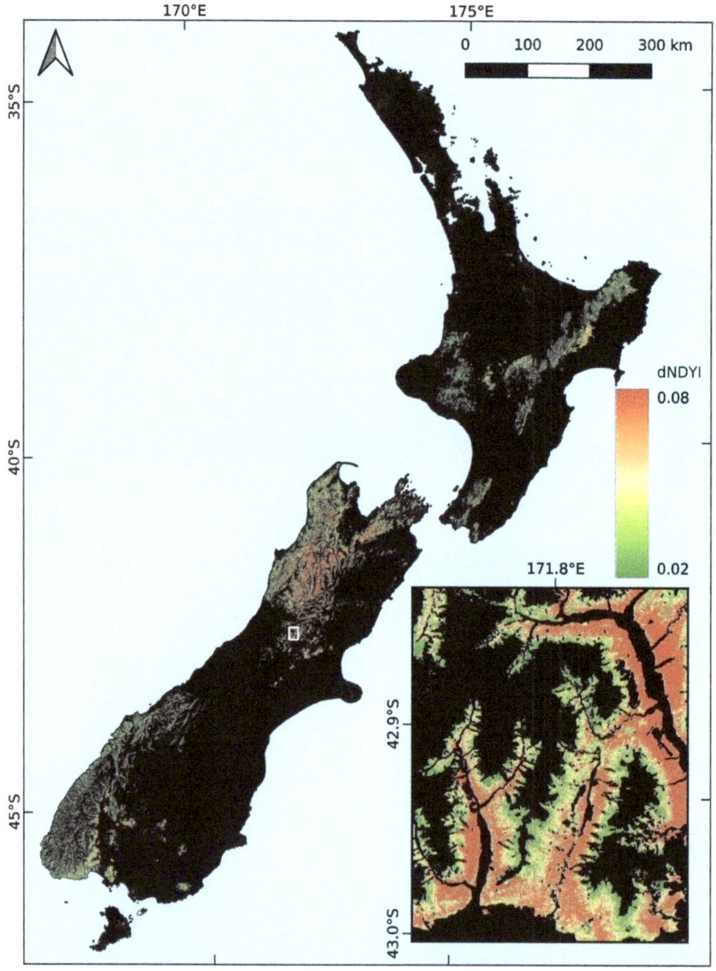

Figure 3. The maximum ΔNDYI (from the model) for spring 2018 in areas of known southern beech forest in New Zealand. Green denotes areas of low (<0.02) maximum ΔNDYI, while red is high (>0.08) and indicates heavy beech flowering. The inset shows the Hawdon and Poulter Valleys near Arthur's Pass (1:250,000 at 42.95°S, 171.82°E; see white box).

Figure 4 shows the maps of heavy beech flowering as detected by our method for years 2017–2021. In spring 2018, a lot of heavy beech flowering was detected in the north-west of the South Island, which was synonymous with a "mega-mast" event in that region. In the North Island and the south-west of the South Island, some pockets of heavy flowering were detected. The following year, in spring 2019, the flowering was much reduced in the north-west of the South Island, but in the North Island, a lot of heavy flowering was detected, which was synonymous with another "mega-mast". The south-west of the South Island had pockets of heavy flowering, much the same as in 2018. In years 2020 and 2021, minimal beech flowering was detected in most areas.

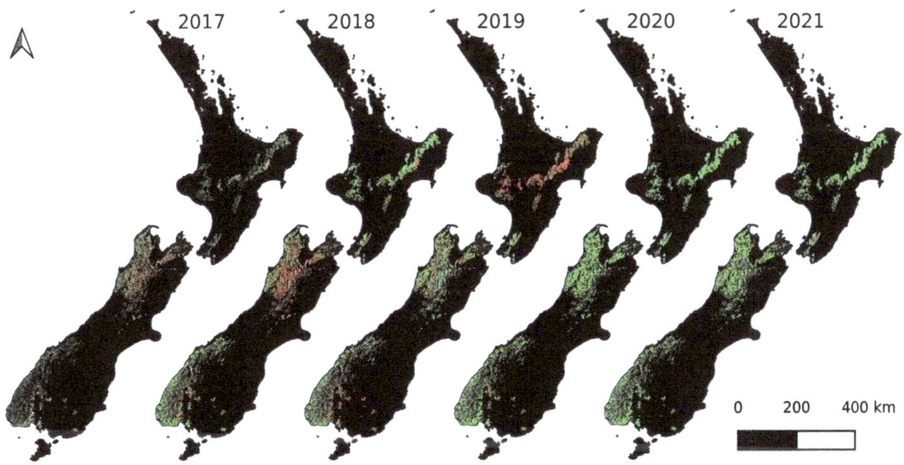

Figure 4. The maps of heavy beech flowering during spring time for four years using Sentinel-2 imagery (2017–2021 inclusive). The classes are "heavy flowering detected" (red), "heavy flowering not detected" (green), and "no cloud-free imagery" (gray).

In the 2018 map, heavy flowering was detected in 27.6% (1,144,382 ha) of the 4.1 million ha of beech forest in New Zealand. Heavy flowering was not detected in 51.2% (2,122,201 ha) of the beech forest. In the remaining 21.2% of the beech forest (878,406 ha), there was no cloud-free imagery in the spring to enable a decision to be reached. In each of the two classes for "heavy flowering detected" and "heavy flowering not detected", we generated 500 random locations at which we compared reference data to map data. The reference data were determined from the visual interpretation of all cloud-free spring imagery for that year. Table 1 shows the confusion matrix of proportions. The overall classification accuracy was 90.8%. The precision (user's accuracy) scores indicate that 90.4% of the area mapped as "heavy flowering detected" was actually heavy flowering. The recall (producer's accuracy) scores indicate that 84.4% of actual "heavy flowering detected" was successfully mapped as "heavy flowering detected". The overall F1-score for "heavy flowering detected" was 0.873 and "heavy flowering not detected" was 0.928. Thus, the ΔNDYI method is likely to underestimate (slightly), rather than overestimate, areas of heavy flowering; however, the scores reflect well on the method overall.

Table 1. A confusion matrix showing detected and not detected heavy flowering (proportion of map) for the ΔNDYI method (Mapped), assessed against a human operator (Reference). The proportions were estimated from a random sample of 500 locations in the "heavy flowering detected" class (weighted by proportion of "heavy flowering detected" in map = 0.35) and a random sample of 500 locations in the "heavy flowering not detected" class (weighted by 0.65).

		Reference Flowering		
		Detected	Not Det.	**Precision**
Mapped flowering	Detected	0.316	0.034	0.904
	Not Det.	0.059	0.592	0.910
	Recall	0.844	0.946	
	F1-Score	0.873	0.928	

4. Discussion

We developed a method that can produce a national map of heavy beech flowering from a temporal sequence of Copernicus Sentinel-2 imagery (Figure 4). The method detects elevated values of a yellow index (NDYI) the are above those normally expected in spring. A ΔNDYI value greater than 0.08 indicates especially heavy flowering; however, these regions can be "grown" into adjacent pixels where ΔNDYI is greater than 0.04 to better capture all heavy flowering. The elevated yellow index is caused by the production of red flowers that obscure the green leaves. The national map of beech flowering can be produced at the end of spring, several months before the subsequent mast event actually occurs and seeds drop to the ground. It can then be provided to DOC, which is the national agency in charge of pest control, to aid with planning additional pest control to be implemented several months later. In the spring of 2018, a nationwide beech mast event was detected and mapped using this method. A manual accuracy assessment determined the heavy flowering map to have an overall accuracy of 90%. The spatial distribution of beech flowering, as mapped by the method, was also consistent with anecdotal observations from DOC field staff.

The national map of beech flowering can be used to provide extra detail to augment the existing ΔT model [7] as it provides a higher spatial resolution of 10 m, as opposed to 5 km. It is also a map of confirmed flowering, which is one less degree of separation from the actual seed fall than the ΔT model, as physical factors, such as carbon availability and soil moisture conditions, also affect flowering and seed productivity [39]. However, one issue with our method is the requirement for cloud-free satellite imagery at critical flowering times. This means that flowering may have been missed in some areas, which effectively makes the map a better indicator of "presence" rather than "absence".

Not all heavy beech flowering in spring results in heavy seed fall in the following autumn. Heavy frost or very wet weather can interfere with seed production [3]. Figure 5 shows how well the maximum ΔNDYI compared to seed counts in trays located on the floor of the beech forest (seed traps are spread throughout beech forests in New Zealand as part of a long-running monitoring program conducted by DOC [4]). Data were restricted to locations with at least eight valid observations in order to obtain representation over the majority of the spring season. There was noise in the data, nevertheless high seed counts generally corresponded with high maximum ΔNDYI ($r^2 = 0.397$). For reference, the ΔT model had r^2 values between 0.331 and 0.556 for the same species range [7] (although that study used the older genus name *Nothofagus*). Reasons for the mismatches between the ΔNDYI and seed count include: cloud coverage still obscuring the flowering event, despite the high number of revisits (low ΔNDYI vs. high seed count); inaccurate trap location (variable impact on relationship); trap location relative to flowering trees combined with wind direction during seed fall (high ΔNDYI vs. low seed count); different beech species (different relationship between ΔNDYI and seed count); climate, adverse weather events, and nutrient availability (lower seed count vs. higher ΔNDYI); and inaccuracies in the method (addressed in accuracy assessment). We recommend that the national map of flowering/not flowering be regarded as a map of potential high seed fall for initial planning purposes, which is to be confirmed later with additional information, such as selected field observations.

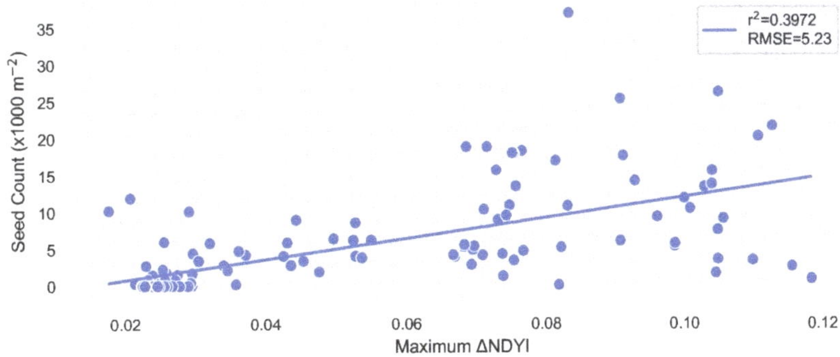

Figure 5. The relationship between the maximum ΔNDYI (spring 2018) and the number of seeds collected from seed traps in the permanent trap network (autumn/winter 2019) for the 2018/2019 masting season. The locations are filtered to exclude those with fewer than eight valid satellite observations.

One way to address the paucity of valid observations would be to add more data sources. As the technique developed in this study relies only on red, green, blue, and NIR (for quality control) wavelengths, it should be possible to include data from commercial satellite constellations that have higher revisit rates but lower spectral ranges or resolutions, such as the Planet "Dove" constellation. Adding freely available Landsat 8 data could also increase the probability of obtaining valid observations at critical times. Targeted aerial imaging campaigns could also provide valuable information in areas of known data paucity, particularly when they are informed by observations from field staff. This study has shown that the resolution requirement is low, by aerial imaging standards, which would allow for higher flight altitudes and larger image footprints, thereby substantially reducing cost. Multiple studies have shown that the fusion of these separate data sources is useful in remote sensing [11,14,16,20,40,41], though the spatial complexity and rugged terrain of the beech forests in New Zealand is likely to reduce the utility of the coarse resolution MODIS optical imagery.

This study successfully mapped the presence of heavy flowering in beech trees at a large scale (greater than 4 million ha) using a visible change in canopy color. A similar study by Garcia et al. [21] was less successful, but did show that moisture-based indices in the lead-up to a flowering or seed/cone event could provide additional information. Fernández-Martínez et al. [19] were also successful in predicting mast events using a combination of the enhanced vegetation index (EVI) and weather data during spring. A number of challenges exist in the context of detecting mast events and the ΔNDYI approach attempts to minimize these. The NDYI index was chosen to specifically target the red and green image bands, avoiding the red edge and near-infrared bands that also respond to vegetation conditions and thus, increase noise. The effectiveness of the multi-year sine and cosine model for modeling the typical behaviour of NDYI, the utilization of extreme ΔNDYI values as "seeds" for regions that grow into areas of lower ΔNDYI values, and the ability to tune the spectral value constraints all contribute to the effectiveness of our approach. To further improve the performance of the ΔNDYI method, it would be worth investigating the use of supporting indices, such as Garcia et al. [21] and Fernández-Martínez et al. [19], in addition to adding extra data sources. Further work into distinguishing different beech species would also add greater value for DOC as those with larger seeds (red and hard beech) have a disproportionately larger impact on rodent irruption.

The temporal analysis of Sentinel-2 satellite imagery has proven to be successful at detecting heavy flowering in New Zealand beech forests. To achieve this, cloud clearing had to be accurate (because the yellow index is sensitive to missed cloud) and automated (because many images are required). The automation of the cloud clearing [31] and other processing means that beech flowering maps can be produced in a timely and cost-effective way. In future, we plan to produce a national map of heavy beech flowering at the end of each spring. This would allow for several months of analysis to plan the extra pest control that would be required in the autumn, thereby improving targeted pest control in masting areas and leading to better outcomes for native fauna.

5. Conclusions

This study used Sentinel-2 top of atmosphere (TOA) imagery to detect and map atypical yellowing that was associated with the heavy flowering of southern beech (*Fuscospora* and *Lophozonia*) forests in New Zealand over 4.1 million ha at an unprecedented 10 m spatial resolution. This was achieved by modeling a normalized difference yellowing index (NDYI) over 5 years of observation and investigating any deviations from the expected values during spring months (September–November). A "threshold" ΔNDYI value of 0.08 was used to identify areas of heavy flowering, with connected areas of ΔNDYI > 0.04 also likely flowering. The method was automated and could be run for all of New Zealand in less than a day on a cluster of approximately 1000 CPU cores. Using Sentinel-2 imagery, the method typically provided information on heavy flowering for 80% of the beech forests in New Zealand with a high overall classification accuracy of 90.8%, resulting in useful information for the planning of national-scale pest control efforts.

Author Contributions: Conceptualization, B.J., J.R.D., T.G. and J.S.; data curation, B.J., J.D.S. and J.S.; formal analysis, B.J.; funding acquisition, J.R.D.; investigation, B.J., J.R.D. and T.G.; methodology, B.J., J.R.D., J.D.S. and T.G.; project administration, J.R.D.; resources, J.R.D.; software, B.J., J.D.S. and J.S.; supervision, J.R.D.; visualization, B.J.; writing—original draft, B.J.; writing—review and editing, J.R.D., J.S. and T.G. All authors have read and agreed to the published version of the manuscript.

Funding: This research was funded by the New Zealand Ministry of Business, Innovation and Employment (MBIE) Endeavour Fund, under the Advanced Remote Sensing of Aotearoa research program (C09X1709).

Data Availability Statement: Data available on reasonable request.

Acknowledgments: The authors would like to thank the New Zealand Department of Conservation (DOC) for providing advice and validation data.

Conflicts of Interest: The authors declare no conflict of interest. The funders had no role in the design of the study, in the collection, analyses or interpretation of data, in the writing of the manuscript or in the decision to publish the results.

Abbreviations

The following abbreviations are used in this manuscript:

DOC	Department of Conservation
EBI	Enhanced Bloom Index
ESA	European Space Agency
EVI	Enhanced Vegetation Index
GRVI	Green-Red Vegetation Index
NDVI	Normalized Difference Vegetation Index
NDYI	Normalized Difference Yellowing Index
NZTM	New Zealand Transverse Mercator
TOA	Top of Atmosphere

References

1. Heenan, P.B.; Smissen, R.D. Revised circumscription of Nothofagus and recognition of the segregate genera Fuscospora, Lophozonia, and Trisyngyne (Nothofagaceae). *Phytotaxa* **2013**, *146*, 1–31. [CrossRef]
2. Shepherd, J.R.D.; Ausseil, A.G.; Dymond, J.R. *EcoSat Forests: A 1: 750,000 Scale Map of Indigenous Forest Classes in New Zealand*; Manaaki Whenua Press: Lincoln, New Zealand, 2005.
3. Wardle, J.A. *The New Zealand Beeches. Ecology, Utilisation and Management*; New Zealand Forest Service: Wellington, New Zelaand, 1984; p. 477.
4. Elliott, G.; Kemp, J. Large-scale pest control in New Zealand beech forests. *Ecol. Manag. Restor.* **2016**, *17*, 200–209. [CrossRef]
5. Ruscoe, W.A.; Pech, R.P. Rodent Outbreaks: Ecology and Impacts. In *Rodent Outbreaks: Ecology and Impacts*; Singleton, G., Belmain, S., Brown, P., Hardy, B., Eds.; Internationl Rice Research Institute: Los Baños, Philippines, 2010; pp. 239–251.
6. King, C.M.; Powell, R.A. Managing an invasive predator pre-adapted to a pulsed resource: A model of stoat (Mustela erminea) irruptions in New Zealand beech forests. *Biol. Invasions* **2011**, *13*, 3039–3055. [CrossRef]
7. Kelly, D.; Geldenhuis, A.; James, A.; Penelope Holland, E.; Plank, M.J.; Brockie, R.E.; Cowan, P.E.; Harper, G.A.; Lee, W.G.; Maitland, M.J.; et al. Of mast and mean: Differential-temperature cue makes mast seeding insensitive to climate change. *Ecol. Lett.* **2013**, *16*, 90–98. [CrossRef] [PubMed]
8. Allen, R.B.; Hurst, J.M.; Portier, J.; Richardson, S.J. Elevation-dependent responses of tree mast seeding to climate change over 45 years. *Ecol. Evol.* **2014**, *4*, 3525–3537. [CrossRef] [PubMed]
9. Bogdziewicz, M.; Kelly, D.; Thomas, P.A.; Lageard, J.G.; Hacket-Pain, A. Climate warming disrupts mast seeding and its fitness benefits in European beech. *Nat. Plants* **2020**, *6*, 88–94. [CrossRef]
10. Bogdziewicz, M.; Hacket-Pain, A.; Kelly, D.; Thomas, P.A.; Lageard, J.; Tanentzap, A.J. Climate warming causes mast seeding to break down by reducing sensitivity to weather cues. *Glob. Chang. Biol.* **2021**, *27*, 1952–1961 . [CrossRef]
11. Bolton, D.K.; Gray, J.M.; Melaas, E.K.; Moon, M.; Eklundh, L.; Friedl, M.A. Continental-scale land surface phenology from harmonized Landsat 8 and Sentinel-2 imagery. *Remote Sens. Environ.* **2020**, *240*, 111685. [CrossRef]
12. Misra, G.; Cawkwell, F.; Wingler, A. Status of phenological research using sentinel-2 data: A review. *Remote Sens.* **2020**, *12*, 2760. [CrossRef]
13. Zeng, L.; Wardlow, B.D.; Xiang, D.; Hu, S.; Li, D. Remote Sensing of Environment A review of vegetation phenological metrics extraction using time-series, multispectral satellite data. *Remote Sens. Environ.* **2020**, *237*, 111511. [CrossRef]
14. Dixon, D.J.; Callow, J.N.; Duncan, J.M.; Setterfield, S.A.; Pauli, N. Satellite prediction of forest flowering phenology. *Remote Sens. Environ.* **2021**, *255*, 112197. [CrossRef]
15. Moon, M.; Seyednasrollah, B.; Richardson, A.D.; Friedl, M.A. Using time series of MODIS land surface phenology to model temperature and photoperiod controls on spring greenup in North American deciduous forests. *Remote Sens. Environ.* **2021**, *260*, 112466. [CrossRef]
16. Moon, M.; Richardson, A.D.; Friedl, M.A. Multiscale assessment of land surface phenology from harmonized Landsat 8 and Sentinel-2, PlanetScope, and PhenoCam imagery. *Remote Sens. Environ.* **2021**, *266*, 112716. [CrossRef]
17. Browning, D.M.; Russell, E.S.; Ponce-Campos, G.E.; Kaplan, N.; Richardson, A.D.; Seyednasrollah, B.; Spiegal, S.; Saliendra, N.; Alfieri, J.G.; Baker, J.; et al. Monitoring agroecosystem productivity and phenology at a national scale: A metric assessment framework. *Ecol. Indic.* **2021**, *131*, 108147. [CrossRef]
18. Atkinson, P.M.; Jeganathan, C.; Dash, J.; Atzberger, C. Inter-comparison of four models for smoothing satellite sensor time-series data to estimate vegetation phenology. *Remote Sens. Environ.* **2012**, *123*, 400–417. [CrossRef]
19. Fernández-Martínez, M.; Garbulsky, M.; Peñuelas, J.; Peguero, G.; Espelta, J.M. Temporal trends in the enhanced vegetation index and spring weather predict seed production in Mediterranean oaks. *Plant Ecol.* **2015**, *216*, 1061–1072. [CrossRef]
20. Cheng, Y.; Vrieling, A.; Fava, F.; Meroni, M.; Marshall, M.; Gachoki, S. Phenology of short vegetation cycles in a Kenyan rangeland from PlanetScope and Sentinel-2. *Remote Sens. Environ.* **2020**, *248*, 112004. [CrossRef]
21. Garcia, M.; Zuckerberg, B.; Lamontagne, J.M.; Townsend, P.A. Landsat-based detection of mast events in white spruce (Picea glauca) forests. *Remote Sens. Environ.* **2021**, *254*, 112278. [CrossRef]
22. Noumonvi, K.D.; Oblišar, G.; Žust, A.; Vilhar, U. Empirical approach for modelling tree phenology in mixed forests using remote sensing. *Remote Sens.* **2021**, *13*, 3015. [CrossRef]
23. Asokan, A.; Anitha, J. Change detection techniques for remote sensing applications: A survey. *Earth Sci. Inform.* **2019**, *12*, 143–160. [CrossRef]
24. Panuju, D.R.; Paull, D.J.; Griffin, A.L. Change detection techniques based on multispectral images for investigating land cover dynamics. *Remote Sens.* **2020**, *12*, 1781. [CrossRef]
25. Chen, B.; Jin, Y.; Brown, P. An enhanced bloom index for quantifying floral phenology using multi-scale remote sensing observations. *ISPRS J. Photogramm. Remote Sens.* **2019**, *156*, 108–120. [CrossRef]
26. Dymond, J.R.; Shepherd, J.R.D. The spatial distribution of indigenous forest and its composition in the Wellington region, New Zealand, from ETM+ satellite imagery. *Remote Sens. Environ.* **2004**, *90*, 116–125. [CrossRef]
27. Landcare Research Ltd. *EcoSat Forests (North Island)*; Landcare Research Ltd.: Lincoln, New Zealand, 2014. [CrossRef]
28. Landcare Research Ltd. *EcoSat Forest (South Island)*; Landcare Research Ltd.: Lincoln, New Zealand, 2014. [CrossRef]
29. Department of Conservation. *Flowering and Fruit Production*; Department of Conservation: Wellington, New Zealand, 2019.

30. Department of Conservation. *Department of Conservation Te Papa Atawhai Annual Report 2019*; Technical Report; Department of Conservation: Wellington, New Zealand, 2019.
31. Shepherd, J.R.D.; Schindler, J.; Dymond, J.R. Automated Mosaicking of Sentinel-2 Satellite Imagery. *Remote Sens.* **2020**, *12*, 3680. [CrossRef]
32. Rouse, W.; Haas, H.; Deering, W. Monitoring vegetation systems in the Great Plains with ERTS. In *Goddard Space Flight Center 3d ERTS-1 Symposium*; NASA: Greenbelt, MD, USA, 1974.
33. Tucker, C.J. Red and Photographic Infrared l, lnear Combinations for Monitoring Vegetation. *Remote Sens. Environ.* **1979**, *8*, 127–150. [CrossRef]
34. Motohka, T.; Nasahara, K.N.; Oguma, H.; Tsuchida, S. Applicability of Green-Red Vegetation Index for remote sensing of vegetation phenology. *Remote Sens.* **2010**, *2*, 2369–2387. [CrossRef]
35. Zhu, Z.; Woodcock, C.E. Automated cloud, cloud shadow, and snow detection in multitemporal Landsat data: An algorithm designed specifically for monitoring land cover change. *Remote Sens. Environ.* **2014**, *152*, 217–234. [CrossRef]
36. Shepherd, J.R.D.; Dymond, J.R.; Cuff, J.R. Monitoring scrub weed change in the Canterbury region using satellite imagery. *N. Z. Plant Prot.* **2007**, *60*, 137–140. [CrossRef]
37. Card, D.H. Using Known Map Category Marginal Frequencies to Improve Estimates of Thematic Map Accuracy. *Photgrammetric Eng. Remote Sens.* **1982**, *48*, 431–439.
38. Maxwell, A.E.; Warner, T.A.; Guillén, L.A. Accuracy assessment in convolutional neural network-based deep learning remote sensing studies—Part 1: Literature review. *Remote Sens.* **2021**, *13*, 2450. [CrossRef]
39. Uscoe, W.E.A.R.; Latt, K.E.H.P.; Richardson, S.J.; Allen, R.B.; Whitehead, D.; Carswell, F.E.; Ruscoe, W.A.; Platt, K.H. Climate and net carbon availability determine temporal patterns of seed production by Nothofagus. *Ecology* **2005**, *86*, 972–981.
40. Thapa, S.; Millan, V.E.G.; Eklundh, L. Assessing forest phenology: A multi-scale comparison of near-surface (UAV, spectral reflectance sensor, phenocam) and satellite (MODIS, sentinel-2) remote sensing. *Remote Sens.* **2021**, *13*, 1597. [CrossRef]
41. Peng, D.; Wang, Y.; Xian, G.; Huete, A.R.; Huang, W.; Shen, M.; Wang, F.; Yu, L.; Liu, L.; Xie, Q.; et al. Investigation of land surface phenology detections in shrublands using multiple scale satellite data. *Remote Sens. Environ.* **2021**, *252*, 112133. [CrossRef]

land

Article

Vegetation Growth Trends of Grasslands and Impact Factors in the Three Rivers Headwater Region

Xiaoping Sun [1] and Yang Xiao [2,*]

1 Jiangsu Key Laboratory for Bioresources of Saline Soils, Jiangsu Synthetic Innovation Center for Coastal Bio-Agriculture, Jiangsu Provincial Key Laboratory of Coastal Wetland Bioresources and Environmental Protection, School of Wetland, Yancheng Teachers University, Yancheng 224001, China
2 China-Croatia "Belt and Road" Joint Laboratory on Biodiversity and Ecosystem Services, CAS Key Laboratory of Mountain Ecological Restoration and Bioresource Utilization & Ecological Restoration and Biodiversity Conservation Key Laboratory of Sichuan Province, Chengdu Institute of Biology, Chinese Academy of Sciences, Chengdu 610041, China
* Correspondence: xiaoyang@cib.ac.cn

Abstract: Areas of grassland improvement and degradation were mapped and assessed to identify the driving forces of change in vegetation cover in the Three Rivers headwater region of Qinghai, China. Based on linear regression at the pixel level, we analyzed the vegetation dynamics of the grasslands of this region using MODIS NDVI data sets from 2000 to 2010. Correlation coefficients were computed to quantitatively characterize the long-term interrelationship between vegetation NDVI and precipitation/temperature variability during this period. The use of time series residuals of the NDVI/precipitation linear regression to normalize the effect of precipitation on vegetation productivity and to identify long-term degradation was extended to the local scale. Results showed that significant improvements occurred in 26.4% of the grassland area in the Three Rivers Headwater region between 2000 and 2010. The study area, which represents about 86.4% of the total grassland area of this headwater region, showed a general trend of improvement with no obvious trend of degradation.

Keywords: NDVI; grassland; MODIS; precipitation variability; human activities

Citation: Sun, X.; Xiao, Y. Vegetation Growth Trends of Grasslands and Impact Factors in the Three Rivers Headwater Region. *Land* **2022**, *11*, 2201. https://doi.org/10.3390/land11122201

Academic Editors: Matteo Convertino and Jie Li

Received: 2 November 2022
Accepted: 29 November 2022
Published: 4 December 2022

1. Introduction

The impact of climate change is multi-scale, all-round and multi-level, with both positive and negative impacts, but its negative impacts are more concerned. After entering the 21st century, climate change has gradually become a comprehensive problem affecting the whole world. As a result of climate change, some grasslands around the world are already experiencing a decline in primary productivity and biodiversity, which is caused by man-made activities and also caused by the climate. These two causes interact and produce a superimposed effect. On the one hand, due to the dry climate, which affects the adaptability of the grassland ecosystem itself, human activities have largely changed the native grassland, which also leads to the decline of the adaptability of the grassland ecosystem itself. A large number of new varieties introduced by artificial and semi-artificial pastures need strict management and protection to create higher pasture yield. Under the climate change model, frequent pasture management measures exert great pressure on the grassland ecosystem. On the other hand, climate change makes the grassland ecosystem itself face more natural disturbances. Climate change affects precipitation and grassland temperature, which causes instability such as year-round drought, summer floods, and winter snowfall. For instance, Orusa and Mondino studied climate change effects on rangelands, and found that phenological and evapotranspiration-related processes and snowpack melting time have been dramatically changed in the last two decades in Aosta Valley (Orusa and Mondino, 2021). Located in the hinterland of the Qinghai-Tibet Plateau in southern Qinghai, the source area of the three rivers refers to the source areas of the Yangtze, Yellow, and Lancang Rivers [1]. This headwater area is the highest, largest, and most concentrated wetland in the world and has the most abundant biodiversity and the most sensitive ecosystem in China. The three rivers supply about 60 billion m^3 of water each year, and the partial water of the Yangtze River, Yellow River, and Lancang River comes from the Three Rivers Headwater Region, thus establishing the area as a veritable "Chinese water tower". The Three Rivers headwater region is a unique area that affects the development of the west; once damaged, recovery is difficult because of the harsh conditions found there and the area's fragile ecosystems [2].

Grassland is the dominant ecosystem in the Three Rivers headwater region, and grassland animal husbandry is the leading industry. One of the main supply functions of this grassland ecosystem is providing herbage allowance for animal husbandry production, which then provides direct benefits to the people through the production of grass-fed livestock [3,4]. In recent decades, due to the impact of the uncontrolled exploitation and overuse by humans, serious grassland degradation has occurred [5]. By the early part of this century, degraded grassland occupied 26–46% of available grassland, and this has had severe impacts on the ecological environment, its security, and on the sustainable development of grassland animal husbandry of the region. To address this problem, the State Council in 2005 approved the Ecosystem Conservation project in the Three Rivers headwater region with the aim of implementing ecological restoration [6]. To design a scientific strategy for the recovery, management, and utilization of the grassland and to effectively evaluate the project [7], analyzing the dynamic change of grassland productivity before and after the start of regional restoration work is essential. This allows an assessment of the trends of the supply function of herbage allowance and the natural and cultural drivers leading to the changes of the grassland ecosystem [8].

A vegetation index refers to ground vegetation coverage gathered through satellite remote sensing; it is a comprehensive, abstract, and indirect method [9] that is accomplished either through empirical modeling or mixed-pixel decomposition [10,11]. An empirical model is applied to achieve vegetation coverage over a large area through the correspondence between the measured data of vegetation coverage in the sample and the vegetation index [12]; however, the application of this method is subject to temporal conditions. With mixed-pixel decomposition, pixel information gathered by remote sensing is simplified into information or non-information on vegetation, and vegetation coverage is estimated from the proportion of vegetation information [13,14]. Studies have shown that this method is not subject to the latest data, and so it is generally applied to dynamic monitoring of remote sensing of vegetation coverage [15].

In this paper, we integrate various remote sensing analytical techniques (trend analysis, correlation analysis, and residual analysis with consideration of hysteresis effect of rainfall on vegetation) to map and assess grassland improvement and degradation areas to identify the driving forces of change in vegetation cover in the Three Rivers headwater region.

2. Study Area

The Three Rivers headwater region (TRHR) refers to the source areas of the Yangtze River, the Yellow River, and the Lancang River [16,17]. The TRHR is the highest, the largest, and the most concentrated wetland in the world, while also being the region with the most abundant biodiversity and the most sensitive ecosystem owing to its varied topography, complex hydrographic network, and numerous lakes. Wetland and meadow plants are the main types of vegetation in the TRHR, and these play a vital role in water conservation, runoff mitigation, and biodiversity maintenance [18].

Additionally, as the primary water source of the Yangtze, Yellow, and Lancang rivers, the Three Rivers headwater region, named "China water tower", is vital for maintaining the water security of 548 million people on the downstream Yellow River basin, Yangtze River basin and Lancang River basin. However, since 1990s, the ecosystem of the Three Rivers headwater region has shown a severe trend of degradation, which seriously threatens the production and living of the local residents in the downstream areas. Therefore, the Chinese government has started to implement a series of ecological policies to curb the degradation of the ecosystem (Figure 1).

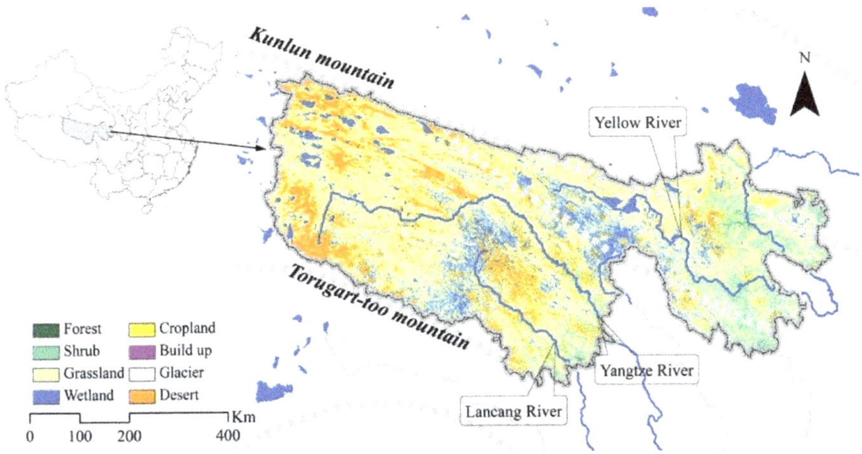

Figure 1. Grassland distribution of the Three Rivers headwater region (The gray bold dashed lines represent the mountains with a name beside it; the blue lines are the rivers; spatial reference: Albers conical equal area; datum: D WGS 1984).

3. Materials and Methods

3.1. Data Sets and Pre-Processing

MODIS data: The normalized difference vegetation index (NDVI) is a normalized ratio of red and near-infrared reflectance. It is used as a terrestrial vegetation growth and conditions proxy, being sensitive to chlorophyll-related plant canopy structural variations and being closely correlated to the fraction of potential photosynthesis and the physiological status of vegetation [19,20]. In this study, we used the moderate resolution imaging spectroradiometer (MODIS) MOD13Q1 time series data set from 2000 to 2010 obtained from the Land Processes Distributed Active Archive Center (LP DAAC). This data set has a spatial resolution of 250 m and temporal resolution of 16 days, an improved sensitivity to vegetation, and a reduced influence of external factors (such as atmosphere, observation angle, sun azimuth, and cloud cover), which have been verified using stable desert control points.

The MODIS data set was geocoded to the universal transverse Mercator (UTM) coordinate system using the MODIS reprojection tool. As these products may be affected by cloud cover, atmosphere, or ice/snow cover, we first reconstructed the NDVI time series data set using the asymmetric Gaussian function filter in TIMESAT 2.3 software program to reduce noise and improve data quality during the data pre-processing procedure [21]. The 16-day MODIS-NDVI was then compiled into monthly NDVI data by applying maximum value compositing (by overlaying multiple raster maps, the value of each raster cell is then taken as the largest of the multiple maps), which was processed in the interactive data language (IDL, https://www.harrisgeospatial.com, 30 November 2022).

Meteorological data: Meteorological data from 2000 to 2010 (including monthly scale precipitation and temperature) were obtained from the Chinese National Metrological Information Center/China Meteorological Administration. Monthly meteorological data derived from station-based information were interpolated to the whole research area at a spatial resolution of 250 m using the kriging interpolation method. In addition, the monthly temperature was calibrated using the digital elevation model, and the coefficient was reduced by a 0.47 degree/100 m increase in elevation. This coefficient was obtained by linear regression between elevation and temperature.

Grassland data: Information related to grassland distribution in the study area was obtained from an ecosystem map of China that was generated and interpreted based on 1:100,000 Landsat TM satellite remote sensing products from the Data Center for Resources and Environmental Sciences, Chinese Academy of Sciences. (Figure 1). Sources of cartographic data and statistics are listed in Table 1.

Table 1. Sources of principal data.

Data Name	Data Resolution	Data Source
MODIS-NDVI	250 m (monthly)	Land Processes Distributed Active Archive Center (https://lpdaac.usgs.gov, 30 November 2022)
SRTM	90 m	CGIAR Consortium for Spatial Information (http://srtm.csi.cgiar.org/, 30 November 2022)
Precipitation, temperature	146 points (monthly)	Chinese National Metrological Information Center/China Meteorological Administration (http://data.cma.cn, 30 November 2022)
Ecosystem map	90 m (yearly)	Resource and Environment Science and Data Center, Chinese Academy of Sciences (https://www.resdc.cn/, 30 November 2022)

3.2. Methods

(1) Trend analysis

To assess variation trends of NDVI and climate (precipitation and temperature) throughout the 2000–2010 study period, we used a linear least-squares regression model to obtain the changing trends of every pixel by fitting a linear equation of NDVI or climate variables as a function of the variable of year to obtain an image of changing slopes [22]. The linear least-squares regression method, which is a commonly used method in trend analysis [23], was applied as follows:

$$y = a + b \times t + \varepsilon \quad (1)$$

where y represents NDVI or climate variables; t is year; a and b are fitted variables (b is the slope as a proxy of trend and a is the intercept); and ε is the residual error. If $b > 0$, there is an increasing trend of NDVI or climate; conversely, if $b < 0$, there is a decreasing trend. $p < 0.05$ was considered a significant change for both increasing and decreasing trends (Table 2).

Table 2. Evaluation standard of trend significance.

Variation Trend	*b* Value Range	*p* Value Range
Significant decrease	$b < 0$	$p \leq 0.05$
Significant increase	$b > 0$	$p \leq 0.05$
No significant change		$p > 0.05$

In addition, the estimation of parameters a and b uses the least square method, ε_i is a random error, the fitting value of parameters a and b is expressed as:

$$\hat{b} = \frac{\sum_{i=1}^{n}(t_i - \bar{t})(y_i - \bar{y})}{\sum_{i=1}^{n}(t_i - \bar{t})^2} \quad (2)$$

$$\hat{a} = \bar{y} - \hat{b} \times \bar{t} \quad (3)$$

The calculation process of changing slopes per pixel was programmed with interactive data language (IDL).

(2) Correlation analysis

We used the Pearson correlation coefficient to explore the relationship between trends of NDVI and climate as follows:

$$r = \frac{\sum_{i=1}^{n}(x_i - \overline{x})(y_i - \overline{y})}{\sqrt{\sum_{i=1}^{n}(x_i - \overline{x})^2}\sqrt{\sum_{i=1}^{n}(y_i - \overline{y})^2}} \quad (4)$$

where r represents the linear correlation coefficient; x is the NDVI variable; y is the precipitation or temperature variable; and n is the number of variables [24].

To evaluate the influence of anthropogenic activities on the trend in NDVI, we used partial correlation between the slope of the NDVI and anthropogenic factors (i.e., population growth and a change in livestock numbers). We also employed stepwise regression between the trend in NDVI and the impact factors (precipitation, temperature, population, and livestock numbers) to assess their relative contribution in influencing spatial characteristics of the NDVI trend [25].

(3) Residual analysis

Residual NDVI is the difference between the observed value y_i (observed NDVI) of the dependent variable and the predicted value $\hat{y}_i$ (predicted NDVI) calculated according to the estimated regression equation, denoted by e. It reflects the error caused by using the estimated regression equation to predict the dependent variable y_i. The residual of the ith observation is:

$$e_i = y_i - \hat{y}_i \quad (5)$$

Residual analysis is used to investigate trends in residual differences (residual NDVI) using a regression model involving. The trends of predicted NDVI were interpreted as climate-induced changes (rainfall as the explanatory variable [26]), while trends in residual NDVI (e_i) were interpreted as human-induced changes [26].

The residual analysis mainly involved the following steps. First of all, a line regression model between observed NDVI and precipitation factor was used to calculate the residual difference (e) between observed NDVI and predicted NDVI (regression result). Then, the trend analysis of residuals as a function of time was processed to investigate the human-induced vegetation degradation (Table 3). Pixels exhibiting marginal decreases (i.e., <5%) in 2000–2010 were excluded or considered as stable because they reflected the potential uncertainties caused by differences due to image dates within the season/month or image calibration processes.

Table 3. Indicators and definitions related to residual analysis methods.

Degradation Trends of Observed NDVI	Trends of Predicted NDVI	Slope of Residual	Definition Description
<0	>0	<0	Human-induced vegetation degradation
	<0	>0	Climate-induced vegetation degradation
	<0	<0	Both climate- and human-induced vegetation degradation
	>0	>0	Uncertainty error

Previous studies reported that annual maximum NDVI representing the growth of grasslands is strongly correlated with climatic variables [26]. Thus, $NDVI_{max}$ as the highest NDVI value can be used to gauge grassland vegetation growth in this study. Based the residual analysis, measured $NDVI_{max}$ values showed both positive and negative deviations from the fitting curve on the $NDVI_{max}$/rainfall linear regression, suggesting that vegetation is not only responsive to rainfall, but also influenced by human activities represented by residuals.

As the rainfall period was most strongly correlated with grassland growth, we calculated the precipitation accumulation periods and lag periods from September to August of next year. And analysis the correlation between the cumulative rainfall and $NDVI_{max}$, to identify the optimal relationship between them in the study area [27]. By prolonging and changing the cumulative period, the process was repeated until all possible combinations were tested, and then the best relevant cumulative period was determined (with the strongest correlation), which is from 1 September to 1 August of next year.

4. Results

4.1. Spatial and Temporal Characteristics of Grassland Variation

Trend analysis showed overall positive NDVI trends for the grasslands (Figure 2). The study area, which is about 84.25% area of the total grassland in the Three Rivers headwater region, showed a continuous trend of improvement from 2000 to 2010.

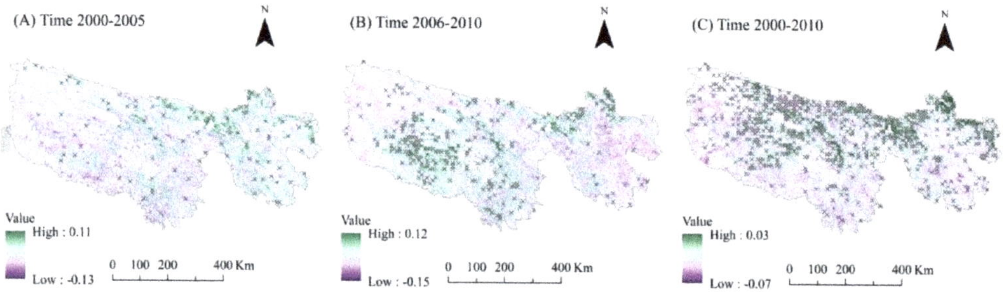

Figure 2. Spatial distribution of NDVI trends in the grassland ecosystem. (**A**) NDVI trend from 2000 to 2005; (**B**) NDVI trend from 2006 to 2010; (**C**) NDVI trend from 2000 to 2010; the fork symbol in the figure indicates the significance of NDVI trend, where p value < 0.05. (Spatial reference: Albers conical equal area; datum: D WGS 1984).

Linear regression analysis of NDVI from 2000 to 2005 indicated that about 37.01% of the total grassland area experienced a declining trend in vegetation production (Figure 2A). The declining patches of vegetation were found mainly in the western and southern pastures; however, this change between 2001 and 2005 was not significant. On the contrary, the long-term productivity of vegetation shows a positive or stable trend in the northern high-altitude pastoral areas affected by fog and haze.

Correlation between grassland degradation and triggered factors:

The correlation between NDVI values and accumulated precipitation revealed a strong positive correlation (r = 0.5 to 0.98) for about 38.41% of the grassland area in the Three Rivers headwater region (Figure 3A). A negative correlation was found in the densely populated southern and eastern parts of the study area, which have low vegetation cover and high levels of human activity.

Moreover, there is little correlation between residual NDVI and rainfall, suggesting that residual analysis can effectively remove climatic factors (precipitation) and effectively reflect the impact of human activities on grassland degradation (Figure 3B).

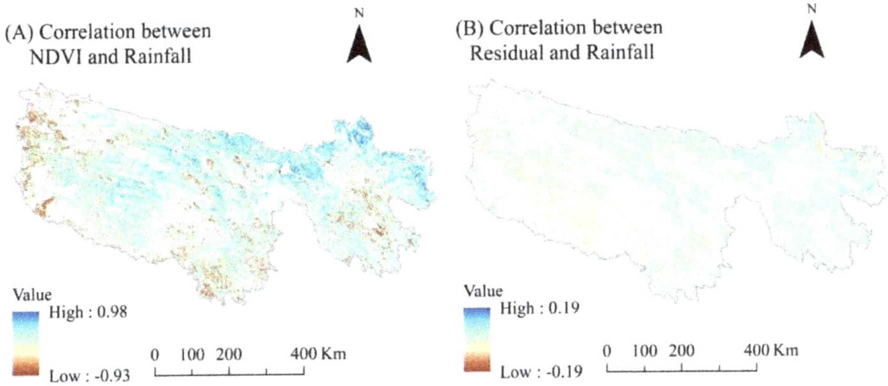

Figure 3. Correlation results based on person correlation analysis. (**A**) Person correlation between NDVI and Rainfall; (**B**) Person correlation between residual NDVI and Rainfall. (spatial reference: Albers conical equal area; datum: D WGS 1984).

4.2. Impacts of Human Activities on Grassland Degradation

Analysis showed that approximately 25.85% of the study area did not show an increase in vegetation growth despite increasing precipitation from 2000 to 2010 (Table 4). The declining patches of vegetation, which can be attributed to human activities, were found mainly in the southern and eastern pastures (Figure 4A). When evaluating the first and second halves of the decade, linear regression analysis of residual NDVI from 2000 to 2005 showed that about 35.72% of the total grassland area experienced a declining trend in vegetation production (Figure 4B). In contrast, about 17.66% of the total grassland area showed a declining trend in vegetation production from 2006 to 2010 (Figure 4C).

Table 4. Proportion of area showing non-climate degradation (units %) and the trend of the residual mean value.

	2000–2010	2000–2005	2006–2010
Non-climate degradation area (%)	25.85	35.72	17.66
Mean of residual slope	−27.35	−88.51	−62.10

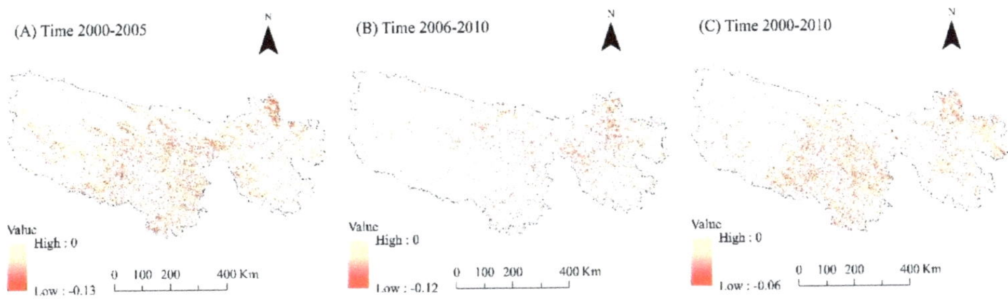

Figure 4. Grassland degradation trend due to human activities (**A**) residual NDVI from 2000 to 2005; (**B**) residual NDVI from 2006 to 2010; (**C**) residual NDVI from 2000 to 2010. (spatial reference: Albers conical equal area; datum: D WGS 1984).

5. Discussion

Our findings showed that an increasing NDVI occurred mainly in the northern plains, while southern and eastern grazing pastures, which are densely populated, showed a small positive trend in long-term vegetation productivity. Degraded patches of vegetation were located on steep slopes (slope > 20°), supporting the findings of Fan et al. (2010) [16]. In addition, our field observations suggested that high wind erosion and intensive human activities were impacting vegetation productivity. Grassland is the most important ecosystem type on the Qinghai Tibet Plateau and the basis of animal husbandry on the plateau. Though a series of grazing grassland ecological protection subsidy incentives such as grassland ecological protection in the construction project, the Qinghai Tibet Plateau grassland conservation effect appeared gradually, but the strength of the human activity factor is different, we also found that declining patches of vegetation were found mainly in the pastures [28].

A large proportion of the total grassland area (80.79%) showed an increasing vegetation trend between 2000 and 2005. Less than 20% of the total grassland showed degradation, and this was patchy rather than continuous. By 2006, the extent of vegetation coverage was even higher. We attribute this improvement to the 2005 approval of the Ecosystem Conservation project in the Three Rivers headwater region [29], which has led to less human disturbance and a focus on vegetation recovery [6,30]. Whether changes in precipitation variability or human activities led to this recovery is not yet clear. However, we found only a limited correlation between precipitation and residual NDVI (Figure 4b), indicating that the trends in vegetation cover were unlikely to be caused as a result of climatic conditions.

To analyze temporal trends in grassland NDVI, we concentrated on the development of the annual maximum NDVI, as proposed by Evans and Geerken (2004) [25]. Vegetation production in a cold alpine environment has been shown to fluctuate strongly according to interannual precipitation and temperature variability [16,31]. However, temperature is periodic and takes years for little change. Therefore, we assume that the correlation between temperature and vegetation growth is unlikely to be used to identify the temporal and spatial trends of grassland, so we ignore temperature. The NDVImax values are calculated and correlated with precipitation, as proposed by Evans and Geerken (2004) and Madonsela et al. (2018) [26,32]. We found that a decreasing NDVI was mainly concentrated in the western region due to enhancing human activity such as over grazing and an increasing NDVI was distributed throughout the rest of the study area due to increasing rainfall and ecological protection policies such as banning grazing [33].

Our findings document the benefits that have occurred since the initiation in 2006 of the ecological restoration project, which included returning grazing land to grassland, the ecological migration project, black soil land recovery (grassland with extreme degradation), and rodent control [16,34]. Non-climatic (human) activities led to serious grassland degradation of the Three Rivers headwater region up until 2005 (Figure 4B), but since 2006, the grassland degradation caused by these activities has decreased and an increase in vegetation cover has occurred. This finding combined with the decreasing residual NDVI indicates that the increasing vegetation cover over time is not related to precipitation variability but to better land management practices.

6. Conclusions

We found an overall positive NDVI trend between 2000 and 2010 for the grasslands in our study area, which cover about 84.25% of the total grassland area in the Three Rivers headwater region. The declining trend in vegetation production between 2000 and 2005, which affected about 37.01% of the total grassland area, appears to have been reversed between 2006 and 2010, most likely by the effectiveness of the government-approved program for ecosystem conservation.

The amount of precipitation in the Three Rivers headwater region has a strong positive correlation with NDVI in about 38.41% of the grassland area. Negative correlations were found in the southern and eastern regions, which have dense populations, low vegetation coverage, and intensive human activity. Our study shows that to promote grassland recovery in Three Rivers headwater region, a number of factors must be taken into account, especially human activities. Our findings also support the need for intensive management of water resources to enhance vegetation and avoid ecological damage in this important ecological area.

Author Contributions: Y.X. conceived and designed the research; X.S. and Y.X. analyzed the data; X.S. and Y.X. wrote the main manuscript; X.S. prepared Figures and Tables. All authors reviewed the manuscript. X.S. conceived and designed the experiments, performed the experiments, contributed reagents/materials/analysis tools, analyzed the data, prepared figures and/or tables, and wrote the main manuscript. Y.X. conceived and designed the experiments, contributed reagents/materials/analysis tools, wrote the main manuscript, authored or reviewed drafts of the paper, and approved the final draft. All authors have read and agreed to the published version of the manuscript.

Funding: This work was funded jointly by the National Natural Science Foundation of China (NSFC) (No: 71904060), the General Project of Natural Science Research of Jiangsu Province Higher Education Institutions (22KJD180005), the Natural Science Foundation of Jiangsu Province (BK20211363), opening grant of Jiangsu Key Laboratory for Bioresources of Saline Soils (JKLBZ202001).

Data Availability Statement: The raw/processed data required to reproduce these findings cannot be shared at this time as the data also form part of an ongoing study.

Conflicts of Interest: The authors declare no competing interest.

References

1. Jiang, C.; Li, D.; Gao, Y.; Liu, W.; Zhang, L. Impact of climate variability and anthropogenic activity on streamflow in the Three Rivers Headwater Region, Tibetan Plateau, China. *Theor. Appl. Climatol.* **2017**, *129*, 667–681. [CrossRef]
2. Alam, M.S.; Lamb, D.W.; Rahman, M.M. A refined method for rapidly determining the relationship between canopy NDVI and the pasture evapotranspiration coefficient. *Comput. Electron. Agric.* **2018**, *147*, 12–17. [CrossRef]
3. Xiong, Q.; Xiao, Y.; Liang, P.; Li, L.; Zhang, L.; Li, T.; Pan, K.; Liu, C. Trends in climate change and human interventions indicate grassland productivity on the Qinghai–Tibetan Plateau from 1980 to 2015. *Ecol. Indic.* **2021**, *129*, 108010. [CrossRef]
4. Miao, C.; Kong, D.; Wu, J.; Duan, Q. Functional degradation of the water–sediment regulation scheme in the lower Yellow River: Spatial and temporal analyses. *Sci. Total Environ.* **2016**, *551*, 16–22. [CrossRef] [PubMed]
5. Gao, Y.; Wang, Y.; Zhang, G.; Xia, J.; Mao, L.; Tang, Z.; Zhou, P. An approach for assessing soil health: A practical guide for optimal ecological management. *Environ. Earth Sci.* **2011**, *65*, 153–159. [CrossRef]
6. Cao, S.; Ma, H.; Yuan, W.; Wang, X. Interaction of ecological and social factors affects vegetation recovery in China. *Biol. Conserv.* **2014**, *180*, 270–277. [CrossRef]
7. An, R.; Wang, H.-L.; Feng, X.-Z.; Wu, H.; Wang, Z.; Wang, Y.; Shen, X.-J.; Lu, C.-H.; Quaye-Ballard, J.A.; Chen, Y.-H.; et al. Monitoring rangeland degradation using a novel local NPP scaling based scheme over the "Three-River Headwaters" region, hinterland of the Qinghai-Tibetan Plateau. *Quat. Int.* **2017**, *444*, 97–114. [CrossRef]
8. Chen, L.; Cao, S. Lack of integrated solutions hinders environmental recovery in China. *Ecol. Eng.* **2013**, *54*, 233–235. [CrossRef]
9. Gao, Y.; He, N.; Yu, G.; Chen, W.; Wang, Q. Long-term effects of different land use types on C, N, and P stoichiometry and storage in subtropical ecosystems: A case study in China. *Ecol. Eng.* **2014**, *67*, 171–181. [CrossRef]
10. Xiao, Y.; Xiong, Q.; Liang, P.; Xiao, Q. Potential Risk to Water Resources under Eco-restoration Policy and Global Change in the Tibetan Plateau. *Environ. Res. Lett.* **2021**, *16*, 094004. [CrossRef]
11. Khan, Z.; Rahimi-Eichi, V.; Haefele, S.; Garnett, T.; Miklavcic, S.J. Estimation of vegetation indices for high-throughput phenotyping of wheat using aerial imaging. *Plant Methods* **2018**, *14*, 20. [CrossRef] [PubMed]
12. Trueman, I.; Mitchell, D.; Besenyei, L. The effects of turf translocation and other environmental variables on the vegetation of a large species-rich mesotrophic grassland. *Ecol. Eng.* **2007**, *31*, 79–91. [CrossRef]
13. Pradhan, S.; Sehgal, V.K.; Bandyopadhyay, K.K.; Sahoo, R.N.; Panigrahi, P.; Parihar, C.M.; Jat, S.L. Comparison of Vegetation Indices from Two Ground Based Sensors. *J. Indian Soc. Remote Sens.* **2017**, *46*, 321–326. [CrossRef]
14. Xiao, Q.; Xiao, Y. Impact of artificial afforestation on the regional water supply balance in Southwest China. *J. Sustain. For.* **2019**, *38*, 427–441. [CrossRef]
15. Patel, N.R.; Padalia, H.; Devadas, R.; Huete, A.; Kumar, A.S.; Murthy, Y.V.N.K. Estimating Net Primary Productivity of Croplands in Indo-Gangetic Plains Using GOME-2 Sun-Induced Fluorescence and MODIS NDVI. *Curr. Sci.* **2018**, *114*, 1333–1337. [CrossRef]
16. Fan, J.-W.; Shao, Q.-Q.; Liu, J.-Y.; Wang, J.-B.; Harris, W.; Chen, Z.-Q.; Zhong, H.-P.; Xu, X.-L.; Liu, R.-G. Assessment of effects of climate change and grazing activity on grassland yield in the Three Rivers Headwaters Region of Qinghai–Tibet Plateau, China. *Environ. Monit. Assess.* **2009**, *170*, 571–584. [CrossRef]
17. Orusa, T.; Mondino, E.B. Exploring Short-Term Climate Change Effects on Rangelands and Broad-Leaved Forests by Free Satellite Data in Aosta Valley (Northwest Italy). *Climate* **2021**, *9*, 47. [CrossRef]
18. Osunmadewa, B.A.; Gebrehiwot, W.Z.; Csaplovics, E.; Adeofun, O.C. Spatio-temporal monitoring of vegetation phenology in the dry sub-humid region of Nigeria using time series of AVHRR NDVI and TAMSAT datasets. *Open Geosci.* **2018**, *10*, 1–11. [CrossRef]

19. Konda, V.G.R.K.; Chejarla, V.R.; Mandla, V.R.; Voleti, V.; Chokkavarapu, N. Correction to: Vegetation damage assessment due to Hudhud cyclone based on NDVI using Landsat-8 satellite imagery. *Arab. J. Geosci.* **2018**, *11*, 67. [CrossRef]
20. Geerken, R.; Ilaiwi, M. Assessment of rangeland degradation and development of a strategy for rehabilitation. *Remote Sens. Environ.* **2004**, *90*, 490–504. [CrossRef]
21. Paudel, K.P.; Andersen, P. Assessing rangeland degradation using multi temporal satellite images and grazing pressure surface model in Upper Mustang, Trans Himalaya, Nepal. *Remote Sens. Environ.* **2010**, *114*, 1845–1855. [CrossRef]
22. Jönsson, P.; Eklundh, L. TIMESAT—A program for analyzing time-series of satellite sensor data. *Comput. Geosci.* **2004**, *30*, 833–845. [CrossRef]
23. Xiao, Y.; Xiao, Q.; Xiong, Q.; Yang, Z. Effects of Ecological Restoration Measures on Soil Erosion Risk in the Three Gorges Reservoir Area Since the 1980s. *Geohealth* **2020**, *4*, e2020GH000274. [CrossRef]
24. Hoagland, S.J.; Beier, P.; Lee, D. Using MODIS NDVI phenoclasses and phenoclusters to characterize wildlife habitat: Mexican spotted owl as a case study. *For. Ecol. Manag.* **2018**, *412*, 80–93. [CrossRef]
25. Xiao, Y.; Xiao, Q. Identifying key areas of ecosystem services potential to improve ecological management in Chongqing City, southwest China. *Environ. Monit. Assess.* **2018**, *190*, 258. [CrossRef] [PubMed]
26. Mondal, S.; Jeganathan, C. Mountain agriculture extraction from time-series MODIS NDVI using dynamic time warping technique. *Int. J. Remote Sens.* **2018**, *39*, 3679–3704. [CrossRef]
27. Evans, J.; Geerken, R. Discrimination between climate and human-induced dryland degradation. *J. Arid Environ.* **2004**, *57*, 535–554. [CrossRef]
28. Xiao, Y.; Xiao, Q.; Sun, X. Ecological Risks Arising from the Impact of Large-scale Afforestation on the Regional Water Supply Balance in Southwest China. *Sci. Rep.* **2020**, *10*, 4150. [CrossRef]
29. Chen, H.; Shao, L.; Zhao, M.; Zhang, X.; Zhang, D. Grassland conservation programs, vegetation rehabilitation and spatial dependency in Inner Mongolia, China. *Land Use Policy* **2017**, *64*, 429–439. [CrossRef]
30. Miao, C.Y.; Ni, J.R.; Borthwick, A.G.L. Recent changes of water discharge and sediment load in the Yellow River basin, China. *Prog. Phys. Geogr.* **2010**, *34*, 541–561. [CrossRef]
31. Xiao, Y.; Xiong, Q.; Pan, K. What Is Left for Our Next Generation? Integrating Ecosystem Services into Regional Policy Planning in the Three Gorges Reservoir Area of China. *Sustainability* **2019**, *11*, 3. [CrossRef]
32. Xiao, Q.; Tao, J.; Xiao, Y.; Qian, F. Monitoring vegetation cover in Chongqing between 2001 and 2010 using remote sensing data. *Environ. Monit. Assess.* **2017**, *189*, 493. [CrossRef] [PubMed]
33. Madonsela, S.; Cho, M.A.; Ramoelo, A.; Mutanga, O.; Naidoo, L. Estimating tree species diversity in the savannah using NDVI and woody canopy cover. *Int. J. Appl. Earth Obs. Geoinf.* **2018**, *66*, 106–115. [CrossRef]
34. Miao, C.; Ni, J.; Borthwick, A.G.; Yang, L. A preliminary estimate of human and natural contributions to the changes in water discharge and sediment load in the Yellow River. *Glob. Planet. Chang.* **2011**, *76*, 196–205. [CrossRef]

land

MDPI

Article

Spatial Pattern of Changing Vegetation Dynamics and Its Driving Factors across the Yangtze River Basin in Chongqing: A Geodetector-Based Study

Bo Yao [1,2,3,†], Lei Ma [1,4,†], Hongtao Si [1,4], Shaohua Li [1,4], Xiangwen Gong [1,2,3,4,*] and Xuyang Wang [2,3,*]

1 Observation and Research Station of Ecological Restoration for Chongqing Typical Mining Areas, Ministry of Natural Resources, Chongqing Institute of Geology and Mineral Resources, Chongqing 401120, China
2 Northwest Institute of Eco-Environment and Resources, Chinese Academy of Sciences, Lanzhou 730000, China
3 University of Chinese Academy of Sciences, Beijing 100049, China
4 Wansheng Mining Area Ecological Environment Protection and Restoration of Chongqing Observation and Research Station, Chongqing Institute of Geology and Mineral Resources, Chongqing 401120, China
* Correspondence: gongxiangwen@nieer.ac.cn (X.G.); xuyangwang@lzb.ac.cn (X.W.)
† These authors contributed equally to this work.

Abstract: Revealing the spatial dynamics of vegetation change in Chongqing and their driving mechanisms is of major value to regional ecological management and conservation. Using several data sets, including the SPOT Normalized Difference Vegetation Index (NDVI), meteorological, soil, digital elevation model (DEM), human population density and others, combined with trend analysis, stability analysis, and geographic detectors, we studied the pattern of temporal and spatial variation in the NDVI and its stability across Chongqing from 2000 to 2019, and quantitatively analyzed the relative contribution of 18 drivers (natural or human variables) that could influence vegetation dynamics. Over the 20-year period, we found that Chongqing region's NDVI had an annual average value of 0.78, and is greater than 0.7 for 93.52% of its total area. Overall, the NDVI increased at a rate of 0.05/10 year, with 81.67% of the areas undergoing significant expansion, primarily in the metropolitan areas of Chongqing's Three Gorges Reservoir Area (TGR) and Wuling Mountain Area (WMA). The main factors influencing vegetation change were human activities, climate, and topography, for which the most influential variables respectively were night light brightness (NLB, 51.9%), annual average air temperature (TEM, 47%), and elevation (ELE, 44.4%). Furthermore, we found that interactions between differing types of factors were stronger than those arising between similar ones; of all pairwise interaction types tested, 92.9% of them were characterized by two-factor enhancement. The three most powerful interactions detected were those for NLB ∩ TEM (62.7%), NLB ∩ annual average atmospheric pressure (PRS, 62.7%), and NLB ∩ ELE (61.9%). Further, we identified the most appropriate kind or range of key elements shaping vegetation development and dynamics. Altogether, our findings can serve as a timely scientific foundation for developing a vegetative resource management strategy for the Yangtze River basin that duly takes into account local climate, terrain, and human activity.

Keywords: normalized difference vegetation index (NDVI); spatial evolution; multi-factor interaction; geographic detector

Citation: Yao, B.; Ma, L.; Si, H.; Li, S.; Gong, X.; Wang, X. Spatial Pattern of Changing Vegetation Dynamics and Its Driving Factors across the Yangtze River Basin in Chongqing: A Geodetector-Based Study. *Land* **2023**, *12*, 269. https://doi.org/10.3390/land12020269

Academic Editors: Matteo Convertino, Jie Li and Nir Krakauer

Received: 21 November 2022
Revised: 20 December 2022
Accepted: 14 January 2023
Published: 17 January 2023

1. Introduction

Vegetation is an important component of terrestrial ecosystems and serves as a link between the atmosphere, water, and soil [1], thus playing a pivotal role in soil conservation, climate regulation, hydrological processes, the carbon cycle, and ecosystem functioning and stability [2,3]. The health of a local ecological environment, such as its water quality, thermal energy, and soil fertility, can also be gauged by its vegetation [4]. Hence, vegetation is often used not only as an indicator of an ecosystem's sensitivity to external disturbances, such as climate change and human activities [5], but also as a comprehensive indicator for characterizing the response and adaptation of a terrestrial ecosystem to environmental change. Accordingly, understanding vegetation's spatio-temporal evolution and the involved driving mechanisms is critical for the regional development of effective vegetation restoration measures and ecological protection policies [6].

Monitoring vegetation dynamics has been a major focus of global change research in recent decades [7]. Because of their unique advantages, namely their large spatial scale, long time series, and short interval period, remote sensing images have become the primary data source for monitoring vegetation change at different scales, especially at multiple spatio-temporal scales [8]. For example, Schultz et al. [9] used a long-time series of Landsat-derived remote sensing imagery to monitor deforestation throughout the tropics. With the continued maturation and development of hyperspectral and thermal infrared remote sensing technologies, the bands of their images are becoming more abundant, making it feasible to use them to study changing spatio-temporal dynamics of terrestrial vegetation. To that end, researchers in China and abroad have proposed more than 100 plant cover indexes, such as the ratio vegetation index (RVI), difference vegetation index (DVI), normalized difference vegetation index (NDVI), enhanced vegetation index (EVI), soil adjusted vegetation index (SAVI), and so forth [10,11], greatly improving the efficiency and accuracy of extracting vegetation information. Currently, of those, the NDVI is recognized as the best indicator of regional vegetation and ecological environment change because its value can convey real information about vegetation's growth status and biomass, among other things; hence, it is widely used in the study of vegetation dynamics [12–14].

The dynamics of vegetation and the involved mechanisms shaping it have drawn much attention in the context of rapid global change [15,16]. Many studies have shown that vegetation dynamics are closely related to a broad suite of natural factors, including climate, terrain, soil, and vegetation types [17,18]. How vegetation responds to climate change is a very complex process, and climatic factors such as precipitation, temperature, and evapotranspiration can jointly affect vegetation dynamics. For instance, Na et al. [17] examined the impact of shifts in extreme air temperature and extreme precipitation indexes on the long-term dynamics of vegetation in inner Mongolia, finding that climate change may explain 68% of the variation in that vegetation's development. The effect of evapotranspiration on vegetation change should not be overlooked either, according to Shuai et al. [18], given the rapid acceleration of surface change. By analyzing the suitable growth conditions of vegetation in the Weihe River Basin, Zhang et al. [19] showed that NDVI is strongly correlated with air temperature, precipitation, evaporation, and soil moisture, with correlation coefficients as high as 0.89, 0.78, 0.71, and 0.65, respectively. The Yangtze River Basin is the largest basin in Asia, and its vegetation cover status and dynamics are fundamental to maintaining the ecological balance of China and its neighboring countries, and perhaps even that of the whole world.

In recent years, great progress has been made in the study of the changing dynamics of NDVI and its influencing factors in the Yangtze River Basin. According to some studies, this basin's NDVI featured an overall upward trend during the years spanning 1982 to 2015, increasing in extent mainly in its middle while decreasing chiefly in its eastern part [20–22]. Furthermore, Qu et al. [20] found that this NDVI trend was more pronounced after 1994 than before. Temperature is the main climatic factor affecting the growth of vegetation in the Yangtze River Basin, while precipitation has a weak effect on it [21]. Other work has reported evidence for lag effects from altered precipitation and temperature regimes on vegetation growth in the Yangtze River Basin, with more than 50% of this growth (on a regional scale) predominantly affected by climate change. Because studies of the whole Yangtze River Basin or portions of it mostly focused on temperature and precipitation, less is known about the possible influences of other climatic factors, in addition to topographic factors and human activities, on vegetation growth and dynamics there [23,24].

Traditional statistical methods, such as linear regression and residual trend analysis, can be applied to reveal the relationship between the monotonous trend of vegetation change and its drivers, but this inference is limited to linear relations [25,26]. However, we know that vegetation growth is often affected by the joint action of multiple factors, so how natural and human factors interact to change vegetation dynamics is unlikely to be a simple linear relationship [27]. Therefore, determining how to accurately quantify the relative contributions of natural and human factors to regional vegetation change and the driving mechanisms involved remains a challenging task [28,29]. The geographic detector model based on spatial stratification heterogeneity theory, introduced by Wang et al. [30,31], provides a reliable and direct methodology to quantify the respective influence of driving factors as well as their interactions. It has three notable advantages: (i) it does not have to strictly follow the assumptions of traditional statistical methods; (ii) it does detect the interaction of two factors, and (3) it does not require a complex parameter setting process [15,32,33]. For example, Zhu et al. [29] quantified the impacts of natural and human factors on changing vegetation dynamics in the middle reaches of the Heihe River by using geographic exploration methods, which revealed that land use conversion type, average annual precipitation, and soil type had the greatest impact. Li et al. [34] quantitatively analyzed the driving factors of grassland vegetation in inner Mongolia from the perspective of spatial stratification heterogeneity, finding that precipitation, livestock density, wind speed, and humans population density were the dominant factors, with these accounting for more than 15% of variation in the data. As such, the geographic detector approach has been successfully applied to quantify the influence of potential driving factors on vegetation dynamics, making it an effective tool for understanding the mechanisms of vegetation change at different spatial scales.

The Chongqing municipality in China is located in the upper reaches of the Yangtze River and in the central zone of the Three Gorges Reservoir Area. It is the last pass of the ecological barrier in the upper reaches of the Yangtze River, and its ecological location is crucial. Therefore, building an important ecological barrier in the upper reaches of the Yangtze River plays an indispensable role in ensuring the ecological balance and homeland security of the entire Yangtze River basin. In recent years, with the intensification of global climate change and human activities, understanding the spatio-temporal dynamic evolution of vegetation in this region and its driving mechanisms has become imperative for the development of reasonable ecosystem protection measures and management in this region. To achieve this aim, based on a time series of SPOT NDVI data, we used trend analysis, stability analysis, and geographic detector methods to fulfill three objectives: (1) to reveal the spatial characteristics and regularities of NDVI-based vegetation dynamics in Chongqing during the years 2000–2019; (2) to quantify the driving mechanisms of natural factors and human activities and their interactions upon vegetation change; and (3) to explore the appropriate types or ranges of the main influencing factors that promote vegetation growth in Chongqing, so as to provide a reference for the implementation of vegetation restoration projects in the Yangtze River Basin and the formulation of sound ecological environmental protection policies.

2. Materials and Methods

2.1. Study Region

Chongqing is located in the transitional zone between the Qinghai-Tibet Plateau and the plain of the middle/lower reaches of the Yangtze River, where it encompasses an area of 8.24×10^4 km^2 (Figure 1a). Aside from being a significant industrial and commercial center in the southwest, a marine and land transportation hub, and the greatest economic hub in the upper reaches of the Yangtze River, Chongqing also serves as a crucial ecological barrier to protect those areas. Implementation of the "one district and two clusters" coordinated development spatial pattern—the major city metropolitan area (MCA), the Three Gorges Reservoir Area town cluster in northeast Chongqing (TGR), and the Wuling Mountain Area town cluster in southeast Chongqing (WMA)—is being accelerated (Figure 1b). In going from south and north to the middle valley area, the topography gradually flattens out, with elevations spanning 63 to 2624 m. The geography varies greatly and is complex, with low mountains and hills in the northwest and center, and Daba Mountain and Wuling Mountain in the southeast (Figure 1c). The Yangtze River, Jialing River, Wujiang River, Fujiang River, Qijiang River, Daning River, and other major rivers flow through this region, which is endowed with abundant surface water resources. Here, a subtropical humid monsoon climate prevails, with annual averages ranging from 4.7–19.7 °C for air temperature, 7.3–21.7 °C for ground temperature, 989–1682 mm for precipitation, 640–1015 mm for evapotranspiration, 79–1646 h for sunshine duration, and 0.8%–84.5% for relative humidity. Leaching soil, primary soil, man-made soil, and iron-bauxite are the main soil types. There are many different types of vegetation, and the vegetation that exists in this region is mostly cultivated crops, shrubs, and other plants (Figure A1).

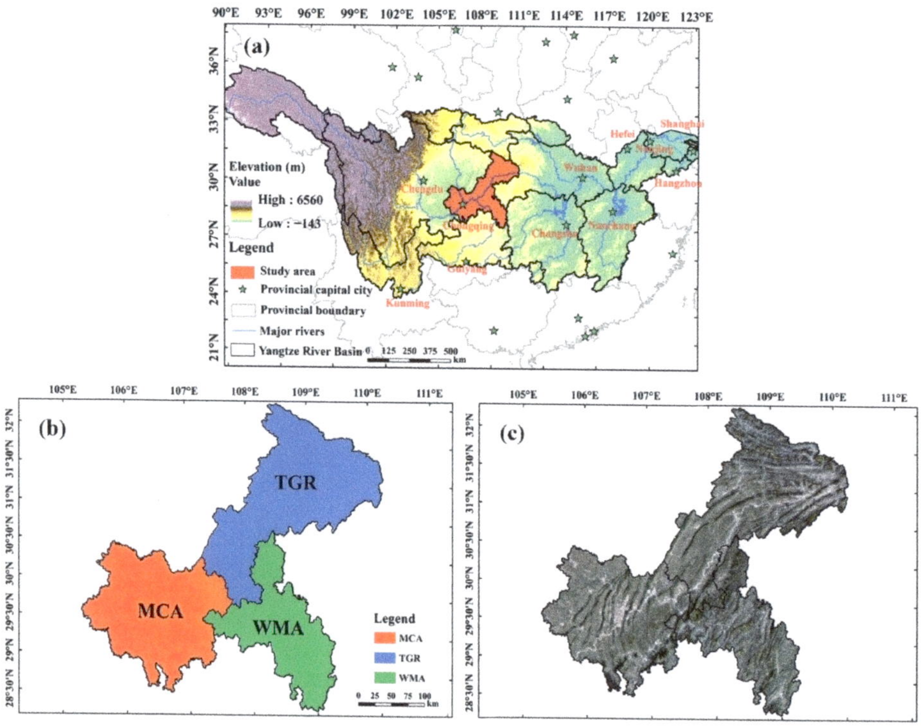

Figure 1. Overview of the study region. (**a**) Its location in the Yangtze River Basin in China; (**b**) the "one district and two clusters" coordinated development spatial pattern; (**c**) original remote sensing image. Notes: MCA, the major city metropolitan area; TGR, Three Gorges Reservoir Area town cluster in northeast Chongqing; and WMA, the Wuling Mountain Area town cluster in southeast Chongqing.

2.2. Data Source and Preprocessing

2.2.1. NDVI Data

The growing status of vegetation on the land surface can be accurately expressed by a vegetation index. In this study, we chose the SPOT NDVI dataset based on the following considerations. (1) Currently, the monitoring of changing vegetation dynamics at various scales has made extensive use of the NDVI time series data derived from SPOT satellite remote sensing imagery [12,35]. (2) The SPOT NDVI dataset with a time span of 2000 to 2019 can be obtained directly from the Resource and Environmental Science Data Center of the Chinese Academy of Sciences (https://www.resdc.cn/, accessed on 18 June 2022) and is free. (3) This data, with a spatial resolution of 1000 m, is consistent with the spatial resolution of mostly other research data (Table 1), and can avoid the influence of data pre-processing processes, such as resampling, on the research results. (4) Since the general quality of this data set is very high based on the maximum value composite (MVC) method, it can accurately reflect the amount and distribution of vegetation in different regions at various geographical and temporal scales; it is now widely used in monitoring regional vegetation change [29,36].

Table 1. The 18 factors considered in this study for their influence on changing vegetation dynamics.

Category	Variable	Time Series	Pixel Resolution	Units	Abbrev.
Climate	Annual average precipitation	2000s/2010s	1000 m	mm	PRE
	Annual average evaporation	2000s/2010s	1000 m	mm	EVP
	Annual average relative humidity	2000s/2010s	1000 m	%	RHU
	Annual average air temperature	2000s/2010s	1000 m	°C	TEM
	Annual average ground temperature	2000s/2010s	1000 m	°C	GST
	Annual sunshine hours	2000s/2010s	1000 m	hour	SSD
	Annual average atmospheric pressure	2000s/2010s	1000 m	hPa	PRS
Soil	Soil type	1995	1000 m	-	SOT
	Soil sand content	2000s	1000 m	%	SSAC
	Soil silt content	2000s	1000 m	%	SSIC
	Soil clay content	2000s	1000 m	%	SCLC
Vegetation	Vegetation type	2000	1000 m	-	VET
Topography	Elevation	2000	250 m	m	ELE
	Slope degree	2000	250 m	°	SLD
Human activity	Land use type	2000/2010/2020	1000 m	-	LUT
	Gross domestic product density	2000/2010/2019	1000 m	10^4 Yuan/km^2	GDP
	Population density	2000/2010/2019	1000 m	persons/km^2	POP
	Night light brightness	2000–2019	1000 m	DN	NLB

2.2.2. Influence Factor Data

Numerous studies have demonstrated that a broad range of factors influence vegetation dynamics [33,37,38]. We concentrated on five components and 18 variables related to climate, soil, vegetation, topography, and human activities (Table 1). The climatic data come from the spatial interpolation data set of the average state of meteorological elements in China [39]. The ANUSPLIN meteorological interpolation software's smoothing spline function was primarily used to obtain the climatic data, which included seven meteorological variables: annual average precipitation (PRE), annual average evaporation (EVP), annual average relative humidity (RHU), annual average air temperature (TEM), annual average ground temperature (GST), annual sunshine hours (SSD), and annual average atmospheric pressure (PRS). Data for soil types was obtained from the 1:1 million Soil Map of the People's Republic of China—created and published by the National Soil Survey Office in 1995—while that for soil sand, silt, and clay content was generated by that soil type map and soil profile information from the second soil survey. Most of the vegetation information came from the "1:1 million Vegetation Atlas of China". The elevation in the terrain data was derived from a 90-m digital elevation model (DEM) and slope data in ArcGIS 10.7 software. Human activities include land use data generated by artificial visual interpretation, this based on American Landsat TM images, and spatial distribution data sets for China's GDP [40] and population [41], which are constantly updated by data producers. All the above data are from the Resource and Environmental Science Data Center of the Chinese Academy of Sciences (https://www.resdc.cn/, accessed on 8 October 2022). Additionally, the National Qinghai-Tibet Plateau Science Data Center (https://data.tpdc.ac.cn, accessed on 19 June 2022) in China provided the long time series data for remote sensing of night light as one indicator of human activity [42].

To meet the input requirements of the geographic detector model, we divided vegetation type into seven categories: coniferous forest, broad-leaved forest, shrub, grass, meadow, cultivated vegetation, and other vegetation. Similarly, we divided land use type into nine categories: cultivated land, woods, shrubs, sparse woodland, other woodland, grassland, water, urban and rural residential land, and unused land. Six categories of soil type data were distinguished: leaching soil, primary soil, hemihydrate soil, artificial soil, iron bauxite, and non-soil type. Additionally, by applying the natural discontinuity approach [43], the remaining 15 continuous variables were classified into 10 groups. Using ArcGIS 10.7, we cast a 2-km fishnet, to extract the NDVI and the effect variables to the point, and then applied the geographic detector's calculation after removing null values.

2.3. *Methods*

2.3.1. Trend Analysis

To explore the spatial distribution characteristics of the multi-year vegetation NDVI in Chongqing, its yearly average NDVI was calculated from 2000 to 2019. We used linear regression analysis to examine the temporal trend change of NDVI in Chongqing from 2000 to 2019, using pixels as the fundamental unit. Its mathematical equation is:

$$slope = \frac{n\sum_{i=1}^{n} NDVI_i - (\sum_{i=1}^{n} i)(\sum_{i=1}^{n} NDVI_i)}{n\sum_{i=1}^{n} i^2 - (\sum_{i=1}^{n} i)^2} \tag{1}$$

where $slope$ is the magnitude and direction of vegetation change, n is the number of years of studied (n = 20 in this study), i denotes a given year from 2000 onward, and $NDVI_i$ denotes the NDVI value for ith year of a pixel.

2.3.2. Stability Analysis

Each observation's level of variation was measured and evaluated statistically using the coefficient of variation (CV). In this time series, CV can indicate the stability of the NDVI data for Chongqing: stronger stability is inferred by a smaller CV value, and weaker stability by a larger CV value [44]. The CV is calculated this way:

$$CV = \frac{\sqrt{\sum_{i=1}^{n}\left(NDVI_i - \overline{NDVI}\right)^2/(n-1)}}{\overline{NDVI}} \tag{2}$$

where $NDVI_i$ denotes the NDVI value for ith year of a pixel, and $\overline{NDVI}$ is the overall average value of NDVI for the whole study period (2000–2019). To more easily compare and convey the variation in vegetation inferred from NDVI across Chongqing, we separated the CV values into four grades, corresponding to extremely stable vegetation ($CV \leq 0.1$), general stable vegetation ($0.1 < CV \leq 0.2$), general unstable vegetation ($0.2 < CV \leq 0.3$), and extremely unstable vegetation ($CV > 0.3$).

2.3.3. Geodetector Model

Geographic detector is a relatively new spatial statistical technique developed by Wang Jinfeng and colleagues that was introduced in 2010 [30,31]. It is typically used to investigate the regional variability of vegetation change and its drivers, and to quantify how potential interactions of these factors may affect the response variables [15,37,38]. It is based on four modules: factor detector, interaction detector, risk detector, and ecological detector.

(1) Factor detector

Its purpose is to detect the spatial heterogeneity of a dependent variable, in this case vegetation NDVI, and to explore the degree to which candidate influencing factors (i.e., the 18 variables in Table 1) could explain that, this expressed as:

$$q = 1 - \frac{\sum_{h=1}^{L} N_h \sigma_h^2}{N\sigma^2} \tag{3}$$

where q can reflect the degree of interpretation of vegetation dynamics by detection factors; $h = 1, 2, \ldots, L$ is the strata of the dependent variable (vegetation NDVI) or of each influencing factor investigated; N_h and N are respectively the number of samples units in layer h and for the whole region; and σ_h^2 and σ^2 denote the variance of the layer h and the NDVI value of the whole region, respectively. The q statistic can take a value in the range of 0 to 1; the higher its value, the greater the power of its corresponding influencing factor to explain vegetation change.

(2) Interaction detector

This may be used to analyze whether the interaction of any two factors will increase or decrease their respective explanatory power for vegetation change, or whether the effects of either factor upon NDVI are independent of each other. The following five categories can be used to illustrate how the two factors could interact (Table 2).

Table 2. Interaction types of factors that affect changing vegetation dynamics.

Number	Judgments Based	Type of Interaction
1	$q(X1\cap X2) < \mathrm{Min}(q(X1), q(X2))$	Non-linear reduction
2	$\mathrm{Min}(q(X1), q(X2)) < q(X1\cap X2) < \mathrm{Max}(q(X1), q(X2))$	Single-factor non-linear reduction
3	$q(X1\cap X2) > \mathrm{Max}(q(X1), q(X2))$	Two-factor enhancement
4	$q(X1\cap X2) = q(X1) + q(X2)$	Independent
5	$q(X1\cap X2) > q(X1) + q(X2)$	Non-linear enhancement

(3) Ecological detector

This module is primarily used to assess whether there is a statistical difference between the two factors in relation to the spatial distribution of the attribute of interest, here NDVI, which is often tested using the F-ratio statistic:

$$F = \frac{N_{x1}(N_{x2} - 1)SSW_{x1}}{N_{x2}(N_{x1} - 1)SSW_{x2}} \tag{4}$$

where N_{x1} and N_{x2} are the total sample size of each of the two factors; and SSW_{x1} and SSW_{x2} denote the summed intra-layer variance formed by $x1$ and $x2$, respectively. If the null hypothesis of $H_0 : SSW_{x1} = SSW_{x2}$ is rejected at the alpha level of significance, then a significant difference between the two factors is inferred for how they influence the spatial distribution of NDVI.

(4) Risk detector

It is frequently used to determine whether there is a statistically significant difference in the mean value of attributes between the two subregions. To test this, the Student's t-test is typically used.

The Geodetector Software in Excel was used as the geographic probe in this study. It is freely available for download online. For more details about Geodetector modeling, please see http://geodetector.cn/, accessed on 10 June 2022.

3. Results

3.1. NDVI's Interannual Variation

As Figure 2 shows, Chongqing's vegetation tended to increase over time, but some regional differences at various geographical scales were evident. From 2000 to 2019, the NDVI increased strongly, at a rate of 0.05/10 year, reaching its maximum value (0.83) in 2017 and its minimum value (0.71) in 2000. Examining the interannual dynamics, we see that the rate of NDVI increase for 2011–2019 was 1.80, 1.33, and 1.43 times greater than that for the 2000–2010 period in the MCA, WMA, and TGR subregions, respectively. This revealed that vegetation restoration was considerably more effective during 2011 to 2019 than 2000 to 2010. Spatially, the NDVI increased at a faster rate in the WMA (0.07/10 year) and TGR (0.06/10 year) than in the MCA (0.03/10 year).

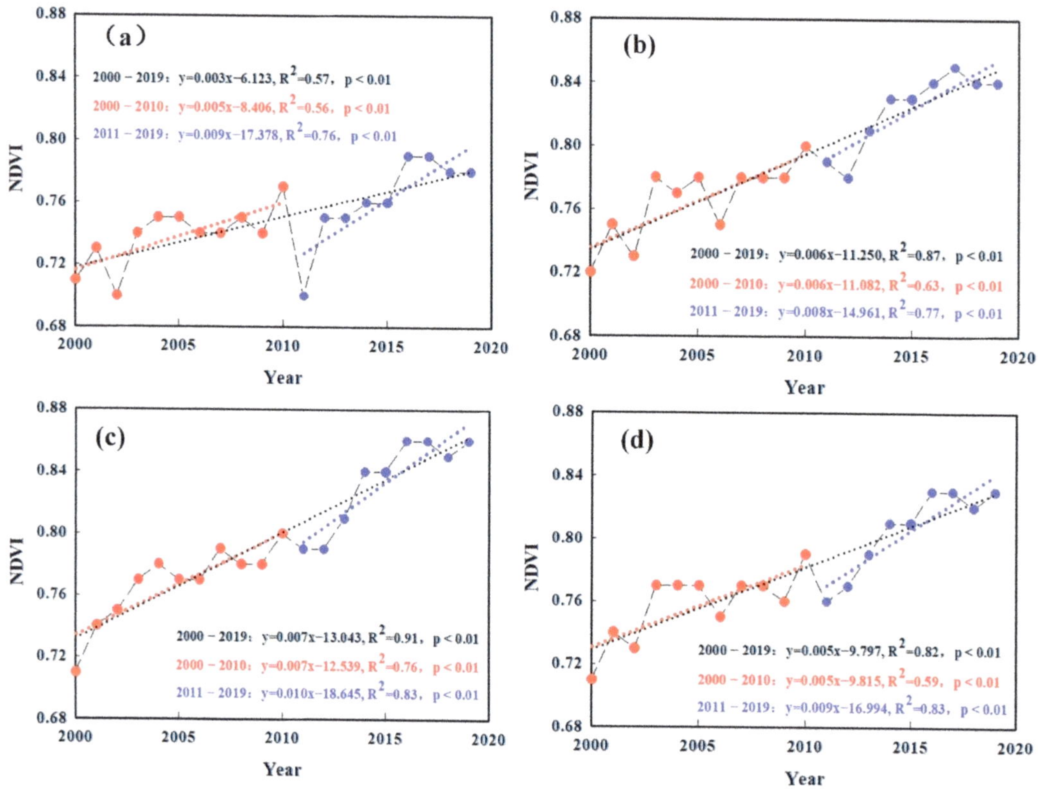

Figure 2. Interannual dynamics of vegetation NDVI in different regions of Chongqing, China, from 2000 to 2019. (**a**) The major city metropolitan area (MCA); (**b**) the Three Gorges Reservoir Area town cluster in northeast Chongqing (TGR); (**c**) the Wuling Mountain Area town cluster in southeast Chongqing (WMA); (**d**) all of Chongqing.

3.2. NDVI's Spatial Distribution

The regional distribution characteristics of NDVI in Chongqing from 2000 to 2019 are depicted in Figure 3 and Table 3. In 2000, 2010, and 2019, the values for NDVI were mainly in the range of 0.6–0.8, >0.7, and >0.7, respectively, with these respectively accounting for 92.21%, 96.91%, and 93.39% of Chongqing's total area. Only 3.35%, 0.63%, and 3.42% of the Chongqing area had NDVI values below 0.6. The percentage of its land area with an NDVI > 0.8 expanded substantially, from 4.45% in 2000 to 75.88% in 2019, a net increase of 71.43%. Chongqing's average NDVI over the entire 20-year period (2000 to 2019) was 0.78, with values primarily distributed between 0.70 and 0.80 that characterized 59.6% of its entire area. The multi-year average of NDVI in the WMA, TGR, and MCA was 0.80, 0.79, and 0.75, respectively.

The regional distribution of trends in the NDVI's change over time in Chongqing is depicted in Figure 4a and Table 4. Those areas distinguished by obvious vegetation restoration (i.e., NDVI rate of increase > 0.07/10 year) together accounted for 28.37% of Chongqing's territory, being mainly distributed in the TGR (42.8%) and WMA (31.05%). Roughly 1.49% of Chongqing's total area consisted of declining NDVI (i.e., a changing slope of less than −0.01/year), this primarily concentrated in the MCA. We found areas with an extremely significant recovery of NDVI as high as 75.19%; these were chiefly concentrated in the WMA and TGR. The parts of Chongqing featuring extremely significant and significant degradation areas, respectively, amounted to just 1.94% and 0.85% of its total area, being mainly concentrated in the MCA (Figure 4b and Table 4). Overall, 52.19% of the study area's vegetation dynamics are in an extremely stable state. Where extremely unstable and general unstable vegetation dynamics did occur, this only affected 1.38% and 3.61% of the total area, principally in the MCA and along either side of the Yangtze River (Figure 4c and Table 4).

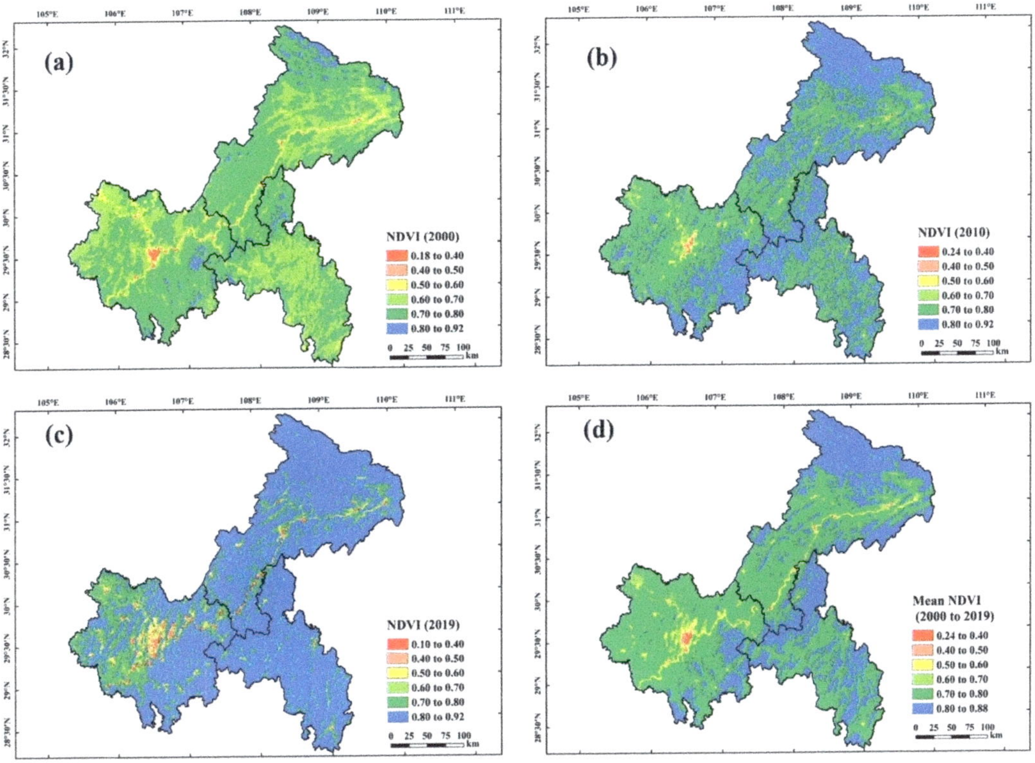

Figure 3. Spatial distribution of vegetation NDVI in Chongqing, China: (**a**) in 2000, (**b**) in 2010, (**c**) in 2019, and (**d**) for the entire 20-year study period of 2000–2019.

Table 3. Spatial distribution characteristics of vegetation NDVI in Chongqing, China, from 2000 to 2019.

Year	2000		2010		2019		2000–2019	
NDVI	Area (km²)	Proportion (%)	Area (km²)	Proportion (%)	Area (km²)	Proportion (%)	Area (km²)	Proportion (%)
<0.4	214	0.26	49	0.06	575	0.7	361	0.44
0.4–0.5	634	0.77	140	0.17	1008	1.22	374	0.45
0.5–0.6	1912	2.32	326	0.40	1239	1.5	−672	−0.82
0.6–0.7	26,547	32.22	2032	2.47	2619	3.18	−23,928	−29.04
0.7–0.8	49,430	59.99	42,332	51.37	14,431	17.51	−34,999	−42.48
>0.8	3665	4.45	37,523	45.54	62,530	75.88	58,865	71.43

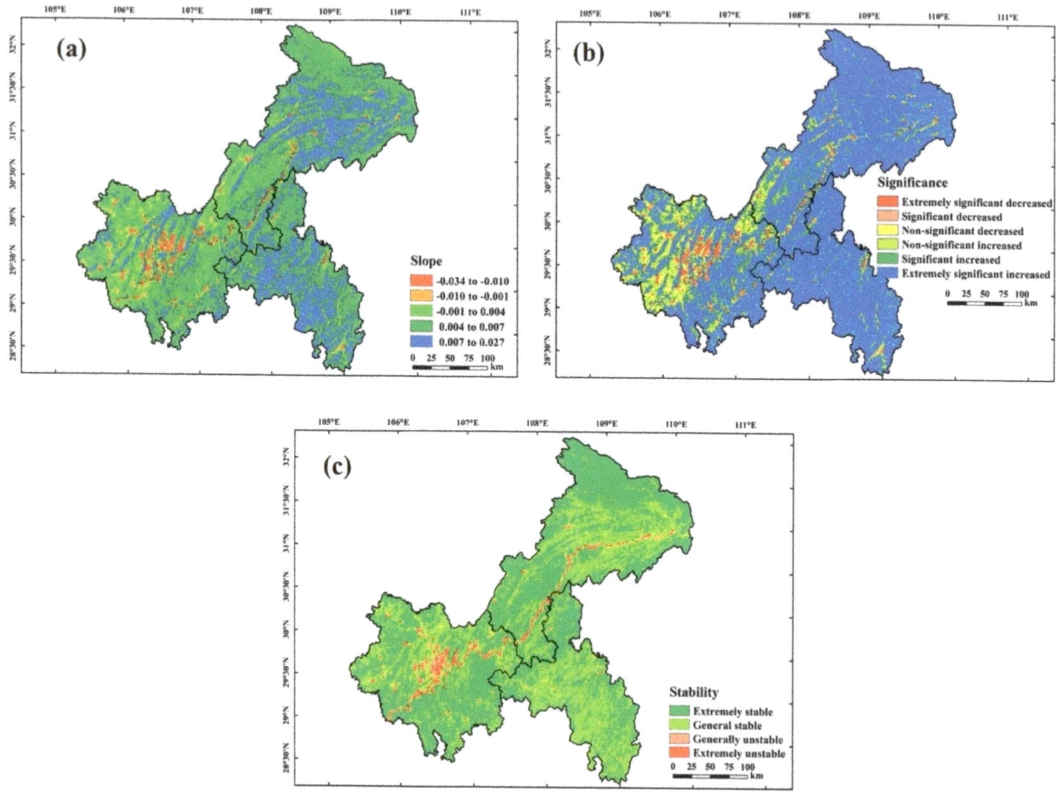

Figure 4. Spatial distributions for (**a**) slope in changing vegetation NDVI dynamics, and their (**b**) significance testing and (**c**) stability analysis, in Chongqing, China, over the entire study period of 2000 to 2019. Notes: we define the significance level as follows: $p < 0.01$ represents extremely significant; $p < 0.05$ represents significant.

Table 4. The slope, significance, and stability of vegetation NDVI in Chongqing, China, from 2000 to 2019.

Type	Range	MCA	TRG	WMA	Chongqing
Slope	−0.034 to −0.010	3.36%	0.62%	0.31%	1.49%
	−0.010 to −0.001	7.93%	2.26%	1.44%	4.03%
	−0.001 to 0.004	35.88%	16.84%	8.36%	21.40%
	0.004 to 0.007	37.67%	49.23%	47.08%	44.71%
	0.007 to 0.027	15.17%	31.05%	42.80%	28.37%
Significance	Extremely significant decrease	4.40%	0.76%	0.43%	1.94%
	Significant decrease	1.65%	0.49%	0.32%	0.85%
	Non-significant decrease	8.29%	2.37%	1.52%	4.22%
	Non-significant increase	22.26%	6.67%	3.56%	11.32%
	Significant increase	10.74%	4.80%	3.23%	6.48%
	Extremely significant increase	52.65%	84.90%	90.95%	75.19%
Stability	Extremely stable	52.54%	57.37%	42.74%	52.19%
	Generally stable	38.03%	39.12%	56.10%	42.82%
	Generally unstable	6.65%	2.59%	0.99%	3.61%
	Extremely unstable	2.78%	0.92%	0.17%	1.38%

Notes: MCA, the major city metropolitan area; TGR, the Three Gorges Reservoir Area town cluster in northeast Chongqing; WMA, the Wuling Mountain Area town cluster in southeast Chongqing. We define the significance level as follows: $p < 0.01$ represents extremely significant; $p < 0.05$ represents significant.

3.3. Single-factor driven analysis

By using the factor detection module, each factor's q statistic was generated to uncover its relative impact on changing vegetation dynamics (Table 5). These results revealed differential impacts of numerous factors among Chongqing as a whole and its three sub-regions, MCA, WMA, and TGR. In the MCA, vegetation change was most influenced by night light brightness (NLB, 0.406), population density (POP, 0.302), atmospheric pressure (PRS, 0.263), and elevation (ELE, 0.258); accordingly, this implied it was mainly affected by human activities and topography. In the TGR, vegetation change was best explained by air temperature (TEM, 0.544), atmospheric pressure (PRS, 0.536), ground temperature (GST, 0.529), and elevation (ELE, 0.511), suggesting it was mainly affected by climate and topography. In the WMA, vegetation change was mainly affected by air temperature (TEM, 0.330), ground temperature (GST, 0.330), PRS (atmospheric pressure, 0.328), and relative humidity (RHU, 0.308), indicating climate was largely responsible.

Table 5. The q statistics value of driving factors of changing vegetation NDVI dynamics in Chongqing, China, at different scales.

Category	Variable	Abbrev.	MCA	TRG	WMA	Chongqing
Climate	Annual average precipitation	PRE	0.121 **	0.363 **	0.102	0.303 **
	Annual average evaporation	EVP	0.104 **	0.453 **	0.220 **	0.297 **
	Annual average relative humidity	RHU	0.188 **	0.171 **	0.308 **	0.168 **
	Annual average air temperature	TEM	0.258 **	0.544 **	0.33 **	0.470 **
	Annual average ground temperature	GST	0.243 **	0.529 **	0.33 **	0.457 **
	Annual sunshine hours	SSD	0.013	0.255 **	0.121 **	0.150 **
	Annual average atmospheric pressure	PRS	0.263 **	0.536 **	0.328 **	0.458 **
Soil	Soil type	SOT	0.089 **	0.260 **	0.090	0.227 **
	Soil sand content	SSAC	0.053 **	0.200 **	0.043	0.152 **
	Soil silt content	SSIC	0.092 **	0.099 **	0.084	0.092 **
	Soil clay content	SCLC	0.071	0.163 **	0.101	0.153 **
Vegetation	Vegetation type	VET	0.059 *	0.109 **	0.061	0.148 **
Topography	Elevation	ELE	0.258 **	0.511 **	0.287 **	0.444 **
	Slope degree	SLD	0.148 **	0.135 **	0.069	0.214 **
Human activity	Land use type	LUT	0.215 **	0.206 **	0.101	0.234 **
	Gross domestic product	GDP	0.186 **	0.096	0.092	0.197 **
	Population density	POP	0.302 **	0.296 **	0.270 *	0.370 **
	Night light brightness	NLB	0.406 **	0.187 **	0.139	0.519 **

Notes: * and ** indicate significant coefficients at $p < 0.05$ and $p < 0.01$, respectively. MCA, the major city metropolitan area; TGR, the Three Gorges Reservoir Area town cluster in northeast Chongqing; WMA, the Wuling Mountain Area town cluster in southeast Chongqing.

For Chongqing's territory, each factor's level (q value) of influence upon the NDVI weakened in this descending rank order: NLB (0.519), TEM (0.470), PRS (0.458), GST (0.457), ELE (0.444), POP (0.370), PRE (0.303), EVP (0.297), land use type (LUT, 0.234), soil type (SOT, 0.227), slope degree (SLD, 0.214), GDP (0.197), RHU (0.168), soil clay content (SCLC, 0.153), soil sand content (SSAC, 0.152), SSD (0.150), vegetation type (VET, 0.148), and soil silt content (SSIC, 0.092). Evidently, a mix of human activities, climate, and topography were the key factor variables that drove the changing vegetation dynamics of Chongqing, whereas the influence of soil and vegetation factors was relatively weak.

3.4. Two-Factor Driven Analysis

By using the interaction detector module, it was possible to calculate how all paired variables could affect changing vegetation dynamics (Table 6). We discovered that the factors influencing vegetation in Chongqing interacted in three different ways: via single-factor nonlinear weakening, nonlinear enhancement, and two-factor enhancement. Among all pairwise interactions, 159 pairs (92.9%) showed two-factor enhancement, indicating this form predominantly drove spatio-temporal changes in vegetation in a complex manner, being affected by the interaction of many factors.

Table 6. Interaction detector results for 18 influencing factors (variables) of changing vegetation NDVI dynamics in Chongqing, China.

Factors	PRE	EVP	RHU	TEM	GST	SSD	PRS	SOT	SSAC	SSIC	SCLC	VET	ELE	SLD	LUT	GDP	POP	NLB
PRE	0.303																	
EVP	0.400	0.297																
RHU	0.403	0.412	0.168															
TEM	0.492	0.490	0.488	0.470														
GST	0.478	0.483	0.480	0.472	0.457													
SSD	0.367	0.381	0.447	0.484	0.473	0.150												
PRS	0.489	0.484	0.478	0.477	0.472	0.476	0.458											
SOT	0.365	0.382	0.322	0.478	0.464	0.292	0.472	0.227										
SSAC	0.353	0.350	0.285	0.480	0.469	0.247	0.470	0.259	0.152									
SSIC	0.355	0.359	0.256	0.486	0.474	0.261	0.473	0.275	0.232	0.091								
SCLC	0.348	0.351	0.282	0.483	0.469	0.260	0.470	0.249	0.232	0.180	0.153							
VET	0.342	0.364	0.265	0.476	0.463	0.277	0.471	0.277	0.229	0.224	0.233	0.148						
ELE	0.478	0.473	0.471	0.486	0.479	0.463	0.475	0.460	0.458	0.463	0.459	0.459	0.444					
SLD	0.387	0.400	0.366	0.491	0.479	0.288	0.481	0.316	0.273	0.273	0.280	0.284	0.472	0.214				
LUT	0.430	0.441	0.369	0.545	0.532	0.347	0.532	0.362	0.323	0.317	0.320	0.306	0.526	0.351	0.234			
GDP	0.424	0.419	0.337	0.551	0.539	0.310	0.541	0.357	0.320	0.271	0.314	0.296	0.529	0.340	0.333	0.197		
POP	0.496	0.498	0.447	0.558	0.546	0.444	0.549	0.423	0.412	0.409	0.404	0.395	0.542	0.414	0.416	0.408	0.370	
NLB	0.551	0.536	0.473	0.627	0.619	0.452	0.627	0.495	0.463	0.429	0.458	0.434	0.619	0.484	0.478	0.419	0.530	0.519

Notes: Blue represents a two-factor enhancement, green represents a non-linear enhancement, and orange represents a single-factor non-linear reduction. More information can be found in Table 2. For details about the factor abbreviations, please see Table 1.

The average value of each interacting factor was next examined. The factor's level (q value) of influence on the NDVI weakened in this descending rank order: ELE (0.522), NLB (0.519), TEM (0.504), PRS (0.502), GST (0.495), POP (0.450), EVP (0.419), PRE (0.415), RHU (0.371), LUT (0.365), SLD (0.361), GDP (0.341), SOT (0.336), VET (0.332), SSD (0.328), SCLC (0.328), SSAC (0.309), and SSIC (0.295). This demonstrated that although soil type and vegetation type can exert some influence, it was still minor compared to human activities, climatic variables, and topographic conditions. Within these similar categories, the strongest prevailing interactions were found for the paired variables: POP ∩ NLB (0.530), TEM ∩ PRE (0.492), SLD ∩ ELE (0.472). Overall, however, between differing types of factors, the strongest dominant interaction factors were the TEM ∩ NLB (0.627), PRS ∩ NLB (0.627), and ELE ∩ NLB (0.619). We found that interactions between differing types of factors were stronger than those arising between similar ones.

3.5. Ecological Detector Analysis

Whether the effects of interactions between two factors on vegetation NDVI differ significantly can be evaluated using the ecological detector module. As seen in Table 7, there were significant differences ($p < 0.05$) in the explanatory power of nearly half (46.4%) of the factor combinations for NDVI. The following scenarios exhibited notable variation in how two factors affected the geographical differentiation of changing vegetation NDVI dynamics in Chongqing: among all climatic variables, TEM ∩ factors (PRE, EVP, RHU), GST ∩ factors (PRE, EVP, RHU), PRS ∩ factors (PRE, EVP, RHU, SSD); among all soil variables, SOT ∩ factors (RHU, SSD), SCLC ∩ SSIC; in the vegetation variables, VET and SSIC; among all human activities variables, LUT ∩ factors (RHU, SSD, SSAC, SSIC, SCLC, VET, SLD), GDP ∩ factors (RHU, SSD, SSAC, SSIC, SCLC, VET), POP ∩ factors (PRE, EVP, RHU, SSD, SOT, SSAC, SSIC, SCLC, VET, SLD, LUT, GDP), NLB and all factors. Additionally, there was no discernible difference between the impacts of the other interactions between two factors on the NDVI's regional differentiation across Chongqing.

Table 7. Statistical tests of 18 influencing factors of changing vegetation NDVI dynamics in Chongqing, China, based on the ecological detector module (significant at $p < 0.05$).

Factors	PRE	EVP	RHU	TEM	GST	SSD	PRS	SOT	SSAC	SSIC	SCLC	VET	ELE	SLD	LUT	GDP	POP	NLB
PRE																		
EVP	N																	
RHU	N	N																
TEM	Y	Y	Y															
GST	Y	Y	Y	N														
SSD	N	N	N	N	N													
PRS	Y	Y	Y	N	N	Y												
SOT	N	N	Y	N	N	Y	N											
SSAC	N	N	N	N	N	N	N	N										
SSIC	N	N	N	N	N	N	N	N	N									
SCLC	N	N	N	N	N	N	N	N	N	Y								
VET	N	N	N	N	N	N	N	N	N	Y	N							
ELE	Y	Y	Y	N	N	Y	N	Y	Y	Y	Y	Y						
SLD	N	N	Y	N	N	Y	N	N	Y	Y	Y	Y	N					
LUT	N	N	Y	N	N	Y	N	N	Y	Y	Y	Y	N	Y				
GDP	N	N	Y	N	N	Y	N	N	Y	Y	Y	Y	N	N	N			
POP	Y	Y	Y	N	N	Y	N	Y	Y	Y	Y	Y	N	Y	Y	Y		
NLB	Y	Y	Y	Y	Y	Y	Y	Y	Y	Y	Y	Y	Y	Y	Y	Y	Y	

Notes: The 'Y' indicates significant difference in the effects of the two factors on changed vegetation NDVI, whereas the 'N' denotes no significant difference detected. For information about the factor abbreviations, please refer to Table 1.

3.6. Types or Range of Suitable Influencing Factors

We presumed that the factor of type or range with a higher NDVI would be better suited for vegetation growth when the risk detector assesses how vegetation changes in response to various factors. As seen in Table 8, in terms of meteorological conditions, PRE, RHU, and SSD tended to increase as the interval increased, whose most suitable ranges were 1538~1682 mm, 81.9%~84.5% and 1526~1646 h, respectively. Conversely, EVP, TEM, GST, and PRS tended to decrease as the interval increased, for which the most suitable range was 640~715mm, 4.7~7.8 °C, 7.3~10.4 °C, 742~802 hPa.

Table 8. The suitable range or type of 18 factors influencing the changing vegetation NDVI dynamics in Chongqing, China.

Category	Variable	Abbrev.	Units	Suitable Range or Type	NDVI Mean
Climate	Annual average precipitation	PRE	mm	1538 to 1682	0.851
	Annual average evaporation	EVP	mm	640 to 715	0.847
	Annual average relative humidity	RHU	%	81.9 to 84.5	0.833
	Annual average air temperature	TEM	°C	4.7 to 7.8	0.856
	Annual average ground temperature	GST	°C	7.3 to 10.4	0.855
	Annual sunshine hours	SSD	hour	1526 to 1646	0.855
	Annual average atmospheric pressure	PRS	hPa	742 to 802	0.854
Soil	Soil type	SOT	-	Semi-leached soil	0.842
	Soil sand content	SSAC	%	33 to 34	0.85
	Soil silt content	SSIC	%	38 to 42	0.835
	Soil clay content	SCLC	%	12 to 16	0.84
Vegetation	Vegetation type	VET	-	Broad-leaved forest	0.828
Topography	Elevation	ELE	m	2000 to 2624	0.854
	Slope degree	SLD	°	39.2 to 56.2	0.824
Human activity	Land use type	LUT	-	Woodland	0.808
	Gross domestic product density	GDP	10^4 Yuan/km^2	0 to 1954	0.788
	Population density	POP	persons/km^2	0 to 143	0.809
	Night light brightness	NLB	DN	0 to 1.6	0.793

In terms of soil conditions, the most suitable SOT was semi-leached soil, and the most suitable ranges for SSAC, SSIC and SCLC were 33%~34%, 38%~42%, and 12%~16%, respectively. In terms of vegetation types, it was most suitable to grow broad-leaved forest. Regarding topography, across Chongqing, NDVI increased with the SLD and ELE, these being most suitable in the range of 39.2~56.2° and 2000~2624 m, respectively.

In terms of human activities, woodlands were the most conducive land use type for vegetation growth, and the NDVI was highest in areas with low GDP, POP, and NLB, meaning that these were most suitable when in the range 0~1954 $\times$ 10^4 Yuan/km^2, 0~143 person/km^2, and 0~1.6 DN.

4. Discussion

4.1. NDVI's Spatio-Temporal Changes

The results of this study demonstrated an upward trend in the vegetation NDVI in Chongqing between 2000 and 2019 (Figure 2), with its vegetation conditions found greatly improved at various temporal and geographical scales (Figure 4). These recovery areas, mainly situated in the southeast and northeast parts of Chongqing, together expanded to 75.19% of its territory from 2000 to 2019, a result consistent with the findings of Xiao et al. [45] and Zhang et al. [23]. This may be attributed to the ecological conservation projects implemented by the government. For example, based on Landsat and MODIS data, Li et al. [46] showed that ecological engineering measures in the Three Gorges Reservoir area, such as the ecological migration project, the ecological protection and restoration project, and Grain for Green, played a positive role in ecological restoration and effectively improved local vegetation coverage. Li Z. and Li X. [47] reported that human activities, such as agricultural production, cultivated land protection, and vegetation ecological construction, were the primary factors responsible for vegetation growth and expansion in Chongqing. Work by Liu et al. [48] quantified the relative contribution rates of human activities and climate to vegetation change in Chongqing as 90.96% and 9.04%, respectively, revealing the overwhelmingly dominant role of human activities. Those areas with significant degradation and instability of vegetation NDVI are mainly concentrated in the major metropolitan area, near water, and some surrounding areas (Figure 4), a pattern basically consistent with the findings of Zhu et al. [49]. According to Li et al., the expansion of human urban construction land across the world's cities is the main reason for the downward trend of regional vegetation NDVI, and the process of urbanization is directly and indirectly having adverse impacts on global urban vegetation growth [50,51].

4.2. NDVI Response to Driving Factors

Previous research has demonstrated that both natural and human activities can distort and modulate the temporal and spatial variation in vegetation dynamics [15,38], making it difficult to investigate the mechanisms underlying the spatial heterogeneity of vegetation [34]. We found that the interaction of most dual-factor variables increased their degree of influence on vegetation change, the latter often affected by the interaction of multiple factors (Table 6). In short, the interaction of two factors is more important to vegetation change than each factor alone, an outcome consistent with previous studies [15,38,52]. Therefore, when considering the change of NDVI, we need to fully consider the interaction between factors. For example, we found that the suitable range of annual mean temperature for vegetation growth in the Chongqing area is 4.7–7.8 °C (Table 8), which was located at higher altitudes (Figure A1, TEM and ELE). Therefore, it is actually a trade-off considering multiple factors, rather than the most suitable temperature for plant physiology. This may be closely related to our hypothesis that the range of factors affecting the maximum NDVI value was the range of the most suitable growth factors.

Our results suggest that the effects of human activity on vegetation change should not be disregarded (Table 5). The explanatory power of nocturnal light brightness (NLB) for vegetation change reached as high as 51.9% (Table 5), confirming that human activities heavily impact changing vegetation dynamics in Chongqing and are of paramount concern.

These findings are in line with those of Liu et al. [48], who estimated that human activities contributed as much as 90.96% to vegetation change in Chongqing. In general, there is a positive correlation between NLB and socio-economic factors, meaning that NLB can effectively express the intensity of human activities such as urbanization level, population density, and GDP [53,54]. Indeed, human driving factors, both population density and GDP, often emerge as the dominant ones affecting regional vegetation change. For example, Sun et al. [55] found that agricultural vegetation NDVI is very sensitive to economic as well as population growth, which may lead to changes in vegetation NDVI in Chongqing given its extensive distribution of cultivated land (Figure 2). Herrero et al. [56] reported a significant negative correlation between population density and NDVI around Southern African national parks during the 21st century (2000–2016). We found that NDVI tends to be augmented in woodland, and by a lower GDP, POP, and NLB (Table 8). This may be due to sparse and small populations and small-scale economies in certain areas, which are less apt to incur damage to vegetation from humans. This suggests trade-offs likely loom between future ecological and economic development, but devising sustainable human interventions may contribute to promoting vegetation recovery and diversity, thereby restoring the ecological balance in the study area.

Climatic factors are generally considered critical to the growth and distribution of vegetation [57,58]. Among these, we found that the explanatory power of annual mean temperature, annual mean pressure, annual mean ground temperature and annual mean precipitation for vegetation change weakened in that order (Table 5). Hence, the influence of air temperature on changing vegetation dynamics in the studied region was greater than that of precipitation. These results are consistent with those of Zhang et al. [59] and Liu et al. [60], and can be explained by Chongqing's location in the upper reaches of the Yangtze River and the central zone of the Three Gorges Reservoir Area. The water needed for vegetation growth here is sufficient, so temperature probably becomes a more pertinent factor than precipitation in modulating vegetation growth and dynamics. Under the premise of sufficient rainfall, a rising temperature can enhance plant photosynthesis, which should favor the growth of most plant species. However, the influence of climatic factors on the changing dynamics of vegetation growth often harbors a threshold effect [61–63]. For example, in Chongqing, the area with sufficient precipitation and annual sunshine duration will most favor the growth of its vegetation, while the area with higher temperature and increased evaporation is more likely to limit that growth in vegetation (Table 8). This may be attributed to the humid subtropical monsoon climate of Chongqing, which has hydrothermal conditions suitable for growing vegetation. When at a low level, temperature often becomes a limiting factor for the plant's physiological processes; hence, an appropriate temperature rise can promote photosynthesis and accelerate the absorption of soil nutrients, thus promoting the growth of vegetation [64]. However, once temperatures exceed the tolerable range of plant species, extreme heat increases transpiration and respiration rates, which accelerate dry matter consumption and soil water losses, leading to reduced photosynthesis and nutrient uptake and transport, which is clearly detrimental to vegetation growth [65].

In terms of topography, with an increase in elevation or slope, the vegetation NDVI in Chongqing gradually increased as well in tandem (Table 8), a trend consistent with the study by Zhu et al. [66], who found that Chongqing had a high vegetation coverage in those areas at high elevations (>1200 m) and with steeper slopes (>15°). Our study showed that ELE explained 44.4% of the vegetation change, likely because it determines the flow and stability of surface materials, modulates the spatial distribution of air temperature and water, and alters vegetation dynamics via temperature, precipitation, soil moisture, soil nutrients, and other factors [67–69]. In fact, temperature tends to have a greater effect on vegetation growth at higher elevations than at lower elevations. For example, Pan et al. [70] used Landsat NDVI and climate data from 1992 to 2020 to explore the impact of topography on vegetation change on the Qinghai-Tibet Plateau. Their results showed that precipitation had a greater impact than air temperature upon vegetation growth in the region lying below 3000 m, and vice versa in the region above 3000 m. Thus, in high-elevation areas, temperature may be the main factor limiting the growth of vegetation [71]; low temperatures often limit the growth of plants by reducing their photosynthesis, soil nutrient absorption rate, and delaying key phenological events, among other impacts [72]. In addition, with rising elevation, the corresponding reduced water availability may also limit vegetation recruitment and growth.

4.3. Caveat and Future Work

Vegetation dynamics are closely related to a variety of factors [37,38,70]. Although this work fully considered the impact of 18 influencing factors, including climate, soil, vegetation, topography, and human activities, upon vegetation change, which helps to further improve our understanding of its driving mechanism, some limitations and uncertainties persist. In terms of method, these are as follows. (1) We found that elucidating the driving mechanisms of vegetation dynamics depends on spatial scale, so we need to consider further the main vegetation drivers involved at multi-scale spatial scales to further reduce the uncertainty concerning how they impact vegetation dynamics. (2) Although Geodetector has realized the measurement, significance test, and attribute analysis of spatial differentiation, it also has limitations in discussing the interactions with temporal vegetation dynamics. (3) The most significant of these is that it cannot simultaneously evaluate the joint influence of multiple variables on changing dynamics of vegetation. Therefore, in future work, we plan to explore the nonlinear driving mechanism of multiple factors on vegetation dynamics. In terms of method, among the 18 variables, the non-time variable data (vegetation type, soil type, soil sand content, soil silt content, soil clay content, elevation, slope degree, etc.) is often difficult to change in a certain period of time, which makes it difficult to understand the driving mechanism of NDVI change from the perspective of time change.

5. Conclusions

This study illustrated the dynamic trends in NDVI's temporal and geographical variability in Chongqing from 2000 through 2019. We discovered that whereas the majority of Chongqing's vegetation recovery area—75.19%—was located in the WMA and TRG, the majority area of the vegetation degradation and lower stability was located within the MCA. As a result, in the future, we need to concentrate on and increase vegetation management and restoration in the MCA. The influencing factors associated with human activities, climate, and topography upon changing vegetation dynamics cannot be ignored. Among all 18 factors considered, NLB (51.9%), TEM (47%), PRS (45.8%), GST (45.7%), ELE (44.4%), POP (37%), and PRE (30.3%) were the main single factors affecting vegetation change, and the relative impacts on vegetation change gradually lessened. We discovered that it was most often (92.9% of all cases) achieved by synergetic interactions between factors (two-factor enhancement)—that is, the combination of two factors has a greater impact on vegetation change than either single component has, and the interaction of differing factors has a greater impact than that of similar factors. For Chongqing, we were able to discern the range of favorable meteorological conditions, adequate precipitation, and yearly sunshine hours that promote vegetation growth there, whereas increased evaporation and rising temperature were more likely to hinder it. In terms of terrain, the Chongqing area's NDVI steadily rises with increasing elevation and slope. In terms of human activity, those areas in the woods and with lower GDP, POP, and NLB were more favorable for sustaining vegetation growth and dynamics. These results could serve as a foundation for improving the management and regeneration of vegetation in the upper parts of the Yangtze River Basin.

Author Contributions: Conceptualization: X.G., B.Y. and S.L. Methodology: X.G., B.Y. and L.M. Software: X.G., X.W. and B.Y. Formal analysis, X.G., B.Y., L.M., X.W. and S.L. Resources: L.M. and H.S. Data curation: X.G. Writing—original draft preparation: B.Y., L.M., X.G. and X.W. Writing—review and editing: X.G., B.Y. and X.W. Visualization: X.W. Supervision: X.G. and X.W. All authors have read and agreed to the published version of the manuscript.

Funding: This research was funded by the Open Subject Funding of Observation and Research Station of Ecological Restoration for Chongqing Typical Mining Areas, Ministry of Natural Resources (grant number CQORS-2021-01), and by a project supported by Chongqing Natural Science Foundation (grant number CSTB2022NSCQ-MSX0233).

Data Availability Statement: Not applicable.

Acknowledgments: We are grateful for the data sets shared with us by the Resource and Environmental Science Data Center of the Chinese Academy of Sciences (https://www.resdc.cn/, accessed on 20 November 2022) and the Qinghai-Tibet Plateau National Scientific Data Center (https://data.tpdc.ac.cn/zh, accessed on 20 November 2022). The authors also express their gratitude to the editors and reviewers for their time and efforts.

Conflicts of Interest: The authors declare no conflict of interest.

Appendix A

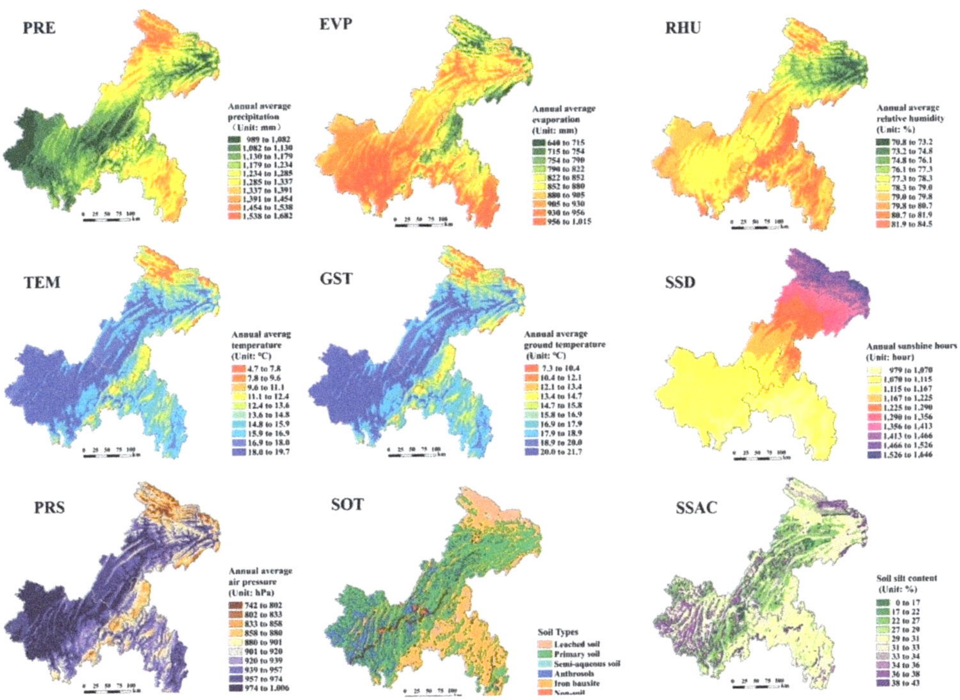

Figure A1. *Cont.*

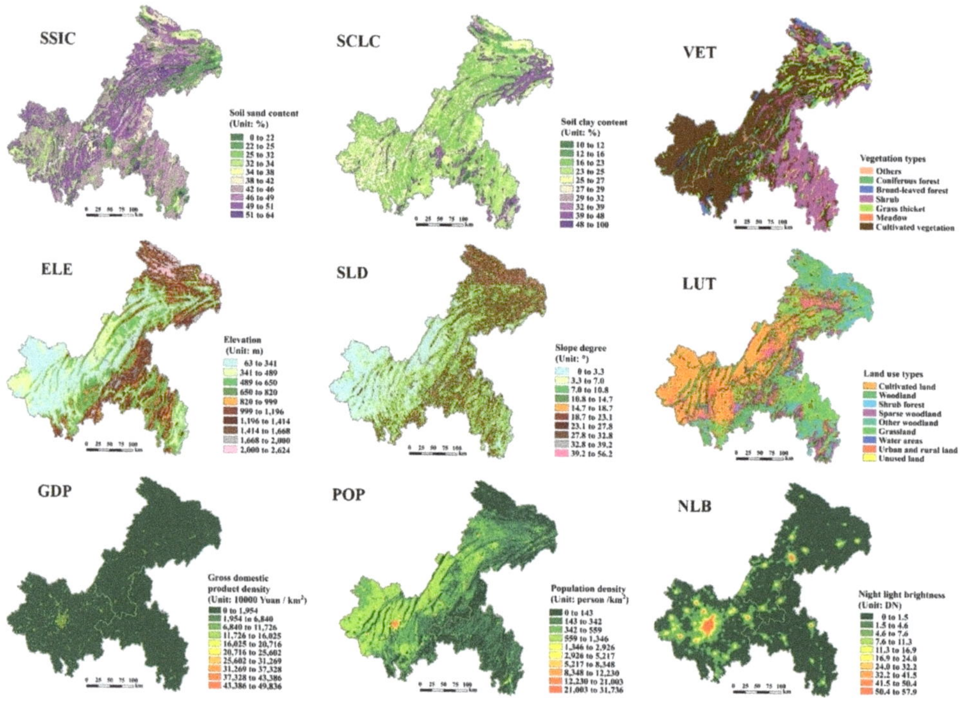

Figure A1. Spatial distributions of all 18 influencing factors (variables) in Chongqing, China. Their abbreviations are detailed in Table 1.

References

1. Deng, Z.; Guan, H.; Hutson, J.; Forster, M.A.; Wang, Y.; Simmons, C.T. A vegetation-focused soil-plant-atmospheric continuum model to study hydrodynamic soil-plant water relations. *Water Resour. Res.* **2017**, *53*, 4965–4983. [CrossRef]
2. Huang, K.; Xia, J.; Wang, Y.; Ahlström, A.; Chen, J.; Cook, R.B.; Cui, E.; Fang, Y.; Fisher, J.B.; Huntzinger, D.N. Enhanced peak growth of global vegetation and its key mechanisms. *Nat. Ecol. Evol.* **2018**, *2*, 1897–1905. [CrossRef]
3. Norton, A.J.; Rayner, P.J.; Wang, Y.P.; Parazoo, N.C.; Baskaran, L.; Briggs, P.R.; Haverd, V.; Doughty, R. Hydrologic connectivity drives extremes and high variability in vegetation productivity across Australian arid and semi-arid ecosystems. *Remote Sens. Environ.* **2022**, *272*, 112937. [CrossRef]
4. Miguez-Macho, G.; Fan, Y. Spatiotemporal origin of soil water taken up by vegetation. *Nature* **2021**, *598*, 624–628. [CrossRef]
5. Seddon, A.W.; Macias-Fauria, M.; Long, P.R.; Benz, D.; Willis, K.J. Sensitivity of global terrestrial ecosystems to climate variability. *Nature* **2016**, *531*, 229–232. [CrossRef]
6. Wang, Y.; Lv, W.; Xue, K.; Wang, S.; Zhang, L.; Hu, R.; Zeng, H.; Xu, X.; Li, Y.; Jiang, L. Grassland changes and adaptive management on the Qinghai–Tibetan Plateau. *Nat. Rev. Earth Environ.* **2022**, *3*, 668–684. [CrossRef]
7. Overpeck, J.T.; Breshears, D.D. The growing challenge of vegetation change. *Science* **2021**, *372*, 786–787. [CrossRef] [PubMed]
8. Zhang, Y.; Migliavacca, M.; Penuelas, J.; Ju, W. Advances in hyperspectral remote sensing of vegetation traits and functions. *Remote Sens. Environ.* **2021**, *252*, 112121. [CrossRef]
9. Schultz, M.; Clevers, J.G.; Carter, S.; Verbesselt, J.; Avitabile, V.; Quang, H.V.; Herold, M. Performance of vegetation indices from Landsat time series in deforestation monitoring. *Int. J. Appl. Earth Obs. Geoinf.* **2016**, *52*, 318–327. [CrossRef]
10. Brando, P.M.; Goetz, S.J.; Baccini, A.; Nepstad, D.C.; Beck, P.S.; Christman, M.C. Seasonal and interannual variability of climate and vegetation indices across the Amazon. *Proc. Natl. Acad. Sci. USA* **2010**, *107*, 14685–14690. [CrossRef]
11. Li, S.; Xu, L.; Jing, Y.; Yin, H.; Li, X.; Guan, X. High-quality vegetation index product generation: A review of NDVI time series reconstruction techniques. *Int. J. Appl. Earth Obs. Geoinf.* **2021**, *105*, 102640. [CrossRef]
12. Jarlan, L.; Mangiarotti, S.; Mougin, E.; Mazzega, P.; Hiernaux, P.; Le Dantec, V. Assimilation of SPOT/VEGETATION NDVI data into a sahelian vegetation dynamics model. *Remote Sens. Environ.* **2008**, *112*, 1381–1394. [CrossRef]
13. Xu, Y.; Zheng, Z.; Guo, Z.; Dou, S.; Huang, W. Dynamic variation in vegetation cover and its influencing factor detection in the Yangtze River Basin from 2000 to 2020. *Environ. Sci.* **2022**, *43*, 3730–3740. (In Chinese)
14. Pan, N.; Feng, X.; Fu, B.; Wang, S.; Ji, F.; Pan, S. Increasing global vegetation browning hidden in overall vegetation greening: Insights from time-varying trends. *Remote Sens. Environ.* **2018**, *214*, 59–72. [CrossRef]

15. Peng, W.; Kuang, T.; Tao, S. Quantifying influences of natural factors on vegetation NDVI changes based on geographical detector in Sichuan, western China. *J. Clean. Prod.* **2019**, *233*, 353–367. [CrossRef]
16. Gong, X.; Li, Y.; Wang, X.; Zhang, Z.; Lian, J.; Ma, L.; Chen, Y.; Li, M.; Si, H.; Cao, W. Quantitative assessment of the contributions of climate change and human activities on vegetation degradation and restoration in typical ecologically fragile areas of China. *Ecol. Indic.* **2022**, *144*, 109536. [CrossRef]
17. Na, L.; Na, R.; Zhang, J.; Tong, S.; Shan, Y.; Ying, H.; Li, X.; Bao, Y. Vegetation dynamics and diverse responses to extreme climate events in different vegetation types of inner mongolia. *Atmosphere* **2018**, *9*, 394. [CrossRef]
18. Shuai, Y.; Tian, Y.; Shao, C.; Huang, J.; Gu, L.; Zhang, Q.; Zhao, R. Potential variation of evapotranspiration induced by typical vegetation changes in Northwest China. *Land* **2022**, *11*, 808. [CrossRef]
19. Zhang, H.; Chang, J.; Zhang, L.; Wang, Y.; Li, Y.; Wang, X. NDVI dynamic changes and their relationship with meteorological factors and soil moisture. *Environ. Earth. Sci.* **2018**, *77*, 582. [CrossRef]
20. Qu, S.; Wang, L.; Lin, A.; Zhu, H.; Yuan, M. What drives the vegetation restoration in Yangtze River basin, China: Climate change or anthropogenic factors? *Ecol. Indic.* **2018**, *90*, 438–450. [CrossRef]
21. Zhang, W.; Wang, L.; Xiang, F.; Qin, W.; Jiang, W. Vegetation dynamics and the relations with climate change at multiple time scales in the Yangtze River and Yellow River Basin, China. *Ecol. Indic.* **2020**, *110*, 105892. [CrossRef]
22. Chen, S.; Wen, Z.; Zhang, S.; Huang, P.; Ma, M.; Zhou, X.; Liao, T.; Wu, S. Effects of long-term and large-scale ecology projects on forest dynamics in Yangtze River Basin, China. *Forest Ecol. Manag.* **2021**, *496*, 119463. [CrossRef]
23. Zhang, Y.; He, Y.; Li, Y.; Jia, L. Spatiotemporal variation and driving forces of NDVI from 1982 to 2015 in the Qinba Mountains, China. *Environ. Sci. Pollut. Res.* **2022**, *29*, 52277–52288. [CrossRef] [PubMed]
24. Wang, C.; Wang, J.; Naudiyal, N.; Wu, N.; Cui, X.; Wei, Y.; Chen, Q. Multiple effects of topographic factors on Spatio-temporal variations of vegetation patterns in the three parallel rivers region, Southeast Qinghai-Tibet Plateau. *Remote Sens.* **2021**, *14*, 151. [CrossRef]
25. Yang, X.; Yang, Q.; Yang, M. Spatio-temporal patterns and driving factors of vegetation change in the Pan-Third Pole Region. *Remote Sens.* **2022**, *14*, 4402. [CrossRef]
26. Cui, L.; Wang, L.; Singh, R.P.; Lai, Z.; Jiang, L.; Yao, R. Association analysis between spatiotemporal variation of vegetation greenness and precipitation/temperature in the Yangtze River Basin (China). *Environ. Sci. Pollut. Res.* **2018**, *25*, 21867–21878. [CrossRef]
27. Yang, L.; Guan, Q.; Lin, J.; Tian, J.; Tan, Z.; Li, H. Evolution of NDVI secular trends and responses to climate change: A perspective from nonlinearity and nonstationarity characteristics. *Remote Sens. Environ.* **2021**, *254*, 112247. [CrossRef]
28. Ji, G.; Song, H.; Wei, H.; Wu, L. Attribution analysis of climate and anthropic factors on runoff and vegetation changes in the source area of the Yangtze River from 1982 to 2016. *Land* **2021**, *10*, 612. [CrossRef]
29. Zhu, L.; Meng, J.; Zhu, L. Applying Geodetector to disentangle the contributions of natural and anthropogenic factors to NDVI variations in the middle reaches of the Heihe River Basin. *Ecol. Indic.* **2020**, *117*, 106545. [CrossRef]
30. Wang, J.F.; Xu, C.D. Geodetector: Principle and prospective. *Acta Georaphica Sin.* **2017**, *72*, 116–134. (In Chinese)
31. Wang, J.F.; Li, X.H.; Christakos, G.; Liao, Y.L.; Zhang, T.; Gu, X.; Zheng, X.Y. Geographical detectors-based health risk assessment and its application in the neural tube defects study of the Heshun Region, China. *Int. J. Geogr. Inf. Sci.* **2010**, *24*, 107–127. [CrossRef]
32. Yang, M.; Zhao, W.; Zhan, Q.; Xiong, D. Spatiotemporal patterns of land surface temperature change in the tibetan plateau based on MODIS/Terra daily product from 2000 to 2018. *IEEE J. Sel. Top. Appl. Earth Obs. Remote Sens.* **2021**, *14*, 6501–6514. [CrossRef]
33. Meng, F.; Luo, M.; Sa, C.; Wang, M.; Bao, Y. Quantitative assessment of the effects of climate, vegetation, soil and groundwater on soil moisture spatiotemporal variability in the Mongolian Plateau. *Sci. Total Environ.* **2022**, *809*, 152198. [CrossRef] [PubMed]
34. Li, S.; Li, X.; Gong, J.; Dang, D.; Dou, H.; Lyu, X. Quantitative analysis of natural and anthropogenic factors influencing vegetation NDVI changes in temperate drylands from a spatial stratified heterogeneity perspective: A case study of Inner Mongolia Grasslands, China. *Remote Sens.* **2022**, *14*, 3320. [CrossRef]
35. Wei, Y.; Zhu, L.; Chen, Y.; Cao, X.; Yu, H. Spatiotemporal variations in drought and vegetation response in Inner Mongolia from 1982 to 2019. *Remote Sens.* **2022**, *14*, 3803. [CrossRef]
36. Chen, Y.; Jiao, J.; Tian, H.; Xu, Q.; Feng, L.; Wang, N.; Bai, L.; Yang, X. Spatial correlation analysis between vegetation NDVI and natural environmental factors based on geographical detector on the Loess Plateau. *Acta Ecol. Sin.* **2022**, *42*, 3569–3580. (In Chinese)
37. Venkatesh, K.; John, R.; Chen, J.; Xiao, J.; Amirkhiz, R.G.; Giannico, V.; Kussainova, M. Optimal ranges of social-environmental drivers and their impacts on vegetation dynamics in Kazakhstan. *Sci. Total Environ.* **2022**, *847*, 157562. [CrossRef]
38. Gao, S.; Dong, G.; Jiang, X.; Nie, T.; Yin, H.; Guo, X. Quantification of natural and anthropogenic driving forces of vegetation changes in the Three-River Headwater Region during 1982–2015 based on Geographical Detector Model. *Remote Sens.* **2021**, *13*, 4175. [CrossRef]
39. Xu, X. *Spatial Interpolation Data Set of Average Status of Meteorological Elements in China*; Scientific Data Registration and Publishing System of Resources and Environment: Beijing, China, 2017.
40. Xu, X. *Chinese GDP Spatial Distribution Kilometer Grid Data Set*; Scientific Data Registration and Publishing System of Resources and Environment: Beijing, China, 2017.

41. Xu, X. *Kilometer Grid Data Set of Spatial Distribution of Population in China*; Scientific Data Registration and Publishing System of Resources and Environment: Beijing, China, 2017.
42. Zhang, L.; Ren, Z.; Chen, B.; Gong, P.; Fu, H.; Xu, B. *A prolonged Artificial Nighttime-Light Dataset of China (1984–2020)*; National Tibetan Plateau Data Center: Beijing, China, 2021.
43. Xu, H.; Wang, Y.; Liu, Y.; Xiao, Y.; Zhou, L. Analysis of the spatial and temporal evolution and driving factors of Rocky Desertification in typical cluster depression areas in the past 30 years based on Google Earth Engine-Taking Xichou County as an example. *Res. Soil Water Conserv.* **2022**, *29*, 407–414. (In Chinese)
44. He, P.; Bi, R.T.; Xu, L.S.; Wang, J.; Cao, C.B. Using geographical detection to analyze responses of vegetation growth to climate change in the Loess Plateau, China. *Chin. J. Appl. Ecol.* **2022**, *33*, 448–456. (In Chinese)
45. Xiao, Q.; Tao, J.; Xiao, Y.; Qian, F. Monitoring vegetation cover in Chongqing between 2001 and 2010 using remote sensing data. *Environ. Monit. Assess.* **2017**, *189*, 493. [CrossRef] [PubMed]
46. Li, F.; Zhou, W.Z.; Shao, Z.L.; Zhou, X.Y. Effects of ecological projects on vegetation in the Three Gorges Area of Chongqing, China. *J. Mt. Sci.* **2022**, *19*, 121–135. [CrossRef]
47. Li, Z.; Li, X. Climatic factors variation and its relation with spatio-temporal changes of vegetation cover in Chongqing, China. *J. Mt. Sci.* **2014**, *32*, 717–724. (In Chinese) [CrossRef]
48. Liu, R.; Xiao, L.; Liu, Z.; Dai, J. Quantifying the relative impacts of climate and human activities on vegetation changes at the regional scale. *Ecol. Indic.* **2018**, *93*, 91–99. [CrossRef]
49. Zhu, L.F.; Xie, S.Y.; Yang, H.; Ma, M.G. Study on the spatial-temporal variability of vegetation coverage based on MODIS-EVI in Chongring. *Acta Ecol. Sin.* **2018**, *38*, 6992–7002. (In Chinese)
50. Zhang, L.; Yang, L.; Zohner, C.M.; Crowther, T.W.; Li, M.; Shen, F.; Guo, M.; Qin, J.; Yao, L.; Zhou, C. Direct and indirect impacts of urbanization on vegetation growth across the world's cities. *Sci. Adv.* **2022**, *8*, eabo0095. [CrossRef]
51. Li, H.; Zhou, Q.; Jiao, H.; Wang, F.; Guo, H.L. Temporal-spatial variation of NDVI in the metropolitan area of Chongqing municipality based on TM. *Res. Soil Water Conserv.* **2015**, *22*, 250–255. (In Chinese)
52. Zhang, M.; Kafy, A.A.; Ren, B.; Zhang, Y.; Tan, S.; Li, J. Application of the optimal parameter Geographic Detector Model in the identification of influencing factors of ecological quality in Guangzhou, China. *Land* **2022**, *11*, 1303. [CrossRef]
53. Levin, N.; Zhang, Q. A global analysis of factors controlling VIIRS nighttime light levels from densely populated areas. *Remote Sens. Environ.* **2017**, *190*, 366–382. [CrossRef]
54. Ustaoglu, E.; Bovkır, R.; Aydınoglu, A. Spatial distribution of GDP based on integrated NPS-VIIRS nighttime light and MODIS EVI data: A case study of Turkey. *Environ. Dev. Sustain.* **2021**, *23*, 10309–10343. [CrossRef]
55. Sun, J.; Li, Y.; Gao, P.; Suo, C.; Xia, B. Analyzing urban ecosystem variation in the City of Dongguan: A stepwise cluster modeling approach. *Environ. Res.* **2018**, *166*, 276–289. [CrossRef] [PubMed]
56. Herrero, H.; Southworth, J.; Muir, C.; Khatami, R.; Bunting, E.; Child, B. An evaluation of vegetation health in and around Southern African National Parks during the 21st century (2000–2016). *Appl. Sci.* **2020**, *10*, 2366. [CrossRef]
57. Afuye, G.A.; Kalumba, A.M.; Orimoloye, I.R. Characterisation of vegetation response to climate change: A review. *Sustainability* **2021**, *13*, 7265. [CrossRef]
58. He, L.; Guo, J.; Yang, W.; Jiang, Q.; Chen, L.; Tang, K. Multifaceted responses of vegetation to average and extreme climate change over global drylands. *Sci. Total Environ.* **2022**, *858*, 159942. [CrossRef]
59. Zhang, F.; Zhang, Z.; Kong, R.; Chang, J.; Tian, J.; Zhu, B.; Jiang, S.; Chen, X.; Xu, C.Y. Changes in forest net primary productivity in the Yangtze River Basin and its relationship with climate change and human activities. *Remote Sens.* **2019**, *11*, 1451. [CrossRef]
60. Liu, C.; Gao, Y.; Li, Y.C.; Yang, S.Q.; Boshun, N. Vegetation coverage change based on NDVI and its relationship with climate factors in Chongqing. *Resour. Environ. Yangtze Basin* **2013**, *22*, 1514–1520. (In Chinese)
61. Zhang, L.; Cong, Z.; Zhang, D.; Li, Q. Response of vegetation dynamics to climatic variables across a precipitation gradient in the Northeast China Transect. *Hydrol. Sci. J.* **2017**, *62*, 1517–1531. [CrossRef]
62. Ma, X.; Bai, H.; Deng, C.; Wu, T. Sensitivity of vegetation on alpine and subalpine timberline in Qinling Mountains to temperature change. *Forests* **2019**, *10*, 1105. [CrossRef]
63. He, J.; Shi, X. Detection of social-ecological drivers and impact thresholds of ecological degradation and ecological restoration in the last three decades. *J. Environ. Manag.* **2022**, *318*, 115513. [CrossRef] [PubMed]
64. Wang, Y.; Hao, L.; Zhao, M.; Xu, X.; Li, C.H. Variation of vegetation NDVI and its response to climatic factors and human activities in Chongqing from 2001 to 2018. *Res. Soil Water Conserv.* **2021**, *28*, 222–229. (In Chinese)
65. Liu, Q.; Su, L.; Xia, Z.; Liu, D.; Xu, W.; Xiao, H. Effects of soil properties and illumination intensities on matric suction of vegetated soil. *Sustainability* **2019**, *11*, 6475. [CrossRef]
66. Zhu, L.; Xie, S.; Yang, H.; Mingguo, M. The response of dynamic change in vegetation coverage to topography in Chongqing based on MODIS EVI. *J. Nat. Resour.* **2017**, *32*, 2023–2033. (In Chinese)
67. Fan, J.; Xu, Y.; Ge, H.; Yang, W. Vegetation growth variation in relation to topography in Horqin Sandy Land. *Ecol. Indic.* **2020**, *113*, 106215. [CrossRef]
68. Del-Toro-Guerrero, F.J.; Kretzschmar, T.; Bullock, S.H. Precipitation and topography modulate vegetation greenness in the mountains of Baja California, México. *Int. J. Biometeorol.* **2019**, *63*, 1425–1435. [CrossRef] [PubMed]
69. Liu, J.; Liu, S.; Tang, X.; Ding, Z.; Ma, M.; Yu, P. The response of land surface temperature changes to the vegetation dynamics in the Yangtze River Basin. *Remote Sens.* **2022**, *14*, 5093. [CrossRef]

70. Pan, Y.; Wang, Y.; Zheng, S.; Huete, A.R.; Shen, M.; Zhang, X.; Huang, J.; He, G.; Yu, L.; Xu, X. Characteristics of greening along altitudinal gradients on the Qinghai–Tibet Plateau based on time-series Landsat images. *Remote Sens.* **2022**, *14*, 2408. [CrossRef]
71. Wang, Y.; Peng, D.; Shen, M.; Xu, X.; Yang, X.; Huang, W.; Yu, L.; Liu, L.; Li, C.; Li, X. Contrasting effects of temperature and precipitation on vegetation greenness along elevation gradients of the Tibetan plateau. *Remote Sens.* **2020**, *12*, 2751. [CrossRef]
72. Mei, L.; Bao, G.; Tong, S.; Yin, S.; Bao, Y.; Jiang, K.; Hong, Y.; Tuya, A.; Huang, X. Elevation-dependent response of spring phenology to climate and its legacy effect on vegetation growth in the mountains of northwest Mongolia. *Ecol. Indic.* **2021**, *126*, 107640. [CrossRef]

remote sensing

MDPI

Article

Temporal Dynamics of the Goose Habitat in the Middle and Lower Reaches of the Yangtze River

Ke He [1,2,†], Jialin Lei [1,†], Yifei Jia [1], Entao Wu [1], Gongqi Sun [1,3], Cai Lu [1], Qing Zeng [1] and Guangchun Lei [1,*]

1 Center for East Asian-Australasian Flyway Studies, Beijing Forestry University, Beijing 100083, China; heke0611@bjfu.edu.cn (K.H.); leijialinbjfu@bjfu.edu.cn (J.L.); jiayifei@bjfu.edu.cn (Y.J.); entaowu@bjfu.edu.cn (E.W.); sungongqi@foxmail.com (G.S.); lucai.wetland@foxmail.com (C.L.); zengqing@bjfu.edu.cn (Q.Z.)

2 Ministry of Education Key Laboratory of Southwest China Wildlife Resource Conservation, China West Normal University, Nanchong 637002, China

3 Academy of Inventory and Planning National Forestry and Grassland Administration, Beijing 100013, China

* Correspondence: guangchun.lei@foxmail.com; Tel.: +86-010-6233-6717

† These authors contributed equally to this work.

Abstract: The middle and lower reaches of the Yangtze River are the most important areas for geese to overwinter in the East Asian–Australasian Flyway, where about 180,000 geese fly to overwinter each year. Over the past 20 years, the region has experienced extensive and rapid land cover changes that may have exceeded the adaptability of geese, and have led to suitable goose habitat area loss, thereby, reducing the stability of the geese population. In order to identify the suitable goose habitat areas in this region, based on ensemble modeling and satellite tracking data, in this study, we simulated the spatial distribution changes in the suitable goose habitat areas over the past 20 years. The results showed that the suitable goose habitat areas had suffered varying degrees of loss, among which, the lesser white-fronted goose had the greatest suitable goose habitat area loss of over 50%. Moreover, we found that wetlands, lakes, and floodplains were the key components of suitable goose habitat areas, and the categories (land use) showed significant differences in different periods ($p < 0.01$). This may be one of the main reasons for the decrease in suitable goose habitat areas. The results of this study provide an important reference for the adaptive management and protection of geese in the middle and lower reaches of the Yangtze River.

Keywords: habitat loss; geese; species distribution models (SDMs); land use change; middle and lower reaches of the Yangtze River

Citation: He, K.; Lei, J.; Jia, Y.; Wu, E.; Sun, G.; Lu, C.; Zeng, Q.; Lei, G. Temporal Dynamics of the Goose Habitat in the Middle and Lower Reaches of the Yangtze River. *Remote Sens.* **2022**, *14*, 1883. https://doi.org/10.3390/rs14081883

Academic Editors: Matteo Convertino and Jie Li

Received: 1 March 2022
Accepted: 7 April 2022
Published: 14 April 2022

1. Introduction

Habitat loss and degradation has been a major cause of wildlife population decline [1–5]. In the nonbreeding season (overwintering period), goose habitat areas mainly consist of floodplains [1,6–9]. While floodplains are highly complex and dynamic ecosystems, they are also among the most threatened ecosystems because they are often dominated by humans and may experience a high intensity of anthropogenic activity [10–14].

The middle and lower reaches of the Yangtze River (MLYR) is one of the most important freshwater ecoregions in the world [15,16]. The numerous lakes connected to the Yangtze River (such as Poyang Lake and Dongting Lake in this region) form a complex river-lake relationship with the Yangtze River, creating extremely rich wetland ecosystem types (nine Ramsar sites of international importance, http://www.ramsar.org/pdf/sitelist.pdf, accessed on 11 October 2021) and providing a habitat for many important and endangered waterbirds [17–19]. The population of geese accounts for about 35% of the total number of waterbirds in the MLYR. This area constitutes the most important overwintering site for geese in the East Asian–Australasian Flyway (EAAF), with nearly 180,000 geese overwintering there every year according to a 2004 survey of birds by [20]. The geese that overwinter in this area include the lesser white-fronted goose (LWFG, Anser erythropus), the greater white-fronted goose (GWFG, Anser albifrons), the bean goose (BG, Anser fabalis), the swan goose (SG, Anser cygnoides), and the greylag goose (GG, Anser anser), among which the LWFG and SG have been recognized as vulnerable by the International Union for Conservation of Nature (IUCN). The LWFG, SG, and GG in the EAAF generally overwinter in this region [21–23], while 20% of GWFGs and 70% of BGs overwinter in this area, and the other geese overwinter in Japan and Korea [24,25]. The population of these five species of geese accounts for more than 99% of the total number of geese in the MLYR (unpublished data from the Center for East Asian–Australasian Flyway Studies).

Waterbirds represent an important environmental indicator group, especially for the status of wetland ecosystems in the MLYR [26]. Over the past 20 years, the MLYR has become one of the regions with the fastest economic growth in China. Human activities in the region have strongly disturbed the hydrological rhythm, especially the unreasonable development and utilization of lakes, wetlands, and floodplains, as well as the cascade development of hydropower stations in the Yangtze River Basin, such as the Three Gorges Dam. As a result, the loss and degradation of goose habitat areas have resulted in a sharp decline in the population of geese in the region [7].

Over the past two decades, species distribution models (SDMs) have been widely used to study species spatial distribution patterns and guide conservation planning [27,28]. SDMs can be adapted to different spatial resolutions, and the available data sources can help researchers to understand the population distribution of species and can provide valuable insights even for species that are rarely studied [29]. Currently, common SDMs include the generalized linear model (GLM) [30], random forest (RF) [31], and maximum entropy (MAXENT) [21]. Each SDM has different characteristics and advantages. Therefore, an increasing number of studies have used ensemble modeling to integrate the advantages of various SDMs to study the spatial distribution of species [27,32]. By combining models with different assumptions and algorithms, the integrated model can provide more robust results than a single model [33].

In this study, SDMs combined with GPS satellite tracking data were used, for the first time, to study the large-scale biogeography of five species of geese in the MLYR, aiming to determine the main environmental variables affecting their habitat areas, and to evaluate their habitat conditions in different periods and the change trends of their habitat areas, which is of great scientific significance to the protection of geese in the research area.

2. Data and Methods

2.1. Study Area

The Yangtze River, the longest river in Asia and the third longest river in the world [34], is unique in its extensive transitory basin wetlands. The wetlands are replenished by summer monsoon rains, bringing nutrient-rich and sediment-rich water, followed by falling water levels in autumn and winter [34]. The MLYR, from the Three Gorges Dam to the estuary, mainly covers most of Hubei, Hunan, Jiangxi, Anhui, Jiangsu, Zhejiang, and Shanghai, as well as some regions of Guangxi and Henan, with a watershed area of about 800,000 km^2 (Figure 1) [35].

2.2. Data and Model

For comparative analysis, we divided the 20 years from 2000 to 2019 into four periods, namely, Period 1 from 2000 to 2004, Period 2 from 2005 to 2009, Period 3 from 2010 to 2014, and Period 4 from 2015 to 2019.

The calculation results of geese in Period 4 were used as the current distribution, and then, the final ensemble model was projected to the past by using the occurrence and environment data of other periods.

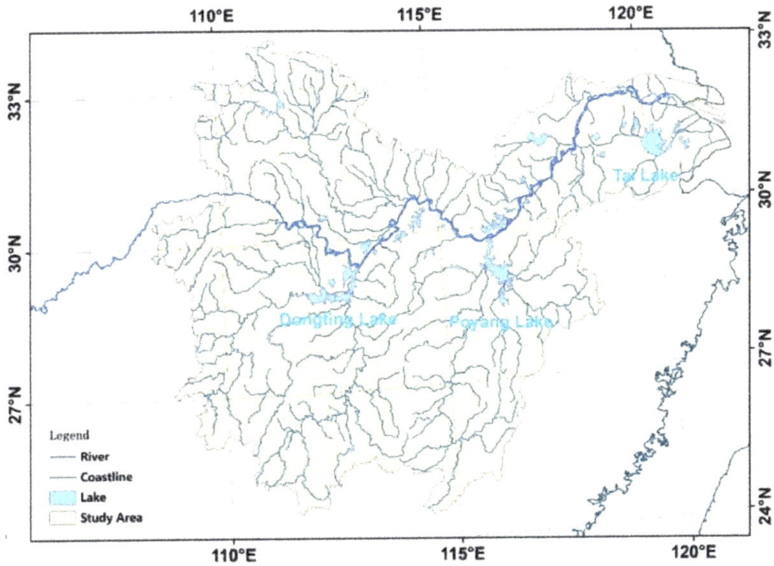

Figure 1. Geographical location of the study area.

2.2.1. Goose Occurrence Data

Since 2015, GPS trackers have been applied to 141 geese, and by 2019, 403,811 goose occurrence data were obtained. Because the MLYR is the overwintering grounds for geese, we only selected the occurrence data from October to December and from January to March. In order to avoid interference with the model due to differences in sites during the migration of geese, we eliminated all sites with velocities greater than 1 [1]. In order to reduce the error of occurrence data in geographical coordinates and reduce the sampling bias effect of the occurrence dataset, the occurrence data were compiled at a spatial resolution of 1 × 1 km [27]. After removing duplicate records within each grid cell, we obtained 2664 presence records to simulate the habitat areas of these five goose species (Table 1).

Table 1. Number of occurrence data points for five species of geese (lesser white-fronted goose (LWFG), greater white-fronted goose (GWFG), bean goose (BG), swan goose (SG), and greylag goose (GG)).

Species	Number of Occurrence Data Points
LWFG	419
GWFG	752
BG	852
SG	487
GG	154
Total	2692

2.2.2. Climate Data

The data of climate variables are important for predicting species distribution, especially for analyses over long time spans [36,37]. The climate variables used in this study were derived from CHELSA (http://chelsa-climate.org, accessed on 5 September 2021) [38,39], mainly using the three variables of monthly rainfall (PRE), monthly average maximum temperature (TMAX), and monthly average minimum temperature (TMIN). The time period was 2000–2018, although the precipitation data for 2018 were missing, and the climate data accuracy was 30 arc seconds (about 1 km^2). In order to effectively assess the goose habitat in the MLYR, only the data during the overwintering period were selected in this study. For example, the averages of October to December of the first year and January to March of the next year were taken as one data point (the rainfall in 2000 was the average of the rainfall from October to December of 1999 and January to March of 2000).

2.2.3. Normalized Difference Vegetation Index (NDVI) and Normalized Difference Water Index (NDWI)

The NDVI and NDWI have been widely used to evaluate the distribution of geese [7,8,40]. Our NDVI and NDWI data were derived from the "Landsat 7 Collection 1 Tier 1 1 8-Day NDVI/NDWI Composite" database in the Earth Engine Data Catalog. At the same time, we used the Google Earth Engine platform (https://code.earthengine.google.com/, accessed on 15 August 2021) to download directly for the period 2000–2019. These data have a resolution of 30 m. In order to ensure the consistency of the data time, we adopted the same processing method as that used for climate data.

2.2.4. Land Use

The use of land use data plays an important role in predicting species distribution at large landscape scales [41–43]. The land use data used in this study were obtained from the Data Center for Resources and Environment of the Chinese Academy of Sciences (http://www.resdc.cn/, accessed on 20 August 2021) and included cultivated land, forest land, grassland, water areas, residential land, and unused land as six primary types, with 20 secondary types. These data have a resolution of 1 km^2. Based on the needs of this study, we selected four primary types and 12 secondary types (Table 2). In this study, the land use data of the first year of each period were selected as the land use data of the period, that is, the land use date of Period 1 was 2000 (it contains 4 years of land use data, 2000, 2005, 2010, and 2015, respectively, Appendix A, Figure A1). Because the land use data were the classification variable and the partial SDM model was not conducive to the classification variable, in order to conduct the quantitative analysis, in this study, we transformed 12 types of land use into a continuous variable using Euclidean distance. For example, wetlands were transformed into the distance to wetland (dis_wl).

Table 2. Land use classification.

Level 1	Level 2	Meaning
Cropland	Paddy Field (dis_pf)	Cropland with a guaranteed water source and irrigation facilities that can be irrigated normally in normal years and used to grow rice, lotus root, and other aquatic crops.
	Upland Field (dis_uf)	Cropland without an irrigation water source or facilities that depends on natural water to grow crops; dry-crop-cultivated land with a water source and irrigation facilities that can be irrigated normally in normal years; cultivated land mainly used for vegetable cultivation.
Water	River (dis_ri)	Land below the perennial water level of rivers and main rivers formed by natural or artificial excavation. Artificial channels include an embankment.
	Lake (dis_la)	Land below the perennial water level in a natural water accumulation area.
	Reservoir (dis_re)	Land below the perennial water level in an artificial water storage area.
	Mudflat (dis_mf)	The tidal zone between the high tide level and the low tide level of the coastal spring tide.
	Floodplain (dis_fp)	Land between the water levels of rivers and lakes in normal seasons and those in flood seasons.

Table 2. *Cont.*

Level 1	Level 2	Meaning
Construction land	Urban Land (dis_ul)	Land used in large, medium, and small cities and built-up areas above the county level.
	Rural Land (dis_rl)	Rural settlements that are independent of cities and towns.
	Other Construction Land (dis_ocl)	Land used for factories and mines, large industrial areas, oil fields, salt fields, quarries, traffic roads, airports, and other construction land uses.
Unused Land	Wetland (dis_wl)	Land with flat and low-lying terrain, poor drainage, long-term moisture, seasonal water accumulation or perennial water accumulation, and surface growth of hygrophytes.
	Bare Land (dis_bl)	Land covered by surface soil where the vegetation coverage is less than 5%.

2.2.5. Elevation Data

The elevation data were obtained from a digital elevation model (DEM) with a 30 m resolution and downloaded from the International Scientific and Technical Data Mirror Site, Computer Network Information Center, Chinese Academy of Sciences (http://www.gscloud.cn, accessed on 15 August 2021) [44]. To match the resolution of climate variables, the DEM data were resampled at 1 km^2 resolution using a bilinear interpolation

2.2.6. Model

We used the stacked species distribution model (SSDM) software package in R software to simulate the suitable goose habitats [45,46]. For this purpose, we used seven species distribution models for the calculations: the GLM, RF, support vector machines (SVM) [47], artificial neural network (ANN) [48], generalized additive model (GAM) [49], classification tree analysis (CTA) [50], and generalized boosting model (GBM) [51].

To evaluate the accuracy of each algorithm, we performed 10 cross-validations for each algorithm; 70% of each dataset was used as training data and the remainder was used to test algorithm performance. The area under the receiver operating characteristic curve (AUC) [27,29,46] was used to evaluate the goodness-of-fit of each model. When the AUC value of the model was greater than 0.9, it was considered to be an excellent fit; when the AUC was 0.9–0.8, it was considered to be a good fit; when the AUC was 0.8–0.7, it was regarded as an acceptable fit; and when the AUC was less than 0.7, the model was regarded as a poor fit [28]. The habitat suitability maps were converted to binary presence absence maps using a threshold that maximums model sensitivity plus specificity [27].

To avoid possible multicollinearity leading to biased model estimates, we tested Pearson correlations between environmental factors and defined the absolute value of the correlation coefficient R > 0.7 as a threshold [27]. Finally, we selected 18 variables, such as land use and climate, among which the correlation between TMAX and TMIN was 0.9. Because these were important indicators for predicting suitable goose habitat, they were also included in the analysis (Appendix A, Figure A2).

All environmental variables for this study were processed using ArcGis 10.6 in order to obtain a uniform resolution and coordinate system. The comparative analysis of all results was completed in R (version 4.1.1) software.

3. Results

3.1. Model Performance and Variable Contribution

It was found that the seven algorithms used for species distribution models had excellent recognition abilities for the five species of geese, and the average AUC values of LWFG, GWEG, SG, and BG were higher than 0.9. The results showed that the models had excellent fits, with AUC values of 0.944 ± 0.002, 0.938 ± 0.002, 0.930 ± 0.002, and

0.920 ± 0.002 for the LWFG, GWEG, SG, and BG models, respectively. The average AUC value of GG was 0.886 ± 0.004, indicating that the model fit was good (Figure 2). Overall, among the seven model algorithms, SVM had the lowest AUC value (0.916 ± 0.031), while GAM had the highest AUC value (0.934 ± 0.028).

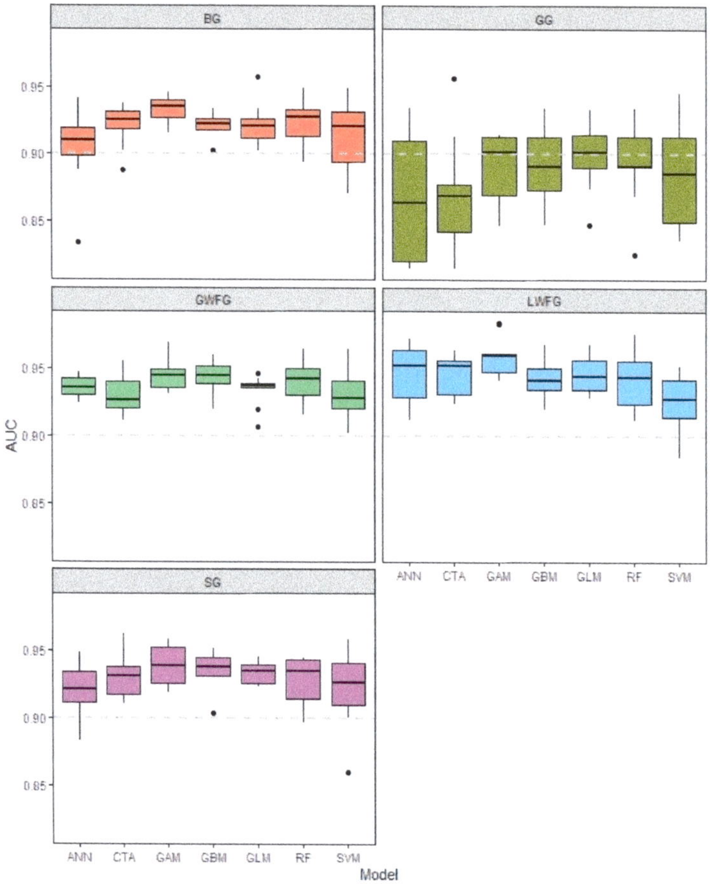

Figure 2. Box plots of interquartile range (IQR), range, and median model performance of the seven modeling algorithms used to predict the habitat suitability of five goose species. The dots are potential outliers which are greater than the 75th percentile + 1.5 IQR or less than the 25th percentile − 1.5 IQR. The medians are represented by thick black lines. Generalized linear model (GLM); random forest (RF); support vector machines (SVM); artificial neural network (ANN); generalized additive model (GAM); classification tree analysis (CTA); generalized boosting model (GBM); lesser white-fronted goose (LWFG); greater white-fronted goose (GWFG); bean goose (BG); swan goose (SG); and greylag goose (GG).

The results showed that land use data contributed the most to the simulation of suitable goose habitat areas, with an average contribution rate of 0.781 ± 0.009, followed by climate data, with a contribution rate of 0.097 ± 0.009. Altitude, NDVI, and NDWI contributed less at 0.053 ± 0.008, 0.045 ± 0.011, and 0.023 ± 0.004, respectively (Appendix B, Table A1). Specifically for each variable, among all 18 variables, the contributions of dis_la, dis_fp, dis_wl, altitude, NDVI, and TMIN were more than 0.05, and the contributions of dis_la, dis_fp, and dis_wl to all the goose habitat areas were more than 0.05 (Figure 3). Although all geese have a high demand for dis_la, dis_fp, and dis_wl, the degree of their specific needs varies. The contributions of dis_wl to the suitable habitat of LWFG, GWFG, and SG geese were 0.294 ± 0.022, 0.238 ± 0.019, and 0.225 ± 0.017, respectively. The contributions of dis_la to the suitable habitat of BG and GG were the highest, at 0.219 ± 0.015 and 0.268 ± 0.018, respectively (Appendix B, Table A1).

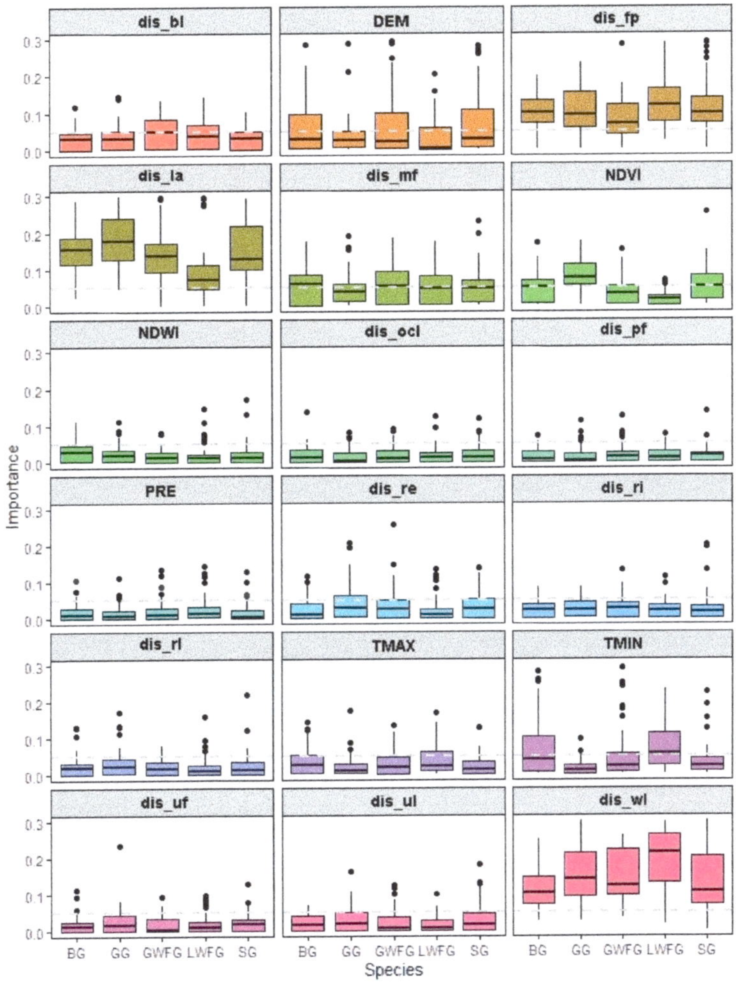

Figure 3. Box plots of interquartile range (IQR), range, and median model performance of algorithms used to predict the variable contribution to the model. The dots are potential outliers which are greater than the 75th percentile + 1.5 IQR or less than the 25th percentile − 1.5 IQR. The medians are represented by thick black lines. Lesser white-fronted goose (LWFG); greater white-fronted goose (GWFG); bean goose (BG); swan goose (SG); and greylag goose (GG).

Because dis_la, dis_fp, and dis_wl are important for predicting suitable goose habitat areas, we compared the differences in these three variables in different periods. This study found that dis_fp and dis_la showed a downward trend from Period 1 to Period 4, while dis_wl showed an upward trend (Appendix B, Table A2). The results showed that the three variables exhibited significant differences in the four periods (ANOVA test, $p < 0.01$, Figure 4).

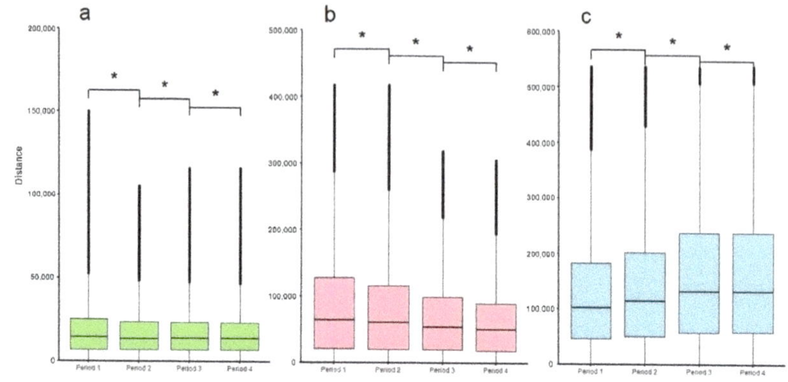

Figure 4. The differences of the three variables in the four periods: (**a**) dis_fp; (**b**) dis_la; (**c**) dis_wl. Period 1 is 2000–2004, Period 2 is 2005–2009, Period 3 is 2010–2014, and Period 4 is 2015–2019. The asterisks indicate a significant difference.

3.2. Suitable Habitat

Using the habitat classification thresholds (LWFG 0.48, GWFG 0.41, BG 0.44, SG 0.43, and GG 0.39) calculated by the SSMD model, we classified the suitable and unsuitable goose habitat areas (Figure 5). The results showed that the largest suitable goose habitat area for all geese was found during Period 1, and the suitable goose habitat areas for all geese were mainly distributed in the Dongting Lake and Poyang Lake areas, as well as the areas near the mainstream of the Yangtze River between the two lakes. In the same period, the suitable goose habitat area of GG was the largest among the five species of geese, while the suitable goose habitat area of LWFG was the smallest. In terms of suitable goose habitat area loss, the suitable goose habitat area of all geese has declined over the past 20 years, with the LWFG losing the most suitable goose habitat area and GG losing the least suitable goose habitat area. Specifically, LWFG and GFWG decreased the most from Period 3 to Period 4, with a loss of 15,905 km^2 (45.85%) and 9217 km^2 (23.26%) of suitable habitat area, respectively; BG and GG lost 7191 km^2 (14.97%) and 2550 km^2 (3.88%) of suitable habitat area from Period 2 to Period 3, respectively. SG lost the most suitable habitat area from Period 1 to Period 2 (3926 km^2, 9.75%, Table 3).

Table 3. The suitable habitat area (km^2) loss and the relative change ratio (%) of suitable habitat between two consecutive periods. Lesser white-fronted goose (LWFG); greater white-fronted goose (GWFG); bean goose (BG); swan goose (SG); and greylag goose (GG).

Species	Suitable Habitat Area/Change	Period 1	Period 1 vs. Period 2	Period 2 vs. Period 3	Period 3 vs. Period 4
LFWG	Suitable habitat	37,872			
	Lost suitable habitat		−2922	−260	−15905
	Relative change ratio		−7.72%	−0.74%	−45.85%
GFWG	Suitable habitat	46,067			
	Lost suitable habitat		−2747	−3699	−9217
	Relative change ratio		−5.96%	−8.54%	−23.26%
BG	Suitable habitat	52,613			
	Lost suitable habitat		−4576	−7191	−4441
	Relative change ratio		−8.70%	−14.97%	−10.87%
SG	Suitable habitat	40,253			
	Lost suitable habitat		−3926	−1591	−2757
	Relative change ratio		−9.75%	−4.38%	−7.94%
GG	Suitable habitat	67,697			
	Lost suitable habitat		−1972	−2550	−1554
	Relative change ratio		−2.91%	−3.88%	−2.46%

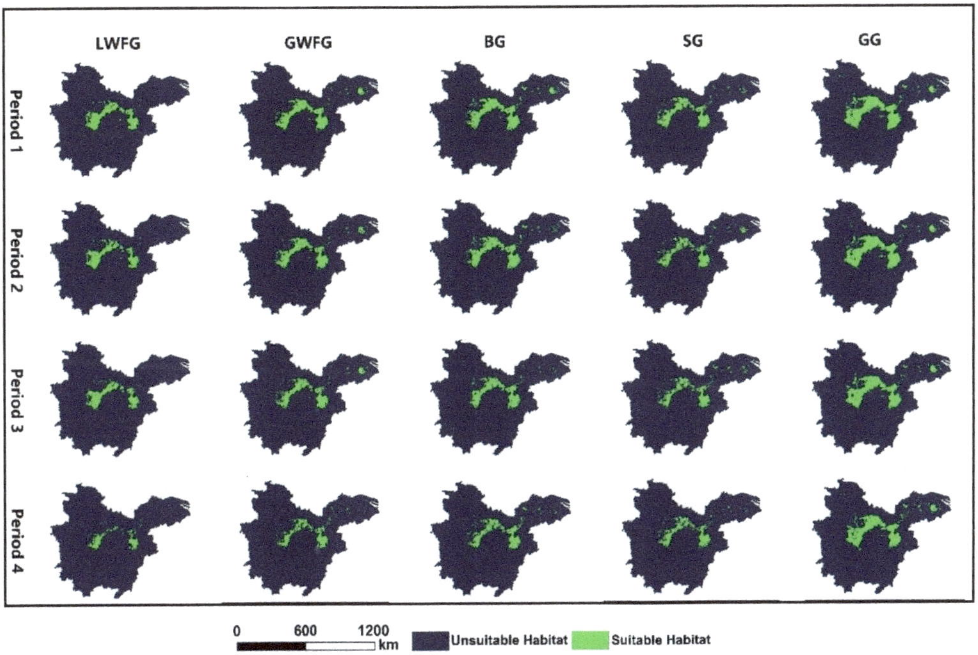

Figure 5. Suitable and unsuitable goose habitat areas in the MLYR for the four periods. Period 1 is 2000–2004, Period 2 is 2005–2009, Period 3 is 2010–2014, and Period 4 is 2015–2019. Lesser white-fronted goose (LWFG); greater white-fronted goose (GWFG); bean goose (BG); swan goose (SG); and greylag goose (GG).

In order to elucidate the overall situation of suitable goose habitat loss, we compared suitable goose habitat areas in two periods: Period 1 and Period 4. The results showed that during the 20 years from 2000 to 2019, the suitable goose habitat area of LWFG increased by 486 km^2 in some areas and decreased by 19,573 km^2 in other areas, with a total suitable habitat area loss of 50.40%, which was the greatest suitable goose habitat area loss among the five species; the suitable goose habitat area of GWFG increased by 1163 km^2 and decreased by 16,826 km^2, with a total suitable habitat area loss of 34.00%; the suitable goose habitat area of BG increased by 817 km^2 and decreased by 17,025 km^2, with a total suitable habitat area loss of 30.81%; the new suitable goose habitat area of SG was 3076 km^2, while it decreased by 8593 km^2, with a total suitable goose habitat area loss of 13.71%; the new suitable goose habitat area of GG was 4688 km^2, while it decreased by 8792 km^2, with a total suitable habitat area loss of 6.24%. GG was the species with the least suitable goose habitat area loss among the five species of geese (Figure 6).

From 2000 to 2019, the peripheries of the suitable goose habitat area for all geese were lost to varying degrees, and the greatest suitable goose habitat area loss was in the marginal areas. The LWFG lost most of its habitat area, except for the Dongting Lake and Poyang Lake areas, especially the suitable habitat area in the northwest of Dongting Lake. Most of the suitable GWFG habitat in the Tai Lake area was lost, and the GWFG habitat in the periphery of the two lakes was also lost. Similar to the GWFG, the BG lost suitable habitat in the Tai Lake area. Moreover, most of the suitable habitat in the peripheries of the two lakes were also lost. Much of the suitable SG habitat in the peripheries of the two lakes was lost as well, but it had also increased in some areas, especially in the lower reaches of the Yangtze River. Among all the geese, the GG had the least suitable goose habitat area loss, which was mainly concentrated in the peripheries of the two lakes, while it had a large increase in suitable goose habitat area in the lower reaches of the Yangtze River (Figure 7).

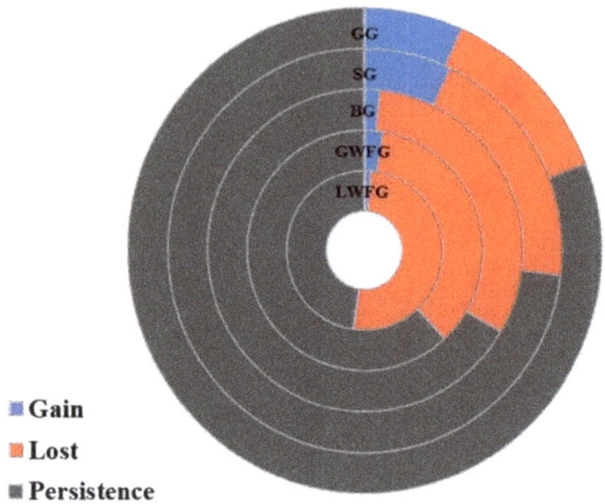

Figure 6. The suitable goose habitat loss over the past 20 years. From inside to outside, lesser white-fronted goose (LWFG); greater white-fronted goose (GWFG); bean goose (BG); swan goose (SG); greylag goose (GG).

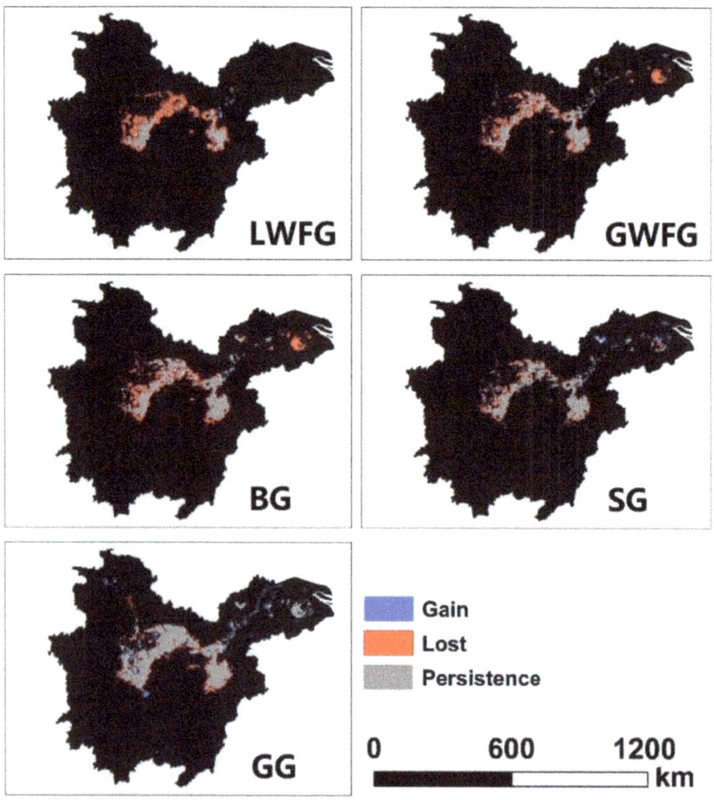

Figure 7. Change in suitable goose habitat during Period 1 and Period 4. Lesser white-fronted goose (LWFG); greater white-fronted goose (GWFG); bean goose (BG); swan goose (SG); and greylag goose (GG).

4. Discussion

SDMs are widely used to address a variety of ecological problems, including predicting the geographic range of species, assessing the impact of biological invasions, and developing conservation strategies [52–54]. However, some studies have shown that a single-species distribution model has greater uncertainty, while an ensemble model can provide more accurate results [55–57]. In this study, seven SDMs (ANN, GAM, GLM, RF, CAT, GBM, and SVM) were used to combine the individual SDMs with the SSDM model, combined with GPS tracking data and environmental data of five goose species from 2015 to 2019. The habitat adaptability and spatial distribution of these geese in the MLYR over the past 20 years were mapped. These findings were basically in line with the scientific consensus on the habitats of these five species, and a large number of areas lost in previous surveys were also found. These results are of great significance for supporting the protection of goose habitats in the MLYR.

The study area is the most important overwintering site for geese on the EAAF. The goose habitat loss in this area means a decrease in food availability and energy reserves, which lead to a series of problems, such as an increase in intraspecific and interspecific competition pressure, a decrease in the subadult survival rate, and an increase in mortality during migration. Over the past 20 years, the suitable goose habitat areas in the MLYR have been reduced to different degrees. Among the five species studied, LWFG, GWFG, and BG lost more than 16,000 km^2 of suitable habitat area (accounting for 50.40%, 34.00%, and 30.81% of their original habitat areas, respectively). However, the suitable habitat areas of SG and GG only decreased by about 8000 km^2 (13.71% and 6.24%, respectively). As compared with LWFG, GWFG, and BG, the suitable habitat area losses for SG and GG have had less impact on these species. Zhang et al. [7] found that large geese showed better adaptability to environmental changes, while small geese had a smaller range of feeding habits due to the length and hardness of their beaks, which made small geese more sensitive to environmental changes. This may be one of the reasons why the LWFG had the greatest suitable habitat area loss and GG had the least suitable habitat area loss. The most direct manifestation of habitat loss is a decline in population size, which has declined significantly for LWFG, GWFG, and BG over the past 20 years. In particular, the eastern population of the LWFG (all of which overwinter in the MLYR) decreased from 65,000 geese in the 1980s to 4020 geese in 2020 [58], and its suitable habitat was the most reduced of all five species. The suitable habitat area losses of SG and GG were less, and their populations were relatively stable in the MLYR. This also reflects that habitat area plays an extremely important role in the stability of goose population.

Our results show that the suitable goose habitat areas in the MLYR are shrinking to the area around the two lakes (Figure 7). Many reports in the literature have also reflected this finding [7,22,23,26]. The reason for this phenomenon may be that the two lakes are the two largest freshwater lakes in China and provide abundant food resources for geese. In addition, there are many nature reserves in the region, with large, protected areas and low human disturbance, which makes the habitat of this region better than that of other regions. The vast waters and floodplains in the two lakes areas are also the largest and most complete natural wetlands in the MLYR, and geese overwintering in China prefer natural wetlands [59,60].

A large number of research results have shown that wetlands, floodplains, and lakes were the main components of goose habitat areas in the MLYR [22,23,25,26,59]. In the present study, it was found that dis_wl, dis_fp, and dis_la were very important for predicting the suitable habitat areas of the five species of geese. However, these three variables have changed dramatically over the past two decades, and there were significant differences in the four periods (Figure 4). These changes have posed significant challenges for geese and have led to the loss of suitable habitat areas for geese with poor adaptability. These three variables are closely related to the hydrological rhythm of the Yangtze River [35,61], and the MLYR is a typical case [7]. During the first 20 years of the 20th century, the economy of the MLYR developed rapidly. To satisfy the high demand for electricity, the Yangtze

River was used for hydropower generation. As of 2019, there were 159 hydropower stations with an annual power generation of more than 300 thousand kilowatts in the Yangtze River Basin [62]. This high quantity of hydropower stations has significantly changed the hydrological rhythm of the Yangtze River Basin, leading to changes in wetlands, floodplains, and lakes [35] that have been disastrous for geese. Some studies have concluded that the construction of hydropower stations has destroyed the natural hydrological rhythm, resulting in early or delayed recession of floodplains in the MLYR, thus, affecting the timing and trends of food growth. These phenological changes no longer match the time when geese arrive at overwintering sites, and therefore geese are unable to obtain food in their original habitat areas [7]. The MLYR is the main grain-producing area in China, with well-developed agriculture; the lake areas in the region are shrinking due to reclamation and the demand for agricultural irrigation. The region also exhibits rapid economic development, a large population, rapid expansion of urban areas, rapid growth of tourism and other tertiary industries, and a large number of natural wetlands that have been exploited, which may be one of the reasons for the greater loss of suitable habitat margins for geese [63,64].

This study found that the use of land use data played an important role in simulating suitable goose habitat in the MLYR, with an average contribution rate as high as 0.781 ± 0.009 (Appendix B, Table A1). The contributions of climate (PRE, TMAX, and TMIN), DEM, NDVI, and NDWI were relatively small at 0.032 ± 0.004, 0.053 ± 0.008, and 0.045 ± 0.011, and 0.023 ± 0.004, respectively (Appendix B, Table A1). This was different from the results of many studies on goose habitat areas, some of which found that hydrological changes were the key factors in changes in goose distribution [65], and some scholars have indicated that food resources were important limiting factors [66,67]. These contrasting results are due to the differences in the scale of the study areas, and the use of land use data plays an extremely important role in predicting the distribution of species at a large landscape scale [41–43]. Studies that have suggested food resources or hydrological changes were the key factors affecting the distribution of geese have mostly been based on small spatial scales, and our research area covers the whole middle and lower reaches of the Yangtze River. Therefore, the difference in spatial scales is an important reason for the differences between our results and those of other studies.

Over the past 20 years, the habitat of wild geese overwintering in the MLYR has experienced different degrees of loss, and suitable goose habitat area has been significantly reduced, resulting in a significant decline in their population. This decline has mainly been caused by human activities. The MLYR is among the areas with the fastest economic development in China, but this rapid economic growth has led to environmental deterioration, especially excessive utilization and development of water resources, which has led to shrinkage of lakes, a reduction in floodplains, and loss of wetlands and other factors that are crucial to goose habitat [14,17,18,35]. The Chinese government has taken many measures to protect and restore the environment in recent years, such as returning farmland to wetlands and the Yangtze River protection strategy, and although the environment in some areas has been improved, the geese in the MLYR still face enormous challenges.

5. Conclusions

Based on SSDM and GPS tracking data, in this study, we analyzed the changes in the suitable goose habitat areas in the MLYR from 2000 to 2019. The results showed that the suitable goose habitat areas in this region had experienced varying degrees of loss, and that the suitable goose habitat area was significantly reduced. The LWFG had the greatest suitable habitat area loss (over 50%), while the GG had the least suitable habitat area loss (6.24%). GWFG, BG, and SG suitable habitat areas were reduced by 34.00%, 30.81% and 13.71%, respectively. The widespread and rapid changes in land use were one of the main reasons, especially the changes in floodplains, lakes, and wetlands. These analyses show that land use is an important factor in studying the spatial and temporal changes of suitable goose habitat areas on a large scale, which is of great significance to the protection and management of goose habitats.

Author Contributions: Conceptualization, K.H., J.L., Y.J. and G.L.; methodology, K.H. and G.L.; validation, K.H., Q.Z. and G.L.; investigation, K.H., J.L., Y.J., E.W., G.S. and Q.Z.; writing—original draft preparation, K.H. and G.L.; writing—review and editing, K.H. and G.L.; supervision, G.L. and C.L.; project administration, G.L. All authors have read and agreed to the published version of the manuscript.

Funding: This work was supported by National Natural Science Foundation of China (no. 31971400) and the National Key Research and Development Program of China (2017YFC0405303).

Data Availability Statement: Not available.

Acknowledgments: We would like to thank researcher Qiang Dai at the Chengdu Institute of Biology Chinese Academy of Sciences for assistance with models and data analysis.

Conflicts of Interest: The authors declare no conflict of interest.

Appendix A

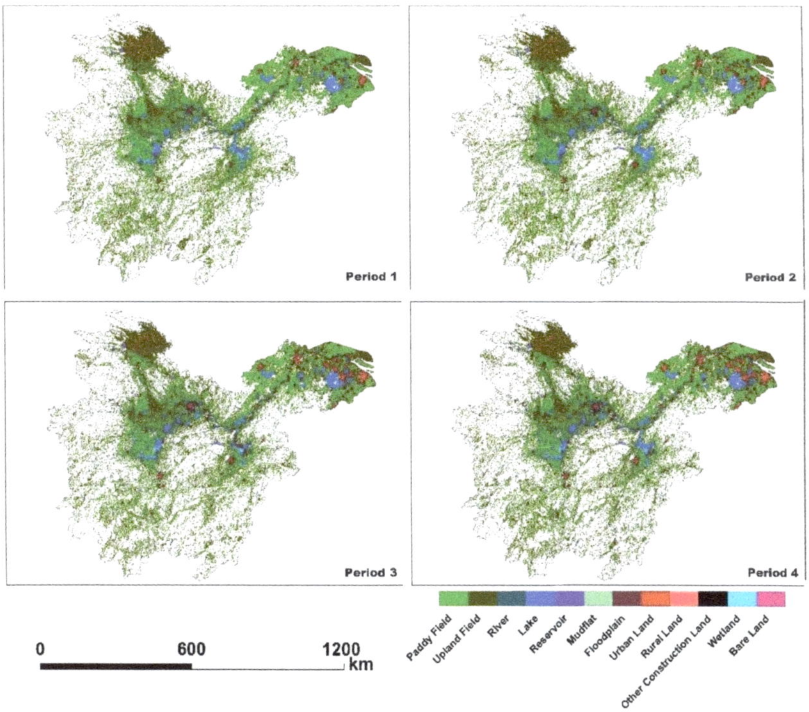

Figure A1. Land use in four periods. Period 1 is 2000; Period 2 is 2005; Period 3 is 2010; Period 4 is 2015.

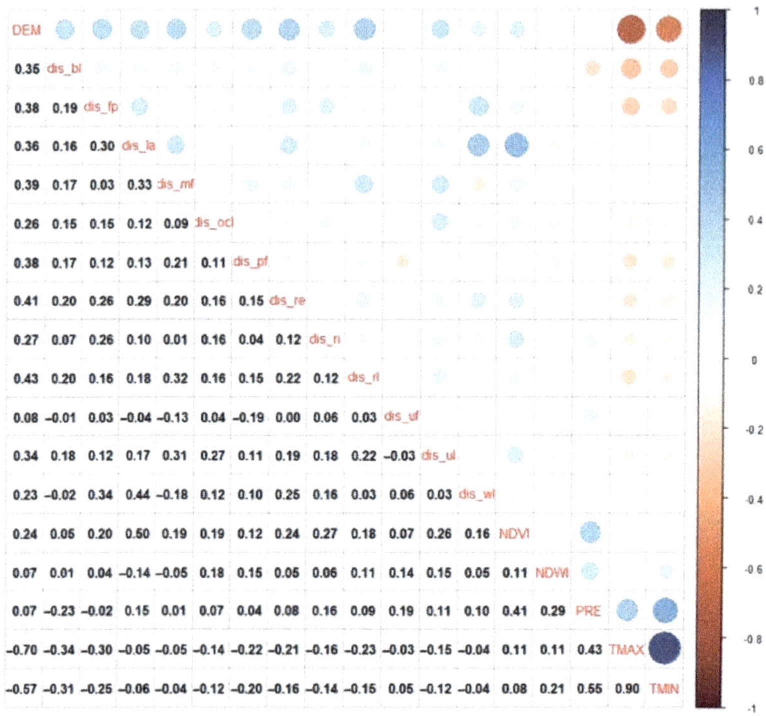

Figure A2. Correlation of 18 variables. (BL, bare land; DEM, altitude; FP, floodplains; LA, lake; MF, mudflat; NDVI, normalized difference vegetation index; NDWI, normalized difference water index; OCL, other construction land; PF, paddy field; PRE: monthly precipitation; RE, reservoir; RI, river; RL, rural land; TMAX, monthly mean maximum temperature; TMIN, monthly mean minimum temperature; UF, upland field; UL, urban land; WL, wetland). Red points represent negative correlations and blue points represent positive correlations.

Appendix B

Table A1. Variable importance of 18 environmental variables of the GAM algorithm.

Variables	LWFG	GWFG	BG	SG	GG
Dis_pf	0.012 ± 0.009	0.020 ± 0.010	0.011 ± 0.012	0.017 ± 0.007	0.008 ± 0.009
Dis_uf	0.015 ± 0.010	0.014 ± 0.016	0.019 ± 0.008	0.026 ± 0.011	0.018 ± 0.004
Dis_ri	0.032 ± 0.013	0.029 ± 0.006	0.020 ± 0.010	0.012 ± 0.012	0.022 ± 0.013
Dis_la	0.065 ± 0.021	0.156 ± 0.051	0.158 ± 0.028	0.167 ± 0.051	0.277 ± 0.097
Dis_re	0.015 ± 0.011	0.036 ± 0.016	0.014 ± 0.013	0.037 ± 0.017	0.032 ± 0.031
Dis_mf	0.086 ± 0.026	0.109 ± 0.025	0.066 ± 0.021	0.062 ± 0.023	0.046 ± 0.022
Dis_fp	0.125 ± 0.036	0.116 ± 0.044	0.137 ± 0.031	0.144 ± 0.046	0.131 ± 0.060
Dis_ul	0.014 ± 0.009	0.012 ± 0.012	0.025 ± 0.013	0.027 ± 0.014	0.013 ± 0.011
Dis_rl	0.016 ± 0.007	0.026 ± 0.012	0.021 ± 0.007	0.025 ± 0.013	0.032 ± 0.029
Dis_ocl	0.013 ± 0.010	0.022 ± 0.013	0.016 ± 0.013	0.017 ± 0.008	0.010 ± 0.011
Dis_wl	0.259 ± 0.045	0.188 ± 0.053	0.088 ± 0.031	0.217 ± 0.079	0.175 ± 0.065
Dis_bl	0.069 ± 0.018	0.078 ± 0.015	0.044 ± 0.015	0.044 ± 0.025	0.024 ± 0.015
DEM	0.019 ± 0.026	0.036 ± 0.047	0.041 ± 0.037	0.036 ± 0.050	0.021 ± 0.026
NDVI	0.022 ± 0.007	0.040 ± 0.014	0.060 ± 0.010	0.059 ± 0.028	0.112 ± 0.035
NDWI	0.020 ± 0.008	0.021 ± 0.012	0.041 ± 0.019	0.024 ± 0.017	0.027 ± 0.025
PRE	0.014 ± 0.011	0.019 ± 0.010	0.038 ± 0.015	0.010 ± 0.008	0.010 ± 0.010
TMAX	0.082 ± 0.032	0.042 ± 0.021	0.094 ± 0.036	0.037 ± 0.021	0.023 ± 0.017
TMIN	0.123 ± 0.031	0.035 ± 0.028	0.108 ± 0.041	0.040 ± 0.017	0.019 ± 0.011

Table A2. The average of the three variables in the four periods.

Period	Distance to Wetland	Distance to Lake	Distance to Floodplain
Period 1	128,104.553 ± 125.594	88,059.992 ± 101.239	18,951.819 ± 20.392
Period 2	137,343.562 ± 128.758	83,960.312 ± 99.655	17,731.429 ± 18.043
Period 3	156,412.35 ± 139.942	69,673.015 ± 74.446	17,362.222 ± 16.905
Period 4	156,595.611 ± 140.027	61,968.91 ± 63.665	17,159.595 ± 16.493

References

1. Xia, S.; Yu, D.; Cui, P.; Duan, H.; Teng, J.; Yu, X. Suitable-habitat dynamics for wintering geese in China's largest freshwater lake. *Glob. Ecol. Conserv.* **2021**, *27*, e01528.
2. Wang, C.; Liu, H.; Li, Y.; Dong, B.; Qiu, C.; Yang, J.; Zong, Y.; Chen, H.; Zhao, Y.; Zhang, Y. Study on habitat suitability and environmental variable thresholds of rare waterbirds. *Sci. Total Environ.* **2021**, *785*, 147316. [CrossRef] [PubMed]
3. Swift, T.L.; Hannon, S.J. Critical thresholds associated with habitat loss: A review of the concepts, evidence, and applications. *Biol. Rev. Camb. Philos. Soc.* **2010**, *85*, 35–53. [CrossRef]
4. Hanski, I.; Ovaskainen, O. Extinction Debt at Extinction Threshold. *Conserv. Biol.* **2002**, *16*, 666–673. [CrossRef]
5. Fahrig, L. Relative Effects of Habitat Loss and Fragmentation on Population Extinction. *J. Wildl. Manag.* **1997**, *61*, 603–610. [CrossRef]
6. Bayliss, P.; Ligtermoet, E. Seasonal habitats, decadal trends in abundance and cultural values of magpie geese (*Anseranus semipalmata*) on coastal floodplains in the Kakadu Region, northern Australia. *Mar. Freshw. Res.* **2018**, *69*, 1079–1091. [CrossRef]
7. Zhang, P.; Zou, Y.; Xie, Y.; Zhang, H.; Liu, X.; Gao, D.; Yi, F. Shifts in distribution of herbivorous geese relative to hydrological variation in East Dongting Lake wetland, China. *Sci. Total Environ.* **2018**, *636*, 30–38. [CrossRef]
8. Zhang, P.; Zou, Y.; Xie, Y.; Zhang, S.; Zhu, F.; Chen, X.; Li, F.; Deng, Z.; Yao, Y.; Song, Y. Phenological mismatch caused by water regime change may explain the population variation of the vulnerable lesser white-fronted goose in east Dongting Lake, China. *Ecol. Indic.* **2021**, *127*, 107776. [CrossRef]
9. Traill, L.W.; Brook, B.W. An aggregative response of the tropical Australian magpie goose (*Anseranas semipalmata*) to seasonal floodplains. *J. Trop. Ecol.* **2011**, *27*, 171–180. [CrossRef]
10. Lu, C.; Jia, Y.; Jing, L.; Zeng, Q.; Lei, J.; Zhang, S.; Lei, G.; Wen, L. Shifts in river-floodplain relationship reveal the impacts of river regulation: A case study of Dongting Lake in China. *J. Hydrol.* **2018**, *559*, 932–941. [CrossRef]
11. Zeng, Q.; Lu, C.; Li, G.; Guo, Z.; Wen, L.; Lei, G. Impact of a dam on wintering waterbirds' habitat use. *Environ. Conserv.* **2018**, *45*, 307–314. [CrossRef]
12. Guan, L.; Wen, L.; Feng, D.D.; Zhang, H.; Lei, G.C. Delayed Flood Recession in Central Yangtze Floodplains Can Cause Significant Food Shortages for Wintering Geese: Results of Inundation Experiment. *Environ. Manag.* **2014**, *54*, 1331–1341. [CrossRef] [PubMed]
13. Feng, D.; Guan, L.; Shi, L.; Zeng, Q.; Liu, X.; Zhang, H.; Lei, G. Impact of autumn hydrologic regime on plants in beach and distribution of populations of wintering lesser white-fronted goose in East Dongting Lake. *Wetl. Sci.* **2014**, *12*, 491–498.
14. Tockner, K.; Stanford, J.A. Riverine flood plains: Present state and future trends. *Environ. Conserv.* **2002**, *29*, 308–330. [CrossRef]
15. Wang, H.; Liu, X.; Wang, H. The Yangtze River Floodplain: Threats and Rehabilitation. In *Fishery Resources, Environment, and Conservation in the Mississippi and Yangtze (Changjiang) River Basins*; American Fisheries Society: New York, NY, USA, 2016.
16. WWF. *Living Yangtze Report 2020*; World Wide Fund for Nature: Gland, Switzerland, 2020.
17. Bao, D.; Wang, X.; Lu, X. Wetland protection basing on basin ecological management in the middle—Lower reaches of the Yangtze River. *Wetl. Sci.* **2006**, *4*, 96–100.
18. Cui, L.; Gao, C.; Zhao, X.; Ma, Q.; Zhang, M.; Li, W.; Song, H.; Wang, Y.; Li, S.; Zhang, Y. Dynamics of the lakes in the middle and lower reaches of the Yangtze River basin, China, since late nineteenth century. *Environ. Monit. Assess.* **2013**, *185*, 4005–4018. [CrossRef]
19. Barter, M.; Cao, L.; Chen, L.; Lei, G. Results of a survey for waterbirds in the lower Yangtze floodplain, China, in January–February 2004. *Forktail* **2005**, *21*, 1.
20. Yan, M.; Yi, K.; Zhang, J.; Batbayar, N.; Xu, Z.; Liu, G.; Hu, B.; Zheng, B.; Antonov, A.; Goroshko, O. Flyway connectivity and population status of the Greylag Goose *Anser anser* in East Asia. *Wildfowl* **2020**, *6*, 157–180.
21. Tian, H.; Solovyeva, D.; Danilov, G.; Vartanyan, S.; Wen, L.; Lei, J.; Lu, C.; Bridgewater, P.; Lei, G.; Zeng, Q. Combining modern tracking data and historical records improves understanding of the summer habitats of the Eastern Lesser White-fronted Goose *Anser erythropus*. *Ecol. Evol.* **2021**, *11*, 4126–4139. [CrossRef]
22. Wang, X.; Fox, A.D.; Cong, P.; Barter, M.; Cao, L. Changes in the distribution and abundance of wintering Lesser White-fronted Geese *Anser erythropus* in eastern China. *Bird Conserv. Int.* **2012**, *22*, 128–134. [CrossRef]
23. Zhang, Y.; Cao, L.; Barter, M.; Fox, A.D.; Zhao, M.; Meng, F.; Shi, H.; Jiang, Y.; Zhu, W. Changing distribution and abundance of Swan Goose *Anser cygnoides* in the Yangtze River floodplain: The likely loss of a very important wintering site. *Bird Conserv. Int.* **2011**, *21*, 36–48. [CrossRef]

24. Li, C.; Zhao, Q.; Solovyeva, D.; Lameris, T.; Batbayar, N.; Bysykatova-Harmey, I.; Lee, H.; Emelyanov, V.; Rozenfeld, S.B.; Park, J.; et al. Population trends and migration routes of the East Asian Bean Goose *Anser fabalis middendorffii* and *A. f. serrirostris*. *Wildfowl* **2020**, *70*, 124–156.
25. Zhao, M.; Cong, P.; Barter, M.; Fox, A.D.; Cao, L. The changing abundance and distribution of Greater White-fronted Geese *Anser albifrons* in the Yangtze River floodplain: Impacts of recent hydrological changes. *Bird Conserv. Int.* **2012**, *22*, 135–143. [CrossRef]
26. Wang, W.; Fraser, J.D.; Chen, J. Wintering waterbirds in the middle and lower Yangtze River floodplain: Changes in abundance and distribution. *Bird Conserv. Int.* **2017**, *27*, 167–186. [CrossRef]
27. Long, T.; Tang, J.; Pilfold, N.W.; Zhao, X.; Dong, T. Predicting range shifts of *Davidia involucrata* Ball. under future climate change. *Ecol. Evol.* **2021**, *11*, 12779–12789. [CrossRef] [PubMed]
28. Qing, J.; Yang, Z.; He, K.; Zhang, Z.; Gu, X.; Yang, X.; Zhang, W.; Yang, B.; Qi, D.; Dai, Q. The minimum area requirements (MAR) for giant panda: An empirical study. *Sci. Rep.* **2016**, *6*, 37715. [CrossRef] [PubMed]
29. Chen, Y.; Shan, X.; Ovando, D.; Yang, T.; Dai, F.; Jin, X. Predicting current and future global distribution of black rockfish (*Sebastes schlegelii*) under changing climate. *Ecol. Indic.* **2021**, *128*, 107799. [CrossRef]
30. McCullagh, P.; Nelder, J.A. *Generalized Linear Models*; Routledge: New York, NY, USA, 2019.
31. Breiman, L. Random forests. *Mach. Learn.* **2001**, *45*, 5–32. [CrossRef]
32. Ilanloo, S.S.; Ashrafi, S.; Shabani, A.A. Modeling Habitat Suitability of the Red-backed Shrike (*Lanius collurio*) in the Irano-Anatolian Biodiversity Hotspot. *J. Zool. Res.* **2021**, *3*, 1–8. [CrossRef]
33. Zhang, Z.; Xu, S.; Capinha, C.; Weterings, R.; Gao, T. Using species distribution model to predict the impact of climate change on the potential distribution of Japanese whiting *Sillago japonica*. *Ecol. Indic.* **2019**, *104*, 333–340. [CrossRef]
34. Chen, J.; Wang, F.; Xia, X.; Zhang, L. Major element chemistry of the Changjiang (Yangtze River). *Chem. Geol.* **2002**, *187*, 231–255. [CrossRef]
35. Gao, B.; Yang, D.; Yang, H. Impact of the Three Gorges Dam on flow regime in the middle and lower Yangtze River. *Quat. Int.* **2013**, *304*, 43–50. [CrossRef]
36. Hampe, A.; Petit, R.J. Conserving biodiversity under climate change: The rear edge matters. *Ecol. Lett.* **2005**, *8*, 461–467. [CrossRef] [PubMed]
37. Jackson, S.T.; Sax, D.F. Balancing biodiversity in a changing environment: Extinction debt, immigration credit and species turnover. *Trends Ecol. Evol.* **2010**, *25*, 153–160. [CrossRef]
38. Booth, T.H.; Nix, H.A.; Busby, J.R.; Hutchinson, M.F.; Franklin, J. BIOCLIM: The first species distribution modelling package, its early applications and relevance to most current MaxEnt studies. *Divers. Distrib.* **2013**, *20*, 1–9. [CrossRef]
39. Karger, D.N.; Conrad, O.; Böhner, J.; Kawohl, T.; Kreft, H.; Soria-Auza, R.W.; Zimmermann, N.E.; Linder, H.P.; Kessler, M. Climatologies at high resolution for the earth's land surface areas. *Sci Data.* **2017**, *4*, 170122. [CrossRef]
40. Doiron, M.; Legagneux, P.; Gauthier, G.; Lévesque, E. Broad-scale satellite Normalized Difference Vegetation Index data predict plant biomass and peak date of nitrogen concentration in Arctic tundra vegetation. *Appl. Veg. Sci.* **2013**, *16*, 343–351. [CrossRef]
41. Luoto, M.; Virkkala, R.; Heikkinen, R.K. The role of land cover in bioclimatic models depends on spatial resolution. *Glob. Ecol. Biogeogr.* **2007**, *16*, 34–42. [CrossRef]
42. Thuiller, W.; Araujo, M.B.; Lavorel, S. Do we need land-cover data to model species distributions in Europe? *J. Biogeogr.* **2004**, *31*, 353–361. [CrossRef]
43. Wilson, J.W.; Sexton, J.O.; Jobe, R.T.; Haddad, N.M. The relative contribution of terrain, land cover, and vegetation structure indices to species distribution models. *Biol. Conserv.* **2013**, *164*, 170–176. [CrossRef]
44. He, K.; Dai, Q.; Gu, X.; Zhang, Z.; Zhou, J.; Qi, D.; Gu, X.; Yang, X.; Zhang, W.; Yang, B.; et al. Effects of roads on giant panda distribution: A mountain range scale evaluation. *Sci. Rep.* **2019**, *9*, 1110. [CrossRef] [PubMed]
45. Schmitt, S.; Pouteau, R.; Justeau, D.; De Boissieu, F.; Birnbaum, P. ssdm: An r package to predict distribution of species richness and composition based on stacked species distribution models. *Methods Ecol. Evol.* **2017**, *8*, 1795–1803. [CrossRef]
46. Struebig, M.J.; Linkie, M.; Deere, N.J.; Martyr, D.J.; Millyanawati, B.; Faulkner, S.C.; Le Comber, S.C.; Mangunjaya, F.M.; Leader-Williams, N.; McKay, J.E. Addressing human-tiger conflict using socio-ecological information on tolerance and risk. *Nat. Commun.* **2018**, *9*, 3455. [CrossRef]
47. Steinwart, I.; Christmann, A. *Support Vector Machines*; Springer Science & Business Media: Berlin/Heidelberg, Germany, 2008.
48. Ripley, B.D. *Pattern Recognition and Neural Networks*; Cambridge University Press: Cambridge, UK, 2007.
49. Hastie, T.; Tibshirani, R. Generalized additive models: Some applications. *J. Am. Stat. Assoc.* **1987**, *82*, 371–386. [CrossRef]
50. Gray, J.B.; Fan, G. Classification tree analysis using TARGET. *Comput. Stat. Data Anal.* **2008**, *52*, 1362–1372. [CrossRef]
51. Suggala, A.; Liu, B.; Ravikumar, P. Generalized boosting. *Adv. Neural Inf. Process. Syst.* **2020**, *33*, 8787–8797.
52. Eberhard, F.E.; Cunze, S.; Kochmann, J.; Klimpel, S. Modelling the climatic suitability of Chagas disease vectors on a global scale. *eLife* **2020**, *9*, e52072. [CrossRef]
53. Elith, J.; Graham, C.H.; Anderson, R.P.; Dudík, M.; Ferrier, S.; Guisan, A.; Hijmans, R.J.; Huettmann, F.; Leathwick, J.R.; Lehmann, A. Novel methods improve prediction of species' distributions from occurrence data. *Ecography* **2006**, *29*, 129–151. [CrossRef]
54. Erauskin-Extramiana, M.; Arrizabalaga, H.; Hobday, A.J.; Cabre, A.; Ibaibarriaga, L.; Arregui, I.; Murua, H.; Chust, G. Large-scale distribution of tuna species in a warming ocean. *Glob. Chang. Biol.* **2019**, *25*, 2043–2060. [CrossRef]
55. Kanagaraj, R.; Araujo, M.B.; Barman, R.; Davidar, P.; De, R.; Digal, D.K.; Gopi, G.; Johnsingh, A.; Kakati, K.; Kramer-Schadt, S. Predicting range shifts of Asian elephants under global change. *Divers. Distrib.* **2019**, *25*, 822–838. [CrossRef]

56. Zhang, P.; Dong, X.; Grenouillet, G.; Lek, S.; Zheng, Y.; Chang, J. Species range shifts in response to climate change and human pressure for the world's largest amphibian. *Sci. Total Environ.* **2020**, *735*, 139543. [CrossRef] [PubMed]
57. Zhang, Z.; Mammola, S.; Liang, Z.; Capinha, C.; Wei, Q.; Wu, Y.; Zhou, J.; Wang, C. Future climate change will severely reduce habitat suitability of the Critically Endangered Chinese giant salamander. *Freshw. Biol.* **2020**, *65*, 971–980. [CrossRef]
58. Ao, P.R.; Wang, X.; Solovyeva, D.; Meng, F.J.; Ikeuchi, T.; Shimada, T.; Park, J.; Gao, D.; Liu, G.; Hu, B.; et al. Rapid decline of the geographically restricted and globally threatened Eastern Palearctic Lesser White-fronted Goose *Anser erythropus*. *Wildfowl* **2020**, *9*, 206–243.
59. Yu, H.; Wang, X.; Cao, L.; Zhang, L.; Jia, Q.; Lee, H.; Xu, Z.; Liu, G.; Xu, W.; Hu, B. Are declining populations of wild geese in China 'prisoners' of their natural habitats? *Curr. Biol.* **2017**, *27*, R376–R377. [CrossRef] [PubMed]
60. Zhao, Q.; Wang, X.; Cao, L.; Fox, A.D. Why Chinese wintering geese hesitate to exploit farmland. *IBIS* **2018**, *160*, 703–705. [CrossRef]
61. Xie, C.; Huang, X.; Mu, H.; Yin, W. Impacts of land-use changes on the lakes across the Yangtze floodplain in China. *Environ. Sci. Technol.* **2017**, *51*, 3669–3677. [CrossRef]
62. Penghao, C.; Pingkuo, L.; Hua, P. Prospects of hydropower industry in the Yangtze River Basin: China's green energy choice. *Renew. Energy* **2019**, *131*, 1168–1185. [CrossRef]
63. Cyranoski, D. Putting China's wetlands on the map. *Nature* **2009**, *458*, 134. [CrossRef]
64. Gong, P.; Niu, Z.; Cheng, X.; Zhao, K.; Zhou, D.; Guo, J.; Liang, L.; Wang, X.; Li, D.; Huang, H. China's wetland change (1990–2000) determined by remote sensing. *Sci. China Earth Sci.* **2010**, *53*, 1036–1042. [CrossRef]
65. Teng, J.; Xia, S.; Liu, Y.; Yu, X.; Zhao, C. Assessing habitat suitability for wintering geese by using Normalized Difference Water Index (NDWI) in a large floodplain wetland, China. *Ecol. Indic.* **2021**, *122*, 107260. [CrossRef]
66. Lei, J.; Jia, Y.; Wang, Y.; Lei, G.; Lu, C.; Saintilan, N.; Wen, L. Behavioural plasticity and trophic niche shift: How wintering geese respond to habitat alteration. *Freshw. Biol.* **2019**, *64*, 1183–1195. [CrossRef]
67. Zou, Y.; Tang, Y.; Xie, Y.; Zhao, Q.; Zhang, H. Response of herbivorous geese to wintering habitat changes: Conservation insights from long-term population monitoring in the East Dongting Lake, China. *Reg. Environ. Change* **2017**, *17*, 879–888. [CrossRef]

environments

Article

Incorporating Industrial and Climatic Covariates into Analyses of Fish Health Indicators Measured in a Stream in Canada's Oil Sands Region

Tim J. Arciszewski [1,*], Erin J. Ussery [2] and Mark E. McMaster [2]

[1] Alberta Environment and Parks, Calgary, AB T2L 1Y1, Canada
[2] Environment and Climate Change Canada Burlington, Burlington, ON L7S 1A1, Canada; erin.ussery@ec.gc.ca (E.J.U.); mark.mcmaster@ec.gc.ca (M.E.M.)
* Correspondence: tima@unb.ca

Abstract: Industrial and other human activities in Canada's oil sands region (OSR) influence the environment. However, these impacts can be challenging to separate from natural stresses in flowing waters by comparing upstream reference sites to downstream exposure locations. For example, health indicators of lake chub (*Couesius plumbeus*) compared between locations in the Ells River (Upper and Lower) in 2013 to 2015 and 2018 demonstrated statistical differences. To further examine the potential sources of variation in fish, we also analyzed data at sites over time. When fish captured in 2018 were compared to pooled reference years (2013–2015), results indicated multiple differences in fish, but most of the differences disappeared when environmental covariates were included in the Elastic Net (EN) regularized regression models. However, when industrial covariates were included separately in the EN, the large differences in 2018 also disappeared, also suggesting the potential influence of these covariables on the health of fish. Further ENs incorporating both environmental and industrial covariates along with other variables which may describe industrial and natural influences, such as spring or summer precipitation and summer wind speeds and distance-based penalty factors, also support some of the suspected and potential mechanisms of impact. Further exploratory analyses simulating changes from zero and the mean (industrial) activity levels using the regression equations respectively suggest effects exceeding established critical effect sizes (CES) for fish measurements may already be present or effects may occur with small future changes in some industrial activities. Additional simulations also suggest that changing regional hydrological and thermal regimes in the future may also cause changes in fish measurements exceeding the CESs. The results of this study suggest the wide applicability of the approach for monitoring the health of fish in the OSR and beyond. The results also suggest follow-up work required to further evaluate the veracity of the suggested relationships identified in this analysis.

Keywords: oil sands; fish; Ells River; elastic net; industry; climate; environmental health

Citation: Arciszewski, T.J.; Ussery, E.J.; McMaster, M.E. Incorporating Industrial and Climatic Covariates into Analyses of Fish Health Indicators Measured in a Stream in Canada's Oil Sands Region. *Environments* **2022**, *9*, 73. https://doi.org/10.3390/environments9060073

Academic Editor: Matteo Convertino

Received: 27 April 2022
Accepted: 7 June 2022
Published: 17 June 2022

1. Introduction

Documenting and assessing the status of biological populations and communities and identifying sources of variation is a priority for ecological monitoring and management programs [1–4], including those in Canada's Oil Sands Region (OSR) [5]. Among the studies performed in the OSR to understand the effects of industrial developments, analyses of fish health indicators have been undertaken in streams in the minable sub-region since the late 1990s [6–8]. Using designs inspired by the Federal Environmental Effects Monitoring program (EEM; [9,10]), these fish studies typically collected sentinel species from an exposed site and statistically evaluate anatomical measurements relative to an upstream or local reference site not exposed to the stressor of interest e.g., [11]. This design has also been used for other indicators and has routinely identified differences at the downstream exposure locations, suggesting the potential influence of industrial activities [12–18].

Although differences in environmental indicators measured at downstream sites compared to upstream references are commonly observed in the OSR, isolating the potential effects of industrial development is often a substantial challenge for monitoring in flowing waters in the region [19,20]. More specifically, while industrial developments have known physical impacts on the landscape and hydrological processes [6,21,22], oil sands mines are preferentially constructed where the bitumen deposits are thickest and the overburden is thinnest [23]. Streams in the areas targeted for development are often influenced by natural exposure to bituminous compounds eroded from the McMurray Formation and other substances derived from other geological strata, including other hydrocarbon-bearing formations present between the study locations [12,15,16,24,25]. Substances originating from deeper rock layers may also be transported to the surface in groundwater in the OSR [26–31]. Although some of the observed changes in environmental indicators in these areas may be associated with facilities, other phenomena may also affect the efficacy of spatial comparisons in the OSR. Contaminants of concern (CoCs) are emitted to the atmosphere from stacks and fugitive sources and many are deposited to the landscape [20,32–36]. While the most intense deposition of materials often occurs within 10–20 km of the sources, e.g., [35], the deposition varies annually [20,32,37,38] and may also occur in areas used as references for studies in streams [39–41]. Additionally, in situ oil sand facilities may also be present upstream of reference sites [16,18], further altering the sensitivity of spatial designs to industrial influences.

The potential overlap of stressors from multiple sources and the confounding of spatial designs often constrains the unequivocal identification of industrial influence in the OSR [19,20]. These challenges of spatial designs in streams from the OSR coupled with the desire to quantify any industrial influence on organisms suggests alternative approaches are required. Although many options are available [18,42], analyses of data collected at individual sites over time may be used to identify industrial influences. These temporal and site-specific analyses can account for natural differences between locations, have been routinely used in the OSR, and have identified likely effects of industrial developments, including some potential influences in streams [13–16,19,25,36,41,43–51].

Although a site-specific approach has been useful in many studies, the design also has challenges. For example, the chosen ecological indicators often vary naturally over time and this may interfere with the identification of any industrial influences. Additionally, pre-industrial baselines may be absent. Whereas some measurement types, such as chemical indicators in lake sediments and peat cores [47,52], may be less sensitive to these challenges, studies in streams require deliberate solutions. To overcome these challenges in flowing waters, generic covariates available from meteorological and hydrological stations can be incorporated to remove their influence on environmental indicators [32,39,53]. Although any residual variation could be qualitatively associated with extraneous sources, descriptors of industrial performance for individual facilities can also be included to identify any potential associations between the indicators and the local development. Although not definitive, this approach may improve the utility of monitoring in the OSR, including identifying industrial influence without a pre-industrial baseline [20,32,39,54–56]

The purpose of this work is to evaluate the status of fish (lake chub; *Couesius plumbeus*) residing in the Ells River relative to industrial activity. To examine the potential effects of industrial influence, several approaches were used and were compared. First, the typical EEM approach was used to identify statistical differences both spatially and temporally using ordinary least squares (OLS) and generalized linear models (GLM) and were compared to the use of regression diagnostics. Second, the effect of including various sets of covariates on the temporal analyses was examined and the elastic net regularized regression (EN). This study suggests the analysis of the ecological data in exposure environments such as the Athabasca tributaries in the OSR may need to integrate environmental and industrial covariates to better account for variability in the data set and to enhance the sensitivity of the analysis to local human activities.

2. Materials and Methods

2.1. Study Area

This study was performed in the Ells River, a tributary of the Athabasca River in Canada's OSR. Two sites, Upper and Lower, along the Ells River were sampled consecutively between 2013 and 2015 and again in 2018 (Figure 1A). The Lower site is in the midst of oil sands development, and the Upper site is located approximately 35 river-km upstream of the Lower site. The Upper site is underlain by the Clearwater Formation, while the Lower site is underlain by the McMurray Formation (Figure 1B). During the initiation of this study (2013), site preparation and construction of the (Total) Joslyn North mine was underway, and the sites were selected to straddle this development (Figure 1C). However, the development of the Joslyn North mine was suspended in May 2014 [57,58]. Most of the development and site preparation in the Ells basin occurred upstream of the Lower location and downstream of the Upper site before the initiation of fish collections (Figure 1C; Supplemental Figure S1).

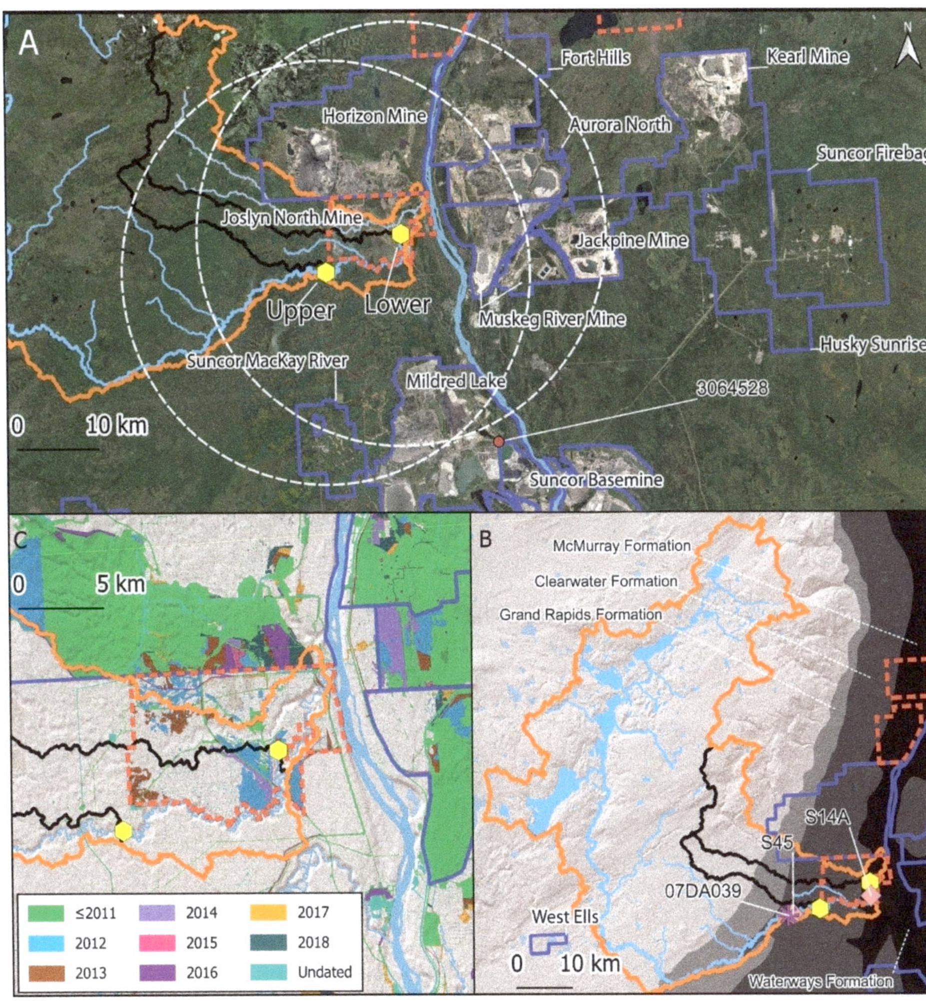

Figure 1. Area of study in Canada's Oil Sands Region; (**A**): fishing locations in the Ells basin (Upper and Lower), project boundaries of facilities operating in 2018 (blue outlines; Fort Hills began production

in 2017 and its data were not included in the current analysis), facilities paused in 2018 (red dashed outlines), 25 km distances from Ells locations (white dashing); orange outline = Ells basin; black outlines show upslope areas for fishing locations; dark red dot shows Mildred Lake meteorological station (3064528); ESRI World Imagery basemap ~ 2019; (**B**): bedrock formations of the Mannville group (Grand Rapids, Clearwater, McMurray) and the Devonian basement (Waterways); all other geological strata not shown; pink diamonds = stream gauges (S45 and S14A); purple star = stream gauge for Upper location in 2018 (07DA039); (**C**): land disturbances in and around the Ells basin by year.

Lake chub were selected as the sentinel species for work in the Ells River. Following the Canadian EEM protocols, ~20 adult female and 20 adult male lake chub were targeted using backpack electrofishers (Smith-Root Type LR-24; Alberta Environment and Parks Fisheries Research Licenses: 13-0445, 14-0456, 15-0456, and 18-0408). September is the typical period of recrudescence for the spring spawning lake chub [59], and fish were collected during this period (22–24 September 2013; 24–26 September 2014; 6–8 October 2015; 19–21 September 2018). The collected fish were euthanized by spinal severance following the Animal Use Protocols approved by Environment and Climate Change Canada's Animal Care Committee (1315, 1415, 1515 and 1815). Sex, body length (BL; ±1 mm), body weight (BW; ±0.001 g), gonad weight (GW; ±0.001 g) and liver weight (LW; ±0.001 g) were recorded. Actual sample sizes range from 12 to 21.

2.2. Analysis of Data

2.2.1. Conventional Statistics and Regression Diagnostics

Conventional analyses of fish health (downstream exposure relative to upstream references) were performed in this study. Site differences in fish health endpoints (GW, LW, BW) were analyzed using the 'lm()' and 'glm()' functions in R to assess differences in slope and intercepts. Ordinary least squares (OLS) analyses used $\log_{10}$-transformed data. Generalized linear models (GLM) were applied using a gamma distribution and a log-link function; visual examination of residuals suggested no substantial deviation from normality, and this technique was used for the remainder of the analyses. All other fish health endpoints were assessed using 1-factor ANOVA followed by Tukey's post hoc test (normality of all data was confirmed using a Shapiro–Wilk test before subsequent statistical analysis; if data were not normally distributed, the data were $\log_{10}$-transformed). An α level of 0.05 was used to determine statistical significance for all tests. These analyses were performed both spatially (within years) and temporally (within sites). To correspond with the regression diagnostics approach, GLMs testing the statistical differences at sites in 2018 compared to a pooled group of fish from 2013 to 2015 were also applied at each location. The results of these analyses are described in detail in the Supplementary Information and Appendix A but are also summarized in the main text.

Additional analyses based on regression diagnostics [60] were also performed in this study. First, GLMs were used to calculate regression equations for GW and LW relative to BW and BW relative to BL at sites over time. From these models, the fish collected from 2013 to 2015 were used to estimate the expected ranges of residuals for fish captured in 2018. Additional regression diagnostics were conducted spatially using the fish captured at the Upper site to predict the GW, LW, and BW of fish for the Lower site.

Bootstrapping was used to calculate both the observed and expected ranges of residuals per year for the fish measurements evaluated here. For the observed residuals, a single bootstrap was used to calculate the central 95% confidence intervals of means. In contrast, a double bootstrap was used to estimate the expected ranges of fish in 2018 compared to the previous 3 years. Details of the single and double bootstrap techniques are found elsewhere [55,61].

2.2.2. Elastic Net Regularized Regressions

The temporal analyses described immediately above examining the health metrics of fish over time identified potentially large differences in several measurements of fish health in 2018 compared to previous years (e.g., Figure 2). In addition, variability within the baseline period of 2013 to 2015 was also apparent among the fish health metrics, including model errors (e.g., Figure 2). Using environmental (ENV) and industrial (IND) covariates, the observed variability and potentially relevant differences prompted additional analyses using the elastic net regularized regression (EN; [62]) to identify potentially relevant drivers of the variability in the fish health metrics. The ENs were applied using the glmnet package in R [63] using penalty ratio (α) of 0.5 and a Gamma distribution with a log-link function. The minimum λ penalty was selected using leave-one-out-cross validation using the 'cv.glmnet()' function in the glmnet package.

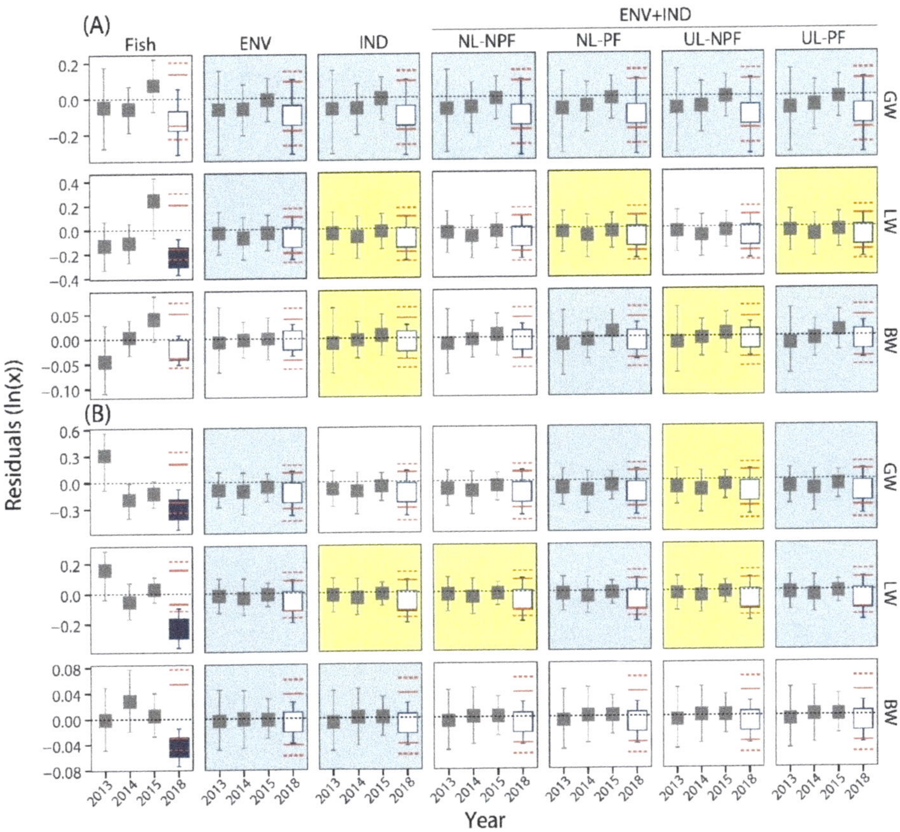

Figure 2. Mean residuals of female endpoints measured at the Upper Ells (**A**) and Lower Ells (**B**) locations; 2013:2015 (gray symbols) are reference years; 2018 is the test year; red horizontal dashed and solid lines show expected ranges of mean residuals in 2018 given the reference data; open symbols suggest no difference in mean residuals; closed symbols suggest a difference in 2018; 'Fish' panes include no environmental or industrial covariates; 'ENV' includes environmental covariates only; 'IND' includes industrial covariates only (with no limits and no penalty factors); ENV+IND with no limits and no penalty factors ('NL-NPF'); ENV+IND with no limits and penalty factors ('NL-PF'); ENV+IND with upper limits and no penalty factors ('UL-NPF'); ENV+IND with upper limits and penalty factors ('UL-PF'); blue shading of panes indicates the most parsimonious models selected by AIC; yellow shading indicates marginally relevant models selected by AIC.

First, ENs were conducted using environmental covariates only. To capture potential natural contributions to indicators of fish health, three environmental covariates were used in this analysis: water temperature (WT), air temperature (AT), and stream discharge (SD; Supplemental Table S1; Supplemental Figure S2). Among these covariates, we included four periods: August 5 until the start of fish sampling, 60 days before fish sampling, summer (June, July, August), and June, July, August, and September until the date of fish capture. Each of these periods corresponds with the numerical descriptors 1–4, with greater values indicating longer periods. Additionally, we also used both means ($\overline{x}$) and medians ($\widetilde{x}$) for each of these periods. For example, 'AT3-$\overline{x}$' represents the mean air temperature from 1 June to 31 August for each respective year of study (Supplemental Table S1; Supplemental Figure S2). Hobo Tidbit temperature recorders were deployed at each Ells River station in each sampling year. In 2015, deployment of the probes did not occur until August 4, and only two estimates of WT were used in this analysis. To compensate for the limited availability of WT, AT data were also obtained and used to calculate possible physical descriptors of environmental conditions in the Ells River. These AT data were obtained from the Mildred Lake meteorological station (Climate ID: 3064528). Stream discharge data were also obtained for the Ells River (2013–2015: S45 and 2018: 07DA039). Given the potential influence of industrial activities at the Joslyn North project, only discharge data from the upstream gauging locations (S45 and 07DA039) were used to approximate natural and unimpeded flows in the Ells River (e.g., [6,64]) for the initial analyses using only the environmental covariates (ENV models); as described below, discharge data from the station near the Lower Ells, S14A, were used in the ENV+IND EN models for the fish captured at that site. Although not an environmental covariate, Julian Day (JD) was also included as a selectable covariate to account for slight differences in the collection of fish among sampling years.

Next, an EN using only industrial covariates was also used. Industrial covariates from eleven facilities (mines: Horizon Mine (HM), Jackpine Mine (JPM), Kearl Mine (KM), Muskeg River Mine (MRM), Syncrude Aurora North (SAN), Suncor Basemine (SBM), and Syncrude Mildred Lake (SML); in situ: Suncor Firebag (SFB), Suncor MacKay River (SMR), Husky Sunrise (HS), and West Ells Sunshine (WES); Figure 1) were obtained from the Alberta Energy Regulator (Supplemental Figure S3; [65,66]). Although they varied by facility, the covariates included fuels combusted (e.g., process (PG-F) and natural gases (NG-F) and petroleum coke (PC-F)), products produced (e.g., crude bitumen (CB-P) and synthetic crude (SC-P)), the mass of mined bitumen (OS-M), stockpiling rate of petroleum coke (PC-SP), and materials flared/wasted (e.g., sulphur (S-FW), diluent naphtha (DN-FW), and crude bitumen (CB-FW)). Steam injection rates (ST) and bitumen production rates (BP) were also obtained for the in situ facilities (HS, SMR, SFB, WES). Data were obtained from mines and in situ facilities with at least two years of operation by 2018 (Figure 1). Summer sums for the industrial facilities were calculated using the reported values from June, July, August, and September. The full list of initial industrial variables is available in the Supplemental Information (Supplemental Table S2).

Land disturbance data was also used. Land disturbance data were obtained from the Alberta Biodiversity Monitoring Institute's 2018 Human footprint data ([67]; Figure 1; Supplemental Figure S1). Not all features in this data layer are dated, but the data serve as a proxy for the proportional increase in land disturbance per year in the watershed areas draining to each sampling location. The sub-watersheds for each stream location were calculated using the 'Upslope Area' function in Quantum GIS (QGIS version 3.16).

Finally, analyses using both environmental and industrial covariates were performed. These analyses used the covariates described above, but also included additional 'mixture' variables to account for the potential cumulative industrial influence, the physical transport processes linking facilities with streams, and other seasonal effects (Supplemental Figure S4; [39,55]). These mixture variables included estimates of mean spring discharge (SP-SD) and air temperature (SP-MT; from Mildred Lake). Spring was estimated as April, May, and June to capture the freshet period. Spring precipitation estimates, mean precipitation per day (SP-MP), number of rain days (SP-RD), and the cumulative total precipitation (SP-TP) were also included. Additional metrics in the summer were also obtained, including mean (e.g., WS4-$\bar{x}$) and upper percentile wind speeds (e.g., WS4-99) from the Bertha Ganter station in the Wood Buffalo Environmental Association air monitoring network and precipitation from the Mildred Lake meteorological station (e.g., P1-$\bar{x}$) for the same periods outlined above. In total, the number of potential environmental, industrial, and mixture covariates included in the ENV+IND models was 147 (Supplemental Tables S1 and S2; Supplemental Figures S1–S3).

In addition to the expanded list of covariates, the ENs were further configured using arguments in the glmnet() function in R [63]. More specifically, an 'upper.limit = 0' argument was used for industrial variables to return only negative coefficients. Secondly, the 'penalty.factor' argument was used to account for the proximity of projects to the sampling locations (Supplemental Table S3). For example, the project boundaries of three operational mines, Horizon, Muskeg River, and Aurora North, are within 10 km of the Lower Ells location, but are physically outside the Ells basin (Figure 1; Supplemental Table S3). A penalty factor (PF) of ((distance (km)/100) + 1) was explored here. The scaled PF was selected during preliminary analyses in which km/10 was likely too punitive, whereas km/1000 had no effect. An exploratory analysis of the effect of varying PFs was also undertaken but showed little effect on the mean squared error (MSE) of the final models (Supplemental Figure S5). However, this preliminary analysis is further considered below (Section 3.3)

Based on the inclusion or exclusion of upper limits and PFs, four ENs were performed when environmental, industrial, and mixture variables were included. These analyses included no limits and no penalty factors (NL-NPF), upper limits of zero and no penalty factors (UL-NPF), no limits and (distance-based) penalty factors (NL-PF), and upper limits of zero and the distance-based penalty factors (UL-PF). To compare the performance of various EN models, the deviance ratio (DR), MSE, and Akaike's Information Criterion (AIC) were calculated [53]. Criteria to evaluate AICs followed generic guidance from the literature [68].

2.2.3. Retrospective and Prospective Model Predictions

Most ENs selected industrial variables, and the regression formulae were also used to estimate the potential current or future effects of these industrial influences and projected climatic and hydrological changes. Among the industrial variables, two sets of predictive models were used. The mean of each industrial variable selected in the best-fit ENs from 2013 to 2015 and 2018 was used to estimate the magnitude of differences causing a change equivalent to the fish Critical Effect Size (CES) used for the morphological ratios of gonado-somatic, liver-somatic, and condition indices and applied to the GW, LW, and BW measurements examined here (±25% for GW and LW and ±10% for BW; [69]). These CESs are used in the EEM program and indicate a potentially high risk to the environment [69]. The same approach was also used to estimate the magnitude of change in the absence of an industrial variable (by setting initial conditions to zero) eliciting a response in a fish endpoint exceeding the established CESs. In addition, potential (and predicted) changes in air temperature (1 to 4 °C) and stream discharge (up to 20% decline; [70,71]) were used to estimate potential future effects of these changes on GW, LW, and BW of lake chub captured in the Ells basin predicted by the various EN models selected by AIC.

3. Results and Discussion

3.1. Analyses without Environmental or Industrial Covariates

As described in detail in the Appendix A, statistical comparisons of fish captured at the Lower Ells compared to fish from the Upper Ells using OLS and GLMs often showed significant differences between sites, and many were larger than the stipulated CESs (Supplemental Tables S3–S6; Supplemental Figures S6–S13). Although the statistical differences between sites were not always present, and the directions or magnitudes of observed differences are not always consistent, two large consecutive differences were observed in GW of females in 2015 and 2018 (Supplemental Table S6) and the GW of males at the Lower Ells was consistently below the value precited from the Upper location (Supplemental Figure S7). These differences parallel other studies in the Ells River showing some differences between upstream and downstream locations [8,14,16,24], but as in some of the previous studies, the role of industrial influences above any differences driven by habitat e.g., [56] are not clear. However, the results would prompt follow-up work in a standard EEM study, including site-specific analyses e.g., [9,72].

As described in the Appendix A, site-specific analyses were performed using traditional EEM statistics (OLS), GLMs, and regression diagnostics. Results from both the traditional EEM OLS analyses and the GLMs generally suggest no consistent differences within sites over time (Supplemental Tables S7 and S8), but some distinctions were also apparent. Among the largest was the common detection of significantly different slopes in GW of female lake chub captured at both sites in GLMs, but not in the OLS (Supplemental Tables S7 and S8). Although the detection of differences in the slope using the GLM, but not the OLS, may be related to the transformation of data before the OLS, overall, there are few temporally persistent patterns suggesting a potential influence of annually variable factors, such as temperature or industrial activity. However, the GW of females captured at the Lower Ells site was statistically greater and larger than the CES of 25% in 2013 compared to all other years (Supplemental Tables S7 and S8).

In this study, we used site-specific analyses over time to determine the health of fish in the Ells River to address the known interpretative limitations of comparing downstream exposure locations to upstream reference sites in the OSR. Corresponding with the remainder of the results described below (Section 3.2), site-specific analyses were also performed using GLMs and regression diagnostics using a pooled reference group of years (2013 to 2015) to determine the potential differences in 2018. The results from the regression (residual) diagnostics mirrored those from the Gamma GLMs performed using the pooled 2013–2015 group compared to 2018 (Figures 2 and 3; Supplemental Table S9). Differences in 2018 compared to 2013–2015 were observed in multiple endpoints, including LW of females from the Upper site and in GW, LW, and BW of females collected at the Lower site (Figure 2). Only one difference identified by the Gamma GLM, the slope of GW of males at the Lower site, was not also identified by regression diagnostics (Figure 3; Supplemental Table S9). However, as described in the Appendix A, tests of intercept may be robust to statistical differences in slope between 0.01 and 0.05 ($p = 0.043$). Some of the temporal differences at sites may be attributable to industrial development, including large declines in GW of females at the Upper site and declines in LW of females at the Upper site in 2018 compared to 2013–2015 (Supplemental Table S9), but the analytical approach used in this portion of the analysis also incorporates temporal variability which may be associated with environmental and/or industrial drivers, suggesting (along with the results of the spatial analyses described above) additional data, analyses, and interpretation are required.

3.2. *Site-Specific Analyses with Covariates*

3.2.1. Which EN Models Performed the Best?

Analyses to include covariates in this study used the EN and included models with only environmental covariates (ENV), only industrial covariates (IND), and those which used both ENV and IND covariates along with various constraints, including ULs of zero for coefficients and PFs for industrial parameters based on the distance of a sampling site to a facility. Among the 12 fish measurements among sites and sexes, the fish-only models for LW and GW of males and females at both sites suggested these models performed the best and as expected that most of the variation in these measurements is influenced by BW of fish ([54]; Supplemental Figures S14 and S15). However, the MSE of the GLM 'fish' models (those including no ENV or IND covariates) were also occasionally the highest, especially in females (Supplemental Figures S14 and S15). In contrast, the BW (vs. BL) models suggested other ENs including ENV and/or IND covariates were preferred

(Supplemental Figures S14 and S15). Irrespective of the AICs suggesting the relative importance of the GLMs including only fish data, the occurrence of large differences in residual variation in 2018 compared to the baseline years (2013 to 2015; Figures 2 and 3) suggested the potential influence of drivers other than BW and BL on lake chub in the Ells River and the potential need for more complex explanatory models [55].

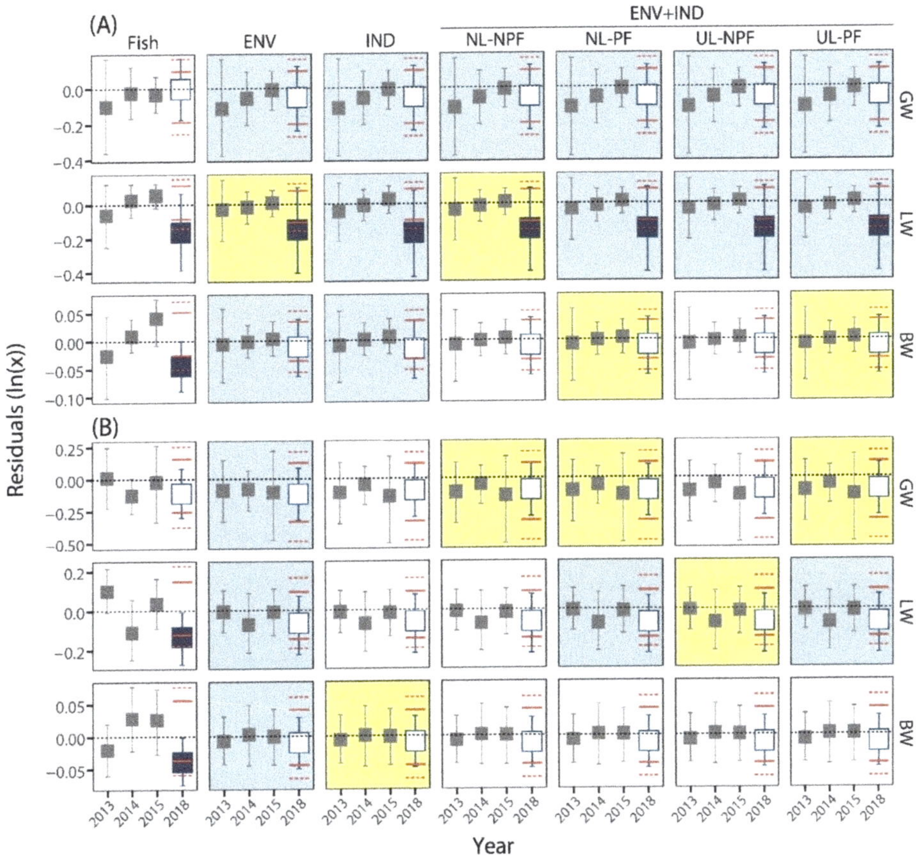

Figure 3. Mean residuals of male endpoints measured at the Upper Ells (**A**) and Lower Ells (**B**) locations; 2013–2015 (gray symbols) are reference years; 2018 is test year; red horizontal dashed and solid lines show expected ranges of mean residuals in 2018 given the reference data; open symbols suggest no difference in mean residuals; closed symbols suggest a difference in 2018; 'Fish' panes include models with no environmental or industrial covariates; 'ENV' includes environmental covariates only; IND includes industrial covariates only (with no limits and no penalty factors); ENV+IND with no limits and no penalty factors (NL-NPF); ENV+IND with no limits and penalty factors (NL-PF); ENV+IND with upper limits and no penalty factors (UL-NPF); ENV+IND with upper limits and penalty factors (UL-PF); blue shading of panes shows the most parsimonious models identified by AIC; yellow shading indicates marginally relevant models selected by AIC.

When only the more complex models with larger sets of initial predictors (i.e., those including environmental and/or industrial covariates) are considered, among the 12 fish measurements among sites and sexes, the ENV model was either the best or among the best explanatory models in 10 and was marginal in an 11th (LW of males at the Upper site; Figures 2 and 3; Supplemental Figures S16 and S17). In the 12th endpoint, BW of females at the Upper site, the AIC of the ENV model was the highest suggesting this model performed the worst (Supplemental Figure S16). Although not always the minimum AIC, the IND models were also commonly selected. For example, the IND models were identified by AIC for GW, LW, and BW of males at the Upper site, BW of females at the Lower site, and GW of females at the Upper site (Supplemental Figures S16 and S17). Other ENs combining ENV and IND variables were also selected by AIC. For example, the UL-PF, NL-PF, and UL-NPF ENV+IND ENs (and the IND model) were among the best for the LW of males at the Upper site (Supplemental Figure S17). Similarly, the UL-PF and the NL-PF ENV+IND models for BW of females at the Upper site had the lowest AIC value (Supplemental Figure S16). The UL-PF and NL-PF models were also selected at the Lower site for the LW of males (Supplemental Figure S17) and GW and LW of females (Supplemental Figure S16). In contrast, the AICs for GW of males and females at the Upper site suggested all models performed similarly (Figures 2 and 3; Supplemental Figures S16 and S17). Finally, although not included in the best set of models, some were deemed marginally relevant by the AIC criteria [68], including several at the Lower site: IND for BW of males, NL-NPF for LW of females and GW of males, and UL-NPF for GW of females and LW of males (Supplemental Figures S16 and S17).

These results suggest some of these more complex models may include variables affecting fish health and may contribute to the declines in GW of females at the Lower site exceeding the CES compared to the Upper site (Supplemental Table S6). The selected models suggest the importance of including environmental covariates, but also the potential for industrial variables to explain the residual variation in the health indicators of lake chub to better focus on stressors of interest [54,55]. However, the analyses also suggested penalty factors may also be useful and are deserving of additional attention.

Although the remainder of this study focuses on the models selected by AIC, the selection of multiple models explaining the reducible error also suggests other criteria may be necessary for implementing the approach used here. For example, while the UL-PF models may represent the most realistic scenario examined in this study, there may be advantages to using the IND models by accounting for the greater likelihood of false positives, including defining testable hypotheses and follow-up studies [72]. Including covariates within the initial analyses and using multiple model forms may also accelerate the development of focused studies to quickly hone the monitoring approaches by further establishing testable and specific hypotheses [73,74].

3.2.2. Are Large Differences Apparent When Environmental and Industrial Covariates Are Included?

All estimates of fish health measured in lake chub captured in the Ells River in 2018 were within the ranges expected from baseline fish collected at each site from 2013 to 2015 when environmental and industrial covariates were included in the models, except LW of males from the Upper site (e.g., Figures 2 and 3). Although the tails of the estimated 95% CIs of the mean annual residuals were occasionally beyond the estimated outer tolerance limits, such as GW of females at the Upper site (Figure 2), these results typically suggest the environmental and industrial factors may be driving some of the variability in the measurements of fish health. In contrast, the mean residual difference in the LW of males from the Upper site was never inside the range of mean residuals irrespective of the covariates and the constraints used (e.g., Figure 3), suggesting that a correlated or causal factor was not included in the set of selectable variables. Despite this result, the correspondence of the GLM and regression diagnostic approach for evaluating sites over time and the common detection of spatial differences (e.g., Figure 3) suggests the utility of this approach. The correspondence between the GLMs and the regression diagnostics further supports the utility of the latter to identify large differences and prompting follow-up analyses or future field studies [72]. Additionally, the approach also highlights the potential sensitivity of these techniques to small influences and the likely utility of the temporal analyses irrespective of the results of spatial comparisons [39]. As discussed below, the site-specific approach also identified the potential influence of industrial activity at both locations, further supporting site-specific analyses.

3.2.3. Variables Selected by Elastic Net Regularized Regression

Influence of Environmental Variation

The results presented already suggest the potential influence of environmental and industrial covariates on the health indicators of lake chub captured in the Ells River, but which variables were selected also provides additional information on the potential drivers of ecological conditions. Among the ENs selected by AIC, all except the GW of males from the Upper site selected additional covariates compared to the independent variables included in the OLS and GLMs (Tables 1 and 2). This result suggests the GW of males at the Upper site is the least sensitive endpoint measured in this study. In contrast, the GW of males at the Lower site suggested the influence of recent AT (Table 2).

Table 1. Environmental and industrial covariates selected by EN for gonad weight (GW), liver wight (LW), and body weight (BW) of female fish captured at the Upper and Lower Ells locations; NL = no limits; UL = upper limits; NPF = no penalty factors; PF = penalty factors; bolded white text with saturated fill color in cells comprise best-fit judged by AIC values; red = positive coefficients; blue = negative coefficients; marginal models indicated as unsaturated fill and black text.

Site	Measure	ENV	IND	ENV+IND NL-NPF	ENV+IND NL-PF	ENV+IND UL-NPF	ENV+IND UL-PF
Upper	GW	AT1-$\tilde{x}$ AT3-$\tilde{x}$ SD3-$\bar{x}$ BW	MRM-CB-FW SBM-PG-FW MRM-DN-F BW	P2-75 MRM-CB-FW SBM-PG-FW BW	P2-75 P3-RD JD BW	P2-75 MRM-CB-FW MRM-NG-F BW	P2-75 P3-RD JD BW
	LW	AT1-$\tilde{x}$ AT3-$\tilde{x}$ JD SD3-$\bar{x}$ BW	MRM-CB-FW HM-PG-FW SBM-PG-FW SML-PC-F MRM-NG-F SAN-CB-P SBM-PG-P BW	P3-RD SP-TP MRM-CB-FW HM-PG-FW SBM-PG-FW MRM-NG-F SAN-CB-P JD BW	P2-75 P3-RD SP-MP SP-TP AT1-$\tilde{x}$ JD SD4-$\bar{x}$ BW	P2-75 P3-RD SP-MP SP-TP MRM-CB-FW MRM-DN-FW MRM-NG-F JD BW	P2-75 P3-RD SP-MP SP-TP AT1-$\tilde{x}$ JD SD4-$\bar{x}$ BW
	BW	JD SD1-$\tilde{x}$ SD2-$\bar{x}$ SD2-$\tilde{x}$ SD3-$\bar{x}$ SD3-$\tilde{x}$ SD4-$\bar{x}$ BL	HM-PG-FW MRM-OS-M SBM-OS-M SAN-CB-P SBM-PG-P BL	P3-RD P4-RD SP-MP SBM-OS-M SAN-CB-P SBM-PG-P SD4-$\bar{x}$ BL	P3-RD SP-MP SP-TP BL	P3-RD P4-RD SP-MP SD3-$\tilde{x}$ SD4-$\bar{x}$ BL	P3-RD SP-MP SP-TP BL
	GW	AT2-$\bar{x}$ AT2-$\tilde{x}$ SD3-$\tilde{x}$ WT1-$\tilde{x}$ BW	SBM-PC-SP SBM-CB-FW KM-DN-FW SBM-PG-F SAN-NG-F SML-NG-F SAN-SC-F SBM-PC-P	P1-99 P2-99 SBM-PC-SP SBM-CB-FW SBM-PG-F SAN-NG-F SML-NG-F SAN-SC-F	P1-99 P2-99 SAN-SC-F AT2-$\bar{x}$ BW	P1-99 P2-99 SBM-PC-SP SAN-NG-F SAN-SC-F SBM-PC-P BW	P1-99 P2-99 SAN-SC-F AT2-$\bar{x}$ BW

Table 1. *Cont.*

Site	Measure	ENV	IND	ENV+IND NL-NPF	ENV+IND NL-PF	ENV+IND UL-NPF	ENV+IND UL-PF
			SBM-CB-P BW	SBM-PC-P WT1-$\tilde{x}$ BW			
Lower	LW	AT1-$\bar{x}$ AT1-$\tilde{x}$ AT2-$\bar{x}$ AT2-$\tilde{x}$ WT1-$\bar{x}$ BW	SBM-CB-FW SBM-PC-F JPM-NG-F SBM-NG-F SML-OS-M JPM-CB-P MRM-CB-P SML-PG-P BW	SBM-PC-F SML-PG-F SBM-NG-F SML-OS-M JPM-CB-P MRM-CB-P AT2-$\bar{x}$ AT2-$\tilde{x}$ BW	MRM-CB-P AT1-$\bar{x}$ AT1-$\tilde{x}$ AT2-$\bar{x}$ AT2-$\tilde{x}$ BW	SML-PG-FW SBM-PC-F HM-DN-F JPM-NG-F JPM-CB-P MRM-CB-P AT1-$\bar{x}$ AT1-$\tilde{x}$ BW	MRM-CB-P AT1-$\bar{x}$ AT1-$\tilde{x}$ AT2-$\bar{x}$ AT2-$\tilde{x}$ BW
	BW	AT1-$\bar{x}$ AT3-$\bar{x}$ AT4-$\bar{x}$ AT4-$\tilde{x}$ BL	HM-DN-FW KM-DN-F HM-OS-M SFB-ST BL	WS1-$\bar{x}$ WS2-$\bar{x}$ WS2-$\tilde{x}$ WS3-$\bar{x}$ WS3-$\tilde{x}$ WS4-$\tilde{x}$ WG3-99 WG4-75 BL	WS1-$\bar{x}$ WS2-$\bar{x}$ WS2-$\tilde{x}$ WS3-$\bar{x}$ WS3-$\tilde{x}$ WS4-$\tilde{x}$ WG3-99 WG4-75 BL	WS1-$\bar{x}$ WS2-$\bar{x}$ WS2-$\tilde{x}$ WS3-$\bar{x}$ WS3-$\tilde{x}$ WS4-$\tilde{x}$ WG3-99 WG4-75 BL	WS1-$\bar{x}$ WS2-$\bar{x}$ WS2-$\tilde{x}$ WS3-$\bar{x}$ WS3-$\tilde{x}$ WS4-$\tilde{x}$ WG3-99 WG4-75 BL

Among the ENV+IND ENs, the environmental or mixture variables tended to be selected more often than the industrial variables (Tables 1 and 2). Across sites, the parameters selected for males and females at the Lower site tended to be more consistent compared to the Upper site (Tables 1 and 2). For example, AT was commonly selected in models for both males and females at the Lower site (Tables 1 and 2), but AT was rarely identified by EN at the Upper site (Table 2).

Among the models of GW and LW for males at the Lower site, only median ATs from the shorter periods were selected: August 4 to the start of fishing (AT1-$\tilde{x}$) and 60 days before the collection of fish (AT2-$\tilde{x}$; Table 2). Although not definitive, there are also some data from female lake chub, suggesting that shorter periods of air temperature are more closely associated with GW and LW and BW with longer periods (Table 2) matching with the environmental biology of fishes [59].

Other spatial patterns among the selected parameters for the metrics of fish health were also identified. For example, estimates of stream discharge (SD) were more commonly selected at the Upper location compared to the Lower site (Tables 1 and 2). When metrics of SD were selected, greater SD tended to reduce fish health measurements, but the single SD measurement selected at the Lower site (female GW; SD3-$\tilde{x}$; Tables 1 and 2), had a positive effect. Although constrained by the late deployment of probes in 2015, the potential influence of water temperature (WT1-$\bar{x}$) was identified by EN in GW and LW of females captured at the Lower site, and, similar to AT and consistent with expectations, higher WT was associated with increases in fish metrics (e.g., Table 1). Although not strictly an 'environmental' covariate, the later collection of fish estimated by the Julian Day (JD) was also associated with increases in some metrics of fish health, including GW and LW of females and BW of males at the Upper site (Tables 1 and 2).

Table 2. Environmental and industrial covariates selected by EN for gonad weight (GW), liver wight (LW), and body weight (BW) of male fish captured at the Upper and Lower Ells locations; NL = no limits; UL = upper limits; NPF = no penalty factors; PF = penalty factors; bolded white text with saturated fill color in cells comprise best-fit judged by AIC values; red = positive coefficients; blue = negative coefficients; marginal models indicated as unsaturated fill and black text.

Site	Measure	ENV	IND	ENV+IND NL-NPF	ENV+IND NL-PF	ENV+IND UL-NPF	ENV+IND UL-PF
Upper	GW	BW	BW	BW	BW	BW	BW
Upper	LW	AT1-$\tilde{x}$ JD SD1-$\bar{x}$ SD4-$\tilde{x}$ BW	SML-PC-F MRM-OS-M SML-PC-P SML-SC-P BW	SP-RD SML-PC-F SML-PC-P SML-SC-P SD1-$\bar{x}$ SD4-$\tilde{x}$ BW	SP-RD SD1-$\bar{x}$ SD4-$\tilde{x}$ BW	SP-RD SD1-$\bar{x}$ SD4-$\tilde{x}$ BW	SP-RD SD1-$\bar{x}$ SD4-$\tilde{x}$ BW
Upper	BW	AT1-$\tilde{x}$ JD SD1-$\bar{x}$ BL	SML-PC-F MRM-OS-M SML-PC-P BL	SP-RD SML-PC-F MRM-OS-M JD SD1-$\bar{x}$ SD2-$\bar{x}$ SD4-$\bar{x}$ SD4-$\tilde{x}$ BL	SP-RD JD SD1-$\bar{x}$ SD1-$\tilde{x}$ SD2-$\bar{x}$ SD4-$\tilde{x}$ BL	SP-RD MRM-NG-F MRM-OS-M JD SD1-$\bar{x}$ SD2-$\bar{x}$ SD4-$\bar{x}$ SD4-$\tilde{x}$ BL	SP-RD JD SD1-$\bar{x}$ SD4-$\bar{x}$ SD4-$\tilde{x}$ BL
Lower	GW	AT1-$\tilde{x}$ AT2-$\tilde{x}$ BW	SBM-CB-FW SBM-DN-FW MRM-DN-F HM-PG-F JPM-NG-F SAN-NG-F BW	SBM-CB-FW SBM-DN-FW SBM-PG-FW MRM-DN-F JPM-NG-F BW	P2-RD MRM-DN-F HM-PG-F AT2-$\tilde{x}$ BW	P2-RD MRM-CB-FW SBM-DN-FW JPM-NG-F SAN-NG-F KM-SC-F AT2-$\tilde{x}$ BW	P2-RD AT2-$\tilde{x}$ WT1-$\tilde{x}$ BW
Lower	LW	AT1-$\tilde{x}$ AT2-$\tilde{x}$ BW	SBM-CB-FW SBM-PC-F JPM-NG-F SAN-NG-F KM-SC-F JPM-CB-P BW	SBM-CB-FW SBM-PC-F JPM-NG-F SAN-NG-F JPM-CB-P AT2-$\tilde{x}$ BW	AT1-$\tilde{x}$ AT2-$\tilde{x}$ BW	SBM-PC-F JPM-NG-F JPM-CB-P AT2-$\tilde{x}$ BW	AT1-$\tilde{x}$ AT2-$\tilde{x}$ BW
Lower	BW	AT3-$\bar{x}$ BL	KM-DN-F SFB-ST BL	WS2-$\tilde{x}$ WS4-$\tilde{x}$ P4-$\bar{x}$ WG2-75 WG3-99 P3-TP P4-TP KM-DN-F AT3-$\bar{x}$ BL	WS2-$\tilde{x}$ WS4-$\tilde{x}$ P4-$\bar{x}$ WG2-75 WG3-99 P3-TP P4-TP AT3-$\bar{x}$ BL	WS2-$\tilde{x}$ WS4-$\tilde{x}$ P4-$\bar{x}$ WG2-75 WG3-99 P3-TP P4-TP KM-DN-F AT3-$\bar{x}$ BL	WS2-$\tilde{x}$ WS4-$\tilde{x}$ P4-$\bar{x}$ WG2-75 WG3-99 P3-TP P4-TP AT3-$\bar{x}$ BL

This study identified potential drivers of annual variability often reflecting known or expected relationships with environmental predictors. For example, sampling of fish later in the year, estimated by Julian Day, tended to increase fish measurements, e.g., [59] as did warmer air and water temperatures [54,55]. In contrast, greater SD tended to reduce the fish health metrics, especially at the Upper site, and may be associated with greater energetic costs of higher flows [54,55]. Although similar results are also reported elsewhere [24,39,56], relationships with natural drivers suggest that effects in fish may occur in the future via expected changes in thermal and hydrological regimes in northeastern Alberta [70,71] and are explored below (See Section 3.2.4). Overall, the analyses reinforce the importance of accounting for background drivers of variability in temporal analyses of fish health (e.g., [53,54]) and how identifying these potential relationships improves on previous approaches [20]. There may, however, also be a role of temperature in greater emissions from some sources, such as tailings ponds and mine faces [75,76], suggesting that other studies may be needed to further improve the monitoring results.

Influence of Industrial Activity

The potential influence of industrial activity on the health indicators of lake chub residing in the Ells River were also identified in this study. The industrial influences were suggested in the IND models, including factors such as HM-DN-FW, HM-OS-M, MRM-DN-F, and SBM-PG-FW (Table 1) and others in the ENV+IND models (Tables 1 and 2). For example, the best fit ENV+IND models for GW of females at the Upper site suggested a negative influence of MRM-CB-FW and MRM-NG-F (Table 1). In contrast, female GW at the Lower site may be negatively influenced by SAN-SC-F (Table 1). Many of these potential relationships occurred only in the NPF ENs, but the potential effects of CB-P at MRM and SC-F at SAN were, respectively, identified in PF models for LW and GW of females at the Lower location (Table 2). Some of the results from the marginal AIC models, such as a negative influence of SBM-PC-F on LW of males (Table 2; UL-NPF) and females (Table 1; NL-NPF, UL-NPF, and IND) at the Lower site and increases in LW of females with greater SBM-NG-F at the Lower site (Table 2; NL-NPF) were also identified.

Although no estimates of relative land disturbance were selected as a relevant factor in any of the ENs (Tables 1 and 2), the selected models did identify the potential roles of multiple types of industrial practices, including flaring and wasting of materials (FW), the combustion of fuels (F), and the production of substances, such as crude bitumen (CB-P) and synthetic crude (SC-P), on indicators of fish health at both locations (Tables 1 and 2). The results also highlight the potential influence of mechanically generated dust, combustion products, and other normal operations along with operational inefficiencies or upsets on ecosystem conditions either suggested, supported, or shown in other studies, e.g., [35,77,78].

While there could be effects of local stressors on fish in the Ells, such as road traffic related to production at HM, all of the industrial facilities identified in the best-fit ENs are physically outside the Ells basin. The potential influence of extrabasin facilities on both sites in the Ells watershed, coupled with no effects of land disturbance, suggests that the primary exposure mechanism is via atmospheric emissions and subsequent deposition of CoCs [20,32,34–38,47,78–80]. These results reinforce findings elsewhere, e.g., [32], including additional analyses suggesting the potential influence of the MRM on benthic invertebrates in the Ells basin [39] and data suggest CoCs are depositing throughout the watershed [37,41,81–83]. Contaminants of concern may be accumulating in headwater lakes in the Ells basin and may originate from oil sands industrial activity [20,41], but additional work on sediments in lakes throughout the region, such as lake SE22 in the Steepbank basin [32,41,47] also suggests that some reference sites may be influenced by industrial activity, albeit to a lesser degree than the downstream locations. Whereas some studies may still detect differences over time despite the potential influence of atmospheric emissions and deposition from industrial development in reference areas [51], accounting for industrial activity at all locations may enhance the sensitivity of monitoring to the stressors of interest. However, industrial emissions may also be associated with enrichment effects [39,84,85], and more work is likely required to determine the veracity of these potentially meaningful associations. The implications of associations between aquatic ecosystems and industrial emissions for future monitoring in Canada's OSR e.g., [32], such as the need to incorporate the results of deposition models [82,86], more detailed records of industrial performance, and/or investigations of more specific effect pathways into an integrated design [87] also likely requires additional work.

The analyses also highlight the likely importance of proximity of facilities to exposure locations. Among facilities identified by AIC in the IND and ENV+IND models, the influence of SML (14 km), MRM (16 km), and SBM (30 km) on fish captured at the Upper site was suggested in the AIC-selected models (Tables 1 and 2). At the Lower site, the influence of HM (5 km), MRM (6 km), SAN (7 km), KM (28 km), and SFB (36 km) were identified in the AIC-selected models (Tables 1 and 2; Supplemental Table S3). The identification of potential relationships between industrial features and indicators of fish health suggests study sites adjacent to industrial facilities may also be influenced by activity from multiple, and potentially remote, operations. However, while the amount of active land disturbance during the study period was low and was not associated with annual variability of fish health metrics, study sites adjacent to facilities may also influenced by activities at those local operations. The influence of land disturbance was apparent in other studies on benthic invertebrates [39] and may be partially responsible for slight temperature differences between sites in 2013 compared to the other study years (Supplemental Figure S18). Greater local land disturbance in watersheds may also lead to a greater influence of atmospherically deposited materials [64], suggesting the potential importance of activity levels or types of land disturbances and the potential need for additional analyses.

Additionally, some results also suggest that background environmental effects may mask industrial influences. The positive effects of AT and negative effects of MRM-CB-FW were concurrently identified in the LW of females from the Lower site (Table 1). This pattern also occurs with the additional 'mixture' variables, and its implications are discussed in the next section.

Influence of 'Mixture' Variables

Assessing the impacts of atmospheric deposition on the health of biota living in streams is also affected by the transport mechanisms linking industrial facilities and exposure environments. For example, the emission and deposition rates of materials from some sources may also be affected by climatic conditions, including wind speeds and precipitation [32,75,76], while these same processes may also affect their post-deposition mobilization and transport to waterbodies [88]. To account for this possibility, additional 'mixture' variables (along with an expanded list of environmental covariates), such as spring and summer precipitation, was also included in this analysis and the ENV+IND models. Among the ENs, these additional environmental covariates were commonly selected (Tables 1 and 2). Among the spring precipitation metrics, SP-MP, SP-TP, and SP-RD were only identified in EN models for lake chub captured at the Upper location (Tables 1 and 2). More specifically, increases in SP-MP and SP-TP were associated with reduced BW in females at the Upper site (Table 1), while greater rain days in the spring (SP-RD) may have reduced the LW of males from the same site (Table 2). In contrast, neither mean spring air temperature (SP-MT) nor mean spring discharge (SP-SD) were selected by any of the ENV+IND ENs (Tables 1 and 2).

Summer wind metrics and summer precipitation were also included as selectable variables in the ENV+IND ENs. The potential effects of summer wind, such as upper percentiles (e.g., WS3-99) and measures of central tendency over various periods, were not included in the best-fit models (Tables 1 and 2). However, both positive and negative effects of greater summer rain were identified in the metrics of fish health. Among females, increases in the P1-99 and P2-99 variables were associated with greater GW of females from the Lower site (Table 1). Increases in the number of rain days from June 1 to August 31 (P3-RD) was associated with increases in GW and BW of females at the Upper site (Table 1), but greater P2-75 was associated with smaller GW of females from the same location (Table 1).

These results suggest the potential influence of summer and spring precipitation on lake chub in the Ells basin supporting earlier hypotheses [64,88]. Although a plausible mechanism linking emissions from industrial facilities and effects on stream biota, the parsing of metrics such as precipitation to 'natural' or 'anthropogenic' causes is not clear. For example, while CoCs deposited on the landscape may migrate to surface water bodies, greater flows in rivers may also increase the natural scouring of the substrate [14,88–92], and, similar to potential interactions of some anthropogenic activities and AT mentioned already, the co-occurrence of drivers may obscure, mask, or counteract some industrial influences [14,88,89,91]. However, the results in this study also suggest that inputs of industrial CoCs in the spring, while they may be difficult to detect, may be more apparent at sites higher in watersheds.

The results also have further implications for how industrial activity may affect the environment. Along with pulses of CoCs during expected freshet, which are often difficult to disentangle from the accompanying in-stream scouring effects [64,89,93], additional complexity may also be present. The results of this study further suggest effects on fish metrics, such as GW of females at the Upper site (Table 1), may be associated with the magnitude (e.g., P2-75) and frequency (e.g., P3-RD) of rain events over slightly differing periods. Furthermore, the selection of precipitation metrics without industrial metrics (e.g., female BW at the Upper site; Table 1) and with industrial predictors showing opposite effects (GW of Lower females; Table 1), such as P1-99, P2-99, and SAN-SC-F suggest the potential for antagonistic effects of multiple factors, including mediating effects of additional transport mechanics, such as landscape retention [88]. However, there is also some evidence in GW of females from the Upper site suggesting synergistic associations between large precipitation events (P2-75) and metrics of industrial activity (MRM-CB-FW; Table 1). Although the effects of precipitation on stream chemistry are well-established [64,94], the data here and elsewhere [55] further suggest potential associations with the status of biological indicators [39]. However, as mentioned already, similar interactions between industrial activity and environmental variables may also extend to air temperature [75,76].

3.2.4. Predicting Exceedances of Critical Effect Sizes

Potential Effects of Industrial Activity

The final portion of this study used the effect sizes of industrial predictors across the ENs to estimate the degree of change in those industrial predictors which may already be associated with effects on fish equal to or exceeding the standard CES used in studies of fish health indicators [69]. When predicting the potential existing effects of industrial activity by setting initial conditions to zero, increases in the mass of mined oil sands at MRM (MRM-OS-M) may currently be associated with declines in BW of males at the Upper site ranging from ~12% to 17%. Similarly, increases in the volume of crude bitumen produced at MRM (MRM-CB-P) may be associated with declines in LW in females from the Lower site ranging from 26% to 38% (Figure 4). In contrast, increases in BW of males at the Upper site ranging from ~41 to 73% may also be associated with greater SML-PC-F (Figure 4). Other effects exceeding the CES may also be occasionally or commonly observed in response to other industrial features, such as a potential influence of SFB-ST on BW of females from the Lower site, declines in GW of females at the Upper site with greater MRM-CB-FW, and increases in LW of males at the Upper site with greater SML-PC-P (Figure 4). In contrast, other effects on fish among the models estimating the onset of industrial activity are typically below the CES, including HM-DN-FW and SBM-PG-FW (Supplemental Figure S19). Although challenging to confirm with existing data, these results suggest the possibility of the existing influence of industrial development on the physiology of sentinel fish beyond the relevant CESs, but also highlight potential counteracting effects of activities and/or climatic conditions.

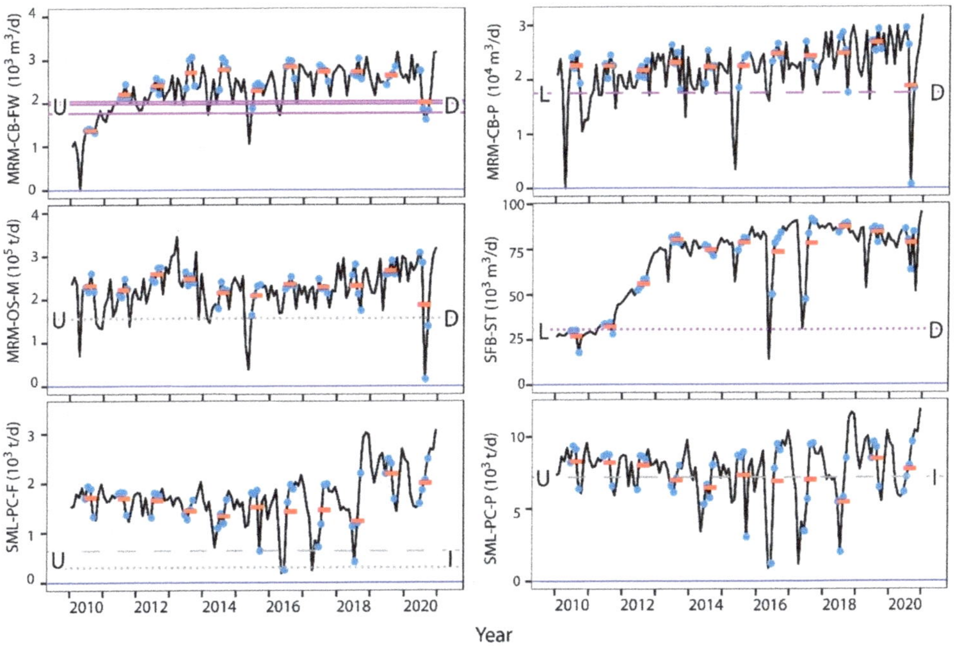

Figure 4. Effect size plots from zero industrial activity (horizontal blue lines) with greater or lesser industrial activities; purple lines = females; dark gray = males; solid lines = gonad weight; dashed lines = liver weight; dotted lines = body weight; L = Lower site; U = Upper site; D = decrease in fish measurement; I = increase in fish measurement; black lines = mean daily industrial metrics from 2010-2020; blue dotes = June, July, August, September; Red lines = mean summer values for industrial endpoints.

Change from the 2013 to 2015 and 2018 mean activity levels of various industrial facilities can also be estimated (Figure 5). When estimates from the mean values of industrial parameters from the summers of 2013–2015 and 2018 are calculated, many models suggested no differences would be expected in fish endpoints with known changes in industrial factors, even when those features are part of the best performing models selected by AIC, such as increases in LW in males from the Upper site in response to changes in SML-SC-P and decreases in GW of females at the Upper site with greater MRM-NG-F (Supplemental Figure S20) or with decreases in female LW at the Lower site with greater MRM-CB-P (Figure 5). Although activity levels at some facilities would need to increase by large degrees to affect fish relative to their status in 2013–2018, such as 53 to 373 times greater oil sand production at HM (HM-OS-M; Supplemental Figure S20), other current activity levels may require smaller increases in activity to potentially affect fish, such as 1.01 times the level of SML-PC-F in 2010 and 6.7 times the level of SML-PC-P in 2018 to increase BW of males at the Upper site (Figure 5) and may occur in the future. However, other predictions also suggest some exceedances may have already occurred. For example, the models predicted changes in BW and LW of males captured at the Upper site in 2019 (and BW in 2020) may have been larger than the CES compared to fish captured earlier in response to the estimated effect of SML-PC-F calculated using the AIC-selected IND model (Figure 5).

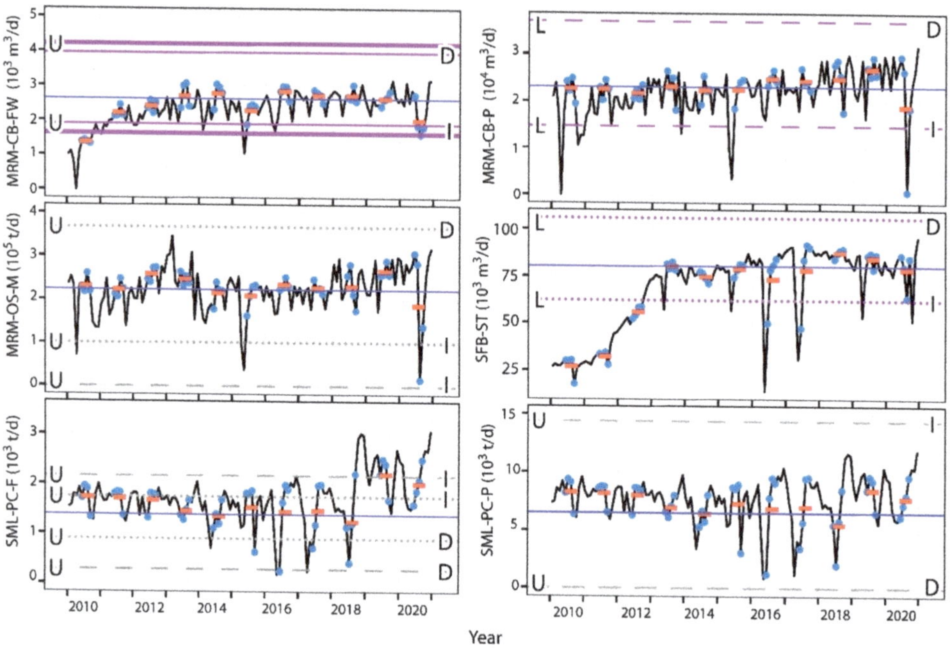

Figure 5. Effect size plots from mean (horizontal blue lines) with greater or lesser industrial activities; purple lines = females; dark gray = males; solid lines = gonad weight; dashed lines = liver weight; dotted lines = body weight; L = Lower site; U = Upper site; D = decrease in fish measurement; I = increase in fish measurement; black lines = monthly industrial values from 2010–2020; blue dotes = June, July, August, September; Red lines = mean summer values for industrial endpoints.

Predicting the impacts of industrial facilities is a goal of the Oil Sands Monitoring program [5]. Although the initial and ongoing influence of industrial development has been demonstrated in some chemical indicators [35,41,47], research in flowing waters has not been performed using undisturbed baseline data, although some potential impacts of industrial development have been identified despite this [50,51]. Although additional data suggest catch rates of lake chub in the Ells basin are increasing over time (Supplemental Figure S21), the analyses here suggest that some impacts of industrial development may be present and may be identified without measured pre-disturbance baselines. These analyses may also be used elsewhere to further simulate pre-disturbance baselines and expected ranges of fish health indicators in the absence of some or all industrial facilities.

Potential Effects of Future Changes in Air Temperature and Stream Discharge

Although industrial impacts may already be present or may occur in the future with changes in activity at the local facilities, physical conditions in water bodies also affects the health status of fish populations, e.g., [95]. Future changes in both air temperature and stream discharge are projected in northern Alberta and these drivers may affect fish [70,71]. When potential changes in summer stream discharge and summer air temperature are used to project future changes in fish health, some effects approaching or exceeding the CESs for GW, LW, and BW may occur in lake chub residing in the Ells River. For example, differences exceeding the CESs may occur in LW and BW of females captured at the Lower site with increases in mean air temperature exceeding ~2 to 3 °C when estimated using the ENV models (Figure 6). Increases approaching 25% are also predicted for GW of female lake chub captured at the Lower site with mean increases of 4 °C in AT, but could be reduced with concurrent and progressively larger declines in SD (Figure 6). Potential effects exceeding the 25% CES for LW of females at the Lower site are also predicted in the UL-PF and NL-PF models if the mean AT increases by ~4 °C (Figure 6). Increases in LW of females at the Upper site and BW of males at the Lower site are also suggested with increases in temperature and/or declines in SD if the ENV models are correct (Figure 6). Similar to the current observations and the results of other studies showing potential responses to warming occurring over decades [41,96], these results suggest background environmental variables are necessary to account for the effects climate change on fish health measurements, especially in long-term monitoring programs [72]. However, as mentioned already, there may also be consequential relationships between greater air temperature and industrial emissions [75,76] or indirect effects of industry embedded in stream discharge measurements, including less mobilization of CoCs from the landscape.

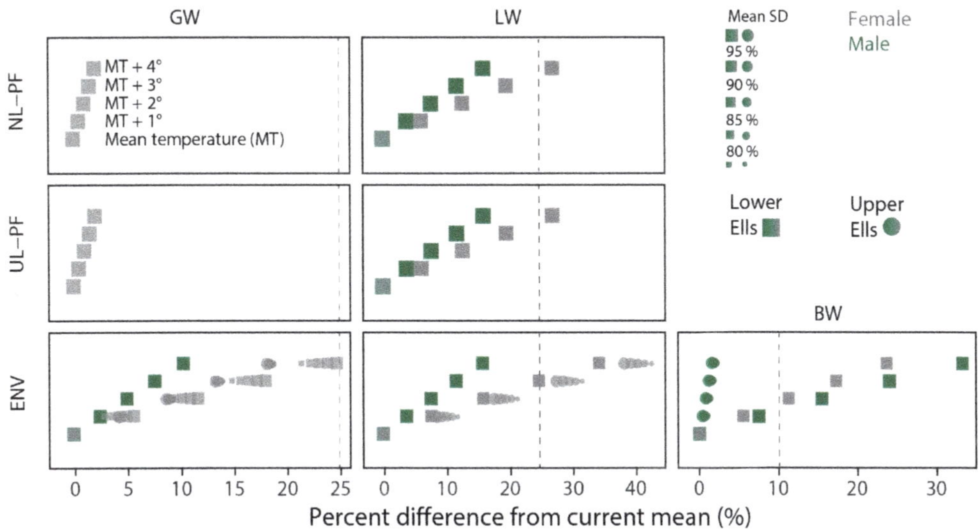

Figure 6. Potential effects of projected future changes in mean air temperature (MT; 1 to 4 °C; vertical scale) and stream discharge (SD; −5% to −20%) in Alberta on male and female lake chub captured in the Ells River basin, 2013–2015 and 2018 only among the best-fit models selected by AIC.

3.3. *Challenges with the Analytical Approach*

Although these analyses suggest some compelling relationships between natural and anthropogenic factors and echo other results [20,32,39,97], many uncertainties remain. First, the magnitude and occurrence of an effect depended on the EN configuration, such as the inclusion of upper limits and/or PFs and the utility of using an upper limit of zero is not clear. Although negative impacts of oil sands industrial activity are often expected and emphasized in reporting [90], some enriching effects of industrial processes are also possible [39,85]. Additionally, zones of exposure from one facility to another may be especially challenging to approximate. Whereas a single PF was used for the main analyses here, additional exploratory analyses used a range (Supplemental Figure S5). The exploratory simulations of varying the PFs suggest different influences of different PFs on the MSE of selected models (Supplemental Figure S5). For example, some suggest PFs which remove all industrial features have the lowest MSE (e.g., female Lower LW), whereas some suggest the opposite (e.g., female Lower GW), and others suggest some other mid-point is optimal (e.g., male Lower GW).

Other uncertainties also remain. Although additional covariates account for some variability in fish endpoints, some residual variation remains suggesting additional factors may be influencing the health measurements of fish. Additionally, some proxies, such as air temperatures from Mildred Lake, may not precisely or accurately describe the conditions affecting study sites in the Ells basin, e.g., [31]. Although identifying natural influences may be used to isolate the impacts of industrial or other anthropogenic drivers [54,55], these analyses are also retrospective and limited to data recorded for parallel purposes. Consequently, any of the established relationships may be causal, and while correlations are relevant for an adaptive monitoring program [72], their explicitness and directness cannot be established with the existing retrospective and exploratory analyses; the results of this study cannot be used to prove the occurrence of industrial impacts on fish in the OSR but may be used to determine the necessity of future monitoring and importantly may direct its form.

3.4. Future Work

The results of this study suggest several patterns that may be relevant to managers, scientists, and or others. A potentially relevant observation in this study was the sensitivity of fish BW to both environmental and industrial covariates (e.g., Figures 2 and 3), echoing other work [39,55]. Among the fish health endpoints examined here, BW can be measured non-lethally [98] and can be used to identify the effects of human activity [99]. These results suggest studies could couple any lethal collections done every 3–4 years with more frequent non-lethal collections of adults to rapidly augment a dataset. Additional data from other species or lake chub from other sites sampled in the Ells River may also be available [8,14,56], but would require techniques suitable for random effects.

In addition to the predictions made using the data from 2010 to 2020, future industrial scheduling may be used to design programs. The Joslyn North project was acquired by Canadian Natural in 2018 and has been incorporated into the mine planning for Horizon [58]. Other changes in operations are also either planned or underway, such as the Mildred Lake expansion in the MacKay basin [100] and the replacement of the coke-fueled boilers at Suncor [101].

Finally, the analyses suggest some potential impacts on lake chub in the Ells River may already be present, diminishing the relevance of identifying differences over time. The potential for existing impacts at the Ells locations, the challenges of comparisons to a spatial reference (although there is some evidence suggesting the GW of females are commonly smaller at the Lower location relative to the upstream site), and the known issues of observational studies [102] suggest manipulative experiments, such as stream-side mesocosms [103] or other analytical approaches, such as structural equation modeling [104] or the more specific use of test and training data may be required to more clearly establish the influence of industrial activity.

4. Conclusions

Oil sands industrial development influences the environment. This study examined health indicators of lake chub captured in the Ells River between 2013 and 2018 and found differences at downstream sites compared to upstream reference locations. The study also identified the potential influence of environmental and industrial variables at both of the Upper and Lower locations. Overall, the analyses suggest environmental and industrial covariates may explain some of the residual variation in fish health metrics. These results suggest these environmental and industrial covariates may be required in future monitoring studies in the OSR to account for background climate signals and to identify potential industrial influence in the absence of pre-disturbance baselines.

Supplementary Materials: The following supporting information can be downloaded at: https://www.mdpi.com/article/10.3390/environments9060073/s1, Supplemental Table S1: Environmental covariates; WT = water temperature; AT = air temperature; SD = stream discharge; P = precipitation; WS = wind speed; WG = wind gusts; $\bar{x}$ = mean; $\tilde{x}$ = median; values of environmental covariates by site shown in Supplemental Figures S1–S3; Supplemental Table S2 Industrial variable names and codes; Supplemental Table S3 Distances from oil sands project boundaries to fish sampling locations; arranged by distance from Lower Ells; Supplemental Table S4 Female Lake Chub metrics (mean ± SD(n)); * denote significant difference at Lower site compared to

Upper site ($p < 0.05$).; Supplemental Table S5 Male Lake Chub metrics (mean ± SD(n)); * denote significant difference at Lower site compared to Upper site ($p < 0.05$).; Supplemental Table S6 *p*-values for OLS and GLM spatial comparisons (Lower Ells vs. Upper Ells) of male and female lake chub collected in the Ells River in 2013–2015, and 2018; Int = intercept; yellow highlighting = slope with *p*-value < 0.05; red highlighting = intercept with *p*-value < 0.05; Supplemental Table S7 *p*-values for OLS temporal comparisons of male and female lake chub collected in the Ells River in 2013–2015, and 2018; Int = intercept; yellow highlighting = slope with *p*-value < 0.05; red highlighting = intercept with *p*-value < 0.05; Supplemental Table S8 *p*-values for GLM temporal comparisons of male and female lake chub collected in the Ells River in 2013–2015, and 2018; Int = intercept; yellow highlighting = slope with *p*-value < 0.05; red highlighting = intercept with *p*-value < 0.05; Supplemental Table S9 GLM *p*-values for grouped Reference years (2013–2015) compared to 2018; yellow highlighting = slope with *p*-value < 0.05; red highlighting = intercept with *p*-value < 0.05; Supplemental Figure S1 Relative land disturbance (%) in watershed above Upper site (grey), above the Lower site (red) and area of lower watershed below the Upper location (blue) over time.; Supplemental Figure S2 Plots of environmental covariates for the Upper and Lower Ells locations; SD values are from the Upper Location; WT = water temperature; AT = air temperature; SD = stream discharge; Supplemental Figure S3 Industrial covariates; values within each pane are the 2013–2015 + 2018 (monthly) means; values standardized to 0–1; Supplemental Figure S4 Extra covariates included in ENV+IND models (and land disturbance), which may capture some industrial influence; Supplemental Figure S5 Mean squared error for 1000 Elastic net model runs using penalty factors based on distance in km divided by 1 to 1000 of Upper and Lower sites to industrial facilities; Supplemental Figure S6 Mean residual gonad weight (GW), liver weight (LW), and body weight (BW) of females (with central 95% confidence interval) at the Upper and Lower Ells locations; residual GW, LW, and BW at the Lower site calculated using models estimated using fish from the Upper location; red lines showing expected range of mean residuals at the Lower site given the sample size for the Lower site; Supplemental Figure S7 Mean residual gonad weight (GW), liver weight (LW), and body weight (BW) of males (with central 95% confidence interval) at the Upper and Lower Ells locations; residual GW, LW, and BW at the Lower site calculated using models estimated using fish from the Upper location; red lines showing expected range of mean residuals at the Lower site given the sample size for the Lower site; Supplemental Figure S8 Gonad weight, liver weight, and body weight of female lake chub relative to anatomical covariates at the Lower and Upper Ells locations in 2013–2015, 2018; Supplemental Figure S9 Gonad weight, liver weight, and body weight of male lake chub relative to anatomical covariates at the Lower and Upper Ells locations by year 2013:2015, 2018; Supplemental Figure S10 Histograms of female lake chub gonad weight, liver weight, and body weight; Supplemental Figure S11 Histograms of male lake chub gonad weight, liver weight, and body weight; Supplemental Figure S12 Gonad weight, liver weight, and body weight of female lake chub relative to anatomical covariates in 2013, 2014, 2015, and 2018 by location (Lower and Upper Ells) Supplemental Figure S13 Gonad weight, liver weight, and body weight of male lake chub relative to anatomical covariates in 2013, 2014, 2015, and 2018 by location (Lower and Upper Ells); Supplemental Figure S14 Akaike's Information Criterion (AIC)s, Mean Squared Errors (MSE), and Deviance Ratios (DR) for female lake chub including fish GLM; Supplemental Figure S15 Akaike's Information Criterion (AIC)s, Mean Squared Errors (MSE), and Deviance Ratios (DR) for male lake chub including fish GLM; Supplemental Figure S16 AICs, MSEs, and deviance ratios for model scenarios among female lake chub at the Upper and Lower locations; Supplemental Figure S17 Akaike's Information Criterion (AIC)s, Mean Squared Errors (MSE), and Deviance Ratios (DR) for model scenarios among male lake chub at the Upper and Lower locations; Supplemental Figure S18 Mean water temperature per day in 2013–2015 and 2018 at the Upper and Lower Ells fishing locations; diagonal black line is 1:1 line; Supplemental Figure S19 Effect Size plots from zero (horizontal blue lines) with greater or lesser industrial activities; purple lines = females; dark gray = males; solid = gonad weight; dashed = liver weight; dotted = body weight; L = Lower site; U = Upper site; D = decrease in fish measurement; I = increase in fish measurement; black lines = monthly industrial values from 2010 to 2020; blue dotes = June, July, August, September; red lines = mean summer values for industrial endpoints; Supplemental Figure S20 Effect Size plots from mean 2013–2015 and 2018 activity levels (horizontal blue lines) with greater or lesser industrial activities; purple lines = females; dark gray = males; solid = gonad weight; dashed = liver weight; dotted = body weight; L = Lower site; U = Upper site; D = decrease in fish measurement; I = increase in fish measurement; black lines = monthly industrial values from 2010–2020; blue dotes = June, July, August, September; red lines =

mean summer values for industrial endpoints; Supplemental Figure S21 Catch-per-unit-effort (CPUE; fish per 100 s of electrofishing) for sites in the Ells Basin (data obtained from the Alberta Fish and Wildlife Management Information System; FWMIS; https://www.alberta.ca/access-fwmis-data.aspx, accessed on 7 March 2022).

Author Contributions: Conceptualization, M.E.M., E.J.U. and T.J.A.; methodology, M.E.M., E.J.U. and T.J.A.; formal analysis, E.J.U. and T.J.A.; data curation, M.E.M., E.J.U. and T.J.A.; writing—original draft preparation, T.J.A.; writing—review and editing, T.J.A., M.E.M. and E.J.U.; visualization, T.J.A. All authors have read and agreed to the published version of the manuscript.

Funding: This research was funded by the Oil Sands Monitoring Program (W-LTM-S-2-2122).

Institutional Review Board Statement: The study was conducted according to the guidelines of the Canadian Council on Animal Care (CCAC) approved Animal Use protocols (1315, 1415, 1515 and 1815) issued by the National Water Research Institute Animal Care Committee.

Data Availability Statement: Most data used in this study are publicly available. Lake chub data: https://www.canada.ca/en/environment-climate-change/services/oil-sands-monitoring.html (accessed on 1 March 2021); stream discharge: https://wateroffice.ec.gc.ca/ (accessed on 15 March 2021); air temperature and precipitation: https://climate.weather.gc.ca/ (accessed on 5 March 2021); Industry data: https://www.aer.ca/providing-information/data-and-reports/activity-and-data (accessed on 10 January 2022); land disturbance data: https://abmi.ca/home/data-analytics/da-top/da-product-overview/Human-Footprint-Products.html (accessed on 5 February 2021); wind data: https://wbea.org/air/wbea-air-monitoring/ (accessed on 14 January 2022); additional data, such as water temperature, is available upon request.

Acknowledgments: This work was funded under the Oil Sands Monitoring program but does not necessarily reflect the position of The Program or its participants. The authors thank Gerald Tetreault, Richard Frank, Sheena Campbell, Jason Miller, Thomas Clark, Abby Wynia, Jessie Cunningham, and Ross Neureuther from Environment and Climate Change Canada (ECCC), Keegan Hicks, Fred Noddin, Kristin Hynes, and Doug Rennie from Alberta Environment and Parks (AEP), Wood Buffalo and Aurora Helicopters (Fabian Moreau), and Jasmin Gee (Hatfield Consultants). The authors also thank the anonymous reviewers for their helpful comments.

Conflicts of Interest: The authors declare no conflict of interest.

Appendix A

Additional results and discussion describing the results of traditional Environmental Effects Monitoring study analyses, including ANOVAs and ANCOVAs for exposed fish (Lower Ells) relative to the reference (Upper Ells) fish.

References

1. Quadroni, S. Monitoring and Management of Inland Waters. *Environments* **2022**, *9*, 48. [CrossRef]
2. Hornbach, D.J. Multi-Year Monitoring of Ecosystem Metabolism in Two Branches of a Cold-Water Stream. *Environments* **2021**, *8*, 19. [CrossRef]
3. Quadroni, S.; Espa, P.; Zaccara, S.; Crosa, G.; Bettinetti, R.; Mastore, M.; Brivio, M.F. Monitoring and Management of Inland Waters: Insights from the Most Inhabited Italian Region. *Environments* **2022**, *9*, 27. [CrossRef]
4. Antognazza, C.M.; Vanetti, I.; de Santis, V.; Bellani, A.; di Francesco, M.; Puzzi, C.M.; Casoni, A.G.; Zaccara, S. Genetic Investigation of Four Beluga Sturgeon (*Huso Huso*, L.) Broodstocks for Its Reintroduction in the Po River Basin. *Environments* **2021**, *8*, 25. [CrossRef]
5. Roberts, D.R.; Hazewinkel, R.O.; Arciszewski, T.J.; Beausoleil, D.; Davidson, C.J.; Horb, E.C.; Sayanda, D.; Wentworth, G.R.; Wyatt, F.; Dubé, M.G. An Integrated Knowledge Synthesis of Regional Ambient Monitoring in Canada's Oil Sands. *Integr. Environ. Assess. Manag.* **2022**, *18*, 428–441. [CrossRef]
6. RAMP (Regional Aquatics Monitoring Program). Regional Aquatics Monitoring in Support of the Joint Oil Sands Monitoring Plan Final 2015 Program Report. 2016. Available online: http://www.ramp-alberta.org/ramp/results/report.aspx (accessed on 6 June 2022).
7. McMaster, M.E.; Tetreault, G.R.; Clark, T.; Bennett, J.; Cunningham, J.; Ussery, E.J.; Evans, M. Baseline White Sucker Health and Reproductive Endpoints for Use in Assessment of Further Development in the Alberta Oil Sands. *Int. J. Environ. Impacts Manag. Mitig. Recovery* **2020**, *3*, 219–237. [CrossRef]

8. Tetreault, G.R.; McMaster, M.E.; Dixon, D.G.; Parrott, J.L. Using Reproductive Endpoints in Small Forage Fish Species to Evaluate the Effects of Athabasca Oil Sands Activities. *Environ. Toxicol. Chem.* **2003**, *22*, 2775–2782. [CrossRef]
9. Walker, S.L.; Hedley, K.; Porter, E. Pulp and Paper Environmental Effects Monitoring in Canada: An Overview. *Water Qual. Res. J. Can.* **2002**, *37*, 7–19. [CrossRef]
10. Ribey, S.C.; Munkittrick, K.R.; McMaster, M.E.; Courtenay, S.; Langlois, C.; Munger, S.; Rosaasen, A.; Whitley, G. Development of a Monitoring Design for Examining Effects in Wild Fish Associated with Discharges from Metal Mines. *Water Qual. Res. J. Can.* **2002**, *37*, 229–249. [CrossRef]
11. Ussery, E.J.; McMaster, M.E.; Servos, M.R.; Miller, D.H.; Munkittrick, K.R. A 30-Year Study of Impacts, Recovery, and Development of Critical Effect Sizes for Endocrine Disruption in White Sucker (*Catostomus Commersonii*) Exposed to Bleached-Kraft Pulp Mill Effluent at Jackfish Bay, Ontario, Canada. *Front. Endocrinol.* **2021**, *12*, 369. [CrossRef]
12. Droppo, I.G.; di Cenzo, P.; Parrott, J.; Power, J. The Alberta Oil Sands Eroded Bitumen/Sediment Transitional Journey: Influence on Sediment Transport Dynamics, PAH Signatures and Toxicological Effect. *Sci. Total Environ.* **2019**, *677*, 718–731. [CrossRef] [PubMed]
13. Evans, M.; Davies, M.; Janzen, K.; Muir, D.; Hazewinkel, R.; Kirk, J.; De Boer, D. PAH Distributions in Sediments in the Oil Sands Monitoring Area and Western Lake Athabasca: Concentration, Composition and Diagnostic Ratios. *Environ. Pollut.* **2016**, *213*, 671–687. [CrossRef] [PubMed]
14. Evans, M.S.; McMaster, M.; Muir, D.C.G.; Parrott, J.; Tetreault, G.R.; Keating, J. Forage Fish and Polycyclic Aromatic Compounds in the Fort McMurray Oil Sands Area: Body Burden Comparisons with Environmental Distributions and Consumption Guidelines. *Environ. Pollut.* **2019**, *255*, 113135. [CrossRef] [PubMed]
15. Headley, J.V.; Crosley, B.; Conly, F.M.; Quagraine, E.K. The Characterization and Distribution of Inorganic Chemicals in Tributary Waters of the Lower Athabasca River, Oilsands Region, Canada. *J. Environ. Sci. Health-Part A Toxic/Hazard. Subst. Environ. Eng.* **2005**, *40*, 1–27. [CrossRef]
16. Culp, J.M.; Glozier, N.E.; Baird, D.J.; Wrona, F.J.; Brua, R.B.; Ritcey, A.L.; Peters, D.L.; Casey, R.; Choung, C.B.; Curry, C.J.; et al. *Assessing Ecosystem Health in Benthic Macroinvertebrate Assemblages of the Athabasca River Main Stem, Tributaries and Peace-Athabasca Delta*; Environment and Climate Change Canada: Edmonton, AB, Canada, 2018. Available online: https://open.alberta.ca/publications/9781460140314 (accessed on 6 June 2022).
17. Pilote, M.; André, C.; Turcotte, P.; Gagné, F.; Gagnon, C. Metal Bioaccumulation and Biomarkers of Effects in Caged Mussels Exposed in the Athabasca Oil Sands Area. *Sci. Total Environ.* **2018**, *610–611*, 377–390. [CrossRef]
18. Tetreault, G.R.; Bennett, C.J.; Clark, T.W.; Keith, H.; Parrott, J.L.; McMaster, M.E. Fish Performance Indicators Adjacent to Oil Sands Activity: Response in Performance Indicators of Slimy Sculpin in the Steepbank River, Alberta, Adjacent to Oil Sands Mining Activity. *Environ. Toxicol. Chem.* **2020**, *39*, 396–409. [CrossRef]
19. Culp, J.M.; Droppo, I.G.; di Cenzo, P.; Alexander, A.C.; Baird, D.J.; Beltaos, S.; Bickerton, G.; Bonsal, B.; Brua, R.B.; Chambers, P.A.; et al. Ecological Effects and Causal Synthesis of Oil Sands Activity Impacts on River Ecosystems: Water Synthesis Review. *Environ. Rev.* **2021**, *29*, 315–327. [CrossRef]
20. Arciszewski, T.J.; Hazewinkel, R.R.O.; Dubé, M.G. A Critical Review of the Ecological Status of Lakes and Rivers from Canada's Oil Sands Region. *Integr. Environ. Assess. Manag.* **2022**, *18*, 361–387. [CrossRef]
21. Chad, S.J.; Barbour, S.L.; McDonnell, J.J.; Gibson, J.J. Using Stable Isotopes to Track Hydrological Processes at an Oil Sands Mine, Alberta, Canada. *J. Hydrol. Reg. Stud.* **2022**, *40*, 101032. [CrossRef]
22. Peters, D.L.; Watt, D.; Devito, K.; Monk, W.A.; Shrestha, R.R.; Baird, D.J. Changes in Geographical Runoff Generation in Regions Affected by Climate and Resource Development: A Case Study of the Athabasca River. *J. Hydrol. Reg. Stud.* **2022**, *39*, 100981. [CrossRef]
23. Hein, F.J.; Cotterill, D.K. The Athabasca Oil Sands—A Regional Geological Perspective, Fort McMurray Area, Alberta, Canada. *Nat. Resour. Res.* **2006**, *15*, 85–102. [CrossRef]
24. Suzanne, C.L. Effects of Natural and Anthropogenic Non-Point Source Disturbances on the Structure and Function of Tributary Ecosystems in the Athabasca Oil Sands Region. Master's Thesis, University of Victoria, Victoria, BC, Canada, 2015.
25. Droppo, I.G.; di Cenzo, P.; Power, J.; Jaskot, C.; Chambers, P.A.; Alexander, A.C.; Kirk, J.; Muir, D. Temporal and Spatial Trends in Riverine Suspended Sediment and Associated Polycyclic Aromatic Compounds (PAC) within the Athabasca Oil Sands Region. *Sci. Total Environ.* **2018**, *626*, 1382–1393. [CrossRef] [PubMed]
26. Hewitt, L.M.; Roy, J.W.; Rowland, S.J.; Bickerton, G.; DeSilva, A.; Headley, J.V.; Milestone, C.B.; Scarlett, A.G.; Brown, S.; Spencer, C.; et al. Advances in Distinguishing Groundwater Influenced by Oil Sands Process-Affected Water (OSPW) from Natural Bitumen-Influenced Groundwaters. *Environ. Sci. Technol.* **2020**, *54*, 1522–1532. [CrossRef]
27. Sun, C.; Shotyk, W.; Cuss, C.W.; Donner, M.W.; Fennell, J.; Javed, M.; Noernberg, T.; Poesch, M.; Pelletier, R.; Sinnatamby, N.; et al. Characterization of Naphthenic Acids and Other Dissolved Organics in Natural Water from the Athabasca Oil Sands Region, Canada. *Environ. Sci. Technol.* **2017**, *51*, 9524–9532. [CrossRef]
28. Ross, M.S.; Pereira, A.D.S.; Fennell, J.; Davies, M.; Johnson, J.; Sliva, L.; Martin, J.W. Quantitative and Qualitative Analysis of Naphthenic Acids in Natural Waters Surrounding the Canadian Oil Sands Industry. *Environ. Sci. Technol.* **2012**, *46*, 12796–12805. [CrossRef]
29. Roy, J.W.; Bickerton, G.; Frank, R.A.; Grapentine, L.; Hewitt, L.M. Assessing Risks of Shallow Riparian Groundwater Quality near an Oil Sands Tailings Pond. *Groundwater* **2016**, *54*, 545–558. [CrossRef]

30. Fennell, J.; Arciszewski, T.J. Current Knowledge of Seepage from Oil Sands Tailings Ponds and Its Environmental Influence in Northeastern Alberta. *Sci. Total Environ.* **2019**, *686*, 968–985. [CrossRef]
31. Gibson, J.J.; Yi, Y.; Birks, S.J. Isotope-Based Partitioning of Streamflow in the Oil Sands Region, Northern Alberta: Towards a Monitoring Strategy for Assessing Flow Sources and Water Quality Controls. *J. Hydrol. Reg. Stud.* **2016**, *5*, 131–148. [CrossRef]
32. Arciszewski, T.J. A Re-Analysis and Review of Elemental and Polycyclic Aromatic Compound Deposition in Snow and Lake Sediments from Canada's Oil Sands Region Integrating Industrial Performance and Climatic Variables. *Sci. Total Environ.* **2022**, *820*, 153254. [CrossRef]
33. Horb, E.C.; Wentworth, G.R.; Makar, P.A.; Liggio, J.; Hayden, K.; Boutzis, E.I.; Beausoleil, D.L.; Hazewinkel, R.O.; Mahaffey, A.C.; Sayanda, D.; et al. A Decadal Synthesis of Atmospheric Emissions, Ambient Air Quality, and Deposition in the Oil Sands Region. *Integr. Environ. Assess. Manag.* **2022**, *18*, 333–360. [CrossRef]
34. Mullan-Boudreau, G.; Belland, R.; Devito, K.; Noernberg, T.; Pelletier, R.; Shotyk, W. Sphagnum Moss as an Indicator of Contemporary Rates of Atmospheric Dust Deposition in the Athabasca Bituminous Sands Region. *Environ. Sci. Technol.* **2017**, *51*, 7422–7431. [CrossRef] [PubMed]
35. Shotyk, W.; Bicalho, B.; Cuss, C.; Donner, M.; Grant-Weaver, I.; Javed, M.B.; Noernberg, T. Trace Elements in the Athabasca Bituminous Sands: A Geochemical Explanation for the Paucity of Environmental Contamination by Chalcophile Elements. *Chem. Geol.* **2021**, *581*, 120392. [CrossRef]
36. Landis, M.S.; Studabaker, W.B.; Patrick Pancras, J.; Graney, J.R.; Puckett, K.; White, E.M.; Edgerton, E.S. Source Apportionment of an Epiphytic Lichen Biomonitor to Elucidate the Sources and Spatial Distribution of Polycyclic Aromatic Hydrocarbons in the Athabasca Oil Sands Region, Alberta, Canada. *Sci. Total Environ.* **2019**, *654*, 1241–1257. [CrossRef] [PubMed]
37. Gopalapillai, Y.; Kirk, J.L.; Landis, M.S.; Muir, D.C.; Cooke, C.A.; Gleason, A.; Ho, A.; Kelly, E.; Schindler, D.; Wang, X.; et al. Source Analysis of Pollutant Elements in Winter Air Deposition in the Athabasca Oil Sands Region: A Temporal and Spatial Study. *ACS Earth Space Chem.* **2019**, *3*, 1656–1668. [CrossRef]
38. Chibwe, L.; Muir, D.C.; Gopalapillai, Y.; Shang, D.; Kirk, J.L.; Manzano, C.A.; Atkinson, B.; Wang, X.; Teixeira, C. Long-Term Spatial and Temporal Trends, and Source Apportionment of Polycyclic Aromatic Compounds in the Athabasca Oil Sands Region. *Environ. Pollut.* **2021**, *268*, 115351. [CrossRef]
39. Arciszewski, T.J. Exploring the Influence of Industrial and Climatic Variables on Communities of Benthic Macroinvertebrates Collected in Streams and Lakes in Canada's Oil Sands Region. *Environments* **2021**, *8*, 123. [CrossRef]
40. Ahad, J.M.E.; Pakdel, H.; Labarre, T.; Cooke, C.A.; Gammon, P.R.; Savard, M.M. Isotopic Analyses Fingerprint Sources of Polycyclic Aromatic Compound-Bearing Dust in Athabasca Oil Sands Region Snowpack. *Environ. Sci. Technol.* **2021**, *55*, 5887–5897. [CrossRef]
41. Kurek, J.; Kirk, J.L.; Muir, D.C.; Wang, X.; Evans, M.S.; Smol, J.P. Legacy of a Half Century of Athabasca Oil Sands Development Recorded by Lake Ecosystems. *Proc. Natl. Acad. Sci. USA* **2013**, *110*, 1761–1766. [CrossRef]
42. Renard, K.G.; Nichols, M.H.; Woolhiser, D.A.; Osborn, H.B. A Brief Background on the U.S. Department of Agriculture Agricultural Research Service Walnut Gulch Experimental Watershed. *Water Resour. Res.* **2008**, *44*, W05S02. [CrossRef]
43. Spiers, G.A.; Dudas, M.J.; Turchenek, L.W. The Chemical and Mineralogical Composition of Soil Parent Materials in Northeast Alberta. *Can. J. Soil Sci.* **1989**, *69*, 721–737. [CrossRef]
44. Akena, A.M. An Intensive Surface Water Quality Study of the Muskeg River Watershed-Vol I. 1979. Alberta Oil Sands Environmental Research Program. Available online: https://doi.org/10.7939/R3NT9F (accessed on 6 June 2022).
45. Davidson, C.J.; Foster, K.R.; Tanna, R.N.R.N. Forest Health Effects Due to Atmospheric Deposition: Findings from Long-Term Forest Health Monitoring in the Athabasca Oil Sands Region. *Sci. Total Environ.* **2020**, *699*, 134277. [CrossRef] [PubMed]
46. Shotyk, W.; Appleby, P.G.; Bicalho, B.; Davies, L.; Froese, D.; Grant-Weaver, I.; Krachler, M.; Magnan, G.; Mullan-Boudreau, G.; Noernberg, T.; et al. Peat Bogs in Northern Alberta, Canada Reveal Decades of Declining Atmospheric Pb Contamination. *Geophys. Res. Lett.* **2016**, *43*, 9964–9974. [CrossRef]
47. Cooke, C.A.; Kirk, J.L.; Muir, D.C.G.; Wiklund, J.A.; Wang, X.; Gleason, A.; Evans, M.S. Spatial and Temporal Patterns in Trace Element Deposition to Lakes in the Athabasca Oil Sands Region (Alberta, Canada). *Environ. Res. Lett.* **2017**, *12*, 124001. [CrossRef]
48. Wiklund, J.A.; Hall, R.I.; Wolfe, B.B.; Edwards, T.W.D.; Farwell, A.J.; Dixon, D.G. Has Alberta Oil Sands Development Increased Far-Field Delivery of Airborne Contaminants to the Peace-Athabasca Delta? *Sci. Total Environ.* **2012**, *433*, 379–382. [CrossRef] [PubMed]
49. Landis, M.S.; Studabaker, W.B.; Pancras, J.P.; Graney, J.R.; White, E.M.; Edgerton, E.S. Source Apportionment of Ambient Fine and Coarse Particulate Matter Polycyclic Aromatic Hydrocarbons at the Bertha Ganter-Fort McKay Community Site in the Oil Sands Region of Alberta, Canada. *Sci. Total Environ.* **2019**, *666*, 540–558. [CrossRef]
50. Schwalb, A.N.; Alexander, A.C.; Paul, A.J.; Cottenie, K.; Rasmussen, J.B. Changes in Migratory Fish Communities and Their Health, Hydrology, and Water Chemistry in Rivers of the Athabasca Oil Sands Region: A Review of Historical and Current Data. *Environ. Rev.* **2015**, *23*, 133–150. [CrossRef]
51. Alexander, A.C.; Chambers, P.A. Assessment of Seven Canadian Rivers in Relation to Stages in Oil Sands Industrial Development, 1972–2010. *Environ. Rev.* **2016**, *24*, 484–494. [CrossRef]
52. Zhang, Y.; Shotyk, W.; Zaccone, C.; Noernberg, T.; Pelletier, R.; Bicalho, B.; Froese, D.G.; Davies, L.; Martin, J.W. Airborne Petcoke Dust Is a Major Source of Polycyclic Aromatic Hydrocarbons in the Athabasca Oil Sands Region. *Environ. Sci. Technol.* **2016**, *50*, 1711–1720. [CrossRef]

53. McMillan, P.G.; Feng, Z.Z.; Deeth, L.E.; Arciszewski, T.J. Improving Monitoring of Fish Health in the Oil Sands Region Using Regularization Techniques and Water Quality Variables. *Sci. Total Environ.* **2022**, *811*, 152301. [CrossRef]
54. Kilgour, B.W.; Munkittrick, K.R.; Hamilton, L.; Proulx, C.L.; Somers, K.M.; Arciszewski, T.J.; McMaster, M.E. Developing Triggers for Environmental Effects Monitoring Programs for Trout-Perch in the Lower Athabasca River (Canada). *Environ. Toxicol. Chem.* **2019**, *38*, 1890–1901. [CrossRef]
55. Arciszewski, T.J.; McMaster, M.E. Potential Influence of Sewage Phosphorus and Wet and Dry Deposition Detected in Fish Collected in the Athabasca River North of Fort McMurray. *Environments* **2021**, *8*, 14. [CrossRef]
56. Wynia, A.G.; Tetreault, G.R.; Clark, T.W.; Cunningham, J.L.; Ussery, E.J.; McMaster, M.E. Fish Assemblage Monitoring in Alberta's Ells River: Baseline Fish and Habitat Variability Prior to Major Development. *Glob. Ecol. Conserv.* **2022**, *34*, e02007. [CrossRef]
57. Anonymous. French Energy Giant Total Shelves the Joslyn Oil Sands Mine Indefinitely. Available online: https://www.oilsandsmagazine.com/news/french-energy-giant-total-shelves-joslyn-oilsands-mine-indefinitely (accessed on 23 April 2022).
58. Anonymous. CNRL Files Application to Integrate Joslyn North into Horizon Mine Plan. Available online: https://www.oilsandsmagazine.com/news/2019/11/20/cnrl-files-application-to-integrate-joslyn-north-into-horizon-mine-plan?rq=joslynnorth (accessed on 21 September 2021).
59. Barrett, T.J.; Munkittrick, K.R. Seasonal Reproductive Patterns and Recommended Sampling Times for Sentinel Fish Species Used in Environmental Effects Monitoring Programs in Canada. *Environ. Rev.* **2010**, *18*, 115–135. [CrossRef]
60. Harrell, F.E. *Regression Modeling Strategies: With Applications to Linear Models, Logistic and Ordinal Regression, and Survival Analysis*; Springer: Berlin/Heidelberg, Germany, 2015; ISBN 3319194259.
61. Arciszewski, T.J.; McMaster, M.E.; Munkittrick, K.R. Long-Term Studies of Fish Health before and after the Closure of a Bleached Kraft Pulp Mill in Northern Ontario, Canada. *Environ. Toxicol. Chem.* **2021**, *40*, 162–176. [CrossRef]
62. Zou, H.; Hastie, T. Regularization and Variable Selection via the Elastic Net. *J. R. Stat. Soc. Ser. B (Stat. Methodol.)* **2005**, *67*, 301–320. [CrossRef]
63. Friedman, J.; Hastie, T.; Tibshirani, R.; Narasimhan, B.; Tay, K.; Simon, N.; Qian, J. Package 'Glmnet'. CRAN R Repository 2021. Available online: https://cran.r-project.org/web/packages/glmnet/index.html (accessed on 2 April 2021).
64. Alexander, A.C.; Levenstein, B.; Sanderson, L.A.; Blukacz-Richards, E.A.; Chambers, P.A. How Does Climate Variability Affect Water Quality Dynamics in Canada's Oil Sands Region? *Sci. Total Environ.* **2020**, *732*, 139062. [CrossRef]
65. AER ST39 | Alberta Energy Regulator. Statistical Report 39. Available online: https://www.aer.ca/providing-information/data-and-reports/statistical-reports/st39 (accessed on 30 May 2021).
66. AER ST53 | Alberta Energy Regulator Statistical Report 53. Available online: https://www.aer.ca/providing-information/data-and-reports/statistical-reports/st53 (accessed on 30 May 2021).
67. ABMI (Alberta Biodiveristy Monitoring Institute). ABMI—Wall-to-Wall Human Footprint Inventory. Available online: https://abmi.ca/home/data-analytics/da-top/da-product-overview/Human-Footprint-Products/HF-inventory.html (accessed on 3 February 2021).
68. Burnham, K.P.; Anderson, D.R. Multimodel Inference: Understanding AIC and BIC in Model Selection. *Sociol. Methods Res.* **2004**, *33*, 261–304. [CrossRef]
69. Munkittrick, K.R.; Arens, C.J.; Lowell, R.B.; Kaminski, G.P. A Review of Potential Methods of Determining Critical Effect Size for Designing Environmental Monitoring Programs. *Environ. Toxicol. Chem.* **2009**, *28*, 1361–1371. [CrossRef]
70. Eum, H.I.; Dibike, Y.; Prowse, T. Climate-Induced Alteration of Hydrologic Indicators in the Athabasca River Basin, Alberta, Canada. *J. Hydrol.* **2017**, *544*, 327–342. [CrossRef]
71. Newton, B.W.; Farjad, B.; Orwin, J.F. Spatial and Temporal Shifts in Historic and Future Temperature and Precipitation Patterns Related to Snow Accumulation and Melt Regimes in Alberta, Canada. *Water* **2021**, *13*, 1013. [CrossRef]
72. Arciszewski, T.J.; Munkittrick, K.R. Development of an Adaptive Monitoring Framework for Long-Term Programs: An Example Using Indicators of Fish Health. *Integr. Environ. Assess. Manag.* **2015**, *11*, 701–718. [CrossRef] [PubMed]
73. Arciszewski, T.J.; Munkittrick, K.R.; Scrimgeour, G.J.; Dubé, M.G.; Wrona, F.J.; Hazewinkel, R.R. Using Adaptive Processes and Adverse Outcome Pathways to Develop Meaningful, Robust, and Actionable Environmental Monitoring Programs. *Integr. Environ. Assess. Manag.* **2017**, *13*, 877–891. [CrossRef] [PubMed]
74. Somers, K.M.; Kilgour, B.W.; Munkittrick, K.R.; Arciszewski, T.J. An Adaptive Environmental Effects Monitoring Framework for Assessing the Influences of Liquid Effluents on Benthos, Water, and Sediments in Aquatic Receiving Environments. *Integr. Environ. Assess. Manag.* **2018**, *14*, 552–566. [CrossRef] [PubMed]
75. Tokarek, T.W.; Odame-Ankrah, C.A.; Huo, J.A.; McLaren, R.; Lee, A.K.Y.; Adam, M.G.; Willis, M.D.; Abbatt, J.P.D.; Mihele, C.; Darlington, A.; et al. Principal Component Analysis of Summertime Ground Site Measurements in the Athabasca Oil Sands with a Focus on Analytically Unresolved Intermediate-Volatility Organic Compounds. *Atmos. Chem. Phys.* **2018**, *18*, 17819–17841. [CrossRef]
76. Moradi, M.; You, Y.; Hung, H.; Li, J.; Park, R.; Alexandrou, N.; Moussa, S.G.; Jantunen, L.; Robitaille, R.; Staebler, R.M. Fugitive Emissions of Polycyclic Aromatic Compounds from an Oil Sands Tailings Pond Based on Fugacity and Inverse Dispersion Flux Calculations. *Environ. Pollut.* **2021**, *269*, 116115. [CrossRef]
77. Gordon, M.; Li, S.M.; Staebler, R.; Darlington, A.; Hayden, K.; O'Brien, J.; Wolde, M. Determining Air Pollutant Emission Rates Based on Mass Balance Using Airborne Measurement Data over the Alberta Oil Sands Operations. *Atmos. Meas. Tech.* **2015**, *8*, 3745–3765. [CrossRef]

78. Mamun, A.A.; Celo, V.; Dabek-Zlotorzynska, E.; Charland, J.P.; Cheng, I.; Zhang, L. Characterization and Source Apportionment of Airborne Particulate Elements in the Athabasca Oil Sands Region. *Sci. Total Environ.* **2021**, *788*, 147748. [CrossRef]
79. McNaughton, C.S.; Vandenberg, J.; Thiede, P. Reanalysis of Aerial Deposition of Metals and Polycyclic Aromatic Compounds to Snow in the Athabasca Oil Sands Region of Alberta Canada. *Sci. Total Environ.* **2019**, *682*, 692–708. [CrossRef]
80. Liggio, J.; Li, S.M.; Hayden, K.; Taha, Y.M.; Stroud, C.; Darlington, A.; Drollette, B.D.; Gordon, M.; Lee, P.; Liu, P.; et al. Oil Sands Operations as a Large Source of Secondary Organic Aerosols. *Nature* **2016**, *534*, 91–94. [CrossRef]
81. Manzano, C.A.; Muir, D.; Kirk, J.; Teixeira, C.; Siu, M.; Wang, X.; Charland, J.P.; Schindler, D.; Kelly, E.; Zhang, Y.F. Temporal Variation in the Deposition of Polycyclic Aromatic Compounds in Snow in the Athabasca Oil Sands Area of Alberta. *Environ. Monit. Assess.* **2016**, *188*, 10–1007. [CrossRef]
82. Cheng, I.; Wen, D.; Zhang, L.; Wu, Z.; Qiu, X.; Yang, F.; Harner, T. Deposition Mapping of Polycyclic Aromatic Compounds in the Oil Sands Region of Alberta, Canada and Linkages to Ecosystem Impacts. *Environ. Sci. Technol.* **2018**, *52*, 12456–12464. [CrossRef] [PubMed]
83. Kirk, J.L.; Muir, D.C.G.G.; Gleason, A.; Wang, X.; Lawson, G.; Frank, R.A.; Lehnherr, I.; Wrona, F. Atmospheric Deposition of Mercury and Methylmercury to Landscapes and Waterbodies of the Athabasca Oil Sands Region. *Environ. Sci. Technol.* **2014**, *48*, 7374–7383. [CrossRef] [PubMed]
84. Wieder, R.K.; Vile, M.A.; Scott, K.D.; Albright, C.M.; Quinn, J.C.; Vitt, D.H. Bog Plant/Lichen Tissue Nitrogen and Sulfur Concentrations as Indicators of Emissions from Oil Sands Development in Alberta, Canada. *Environ. Monit. Assess.* **2021**, *193*, 208. [CrossRef] [PubMed]
85. Wieder, R.K.; Vile, M.A.; Vitt, D.H.; Scott, K.D.; Xu, B.; Quinn, J.C.; Albright, C.M. Can Plant or Lichen Natural Abundance 15N Ratios Indicate the Influence of Oil Sands N Emissions on Bogs? *J. Hydrol. Reg. Stud.* **2022**, *40*, 101030. [CrossRef]
86. Makar, P.A.; Akingunola, A.; Aherne, J.; Cole, A.S.; Aklilu, Y.A.; Zhang, J.; Wong, I.; Hayden, K.; Li, S.M.; Kirk, J.; et al. Estimates of Exceedances of Critical Loads for Acidifying Deposition in Alberta and Saskatchewan. *Atmos. Chem. Phys.* **2018**, *18*, 9897–9927. [CrossRef]
87. Arciszewski, T.J.; Roberts, D.R.; Munkittrick, K.R.; Scrimgeour, G.J. Challenges and Benefits of Approaches Used to Integrate Regional Monitoring Programs. *Front. Environ. Sci.* **2021**, *9*, 256. [CrossRef]
88. Wasiuta, V.; Kirk, J.L.; Chambers, P.A.; Alexander, A.C.; Wyatt, F.R.; Rooney, R.C.; Cooke, C.A. Accumulating Mercury and Methylmercury Burdens in Watersheds Impacted by Oil Sands Pollution. *Environ. Sci. Technol.* **2019**, *53*, 12856–12864. [CrossRef]
89. Birks, S.J.; Cho, S.; Taylor, E.; Yi, Y.; Gibson, J.J. Characterizing the PAHs in Surface Waters and Snow in the Athabasca Region: Implications for Identifying Hydrological Pathways of Atmospheric Deposition. *Sci. Total Environ.* **2017**, *603–604*, 570–583. [CrossRef]
90. Alexander, A.C.; Chambers, P.A.; Jeffries, D.S. Episodic Acidification of 5 Rivers in Canada's Oil Sands during Snowmelt: A 25-Year Record. *Sci. Total Environ.* **2017**, *599–600*, 739–749. [CrossRef]
91. Yi, Y.; Birks, S.J.; Cho, S.; Gibson, J.J. Characterization of Organic Composition in Snow and Surface Waters in the Athabasca Oil Sands Region, Using Ultrahigh Resolution Fourier Transform Mass Spectrometry. *Sci. Total Environ.* **2015**, *518–519*, 148–158. [CrossRef]
92. Ghotbizadeh, M.; Cuss, C.W.; Grant-Weaver, I.; Markov, A.; Noernberg, T.; Ulrich, A.; Shotyk, W. Spatiotemporal Variations of Total and Dissolved Trace Elements and Their Distributions amongst Major Colloidal Forms along and across the Lower Athabasca River. *J. Hydrol. Reg. Stud.* **2022**, *40*, 101029. [CrossRef]
93. Thomas, K.E.; Alexander, A.C.; Chambers, P.A. Contribution of Rain Events to Surface Water Loading in 3 Watersheds in Canada's Alberta Oil Sands Region. *J. Hydrol. Reg. Stud.* **2022**, *40*, 101028. [CrossRef]
94. Emmerton, C.A.; Cooke, C.A.; Hustins, S.; Silins, U.; Emelko, M.B.; Lewis, T.; Kruk, M.K.; Taube, N.; Zhu, D.; Jackson, B.; et al. Severe Western Canadian Wildfire Affects Water Quality Even at Large Basin Scales. *Water Res.* **2020**, *183*, 116071. [CrossRef] [PubMed]
95. Paul, A.J. Environmental Flows and Recruitment of Walleye (*Sander Vitreus*) in the Peace-Athabasca Delta. *Can. J. Fish. Aquat. Sci.* **2013**, *70*, 307–315. [CrossRef]
96. Summers, J.C.; Kurek, J.; Kirk, J.L.; Muir, D.C.G.; Wang, X.; Wiklund, J.A.; Cooke, C.A.; Evans, M.S.; Smol, J.P. Recent Warming, Rather than Industrial Emissions of Bioavailable Nutrients, Is the Dominant Driver of Lake Primary Production Shifts across the Athabasca Oil Sands Region. *PLoS ONE* **2016**, *11*, e0153987. [CrossRef]
97. Wnorowski, A.; Aklilu, Y.-A.; Harner, T.; Schuster, J.; Charland, J.P. Polycyclic Aromatic Compounds in Ambient Air in the Surface Minable Area of Athabasca Oil Sands in Alberta (Canada). *Atmos. Environ.* **2021**, *244*, 117897. [CrossRef]
98. Gray, M.A.; Curry, A.R.; Munkittrick, K.R. Non-Lethal Sampling Methods for Assessing Environmental Impacts Using a Small-Bodied Sentinel Fish Species. *Water Qual. Res. J. Can.* **2002**, *37*, 195–211. [CrossRef]
99. Arciszewski, T.J.; Kidd, K.A.; Munkittrick, K.R. Comparing Responses in the Performance of Sentinel Populations of Stoneflies (Plecoptera) and Slimy Sculpin (*Cottus Cognatus*) Exposed to Enriching Effluents. *Ecotoxicol. Environ. Saf.* **2011**, *74*, 1844–1854. [CrossRef]
100. Anonymous. After a 5-Year Review, Provincial Regulators Approve Syncrude's Plan to Extend Mildred Lake Operations. Available online: https://www.oilsandsmagazine.com/news/2019/7/23/after-a-5-year-review-alberta-regulators-approve-extension-of-mildred-lake-operations (accessed on 22 April 2022).

101. Suncor Suncor Energy Inc. Coke Fired Boiler Replacement Project Project Description. 2017. Available online: https://www.suncor.com/en-ca/what-we-do/oil-sands/coke-boiler-replacement-project (accessed on 3 August 2021).
102. Platt, J.R. Strong Inference: Certain Systematic Methods of Scientific Thinking May Produce Much More Rapid Progress than Others. *Science* **1964**, *146*, 347–353. [CrossRef]
103. Cash, K.J.; Culp, J.M.; Dubé, M.G.; Lowell, R.B.; Glozier, N.E.; Brua, R.B. Integrating Mesocosm Experiments with Field and Laboratory Studies to Generate Weight-of-Evidence Risk Assessments for Ecosystem Health. *Aquat. Ecosyst. Health Manag.* **2003**, *6*, 177–183. [CrossRef]
104. Shipley, B. *Cause and Correlation in Biology: A User's Guide to Path Analysis, Structural Equations and Causal Inference with R*; Cambridge University Press: Cambridge, UK, 2016; ISBN 1107442591.

environments

MDPI

Article

High and Low Air Temperatures and Natural Wildfire Ignitions in the Sierra Nevada Region

Matthew D. Petrie [1,*], Neil P. Savage [1] and Haroon Stephen [2]

1 School of Life Sciences, University of Nevada, Las Vegas, Las Vegas, NV 89154, USA; savagn1@unlv.nevada.edu
2 Civil and Environmental Engineering and Construction, University of Nevada, Las Vegas, Las Vegas, NV 89154, USA; haroon.stephen@unlv.edu
* Correspondence: matthew.petrie@unlv.edu

Abstract: The Sierra Nevada region has experienced substantial wildfire impacts. Uncertainty pertaining to fire risk may be reduced by better understanding how air temperature (Ta: °C) influences wildfire ignitions independently of other factors. We linked lightning-ignited wildfires to Ta patterns across the region from 1992 to 2015 and compared monthly high- and low-air-temperature patterns between ignition and non-ignition locations at local scales (4 km). Regionally, more ignitions occurred in springs with a greater number of high-Ta months and fewer cool Ta months (analyzed separately) and in summers with fewer cool Ta months. Locally, summer ignition locations experienced warmer summer months on a normalized scale than non-ignition locations. The probability of a wildfire ignition was positively associated with a greater number of high-Ta months during and prior to fire seasons. Regionally, springs with a greater number of high-Ta months had more wildfire ignitions. Locally, as individual locations in the region experienced a greater number of high-Ta months preceding and including the fire season, they exhibited substantial increases in spring (+1446%), summer (+365%), and fall (+248%) ignitions. Thus, the frequent occurrence of high-Ta months is positively associated with lightning-ignited wildfires in the Sierra Nevada region.

Keywords: fire risk; forest; climate; California; Nevada

Citation: Petrie, M.D.; Savage, N.P.; Stephen, H. High and Low Air Temperatures and Natural Wildfire Ignitions in the Sierra Nevada Region. *Environments* **2022**, *9*, 96. https://doi.org/10.3390/environments9080096

Academic Editors: Matteo Convertino and Jie Li

Received: 9 June 2022
Accepted: 18 July 2022
Published: 28 July 2022

1. Introduction

Measures of wildfire activity, including the number of large wildfires, burned area extent, and wildfire severity, have been increasing in many locations across the western United States since the 1980s [1–4]. These increases have largely been attributed to the interactive effects of climate warming and the legacy of human wildfire and forest management [1,3]. The average annual burned area in the western US increased at a rate of 355 km^2 per year from 1984 to 2011 [5], and wildfires have caused considerable damage to ecosystems, human health, and human communities in the wildland–urban interface [6,7]. As conditions supporting wildfire—and especially very large wildfires—increase across the western US as a result of climate change, wildfire affected areas are expected to also increase considerably within this region [8–11]. The Sierra Nevada region of California and Nevada is one of the largest wildfire-affected areas of the western US and has experienced a greater than six-fold increase in average annual burned area since 1972 [2,12]. Yet, while wildfire increases are anticipated, fire risk assessments remain difficult to quantify across the diverse causative agents—human actions, climates, landscape and topographic conditions, and over- and understory characteristics of ecosystems—that comprise the Sierra Nevada region [13–16].

Many wildfire ignitions in the western US—including the Sierra Nevada region—are initiated by lightning strikes (e.g., natural wildfire ignitions; Short [17]). The fire risk associated with lightning-ignited wildfires is influenced by multiple factors. Short-term weather patterns influence the frequency and timing of lightning strikes, and lightning strikes vary geographically and topographically [18–21]. At seasonal to interannual timescales, precipitation and temperature influence the ignition probability component of fire risk through their interactive effects on understory vegetation and fuel moisture [19,22], and wildfire has been found to coincide with dry periods and lightning strikes [23]. In the Sierra Nevada region, vegetation composition and structure vary at fine spatial scales, and over interannual to decadal time periods [24,25]. Thus, the components of natural fire risk—lightning occurrence and ignition probability—may often exhibit considerable spatial and temporal variation, and their interaction may in some instances enhance risk and in other instances lessen it [19]. Considerable research suggests that climate change will increase lightning occurrence and ignition probability [8,26,27], although Finney et al. [28] suggest that global lightning density may actually decline in coming decades. Increasing temperatures and earlier spring snowmelt are already lengthening and intensifying ecosystem drying and wildfire activity periods [29,30], and the associated occurrence of extreme climate events, including drought and heat waves, have been linked to regional patterns of wildfire extent and severity [31,32]. These studies point to a future where enhanced lightning occurrence and ignition probability could lead to enhanced fire risk across both space and time.

Wildfire characteristics—ignition, extent, and severity, for example—are enhanced by hot and dry conditions [20] but through different pathways. Once hot and dry conditions have been realized, wildfire ignition is strongly influenced by weather patterns and the favorability of ignition agents [19,27], wildfire severity is strongly influenced by topography, vegetation composition, and vegetation structural characteristics [33,34], and wildfire extent is strongly influenced by wind speed, topography, and the occurrence of precipitation [22,33,35]. Complex and interactive processes such as these may be simplified—if not fully resolved—by compartmentalizing their components independently. To this end, we propose that air temperature (Ta: °C) may have utility for quantifying fire risk. Ta is positively associated with lightning strikes, heat waves, and meteorological drought, such that warmer temperatures may capture variation in both lightning occurrence and ignition probability [20,26]. Recent research suggests that properties of extreme events may capture conditions exceeding normal environmental variation and help to identify changes in fire risk [32]. We therefore postulate that Ta may indirectly capture ignition probability through its strong negative association with fuel moisture (indicating enhanced ignition probability; Abatzoglou and Kolden [36], Flannigan et al. [37]) and through its positive association with dry atmospheric events and meteorological conditions supporting lightning strikes [20,26]. That is, Ta variation—and especially the occurrence of high-Ta periods—is associated with the meteorological and environmental conditions supporting natural wildfire ignition and may be a useful way to compartmentalize and simplify one component of fire risk.

Although considerable research has linked Ta to lightning strikes and wildfire ignitions, it is not clear if analyses of Ta can be used to quantify change to fire risk and at what spatial and temporal scales these changes may be anticipated. If Ta patterns are related to the probability of wildfire ignition, this determination can help to identify how ignition probabilities will change in a warming regional climate and indicate locations where management interventions should be prioritized. In this study, we explored relationships between Ta change, the properties of high (warm) and low (cool) Ta months, and the occurrence of lightning-ignited wildfires in the Sierra Nevada region from 1992 to 2015. During a similar timeframe (1992–2012), this region experienced the most lightning-ignited wildfires in the contiguous US [38]. The objectives of this research were to: (1) determine if regional temperature patterns from 1992 to 2015 in the Sierra Nevada region were related to the number of regional wildfire ignitions in spring, summer, and fall fire seasons; (2) contrast the frequency of occurrence and magnitude of high- and low-Ta months in spring, summer, and fall fire seasons between locations that experienced a wildfire ignition and locations that did not (fire ignition vs. non-ignition locations); and (3) explore to what degree the frequency of high- and low-Ta months prior to spring, summer, and fall fire seasons are related to ignition probability. Our analyses explore Ta–ignition relationships averaged across the region and for individual locations through time. Our analyses also evaluate the magnitude and occurrence of high- and low-Ta months within individual wildfire seasons and also evaluate high- and low-Ta months in preceding seasons, which may indicate sustained conditions (or lack thereof) that amplify wildfire ignition probability. Because wildfires have been strongly tied to hot and dry conditions in this region [14,32], we hypothesized that regional analyses focusing on high Ta would show a greater association with wildfire ignition compared to analyses of low Ta. Similarly, we hypothesized that the magnitude of high-Ta months would have a greater association with wildfire ignitions (due to their general indication of drier conditions and favorable weather for lightning strikes) compared to the frequency of high-Ta months.

2. Site Description

The Sierra Nevada region encompasses an area of 64,544 km^2 ([39]; Figure 1a). This ecologically diverse region is comprised of forest, woodland, and shrubland vegetation, with a plant species composition that is influenced by the surrounding ecoregions: the Cascade–Sierra Mountains to the north, the Great Basin to the east, the Mojave Desert to the south, and the Central Valley to the west [24]. The Sierra Nevada has a Mediterranean climate, with the majority of annual precipitation (~500–2030 mm total) falling between fall and spring, and with dry summers [24]. Average daily temperatures vary across seasons, topography, and elevation (~100–3900 m; United States Forest Service [24], PRISM Climate Group [40]). Vegetation composition also varies across topography and elevation, often at relatively fine spatial scales [14,41]. Frequent, small fires that maintained ecosystem health and structure in this region were reduced due to human fire suppression in the 20th century, which resulted in increasing fuels and vegetation density and contributed to the occurrence of large wildfires over the past 40 years [42]. Vegetation mortality and die-offs are also increasing across the region due to the interplay of and altered fire regime, drought, and pathogen outbreaks, which portend to an uncertain future for the region's ecosystems and the services they provide [21,32].

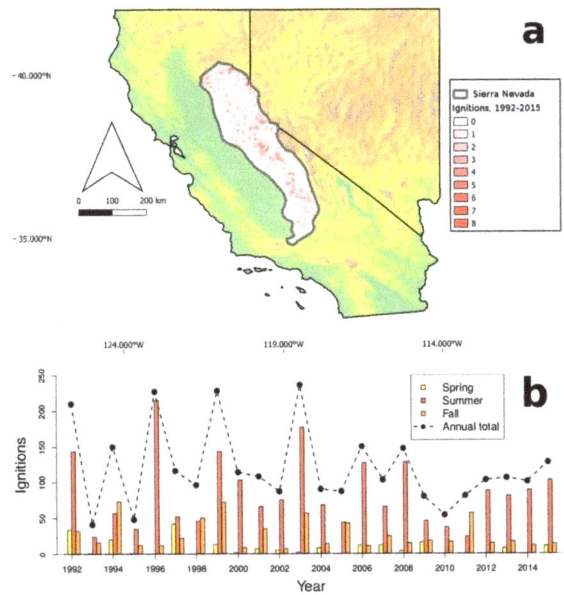

Figure 1. Study area map of the Sierra Nevada region, including the number of fire ignitions in each 4 km grid cell from 1992 to 2015 (Panel **a**), and a barplot illustrating the timeseries of wildfire ignitions in spring, summer, and fall from 1992 to 2015 (Panel **b**). The black line in Panel (**b**) illustrates total wildfire ignitions in each year.

3. Methods

3.1. Weather Estimates and Spatial Wildfire Data

We obtained daily maximum (Ta_{max}) and minimum (Ta_{min}) air temperature estimates at 4 km resolution from 1990 to 2016 from the Parameter-elevation Regressions on Independent Slopes Model (PRISM; PRISM Climate Group [40]), which has been used in other, recent wildfire-focused research [43,44]. We clipped PRISM raster files to the Sierra Nevada region (4034 total 4 km resolution cells). Weather estimates, including PRISM, may experience autocorrelation due to both naturally occurring patterns as well as interpolation procedures, and we found a moderate degree of autocorrelation of Ta_{max} and Ta_{min} for our study area (Figure S1). Fortin and Dale [45] recommend the use of small blocks when autocorrelation is present, and we minimized potential autocorrelation effects by analyzing Ta in each grid cell independently of surrounding cells. We conducted spatial data harmonization using QGIS [46], and we compiled and analyzed all data using R [47].

We obtained point-based wildfire ignition data for the Sierra Nevada region from 1992 to 2015 from the Fire Program Analysis Fire-Occurrence Database [17]. Our analyses included only lightning-ignited wildfires and included fire size classes from >0.10 to ≥2023.43 ha (Class B to Class G), the characteristics of which are provided in the attribute data of the Fire Program Analysis Fire-Occurrence Database [17,48]. We did not include fires ≤ 0.10 ha (Class A; 7897 of 10,788 total fires; 73%), which may occur following lightning strikes even when ignition agents are unfavorable for wildfire spread. These small fires may include ignitions with greater wildfire potential that were quickly contained and suppressed, and there is therefore some uncertainty in whether the fires removed from our analysis were of lower or higher risk.

We assigned point-based wildfire occurrence data to the centroid of the nearest 4 km cell. In our analyses, we compared ignition and non-ignition cells, evaluated seasonally at discrete time steps. For example, a single 4 km cell was designated to be an ignition location in any season when it experienced a wildfire ignition (summer 2010, for example) and was designated a non-ignition location in seasons when it did not. We evaluated 3 fire seasons (spring: March–May; summer: June–August; fall: September–November). Only 10 ignitions during winter met the Class B to Class G criteria of this study, and we therefore did not analyze winter ignitions.

3.2. High- and Low-Temperature Months

We used a peak-over-threshold approach from Coelho et al. [49] to calculate the occurrence and properties of high- and low-Ta months. This is a nonstationary approach that is used to identify anomalous and extreme values at the monthly time step from the distribution of daily values, across a moving time window. This technique is especially useful for identifying unique features in climatic data that have a positive or negative trend in their mean or variance through time (due to climate change, for example). Using this technique, analysis can focus on the properties of daily values occurring above a time-varying threshold, or on variation in the threshold values (see Petrie et al. [50] for an example using temperature simulations and a focus on statistically extreme events). For each 4 km cell in each month, our analyses focused on determining the temperature threshold value designating a high- or low-Ta month. On average, 20–30% of all months in our analyses were high- or low-Ta months. Our reported high- and low-Ta months are therefore best understood as the occurrence and magnitude of the top 20–30% of warm or cool months, instead of extreme events in the top 5% of all values.

To determine the occurrence and threshold magnitude of high-Ta months, we first calculated a 3-year floating mean of daily Ta_{max} values in each month. For example, the 3-year floating mean for a single cell in August 1995 was the mean of 93 daily Ta_{max} values from August, 1994, 1995, and 1996. In this analysis, we found that window length had minimal influence on monthly floating mean values and chose a 3-year window length to maximize the number of observations. After calculating the 3-year floating mean, we then determined the daily Ta_{max} values in the analysis month (August, 1995, in this example) that were greater than the 3-year floating mean value and ranked these daily values from the highest daily value to the lowest. A month with ≥20 daily Ta_{max} values in this category therefore experienced at least 1 day with a Ta_{max} value that exceeded the top 5% of days and was therefore a statistical extreme event for daily values. The observed daily Ta_{max} value ranked sequentially below the top 5% of values (e.g., 1–2 values below the highest daily value) was the threshold value, which designated the boundary air temperature between extreme and non-extreme daily values. Although this threshold value indicates a month with an extreme Ta day, we clarify that this is best viewed as a month located in the top 27% of observed daily temperature values. We calculated high- and low-Ta months separately. To determine the occurrence of low-Ta months, we used the same procedure but evaluated daily values of Ta_{min} below the 3-year floating mean.

Because the distribution of average Ta differed among the cells in our study region, we normalized the magnitude of high- and low-Ta months as an anomaly ((observed–long-term mean)/standard deviation of long-term mean). To allow for analysis of these anomaly values across cells with significant and insignificant trends in Ta magnitude, we detrended the magnitude of high and low Ta for cells that experienced a statistically significant linear trend in the threshold value over the 1991–2015 study period (at least 7 high- or low-Ta months observed; $p < 0.05$).

3.3. Analysis

Our regional analyses explored relationships between temperatures and wildfire ignitions from 1991 to 2015 and relationships between the frequency of occurrence and the magnitude of high- and low-Ta months and the occurrence of wildfire ignitions. Our analyses of individual locations (e.g., 4 km cells) focused on contrasting the frequency of occurrence and the magnitude of high- and low-Ta months between ignition locations and non-ignition locations in spring, summer, and fall fire seasons. We determined significant relationships between the number of high- and low-Ta months and wildfire ignitions using linear correlations (R^2 coefficient of determination; $p < 0.05$), and we determined significant differences in the average magnitude of high- and low-Ta months between ignition and non-ignition locations using ANOVA and Tukey's honest significant differences ($p < 0.05$).

We determined to what degree the occurrence and magnitude of high- and low-Ta months in seasons preceding the fire season were related to differences between ignition and non-ignition locations. To determine if ignition locations experienced a differing magnitude of high- or low-Ta months compared to non-ignition locations, we contrasted the magnitude of high- and low-Ta months between ignition and non-ignition locations in the seasons preceding each fire season, as well as in spring, summer, and fall fire seasons. To determine if ignition locations experienced a greater frequency of high- and low-Ta months compared to non-ignition locations, we counted the total number of high- or low-Ta months (separate analyses for high versus low) in seasons preceding each fire season. Our counts included the 3 seasons prior to each fire season (0 months minimum, 9 months maximum) as well as the 3 prior seasons up to and including spring, summer, and fall fire seasons (0 months minimum, 12 months maximum). We interpreted a positive relationship between the proportion of locations experiencing a wildfire ignition and a higher number of Ta months (either high or low) as evidence that Ta and wildfire ignitions were positively associated. To evaluate these relationships, we grouped all locations experiencing the same number of high- or low-Ta months into discrete categories (0 months, 1 month, 2 months, etc.) and calculated the ignition proportion of each category independently from the number of locations experiencing a wildfire ignition divided by the number not experiencing an ignition. We required at least 8 locations in each category, and we combined observations when necessary (for example, combining 9 and 10 month locations to reach the required number of 8 observations, resulting in a 9–10 category).

4. Results

4.1. Regional Change from 1992 to 2015

The number of wildfire ignitions was lower in the later third of our 1992–2015 study period (2008–2015) compared to the first two-thirds (1992–2007), and summer and fall accounted for 87.8% of the total ignitions (Figure 1b, Table 1b). The magnitude of the high-Ta months was generally lower in the first 8 years of our 24-year study period (1992–1999) compared to the last 16 years, but significant differences between study period intervals were inconsistent and were strongly influenced by a large number of observations (Table 1a, Figure 2). The magnitude of high-Ta months was generally greater in the last 8 years of our study period (2008–2015) compared to the first 16 years (ANOVA and Tukey's honest significant differences, $p < 0.05$; Table 1b). We observed a slight decline in the frequency of high-Ta months in spring and summer in the later third of our 1992–2015 study period (2008–2015) compared to the first two-thirds (1992–2007; Table 1a).

Table 1. Seasonal air temperature (Ta: °C) differences between 8-year time periods from 1992 to 2015 in the Sierra Nevada region. Ta differences were analyzed for high-Ta months (Section **a**) and locations with high-Ta months that experienced a concurrent wildfire ignition (Section **b**). The number of observations in each category is indicated in the table, and significance is indicated by differing letters (determined at $p < 0.05$).

(a) High-Ta months.

Time Period	Spring		Summer		Fall		Winter	
	# obs.	(°C)	# obs.	(°C)	# obs.	(°C)	# obs.	(°C)
1992:1999	29,541	22.8 ^{c}	33,485	31.6 ^{b}	27,878	24.9 ^{c}	20,147	17.1 ^{b}
2000:2007	32,926	24.1 ^{a}	29,990	31.1 ^{c}	29,625	25.6 ^{b}	19,751	17.2 ^{a}
2008:2015	18,027	23.7 ^{b}	19,826	32.5 ^{a}	27,667	28.1 ^{a}	20,828	16.7 ^{c}

(b) High-Ta months, fire ignition locations.

Time period	Spring		Summer		Fall		Winter	
	# obs.	(°C)	# obs.	(°C)	# obs.	(°C)	# obs.	(°C)
1992:1999	83	25.6 ^{b}	318	31.9 ^{ab}	89	28.5 ^{ab}	5	21.4
2000:2007	23	27.6 ^{a}	213	31.2 ^{b}	84	27.4 ^{b}	5	18.4
2008:2015	15	29.3 ^{a}	175	32.1 ^{a}	58	29.9 ^{a}	0	-

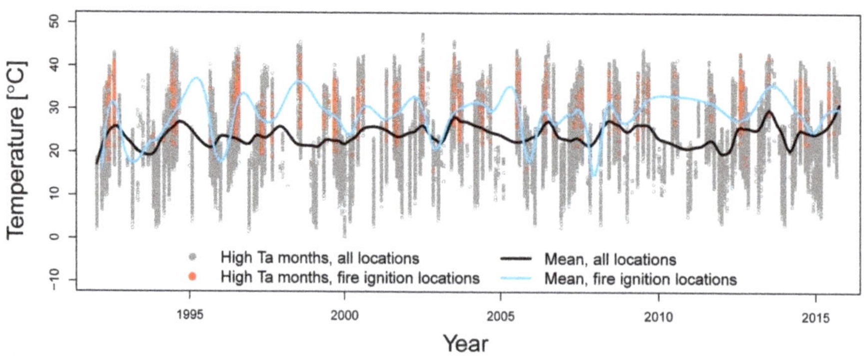

Figure 2. Timeseries of the magnitude of high-air-temperature months (Ta: °C) across the Sierra Nevada region from 1992 to 2015, and the magnitude of high-Ta months for fire ignition locations in the month when ignition occurred. Lines were loess smoothed for illustration.

4.2. The Magnitude of High and Low Temperatures and Wildfire Ignitions

Across the Sierra Nevada region, there was no relationship between the normalized magnitude of high- or low-Ta months and the percentage of locations that experienced a wildfire ignition in the spring, summer, or fall fire seasons (Figures 3a and S2a). At local scales, the ignition locations in summer experienced (on a normalized scale) warmer high-Ta summer months (Figure 4b), and the ignition locations in fall experienced less cool low-Ta fall months (Figure S3c). The differences in the magnitude of high and low Ta in months preceding the fire seasons were inconclusive (Figures 3 and S2).

4.3. The Frequency of High- and Low-Temperature Months and Wildfire Ignitions

Regionally, a greater number of ignitions occurred in springs that experienced a greater proportion of locations experiencing one or more high-Ta months (0.003% ignition percentage at a proportion of 0.0% of locations; 0.4% ignition percentage at 100% of locations; Figure 3b), a lower proportion of the locations experiencing one or more low-Ta months (0.0% ignition percentage at a proportion of 100% of locations; 0.4% ignition percentage at 0.0% of locations; Figure S2b), and in summers with a lower proportion of locations experiencing one or more low-Ta months (1.3% ignition percentage at a proportion of 100% of locations; 3.0% ignition percentage at 0.0% of locations; Figure S2b). At local scales, the ignition locations in spring, summer, and fall experienced a greater number of high-Ta months, both in the fire season and in the previous seasons (Figure 5). In spring and summer, the ignition locations generally experienced fewer low-Ta months (Figure S4).

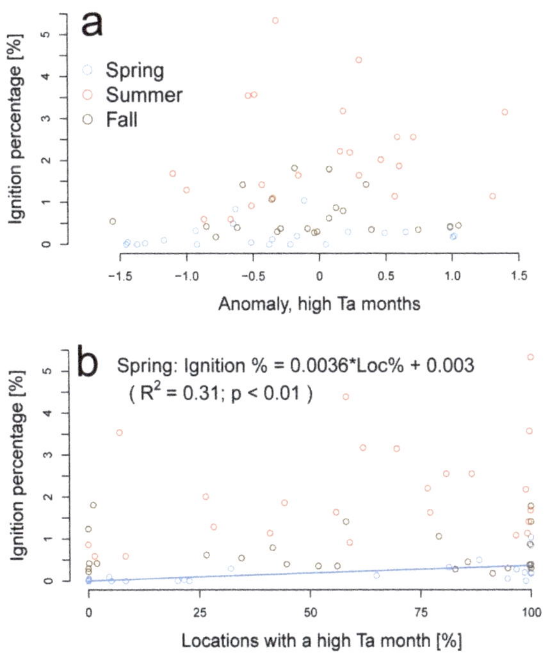

Figure 3. Relationships between the average anomaly of high-air-temperature months (anomaly = observed − long-term mean/standard deviation of long-term mean) and the percentage of locations in the Sierra Nevada region experiencing a wildfire ignition in spring, summer, and fall seasons (Panel **a**), and relationships between the percentage of locations in the Sierra Nevada region experiencing a high-air-temperature month and the percentage of locations in the Sierra Nevada region experiencing a wildfire ignition in spring, summer, and fall seasons (Panel **b**). Significant relationships are shown in each panel (R^2 coefficient of determination; $p < 0.05$).

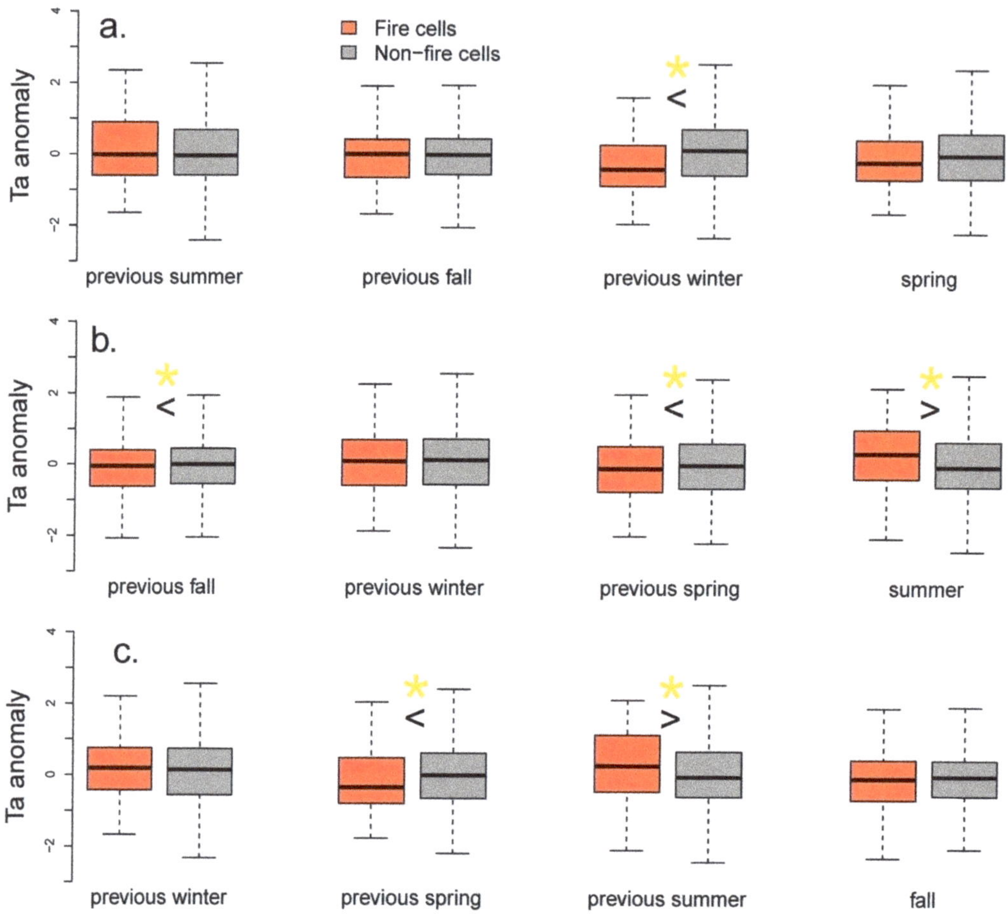

Figure 4. Boxplots illustrating differences in the anomaly of high-air-temperature months (anomaly = observed − long-term mean/standard deviation of long-term mean) between ignition (red) and non-ignition (gray) locations in the 3 seasons previous to and including the spring (Panel **a**), summer (Panel **b**), and fall (Panel **c**) fire seasons. Significant differences and the direction of difference are indicated for each boxplot pair (one-tailed *t*-tests; $p < 0.05$).

The number of high-Ta months occurring prior to and during fire seasons was positively correlated to the percentage of locations experiencing a wildfire ignition (Figure 6). In spring, wildfire ignitions occurred in ~0.08% of the locations that did not experience a high-Ta month, and increased to ~0.5% for the locations experiencing 5–6 of these months in the prior three seasons (of nine possible; Figure 6c), and to ~0.9% of the locations experiencing 7–8 of these months in the prior 3 seasons plus the spring fire season (of 12 possible; Figure 6b). In summer, wildfire ignitions occurred in ~1.2% of the locations that did not experience a high-Ta month, and increased to ~4.2% for locations experiencing 7–8 of these months in the prior three seasons (of nine possible; Figure 6c), and to ~4.4% for locations experiencing 9–10 of these months in the prior 3 seasons plus the summer fire season (of 12 possible; Figure 6d). In fall, wildfire ignitions occurred in ~0.7% of the locations that did not experience a high-Ta month, and increased to ~1.5% for the locations experiencing seven of these months in the prior three seasons (of nine possible), and to

~1.7% for locations experiencing 8 of these months in the prior 3 seasons plus the fall fire season (of 12 possible; Figure 6e,f). We were not able to assess ignition increases for the three preceding seasons in spring due to too few observations (Figure 6a). We did not observe significant relationships between the number of low-Ta months and the percentage of locations experiencing a wildfire ignition (not shown).

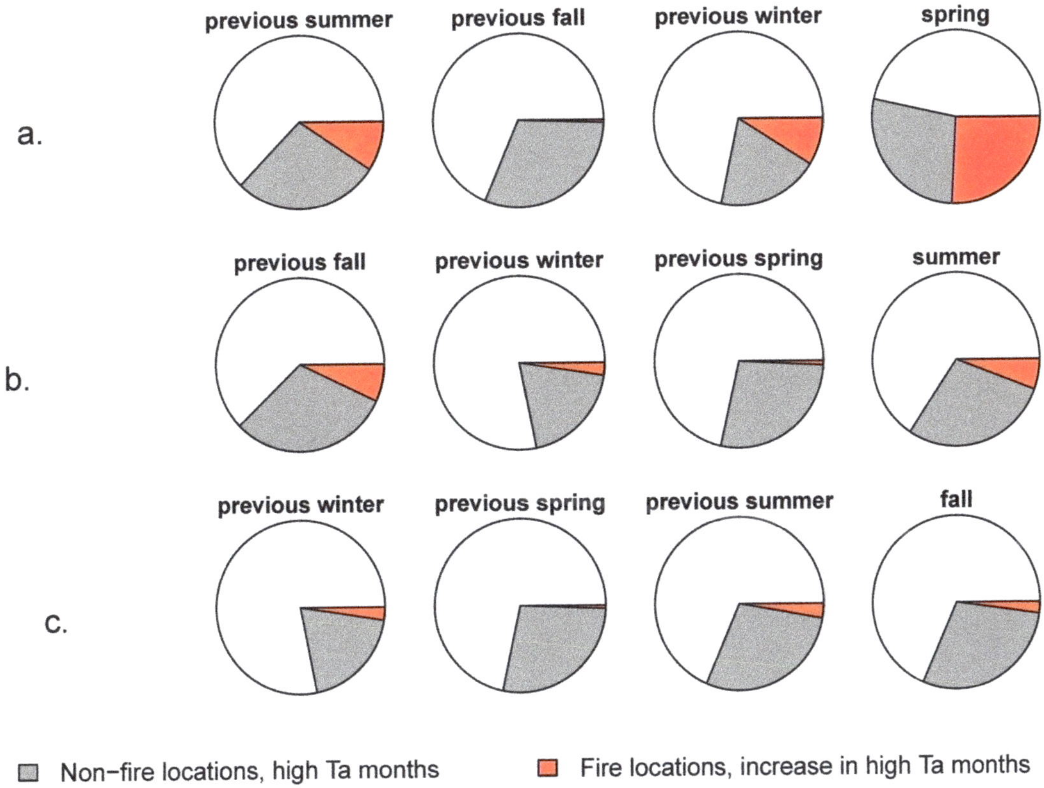

Figure 5. Pie charts illustrating differences in the proportion of ignition and non-ignition locations experiencing a high-air-temperature month in the 3 seasons previous to and including the spring (Panel **a**), summer (Panel **b**), and fall (Panel **c**) fire seasons. The average proportion of non-ignition locations experiencing a high-Ta month in each season is illustrated in gray, and the proportional increase in ignition locations experiencing a high-Ta month in each season is shown in red.

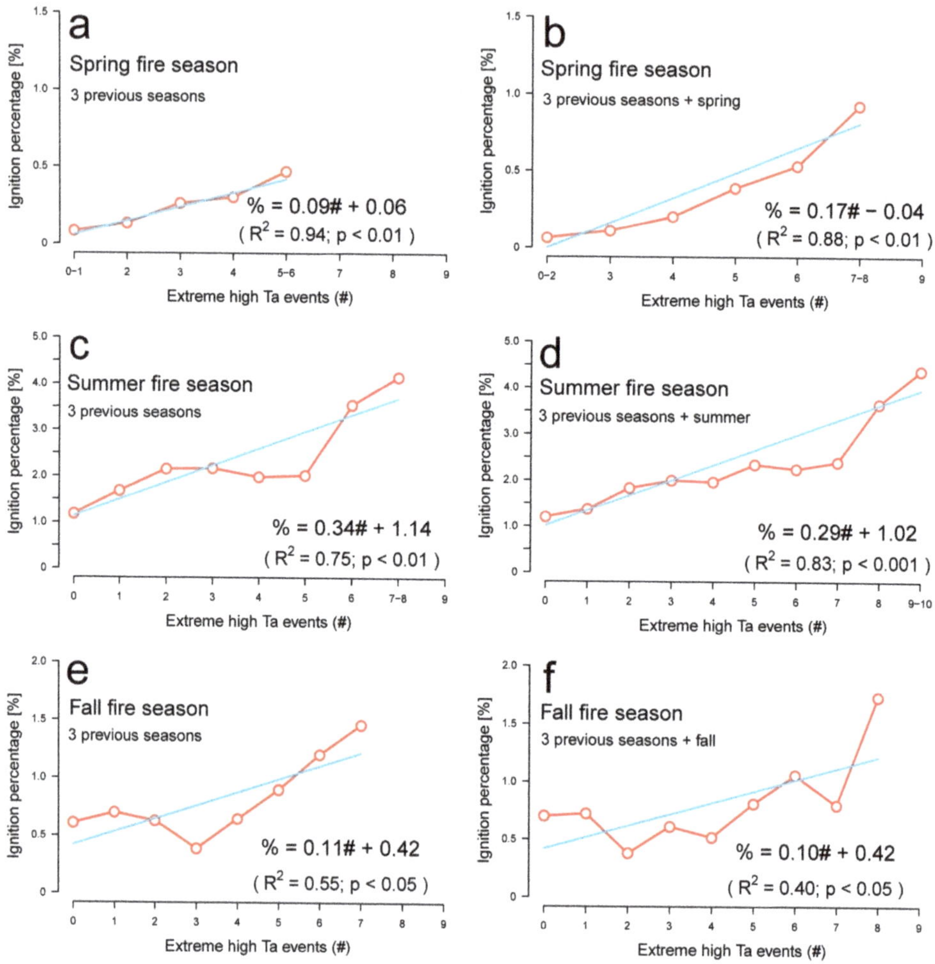

Figure 6. Fire ignition probability for locations experiencing differing numbers of high-air-temperature months during and prior to the fire season, summarized across 1992–2015. Panels (**a**,**c**,**e**) illustrate fire ignition probability in the spring (Panel **a**), summer (Panel **c**), and fall (Panel **e**) fire season in response to high-Ta month occurrence during the 3 prior seasons (3 seasons, 9 months maximum). Panels (**b**,**d**,**f**) illustrate fire ignition probability in the spring (Panel **b**), summer (Panel **d**), and fall (Panel **f**) fire season in response to high-Ta month occurrence during the 3 prior seasons and also the fire season (4 seasons, 12 months maximum). In cases where too few events were observed in a single category, these events were merged with the next lowest category. Because multiple observations were combined in some categories, linear correlations were developed for category number instead of observation number (Panel (**a**) example: 0 observations = category 0; 5–6 observations = category 4).

5. Discussion

The slightly increasing magnitude of Ta across the Sierra Nevada region did not correspond to increased wildfire ignitions, which have been declining across the western US in recent decades [51]. In refutation of our first hypothesis, we did not find associations between the magnitude of high- and/or low-Ta months with wildfire ignitions at the regional scale. That is, the Sierra Nevada region as a whole did not experience variation in wildfire ignitions that can be tied to regionally warmer or cooler months. At the 4 km local scale, however, ignition locations in summer had relatively warmer high-Ta months compared to non-ignition locations, and ignition locations in fall had less cool low-Ta months. This corroborates other work linking high air temperatures to the dry environmental and high-pressure systems that support lightning strikes and wildfire ignition [20,52].

In refutation of our second hypothesis, we found that the frequency of high- and low-Ta months was closely associated with wildfire ignitions. Regionally, more ignitions occurred in springs with a greater number of high and fewer cool Ta months, and in summers with fewer cool Ta months. Locally, ignition locations consistently experienced a greater number of high-Ta months in fire seasons and in the three preceding seasons. As the number of high-Ta months increased from 0 to a maximum of ~7–10, we found a 1446% increase in spring ignitions, a 365% increase in summer ignitions, and a 248% increase in fall ignitions. This considerable increase in wildfire ignition probability suggests that fire risk in the Sierra Nevada region is enhanced by the more frequent occurrence of warm temperatures over multiple months. In summer and fall, an observation of high-Ta months prior to the fire season suggested ignition percentages that were similar to percentages observed in analyses that included observations during the fire season (summer: 4.2% prior, 4.4% prior + fire season; fall: 1.5%, 1.7%). This suggests that the variation in the probability of fire ignitions in summer and fall may in many cases be strongly influenced by the legacy effects of prior seasons.

Spring ignitions were notably emergent under a greater frequency of warm conditions and were enhanced by additional high-Ta months in spring (0.5% prior, 0.9% prior + fire season). Spring ignition locations experienced a greater number of high-Ta months and fewer low-Ta months in the spring and preceding winter, which corroborates previous research linking winter weather to spring temperatures, moisture patterns, and fire activity [29,30]. A higher Ta also decreases snowpack, resulting in more lightning strikes reaching surface fuels [53]. Although our analyses of spring ignitions were limited by relatively low ignition occurrence, our results suggest that consistent movement toward warmer and less cool winter conditions—and not change to episodic periods of the very coldest or warmest winter temperatures—may support the emergence of spring wildfire ignitions in this region.

5.1. How Broadly Is Air Temperature Associated with Wildfire?

Ta patterns have been previously linked to wildfire ignition [54,55], and our results show that observation of the frequency of high-Ta months has utility for the assessment of fire risk. It remains unclear, however, if Ta variation can help to understand wildfire extent and severity, which are underrepresented components of wildfire that differ from more commonly focused on components, such as fire risk and fire hazard [19]. Ta variation may capture conditions shaping wildfire severity and extent due to the influence of weather on wildfire activity [56] and climate on vegetation characteristics [27,57]. Understory fuel moisture influences both wildfire ignition and wildfire spread, and periods of low humidity associated with higher temperatures can increase dry fuel loads [37]. Thus, wildfire extent and severity may in some cases be associated with Ta, but these components are influenced more directly by over- and understory vegetation characteristics, such as tree stand health and density [34,36]. Additionally, Ta variation may have indirect effects on wildfire through other physical processes, including the potential for relatively cool and wet periods to increase wildfire severity by supporting the growth of understory vegetation [41]. Thus, although our results point to one pathway by which high temperatures may be associated with an increase in wildfire ignitions, there are many factors and mechanisms shaping ignitions and other wildfire components, especially over multiyear and decadal timescales.

We note a few caveats associated with the datasets we used in our study that should be considered when evaluating our reported Ta–ignition relationships. First, the accuracy of wildfire ignition coordinates may differ from the actual ignition location [58]. Second, spatial datasets such as PRISM provide an estimate of Ta patterns and do not fully resolve Ta in heterogeneous mountain landscapes [59]. Third, our analyses did not account for multiple ignitions in the same month or season in a single location, for the effect of previous fires on future ignitions [60], and our study area includes locations with limited causative agents (such as low vegetation cover) that may limit ignition.

5.2. Managing for Wildfire

Climate change is increasing the occurrence and extent of large wildfires, and over 4 million hectares of burned area in the western US has been linked to Ta-associated increases in vapor-pressure deficit in recent decades [43]. The frequency and magnitude of high-Ta months will likely increase throughout the western US in coming decades [50]. When vapor-pressure deficit is high, sustained high Ta enables the drying of fuels and may be associated with weeks to months of severe fire weather, and these conditions may often be clustered spatially [23,37]. In our analyses, we used a nonstationary method to calculate high- and low-Ta months, which minimizes the effect of changing mean temperatures through time [49], and we detrended their magnitude when appropriate. It follows that the methodology we employed could provide consistent indication of wildfire ignition probability in a warmer future climate, but we caution that climate change may alter some or many of the physical processes that influence wildfire in ways that do not have a contemporary analog. It is not yet fully clear if the associations between Ta and wildfire ignition that we observed over a 24-year period in the Sierra Nevada region will be consistent in a changing climate and to what degree they may be extended to other semiarid regions that differ in regional climate, topography, and ecosystem characteristics.

The use of Ta in wildfire research is bolstered by the variable's accessibility and predictability, and relationships between Ta and wildfire ignitions can provide practical utility for fire season planning. Specifically, consistent observations of high monthly temperatures preceding spring, summer, and fall fire seasons indicate elevated wildfire ignition probability for locations in the Sierra Nevada region. Future work could improve these results by determining to what degree spatial clustering of high Ta may indicate higher ignition probabilities (see Podschwit and Cullen [23] for a similar focus over very broad spatial areas), although the determination of natural versus artificial spatial autocorrelation across topographically complex areas may complicate this determination. We recommend that the findings of the present study can help land managers efficiently deploy actions that minimize the potential for wildfire ignitions and spread, including vegetation thinning and using natural fires to reduce fuels [61,62]. Specifically, locations that are regularly experiencing high-Ta months and also have high vegetation density and/or high understory fuels should be prioritized for thinning and fuel reduction treatments, especially when these locations are experiencing abnormal high summer temperatures. Locations experiencing an increase in high-Ta months and/or a decline in low-Ta months in spring may be more likely to experience an emergence of spring wildfire ignitions. That is, better anticipation of ignitions can be achieved through analyses of Ta and can be a useful tool to direct resources that help to protect regional ecosystems.

6. Conclusions

In this work, we sought to improve the understanding and anticipation of fire risk in the Sierra Nevada region, USA, by linking wildfire ignitions at regional and local scales to high and low air temperatures (Ta: °C) and the properties of high and low Ta evaluated at the monthly time step. Regionally, more ignitions occurred in springs with a greater number of high-Ta months and fewer low-Ta months, and in summers with fewer low-Ta months. We found that wildfire ignitions were most strongly associated with the frequency of high-Ta months, and wildfire ignition probability increased substantially when a greater number of these months occurred both prior to and during the fire season. Summer ignitions experienced the highest percentage increases (up to 4.4% of the locations) and were enhanced by both the frequency and the magnitude of high-Ta months in summer. Spring ignitions became emergent when they experienced an increase in the occurrence of high-Ta months or a decline in low-Ta months. Thus, the more frequent occurrence of high-

Ta months is positively associated with lightning-ignited wildfires in the Sierra Nevada region. Although there remains some uncertainty pertaining to the physical mechanisms and pathways that underlie these findings, Ta increases have the potential to enhance fire risk by increasing ignition probability and possibly by increasing lightning occurrence. The identification of change to the properties of Ta can help to identify locations and time periods experiencing enhanced fire risk and prioritize management interventions.

Supplementary Materials: The following supporting information can be downloaded at: https://www.mdpi.com/article/10.3390/environments9080096/s1, Figure S1: Spatial autocorrelation results; Figure S2: Regional relationships between ignitions and low Ta months; Figure S3: Low Ta month anomaly between ignition and non-ignition locations; Figure S4: Differences in the proportion of ignition and non-ignition locations experiencing a low Ta month.

Author Contributions: Conceptualization, M.D.P., H.S.; methodology, M.D.P., N.P.S., H.S.; software, M.D.P., N.P.S.; validation, M.D.P., N.P.S.; formal analysis, M.D.P., N.P.S.; investigation, M.D.P., N.P.S.; resources, M.D.P., N.P.S.; data curation, M.D.P., N.P.S.; writing—original draft preparation, M.D.P., N.P.S.; writing—review and editing, M.D.P., H.S.; visualization, M.D.P., N.P.S.; supervision, M.D.P., H.S.; project administration, M.D.P.; All authors have read and agreed to the published version of the manuscript.

Funding: This research received no external funding.

Data Availability Statement: All data and derived data products were obtained from online and publicly available sources, as described in the manuscript.

Conflicts of Interest: The authors declare no conflict of interest.

References

1. Dillon, G.; Holden, Z.; Morgan, P.; Crimmins, M.; Heyerdahl, E.; Luce, C. Both topography and climate affected forest and woodland burn severity in two regions of the western US, 1984 to 2006. *Ecosphere* **2011**, *2*, 1–33. [CrossRef]
2. Miller, J.; Safford, H. Trends in wildfire severity: 1984 to 2010 in the Sierra Nevada, Modoc Plateau, and southern Cascades, California, USA. *Fire Ecol.* **2012**, *8*, 41–57. [CrossRef]
3. Calkin, D.; Thompson, M.; Finney, M. Negative consequences of positive feedbacks in US wildfire management. *For. Ecosyst.* **2015**, *2*, 9. [CrossRef]
4. Doerr, S.; Santin, C. Global trends in wildfire and its impacts: perceptions versus realities in a changing world. *Philos. Trans. R. Soc. B Biol. Sci.* **2016**, *371*, 20150345. [CrossRef] [PubMed]
5. Dennison, P.; Brewer, S.; Arnold, J.; Moritz, M. Large wildfire trends in the western United States, 1984–2011. *Geophys. Res. Lett.* **2014**, *41*, 2928–2933. [CrossRef]
6. Miller, R.; Chambers, J.; Pellant, M. *A Field Guide for Selecting the Most Appropriate Treatment in Sagebrush and Pinon-Juniper Ecosystems in the Great Basin*; USDA Forest Service General Technical Report; U.S. Department of Agriculture, Rocky Mountain Research Station: Ft. Collins, CO, USA, 2014; pp. 1–76.
7. Fann, N.; Alman, B.; Broome, R.; Morgan, G.; Johnston, F.; Pouliot, G.; Rappold, A. The health impacts and economic value of wildland fire episodes in the US: 2008–2012. *Sci. Total. Environ.* **2018**, *610*, 802–809. [CrossRef]
8. Spracklen, D.; Mickley, L.; Logan, J.; Hudman, R.; Yevich, R.; Flannigan, M.; Westerling, A. Impacts of climate change from 2000 to 2050 on wildfire activity and carbonaceous aerosol concentrations in the western United States. *J. Geophys. Res. Atmos.* **2009**, *114*, D20301. [CrossRef]
9. Stavros, E.; Abatzoglou, J.; McKenzie, D.; Larkin, N. Regional projections of the likelihood of very large wildland fires under a changing climate in the contiguous Western United States. *Clim. Change* **2014**, *126*, 455–468. [CrossRef]
10. Barbero, R.; Abatzoglou, J.; Larkin, N.; Kolden, C.; Stocks, B. Climate change presents increased potential for very large fires in the contiguous United States. *Int. J. Wildland Fire* **2015**, *24*, 892–899. [CrossRef]
11. Podschwit, H.; Larkin, N.; Steel, E.; Cullen, A.; Alvarado, E. Multi-Model Forecasts of Very-Large Fire Occurences during the End of the 21st Century. *Climate* **2018**, *6*, 100. [CrossRef]
12. Williams, A.; Abatzoglou, J.; Gershunov, A.; Guzman-Morales, J.; Bishop, D.; Balch, J.; Lettenmaier, D. Observed impacts of anthropogenic climate change on wildfire in California. *Earth's Future* **2019**, *7*, 892–910. [CrossRef]
13. Collins, B.; Kelly, M.; Van Wagtendonk, J.; Stephens, S. Spatial patterns of large natural fires in Sierra Nevada wilderness areas. *Landsc. Ecol.* **2007**, *22*, 545–557. [CrossRef]
14. Liang, S.; Hurteau, M.; Westerling, A. Response of Sierra Nevada forests to projected climate–wildfire interactions. *Glob. Chang. Biol.* **2017**, *23*, 2016–2030. [CrossRef] [PubMed]
15. Syphard, A.; Keeley, J.; Pfaff, A.; Ferschweiler, K. Human presence diminishes the importance of climate in driving fire activity across the United States. *Proc. Natl. Acad. Sci. USA* **2017**, *114*, 13750–13755. [CrossRef]

16. Syphard, A.; Sheehan, T.; Rustigian-Romsos, H.; Ferschweiler, K. Mapping future fire probability under climate change: Does vegetation matter? *PLoS ONE* **2018**, *13*, e0201680. [CrossRef] [PubMed]
17. Short, K. *Spatial Wildfire Occurrence Data for the United States, 1992–2015 [FPA_FOD_20170508]*, 4th ed.; Forest Service Research Data Archive; United States Department of Agriculture: Fort Collins, CO, USA, 2017. [CrossRef]
18. Flannigan, M.; Wotton, B. Chapter 10—Climate, Weather, and Area Burned. In *Forest Fires: Behavior and Ecological Effects*; Academic Press: Cambridge, MA, USA, 2001; pp. 351–373. [CrossRef]
19. Hardy, C. Wildland fire hazard and risk: Problems, definitions, and context. *For. Ecol. Manag.* **2005**, *211*, 73–82. [CrossRef]
20. Littell, J.; Peterson, D.; Riley, K.; Liu, Y.; Luce, C. A review of the relationships between drought and forest fire in the United States. *Glob. Chang. Biol.* **2016**, *22*, 2353–2369. [CrossRef]
21. Stevens, J.; Collins, B.; Miller, J.; North, M.; Stephens, S. Changing spatial patterns of stand-replacing fire in California conifer forests. *For. Ecol. Manag.* **2017**, *406*, 28–36. [CrossRef]
22. Littell, J.; McKenzie, D.; Peterson, D.; Westerling, A. Climate and wildfire area burned in western US ecoprovinces, 1916–2003. *Ecol. Appl.* **2009**, *19*, 1003–1021. [CrossRef]
23. Podschwit, H.; Cullen, A. Patterns and trends in simultaneous wildfire activity in the United States from 1984 to 2015. *Int. J. Wildland Fire* **2020**, *29*, 1057–1071. [CrossRef]
24. United States Forest Service. Ecological Subregions of the United States. 1994. Available online: https://www.fs.fed.us/land/pubs/ecoregions/ch33.html (accessed on 24 June 2020).
25. Hanberry, B. Compositional changes in selected forest ecosystems of the western United States. *Appl. Geogr.* **2014**, *52*, 90–98. [CrossRef]
26. Price, C.; Rind, D. Possible implications of global climate change on global lightning distributions and frequencies. *J. Geophys. Res. Atmos.* **1994**, *99*, 10823–10831. [CrossRef]
27. Bajocco, S.; Koutsias, N.; Ricotta, C. Linking fire ignitions hotspots and fuel phenology: The importance of being seasonal. *Ecol. Indic.* **2017**, *82*, 433–440. [CrossRef]
28. Finney, D.; Doherty, R.; Wild, O.; Stevenson, D.; MacKenzie, I.; Blyth, A. A projected decrease in lightning under climate change. *Nat. Clim. Change* **2018**, *8*, 210-213. [CrossRef]
29. Westerling, A.; Hidalgo, H.; Cayan, D.; Swetnam, T. Warming and earlier spring increase western US forest wildfire activity. *Science* **2006**, *313*, 940–943. [CrossRef] [PubMed]
30. Westerling, A. Increasing western US forest wildfire activity: Sensitivity to changes in the timing of spring. *Philos. Trans. R. Soc. B Biol. Sci.* **2016**, *371*, 20150178. [CrossRef]
31. Liu, Y.; Goodrick, S.; Stanturf, J. Future US wildfire potential trends projected using a dynamically downscaled climate change scenario. *For. Ecol. Manag.* **2013**, *294*, 120–135. [CrossRef]
32. Crockett, J.; Westerling, A. Greater temperature and precipitation extremes intensify Western US droughts, wildfire severity, and Sierra Nevada tree mortality. *J. Clim.* **2018**, *31*, 341–354. [CrossRef]
33. Turner, M.; Romme, W. Landscape dynamics in crown fire ecosystems. *Landsc. Ecol.* **1994**, *9*, 59–77. [CrossRef]
34. Parks, S.; Holsinger, L.; Panunto, M.; Jolly, W.; Dobrowski, S.; Dillon, G. High-severity fire: Evaluating its key drivers and mapping its probability across western US forests. *Environ. Res. Lett.* **2018**, *13*, 044037. [CrossRef]
35. Moritz, M.; Moody, T.; Krawchuk, M.; Hughes, M.; Hall, A. Spatial variation in extreme winds predicts large wildfire locations in chaparral ecosystems. *Geophys. Res. Lett.* **2010**, *37*, L04801. [CrossRef]
36. Abatzoglou, J.; Kolden, C. Relationships between climate and macroscale area burned in the western United States. *Int. J. Wildland Fire* **2013**, *22*, 1003–1020. [CrossRef]
37. Flannigan, M.; Wotton, B.; Marshall, G.; De Groot, W.; Johnston, J.; Jurko, N.; Cantin, A. Fuel moisture sensitivity to temperature and precipitation: Climate change implications. *Clim. Chang.* **2016**, *134*, 59–71. [CrossRef]
38. Balch, J.; Bradley, B.; Abatzoglou, J.; Nagy, R.; Fusco, E.; Mahood, A. Human-started wildfires expand the fire niche across the United States. *Proc. Natl. Acad. Sci. USA* **2017**, *114*, 2946–2951. [CrossRef] [PubMed]
39. United States Geological Survey. Physiographic Divisions of the Conterminous U. S. 2004. Available online: https://water.usgs.gov/GIS/metadata/usgswrd/XML/physio.xml (accessed on 21 October 2019).
40. PRISM Climate Group. Oregon State University, Created 4 February 2004. 2019. Available online: http://www.prism.oregonstate.edu (accessed on 18 December 2019).
41. Miller, J.; Safford, H.; Crimmins, M.; Thode, A. Quantitative evidence for increasing forest fire severity in the Sierra Nevada and southern Cascade Mountains, California and Nevada, USA. *Ecosystems* **2009**, *12*, 16–32. [CrossRef]
42. Stephens, S.; Collins, B.; Fettig, C.; Finney, M.; Hoffman, C.; Knapp, E.; North, M.; Safford, H.; Wayman, R. Drought, tree mortality, and wildfire in forests adapted to frequent fire. *BioScience* **2018**, *68*, 77–88. [CrossRef]
43. Abatzoglou, J.; Williams, A. Impact of anthropogenic climate change on wildfire across western US forests. *Proc. Natl. Acad. Sci. USA* **2016**, *113*, 11770–11775. [CrossRef] [PubMed]
44. Hallema, D.; Sun, G.; Caldwell, P.; Norman, S.; Cohen, E.; Liu, Y.; Bladon, K.; McNulty, S. Burned forests impact water supplies. *Nat. Commun.* **2018**, *9*, 1307. [CrossRef]
45. Fortin, M.; Dale, M. *Spatial Analysis: A Guide for Ecologists*; Cambridge University Press: Cambridge, UK, 2005.
46. QGIS Development Team. QGIS Geographic Information System, Open Source Geospatial Foundation Project. 2021. Available online: http://qgis.osgeo.org (accessed on 24 June 2020).

47. R Development Core Team. *R: A Language and Environment for Statistical Computing*; R Foundation for Statistical Computing: Vienna, Austria, 2021; ISBN 3-900051-07-0.
48. National Wildfire Coordinating Group. Size Class of Fire. 2020. Available online: https://www.nwcg.gov/term/glossary/size-class-of-fire (accessed on 24 June 2020).
49. Coelho, C.A.S.; Ferro, C.A.T.; Stephenson, D.B. Methods for exploring spatial and temporal variability of extreme events in climate data. *J. Clim.* **2008**, *21*, 2072–2092. [CrossRef]
50. Petrie, M.; Bradford, J.; Lauenroth, W.; Schlaepfer, D.; Andrews, C.; Bell, D. Non-analog increases to air, surface and belowground temperature extremes in the 21st century due to climate change. *Clim. Chang.* **2020**, *163*, 2233–2256. [CrossRef]
51. National Interagency Fire Center. Total Wildland Fires and Acres. 2019. Available online: https://www.nifc.gov/fireInfo/fireInfo_stats_totalFires.html (accessed on 26 June 2020).
52. Van Wagtendonk, J.; van Wagtendonk, K.; Thode, A. Factors associated with the severity of intersecting fires in Yosemite National Park, California, USA. *Fire Ecol.* **2012**, *8*, 11–31. [CrossRef]
53. Lutz, J.; van Wagtendonk, J.; Thode, A.; Miller, J.; Franklin, J. Climate, lightning ignitions, and fire severity in Yosemite National Park, California, USA. *Int. J. Wildland Fire* **2009**, *18*, 765–774. [CrossRef]
54. Preisler, H.; Brillinger, D.; Burgan, R.; Benoit, J. Probability based models for estimation of wildfire risk. *Int. J. Wildland Fire* **2004**, *13*, 133–142. [CrossRef]
55. Chang, Y.; Zhu, Z.; Bu, R.; Chen, H.; Feng, Y.; Li, Y.; Hu, Y.; Wang, Z. Predicting fire occurrence patterns with logistic regression in Heilongjiang Province, China. *Landsc. Ecol.* **2013**, *28*, 1989–2004. [CrossRef]
56. Flannigan, M.; Logan, K.; Amiro, B.; Skinner, W.; Stocks, B. Future area burned in Canada. *Clim. Chang.* **2005**, *72*, 1–16. [CrossRef]
57. Marlon, J.; Bartlein, P.; Gavin, D.; Long, C.; Anderson, R.; Briles, C.; Brown, K.; Colombaroli, D.; Hallett, D.; Power, M.; et al. Long-term perspective on wildfires in the western USA. *Proc. Natl. Acad. Sci. USA* **2012**, *109*, E535–E543. [CrossRef] [PubMed]
58. Short, K. A spatial database of wildfires in the United States, 1992–2011. *Earth Syst. Sci. Data* **2014**, *6*, 1–27. [CrossRef]
59. Strachan, S.; Daly, C. Testing the daily PRISM air temperature model on semiarid mountain slopes. *J. Geophys. Res. Atmos.* **2017**, *122*, 5697–5715. [CrossRef]
60. Parks, S.; Miller, C.; Holsinger, L.; Baggett, L.; Bird, B. Wildland fire limits subsequent fire occurrence. *Int. J. Wildland Fire* **2016**, *25*, 182–190. [CrossRef]
61. North, M.; Collins, B.; Stephens, S. Using fire to increase the scale, benefits, and future maintenance of fuels treatments. *J. For.* **2012**, *110*, 392–401. [CrossRef]
62. Stephens, S.; Collins, B.; Biber, E.; Fule, P. US federal fire and forest policy: Emphasizing resilience in dry forests. *Ecosphere* **2016**, *7*, e01584. [CrossRef]

entropy

Article

Multiscale Analysis of Runoff Complexity in the Yanhe Watershed

Xintong Liu [1,2] and Hongrui Zhao [1,2,*]

1 Institute of Transportation Engineering and Geomatics, Department of Civil Engineering, Tsinghua University, Beijing 100084, China
2 3S Center, Tsinghua University, Beijing 100084, China
* Correspondence: zhr@mail.tsinghua.edu.cn

Abstract: Runoff complexity is an important indicator reflecting the sustainability of a watershed ecosystem. In order to explore the multiscale characteristics of runoff complexity and analyze its variation and influencing factors in the Yanhe watershed in China during the period 1991–2020, we established a new analysis method for watershed runoff complexity based on the complete ensemble empirical mode decomposition with adaptive noise (CEEMDAN) method for the decomposition of multiscale characteristics and the refined composite multiscale entropy (*RCMSE*) method for the quantification of the system complexity. The results show that runoff and its components all present multiscale complexity characteristics that are different from random signals, and the intermediate frequency modes contribute the most to runoff complexity. The runoff complexity of the Yanhe watershed has decreased gradually since 1991, and 2010 was a turning point of runoff complexity, when it changed from a decline to an increase, indicating that the ecological sustainability of this basin has improved since 2010, which was mainly related to the ecological restoration measures of the Grain for Green Project. This study expands the research perspective for analyzing the variation characteristics of runoff at the multiscale, and provides a reference for the study of watershed ecological sustainability and ecological management.

Keywords: runoff complexity; *RCMSE*; CEEMDAN; Yanhe watershed; multiscale

Citation: Liu, X.; Zhao, H. Multiscale Analysis of Runoff Complexity in the Yanhe Watershed. *Entropy* **2022**, *24*, 1088. https://doi.org/10.3390/e24081088

Academic Editor: Reik Donner

Received: 18 June 2022
Accepted: 5 August 2022
Published: 7 August 2022

1. Introduction

A watershed is a topographically delineated area formed under the internal forces of the earth and modified by external forces and human activities [1]. Watersheds are hydrological response units, biophysical units, and holistic ecosystems, possessing all the complexities of land surface systems, thereby making them excellent candidates for the practice of earth system science [2–4]. The natural flow regime is considered the primary driving force behind the formation of habitats and distribution, diversity, and abundance of biota, and it is extremely important for maintaining and sustaining riverine ecosystem integrity and its biodiversity [5]. Climate change and human activities are the two main driving factors that affect water cycles and the evolution of water resources. Frequent and intense human activities, such as afforestation and deforestation, grassland conversion, urbanization, and dam construction, determine rainfall redistribution and alter surface runoff, infiltration, groundwater recharge, instream flow, and evapotranspiration processes [6,7]. Runoff, the key component of the hydrological cycle, is directly or indirectly influenced by numerous types of positive and negative feedbacks at various scales, such as rainfall, climate change, human activities, and other surface factors [8,9], so that the runoff of a watershed is a complex system that is nonlinear, nonstationary, and uncertain [10,11]. Complexity is an essential and core feature of a hydrological system [12]. In-depth exploration of the inherent complexity of runoff is of theoretical and practical significance for revealing the instability of hydrological cycle dynamic processes and the self-organization ability of watershed systems.

Complex system science provides a valuable reference for complexity research into runoff systems. Entropy methods derived from information theory are simple algorithms with high sensitivity, strong robustness against noise, and no assumption of the statistical characteristics of the data [13,14]. By regarding the watershed system as a living organism, the concept that 'life feeds on negative entropy' has a profound impact on the study of natural systems [15]. Entropy methods have been widely used in evaluating the complexity of nonlinear and overall hydrological dynamics [11,16,17]. Sample entropy (SE) [18] quantifies the degree of regularity of a time series by evaluating the appearance of repetitive patterns, and has excellent stability and reliability in detecting the randomness and complexity of runoff [19,20]. Complexity is associated with 'meaningful structural richness' [21], but SE essentially comprises the statistical analysis of regularity, without detecting the nonlinear characteristics or quantifying the fractal behaviors of signals [22]. Therefore, an increase in SE is related to an increase in irregularity, but does not necessarily mean an increase in system complexity.

Costa et al. [23] introduced multiscale entropy (MSE) analysis to quantify the complexity of biological systems in cardiology. MSE takes into account the multiple temporal scales by the estimation of SE depending on the coarse-graining time series. MSE is based on the observation that the state of a complex system is far from perfect regularity and complete randomness [21] and reveals the structure of long-range correlation on multiple scales by quantifying the multiscale variability of signals [24]. The complexity is usually related to the ability of life systems to adapt to changing environments, which requires integrated multiscale functions. The results of MSE have been proven to be consistent with the 'complex loss' of ill-conditioned systems [21,24–26]. Similarly, the runoff structure in hydrology also has its own complexity. The more complex the structure, the stronger the self-regulation and restoration ability, which usually means that the watershed is healthier [17]. In general, the original complexity of a hydrological system is close to the maximum that can occur with a long-term evolution of natural conditions, but it may lose its complexity and become an unhealthy watershed system because of human activities, such as soil and water conservation and construction of water conservancy projects [27,28]. Therefore, MSE is also applicable to the complexity study of a runoff system for the measurement of multiscale characteristics of runoff and the system's adaptability to the environment.

However, since coarse-graining procedures greatly shorten the length of time series, MSE may produce inaccurate entropy estimates or induce undefined entropy [29]. Composite multiscale entropy (CMSE) [30] was proposed to improve the accuracy of MSE, but it does not resolve undefined entropy. Wu et al. [29] proposed a refined composite multiscale entropy (*RCMSE*) to improve CMSE, which improves the accuracy of entropy estimation and reduces the probability of generating undefined entropy, making it more suitable for the analysis of runoff data with a limited sequence length.

Due to the interaction between various dynamic mechanisms, runoff time series contain various scales of fluctuations and possess complexity of different time scales. Empirical mode decomposition (EMD) is an adaptive signal decomposition method that was proposed by Huang et al. [31]. It assumes that the data may have many different coexisting modes of oscillations in various scales at the same time, and decompose the original series into these intrinsic modes based on the local characteristic scales of data themselves; these components are called intrinsic mode functions (IMFs). The complete ensemble empirical mode decomposition with adaptive noise (CEEMDAN) [32] is an important improvement on EMD. Compared with most EMD improvement methods, CEEMDAN effectively solves the mode mixing problem and generates complete and noise-free reconstruction. Currently, the CEEMDAN method has been widely used in the signal processing field [33–35], but it has insufficient applications in hydrology. Combining *RCMSE* with CEEMDAN, the characteristics of runoff time series can be understood sufficiently at the micro and macro levels.

The Yanhe watershed is located in the middle of the Loess Plateau in China, which is a landscape that has been significantly affected by climate change and anthropogenic activities [36,37]. There is an urgent need to evaluate the ecosystem sustainability of this region. The overall aim of this paper was to propose a new research method for watershed runoff complexity based on *RCMSE* and CEEMDAN. The mechanism of multiscale runoff complexity and the variation and influencing factors of complexity in the Yanhe watershed over the last 30 years were studied to provide references for the implementation of ecological conservation and watershed management.

2. Materials and Methods

2.1. Study Area

As a primary tributary of the Yellow River, the Yanhe River has a total length of 286.9 km. The Yanhe watershed (36°27′–37°58′ N, 108°41′–110°29′ E) is located in the hinterland of the Loess Plateau, with a total area of 7687 km^2 and an altitude of 491–1787 m, as illustrated in Figure 1. The Yanhe watershed has a typical loess landform with crisscross ravines, loose soil, and poor antierosion ability. This region is a semiarid continental climate zone, with a mean annual precipitation of about 520 mm and a multiyear mean temperature ranging from 8.8 to 10.2 °C. The seasonal distribution of precipitation is quite uneven; more than 75% occurs between June and September as rainstorms. In the past, due to the influence of unreasonable anthropogenic activities and natural factors, the ecosystem in this region was significantly degraded, with a sharp decrease in natural vegetation and severe soil erosion, resulting in serious impacts on regional sustainable development [36]. Consequently, the Grain for Green Project, which includes a series of ecological construction policies, has been carried out in this region since 1999 [37].

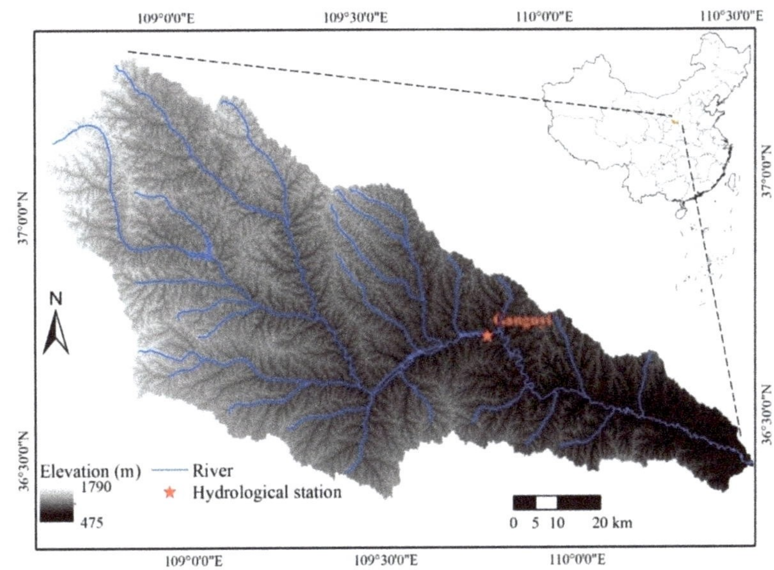

Figure 1. Basic information map of the Yanhe watershed.

2.2. Data Sources

The data used in this study included the following: (1) daily runoff data for the period of 1991–2020 were collected from the Ganguyi hydrological station, which is the control hydrographic station and the hydrological calibration outlet in the research basin with a control area of 5891 km^2, accounting for about 76.6% of the basin area [38,39]. All the runoff data came from the hydrological yearbook of the Yellow River Basin provided by the Yellow River Conservancy Commission of the Ministry of Water Resources [40]; (2) digital elevation model (DEM) data with 30 m resolution, obtained from the Geospatial

Data Cloud [41]; and (3) Yan'an Statistical Yearbook data for 2020, obtained from the Yan'an Bureau of Statistics [42].

2.3. Methods

2.3.1. Refined Composite Multiscale Entropy

Multiscale entropy is an effective method used to measure the complexity of a time series and has been applied in many fields successfully [21,23–26], but as the time scale factor increases, MSE may yield an inaccurate estimation of entropy or induce undefined entropy. Composite multiscale entropy (CMSE) [30] was proposed to improve the accuracy of entropy estimation, but CMSE increases the probability of inducing undefined entropy. In 2014, Wu et al. [29] proposed refined composite multiscale entropy (*RCMSE*) to improve MSE and CMSE for the undefined entropy problem of short time series. The *RCMSE* algorithm consists of the following three procedures:

(1) For a discrete time series $\mathbf{x} = \{x_1, x_2, \ldots, x_N\}$, after the initial normalization of the original series, consecutive coarse-graining procedures are performed at different scales, and the coarse-grained sequence represents the system dynamics at different time scales. For a scale factor τ, the k-th coarse-grained time series is defined as $\boldsymbol{y}_k^{(\tau)} = \left\{y_{k,1}^{(\tau)}, y_{k,2}^{(\tau)}, \ldots, y_{k,p}^{(\tau)}\right\}$, where:

$$y_{k,j}^{(\tau)} = \frac{1}{\tau}\sum\nolimits_{i=(j-1)\tau+k}^{j\tau+k-1} x_i, \qquad 1 \le j \le \frac{N}{\tau}, 1 \le k \le \tau. \tag{1}$$

(2) For all coarse-grained time series of each scale factor τ, the numbers of similar vector pairs $n_{k,\tau}^{m+1}$ and $n_{k,\tau}^{m}$ are computed, where $n_{k,\tau}^{m}$ represents the total number of m-dimensional vector pairs from the k-th coarse-grained time series for a scale factor τ for which the distance between the two vectors is smaller than a predefined tolerance r [18]. Referring to the relevant literature [21], for larger m, both the SE and the coefficient of variation increase dramatically due to the finite number of data points, and for larger r, fewer vectors are distinguishable, so we used $m = 2$ and $r = 0.15\sigma$, where σ denotes the standard deviation (SD) of the original time series, and m and r both remain constant for all scales.

(3) Let $\overline{n}_{k,\tau}^{m+1}$ and $\overline{n}_{k,\tau}^{m}$ represent the mean of $n_{k,\tau}^{m+1}$ and $n_{k,\tau}^{m}$, respectively, for $1 \le k \le \tau$. At a scale factor τ, the *RCMSE* value is defined as the logarithm of the ratio of $\overline{n}_{k,\tau}^{m+1}$ to $\overline{n}_{k,\tau}^{m}$, which is provided as Equation (2):

$$RCMSE(\mathbf{x}, \tau, m, r) = -\ln\left(\frac{\overline{n}_{k,\tau}^{m+1}}{\overline{n}_{k,\tau}^{m}}\right) \tag{2}$$

where $\overline{n}_{k,\tau}^{m+1} = \frac{1}{\tau}\sum_{k=1}^{\tau} n_{k,\tau}^{m+1}$ and $\overline{n}_{k,\tau}^{m} = \frac{1}{\tau}\sum_{k=1}^{\tau} n_{k,\tau}^{m}$. Equation (2) can be simplified as:

$$RCMSE(\mathbf{x}, \tau, m, r) = -\ln\left(\frac{\sum_{k=1}^{\tau} n_{k,\tau}^{m+1}}{\sum_{k=1}^{\tau} n_{k,\tau}^{m}}\right) \tag{3}$$

The *RCMSE* is used to compare the relative complexity based on the following guidelines [21]: (1) If for most scales the entropy measures are higher for one time series than for another, the former is considered more complex than the latter; (2) a monotonic decrease in entropy measures indicates that the original signal only contains information at the smaller scales. Therefore, in the analysis of the complexity of the runoff system, not only the specific entropy values but also their dependence on scales needs to be considered, such as the areas under the *RCMSE* curves and the morphological characteristics of *RCMSE* curves.

2.3.2. Complete Ensemble Empirical Mode Decomposition with Adaptive Noise

Huang et al. [31] proposed the empirical mode decomposition (EMD) to decompose the complex time series into intrinsic mode functions (IMFs). The EMD has great advantages in dealing with nonstationary and nonlinear signals, but it still has a 'mode mixing' problem, which refers to the presence of similar oscillations in different modes or disparate amplitudes in a mode [43]. Therefore, the ensemble EMD (EEMD) [44] adds the Gaussian white noise to eliminate the mode mixing in the EMD. However, along with introducing the Gaussian white noise, the EEMD algorithm cannot completely eliminate Gaussian white noise after signal reconstruction, and it probably generates a different number of IMFs after adding different noise. Consequently, the complete ensemble empirical mode decomposition with adaptive noise (CEEMDAN) [32] was proposed as an improved version of EEMD. The CEEMDAN adds Gaussian white noise to each stage of the decomposition process, and each IMF is calculated by averaging the results, obtaining decomposed components with less noise and more physical meaning [33,34]. The CEEMDAN process proceeds as follows:

(1) Add Gaussian white noise to the original data $x(t)$ to create new time series and use the EMD method [31] to obtain the first IMF, IMF_1, and the first residue, r_1.

(2) The following k-th IMF ($k \geq 2$) IMF_k and residue r_k can be obtained by:

$$\mathrm{IMF}_k\,(t) = \frac{1}{N}\sum_{i=1}^{N} E_1\left(r_{k-1}(t) + \varepsilon_{k-1}E_{k-1}\left(G^i(t)\right)\right) \tag{4}$$

$$r_k(t) = r_{k-1}(t) - \mathrm{IMF}_k(t) \tag{5}$$

where N is the number of ensemble members, that is, the number of different realizations of white Gaussian noise; G^i is the i-th Gaussian white noise to be added; and ε_{k-1} is the signal-to-noise ratio between the additional noise and original signal. Define the operator $E_j(\cdot)$ that produces the jth mode obtained by EMD.

(3) Iterate Step 2 until the obtained residue can no longer be decomposed. The original sequence can be computed as:

$$x(t) = \sum_{k=1}^{K} \mathrm{IMF}_k(t) + R(t) \tag{6}$$

where K is the total number of IMFs, which comprise the characteristics of the original signal at different time scales, and R is the final residue, which clearly shows the trend in the original sequence.

The noise standard deviation was set to 0.2, the number of ensemble members N was 100, and the maximum number of sifting iterations was 500 in this paper, which were typically used in practice.

3. Results

3.1. Multiscale Complexity Characteristics of Runoff

The *RCMSE* under scale factors from 1 to 365 d for the daily runoff data during the period of 1991–2020 for the Ganguyi hydrological station were calculated, as shown in Figure 2. When $\tau < 90$ d, the entropy measure gradually increased, with an increase in τ until it reached the maximum among all 365 scales at about $\tau = 90$ d, and it remained stable when $\tau \in [90, 110]$ d. Then, it decreased rapidly until $\tau = 170$ d, and there was a sudden drop near $\tau = 120$ d. When $\tau \in [170, 210]$ d, it showed a slight increase; in that period, a minimum point of sudden drop appeared again near $\tau = 180$ d. Entropy then decreased gradually after $\tau = 210$ d.

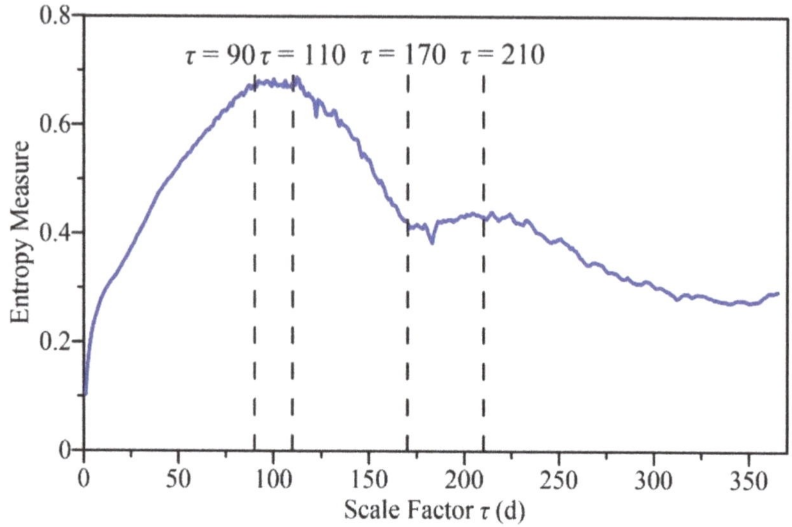

Figure 2. The refined composite multiscale entropy (*RCMSE*) curve for the daily runoff data during the period 1991–2020 for the Ganguyi hydrological station in the Yanhe watershed.

To further analyze the runoff at different temporal scales, the CEEMDAN method was applied to the daily runoff data during the period 1991–2020 for the Ganguyi hydrological station, and 14 IMFs and 1 residual term (RES) were obtained, as shown in Figure 3. To find the statistical characteristics of each IMF, the mean periods were calculated, which were derived by dividing the total number of points by the number of peaks (see Table 1). The fluctuation characteristics of all IMFs were different, and as the IMF number increased, both frequency and amplitude reduced. The mean periods of IMF1–IMF4 were lower than 10 d as high-frequency modes, and most of short-term strong runoff was decomposed into these IMFs. The mean periods of IMF5–IMF10 were between 10 d and 1 year as the intermediate-frequency modes. The periods of IMF11–IMF14 were longer than 1 year as the low-frequency modes, representing the influence of long-term factors. The RES presented a pattern of slow change around the long-term average, which shows that the runoff gradually decreased from 1991 to 2009, and the decline was faster after 1994; then it slowly increased after 2009.

Table 1. Mean periods of intrinsic mode functions (IMFs) for the daily runoff data from 1991 to 2020 for the Ganguyi hydrological station through CEEMDAN.

	IMF1	IMF2	IMF3	IMF4	IMF5	IMF6	IMF7	IMF8	IMF9	IMF10	IMF11	IMF12	IMF13	IMF14
Mean Period/d	2.91	3.62	3.59	6.45	12.2	22.69	41.67	89.09	171.22	342.44	576.74	1095.80	2191.60	3652.67

The *RCMSE* was calculated for all IMFs, as shown in Figure 4. The scales were from 1 d to 90 d because the entropy measures of the original runoff series reached the maximum around τ = 90 d (see Figure 2). Figure 4 shows that the high-frequency modes (IMF1–IMF4) have low entropy values and fluctuations at almost all scales, with low complexity. The entropy values of IMF5–IMF9 gradually increased within a certain range, and then decreased after reaching the maximum. The *RCMSE* curves of IMF10–IMF14 gradually increased under scales 1–90 d due to their large periods and long-range correlations. IMF9 and IMF10 had larger summations of the entropy values over research scales than the others (see Table 2), and maintained a growth trend over a wide range so that they made the greatest contribution to runoff complexity. At small scales, intermediate-frequency and high-frequency modes are the dominant modes of runoff complexity. The contribution of low-frequency modes to runoff complexity increased gradually with increasing temporal scales.

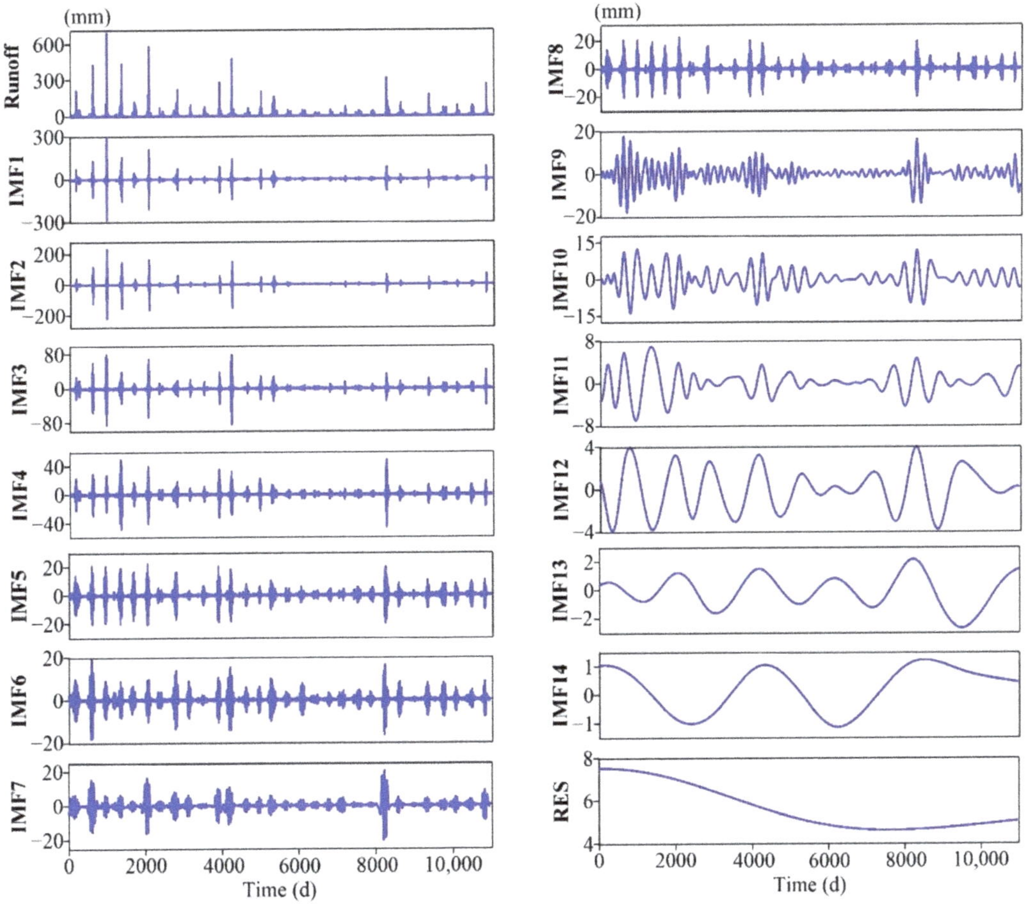

Figure 3. The intrinsic mode functions (IMFs) and residue (RES) for the daily runoff data during the period 1991–2020 for the Ganguyi hydrological station, through complete ensemble empirical mode decomposition with adaptive noise (CEEMDAN).

Table 2. The summations of the entropy values over the scales 1–90 d of *RCMSE* ($RCMSE_{\Sigma}$) of IMFs of the daily runoff data during the period 1991–2020 for the Ganguyi hydrological station through CEEMDAN.

	IMF1	IMF2	IMF3	IMF4	IMF5	IMF6	IMF7	IMF8	IMF9	IMF10	IMF11	IMF12	IMF13	IMF14
$RCMSE_{\Sigma}$	6.09	4.23	6.96	8.76	15.18	25.61	38.82	55.38	73.10	72.91	40.67	35.66	40.96	18.40

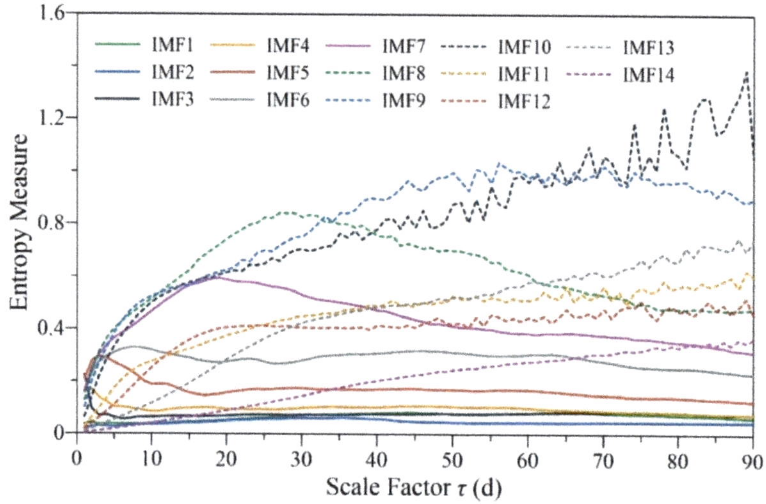

Figure 4. *RCMSE* for IMFs of the daily runoff data during the period 1991–2020 for the Ganguyi hydrological station through CEEMDAN.

We next tested the hypothesis that due to the complexity of runoff series, they cannot be generated by uncorrelated random processes. The complexity of the original runoff series and its IMFs was compared with that of the randomized time series obtained by shuffling the order of original data points. Because, by construction, both the hydrological and the shuffled time series had the same mean, variance, and distribution, any differences in the complexity indexes were caused by differences in the temporal order of the data points and their correlation properties.

The comparison results for the *RCMSE* curves of the original series with the average curves obtained by randomly shuffling 30 times are shown in Figure 5. The *RCMSE* curves of the original and shuffled series all have large differences in numerical size and trend. The *RCMSE* curves of disordered time series should be expressed as entropy values monotonically decreasing with scale factors [21], just like those of the shuffled series of IMF8–IMF14. However, the *RCMSE* curves of the shuffled series of IMF1–IMF7 showed a rapid increase in the initial stage, and then decreased gradually, like those of uncorrelated noise, resulting in a short 'fake complexity' phenomenon. This was mainly due to the seasonal heavy rainfall in the Yanhe watershed, which often leads to an explosion of runoff, and these extreme values were mainly decomposed into the high-frequency modes by the CEEMDAN method, leading to a large fluctuation in these IMFs. The shuffling treatment distributes these extreme values relatively evenly throughout the sequence, resulting in an increase in entropy at small scales, but in the long run, entropy still conforms to the characteristics of uncorrelated noise. Moreover, the shuffled series present smoother *RCMSE* curves without small fluctuations, while those of the original sequences often fluctuate with the scale change, such as IMF10 under scales 60–110 d and IMF5 under scales 1–20 d. The above results show that the original runoff series and IMFs all have unique and high complexity.

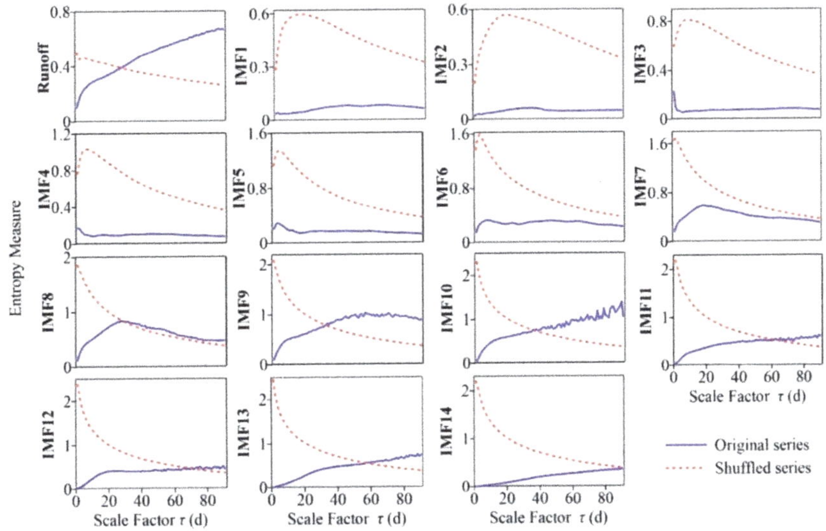

Figure 5. Comparison of the *RCMSE* of the original runoff and IMF series with the corresponding shuffled series.

3.2. Trend Analysis of Runoff Complexity

In order to analyze the temporal evolution of runoff complexity in the Yanhe watershed over the past 30 years, considering that the mean period of the slowest fluctuation, IMF14, is about a decade (see Table 1), the *RCMSE* of the runoff series in sliding windows of 10 years, shifting the window by 1 year, was calculated, and the summations of the entropy values over the scales 1–90 d were taken as complexity, and the fifth year of each time period was set to represent the sequence in Figure 6. This shows that the runoff complexity in the Yanhe watershed presented a downward trend from 1995 to 2010. After 2010, the runoff complexity changed from a decline to an increase.

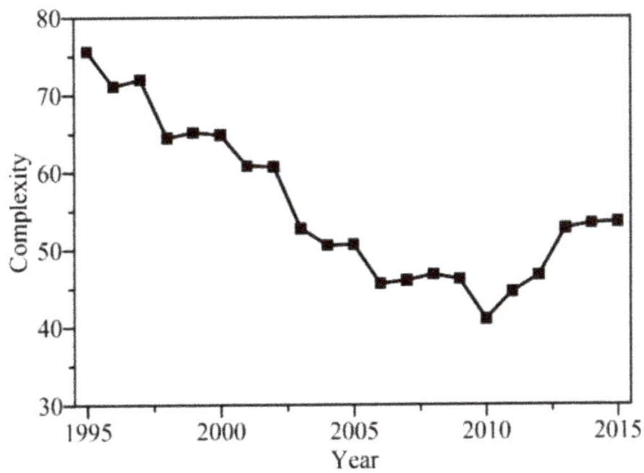

Figure 6. The complexity of runoff in the Yanhe watershed, as shown by sliding windows of 10 years, shifting the window by 1 year.

In order to explore the variation in runoff complexity under different temporal scales, the complexity of each IMF was calculated through *RCMSE* curves in sliding windows of 10 years, shifting the window by 1 year. It should be noted that although the complexity performance at different temporal scales has different characteristics in terms of *RCMSE* curves, when analyzing the variation in each IMF, the summations of the entropy values over the same scale range of 1–90 d can be used as a measure of complexity. To make the results more comparable, we divided the complexity of each sequence by the complexity of the first 10 years, as shown in Figure 7. It can be seen that the complexity of high-frequency modes (IMF1–IMF4) and intermediate-frequency modes (IMF5–IMF10) showed a flat or slightly upward trend before 1997, and gradually decreased from 1997 to 2010. After 2010, the variations in the complexity of high-frequency modes were relatively flat, while those of medium-frequency modes gradually increased, similar to the change trend for the original runoff complexity (see Figure 6). The changes in complexity of the low-frequency modes (IMF11–IMF14) did not show a uniform and obvious varying regularity, and presented little correlation with the variation in runoff complexity. Therefore, it can be concluded that the change in runoff complexity was mainly due to the intermediate-frequency and high-frequency components, and the influence of low-frequency components was slight.

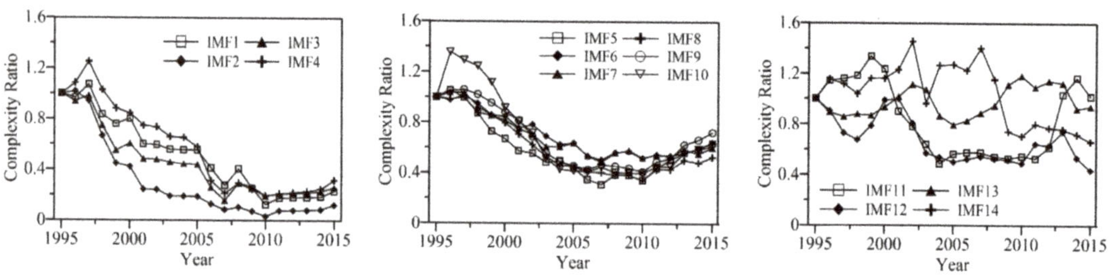

Figure 7. The complexity of IMFs with sliding windows of 10 years and shifting the window by 1 year. (**a**) High-frequency modes (IMF1–IMF4), (**b**) intermediate-frequency modes (IMF5–IMF10), (**c**) low-frequency modes (IMF11–IMF14).

4. Discussion

4.1. Characteristics of Runoff Complexity

In this paper, the watershed system was considered a complex system, assuming the following: (1) the watershed system has complexity characteristics across multiple temporal scales, (2) the complexity reflects its ability of adaptation and regulation in changing environments, and (3) a 'sick' watershed will have reduced adaptability and information carried by output variables. The multiscale entropy method was used to characterize the complexity of the watershed system, and the complexity characteristics of different scale components were studied by empirical mode decomposition. The uncertainty characteristic of runoff had two peak areas within 1 year around the scales of 100 and 210 d, and it reached the maximum at the 90-day scale, which was probably related to the correlation time and the period of possible nonlinear oscillations of runoff series. The runoff was decomposed into 14 IMFs with temporal scales from 3 d to 10 years by CEEMDAN. Each component had separate physical meanings and complexity characteristics that were completely different from random signals. The high-frequency IMFs with short periods and large amplitudes represent short-term fluctuation of runoff, which have the minimum multiscale complexity. The low-frequency IMFs with long periods and small amplitudes signify the components of slow variation, which may be mainly affected by atmospheric circulation or celestial activities. The intermediate-frequency IMFs, especially IMF9 of the half-year scale and IMF10 of the annual scale, made the greatest contribution to runoff complexity. This also indicates that the complexity research based on the *RCMSE* method should refer to not only the specific numerical size of *RCMSE* curves, but also the trend changes and the difference with those of the corresponding random shuffled series.

4.2. Factors Impacting Runoff Complexity

Runoff variation in the hydrological system is affected by multiple factors. Climate change and human activities are two main driving factors that affect water cycles and the evolution of runoff in a watershed [17]. Studies have shown that meteorological factors have affected the runoff complexity of the Yellow River Basin, and precipitation has the greatest impact, followed by evaporation and temperature [17,45]. However, the literature and actual meteorological data do not indicate that meteorological factors underwent abrupt changes around 2010, which was the turning point of runoff complexity (see Figure 6). Therefore, meteorological factors may affect runoff complexity to a certain extent, but are not the main factors.

Human activities in the Yanhe watershed mainly include the Grain for Green Project and urbanization. The impacts of the Grain for Green Project on watershed runoff can be explained based on two aspects. On the one hand, since 1999, a series of ecological construction projects have been carried out in this area. Research shows that the Yanhe watershed has had increases in areas of woodland and grassland and significant decreases in the amount of soil and water loss, and the quality and service function of the ecosystem have improved [36]. Following the ecosystem undergoing a growth period, the improvement effect on runoff complexity may have lagged behind, so the runoff complexity changed from a decline to an increase after 2010. On the other hand, the Grain for Green Project had two phases: 2000–2010 and 2010–2020. In the early stage of the project, rapid progress of returning farmland to forest and grassland was carried out with a lack of scientific planning and demonstration, resulting in low vegetation survival rate and damaged plots [46,47], which may lead to further reduction of runoff complexity. Based on in-depth field research and scientific planning, a series of improvement policies of the Grain for Green Project were issued, such as the 'Notice on Improving the Policy of Grain for Green Project from the State Council' [48], so that these ecological managements in the latter period may have significantly improved the ecological sustainability of the watershed. The annual afforestation area of Yan'an City is shown in Figure 8, which shows that large-range disorderly afforestation in this area has improved significantly since 2004. Moreover, with the increase in the range of afforestation, runoff ecological status is not necessarily improved. Studies have reported that the growth and development of a large number of artificial vegetations have led to an increase in evapotranspiration and a decrease in surface water resources in the Loess Plateau [49]. The complexity of runoff is a comprehensive variable reflecting the ecological status of the basin, which is expected to be an important reference index for ecological sustainability. In addition, Yan'an City has undergone accelerated urbanization in recent years, which may also have affected the runoff complexity to some extent.

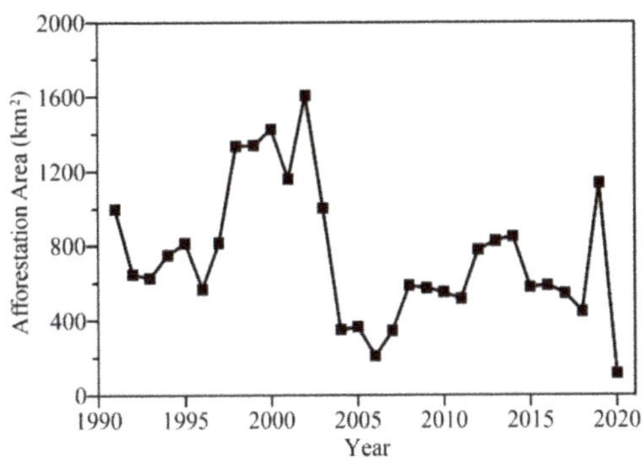

Figure 8. The annual afforestation area of Yan'an City in the period 1991–2020.

5. Conclusions

Complexity has always been the focus and difficulty of watershed system science research, which is closely related to the sustainability of a watershed. Regarding the watershed system as a complex system with multiscale characteristics, in this paper, we established a complexity analysis framework for watershed runoff based on CEEMDAN for the decomposition of multiscale characteristics and *RCMSE* for the quantification of system complexity. The results show that the runoff and each temporal-scale component present completely different complexity characteristics exhibited in the numerical size and trend changes of *RCMSE* curves, which verifies that the runoff sequence has multiscale complexity. The high-frequency components with short periods and large amplitudes represent short-term fluctuation of runoff, which may contribute to understanding the response of runoff to short-term interference, such as rainstorms. The low-frequency components with long periods and small amplitudes signify the components of slow variations, which may be mainly affected by atmospheric circulation or celestial activities. The intermediate-frequency components, especially the components with mean periods of half-year and annual scales, make the greatest contribution to runoff complexity, which are the key components in the study of runoff complexity variation. The runoff complexity of the Yanhe watershed has shown a downward trend since 1991, but with a gradual increase after 2010, indicating that the ecological sustainability of this basin improved after 2010, which was probably related to the ecological restoration measures of the Grain for Green Project, showing that the measures in the past decade have effectively improved the degradation phenomenon of runoff complexity. This study has expanded the research perspective in relation to multiscale runoff complexity and the variation characteristics of runoff systems. It also provides a reference for the evaluation of watershed ecological sustainability and ecological management.

The following issues require attention in the future. In this paper, only runoff data for the Ganguyi hydrological station in the Yanhe watershed were considered; future research should introduce data from more hydrological stations to analyze the spatial multiscale characteristics of runoff complexity. In addition, we only analyzed the causes of runoff complexity change qualitatively; the quantitative contributions of climate change and human activities to runoff changes need to be combined in the future.

Author Contributions: Conceptualization, X.L. and H.Z.; data curation, X.L.; formal analysis, X.L.; methodology, X.L.; supervision, H.Z.; validation, X.L.; writing—original draft, X.L.; writing—review and editing, H.Z. All authors have read and agreed to the published version of the manuscript.

Funding: This work was funded by the National Natural Science Foundation of China, grant number 41971379.

Institutional Review Board Statement: Not applicable.

Informed Consent Statement: Not applicable.

Data Availability Statement: Runoff data come from the hydrological yearbook of the Yellow River Basin from the Yellow River Conservancy Commission of the Ministry of Water Resources (http://yrcc.gov.cn/ (accessed on 12 December 2021)). Digital elevation model (DEM) data are openly available in the Geospatial Data Cloud (http://www.gscloud.cn/ (accessed on 6 December 2021)). The Yan'an Statistical Yearbook of 2020 comes from the Yan'an Bureau of Statistics (http://tjj.yanan.gov.cn/Item/5085.aspx/ (accessed on 6 December 2021)) [42]. The datasets generated during this study are available from the corresponding author on reasonable request.

Acknowledgments: The authors greatly appreciate the insightful advice and comments on the manuscript received from the reviewers who helped improve the quality of this research.

Conflicts of Interest: The authors declare no conflict of interest.

References

1. Lotspeich, F.B. Watersheds as the basic ecosystem: This conceptual framework provides a basis for a natural classification system. *J. Am. Water Resour. Assoc.* **1980**, *16*, 581–586. [CrossRef]
2. Cheng, G.; Li, X. Integrated research methods in watershed science. *Sci. China Earth Sci.* **2015**, *58*, 1159–1168. [CrossRef]
3. Wang, G.; Mang, S.; Cai, H.; Liu, S.; Zhang, Z.; Wang, L.; Innes, J.L. Integrated watershed management: Evolution, development and emerging trends. *J. For. Res.* **2016**, *27*, 967–994. [CrossRef]
4. Pande, C.B. Watershed management and development. In *Sustainable Watershed Development*; Springer: Cham, Switzerland, 2020; pp. 13–26.
5. Singh, R.K.; Jain, M.K. Complexity analyses of Godavari and Krishna river streamflow using the concept of entropy. *Acta Geophys.* **2021**, *69*, 2325–2338. [CrossRef]
6. Wu, J.; Miao, C.; Zhang, X.; Yang, T.; Duan, Q. Detecting the quantitative hydrological response to changes in climate and human activities. *Sci. Total Environ.* **2017**, *586*, 328–337. [CrossRef]
7. Kang, Y.; Gao, J.; Shao, H.; Zhang, Y. Quantitative analysis of hydrological responses to climate variability and land-use change in the hilly-gully region of the Loess Plateau, China. *Water* **2019**, *12*, 82. [CrossRef]
8. Lian, L. Runoff forecasting model based on CEEMD and combination model: A case study in the Manasi River, China. *Water Sci. Technol. Water Supply* **2022**, *22*, 3921–3940. [CrossRef]
9. Zheng, H.; Miao, C.; Zhang, G.; Li, X.; Wang, S.; Wu, J.; Gou, J. Is the runoff coefficient increasing or decreasing after ecological restoration on China's Loess Plateau? *Int. Soil Water Conserv. Res.* **2021**, *9*, 333–343. [CrossRef]
10. Chou, C.M. Complexity analysis of rainfall and runoff time series based on sample entropy in different temporal scales. *Stoch. Environ. Res. Risk Assess.* **2014**, *28*, 1401–1408. [CrossRef]
11. Ma, W.; Yan, K.; Song, S. Analysis of streamflow complexity based on entropies in the Weihe River Basin, China. *Entropy* **2019**, *22*, 38. [CrossRef]
12. Huang, F.; Ochoa, C.G.; Guo, L.; Wu, Y.; Qian, B. Investigating variation characteristics and driving forces of lake water level complexity in a complex river–lake system. *Stoch. Environ. Res. Risk Assess.* **2021**, *35*, 1003–1017. [CrossRef]
13. Zhang, L.; Li, H.; Liu, D.; Fu, Q.; Li, M.; Faiz, M.A.; Khan, M.I.; Li, T. Identification and application of the most suitable entropy model for precipitation complexity measurement. *Atmos. Res.* **2019**, *221*, 88–97. [CrossRef]
14. Pechlivanidis, I.G.; Jackson, B.; McMillan, H.; Gupta, H.V. Robust informational entropy-based descriptors of flow in catchment hydrology. *Hydrol. Sci. J.* **2016**, *61*, 1–18. [CrossRef]
15. Rodríguez, R.A.; Rodrigo, R.; Juan, D.D. Ecology: Science or philately? An interdisciplinary analysis of sustainability by exploring if it is possible to get more and more information by reducing collateral environmental dam-ages. *Sci. Total Environ.* **2017**, *596*, 43–52. [CrossRef] [PubMed]
16. Joo, H.; Jun, H.; Lee, J.; Kim, H.S. Assessment of a stream gauge network using upstream and downstream runoff characteristics and entropy. *Entropy* **2019**, *21*, 673. [CrossRef]
17. Su, X.; Li, X.; Niu, Z.; Liang, X. A new complexity-based three-stage method to comprehensively quantify positive/negative contribution rates of climate change and human activities to changes in runoff in the upper Yellow River. *J. Clean Prod.* **2021**, *287*, 125017. [CrossRef]
18. Richman, J.S.; Moorman, J.R. Physiological time-series analysis using approximate entropy and sample entropy. *Am. J. Physiol.-Heart Circul. Physiol.* **2000**, *278*, 2039–2049. [CrossRef]
19. Sheikh, Z.; Alireza, M.N.; Dawei, H. Climate change impact assessment on low streamflows using cross-entropy methods. *Clim. Res.* **2021**, *85*, 159–176. [CrossRef]
20. Huang, F.; Chunyu, X.; Wang, Y.; Wu, Y.; Qian, B.; Guo, L.; Zhao, D.; Xia, Z. Investigation into multi-temporal scale complexity of streamflows and water levels in the Poyang Lake basin, China. *Entropy* **2017**, *19*, 67. [CrossRef]
21. Costa, M.; Goldberger, A.L.; Peng, C.K. Multiscale entropy analysis of biological signals. *Phys. Rev. E* **2005**, *71*, 021906. [CrossRef]
22. Goldberger, A.L.; Peng, C.K.; Lipsitz, L.A. What is physiologic complexity and how does it change with aging and disease? *Neurobiol. Aging* **2002**, *23*, 23–26. [CrossRef]
23. Costa, M.; Goldberger, A.L.; Peng, C.K. Multiscale entropy analysis of complex physiologic time series. *Phys. Rev. Lett.* **2002**, *89*, 068102. [CrossRef] [PubMed]
24. Kang, H.G.; Costa, M.D.; Priplata, A.A.; Starobinets, O.V.; Goldberger, A.L.; Peng, C.K.; Kiely, D.K.; Cupples, L.A.; Lipsitz, L.A. Frailty and the degradation of complex balance dynamics during a dual-task protocol. *J. Gerontol. Ser. A-Biol. Sci. Med. Sci.* **2009**, *64*, 1304–1311. [CrossRef]
25. Henriques, T.S.; Costa, M.D.; Mathur, P.; Mathur, P.; Davis, R.B.; Mittleman, M.A.; Khabbaz, K.R.; Goldberger, A.L.; Subramaniam, B. Complexity of preoperative blood pressure dynamics: Possible utility in cardiac surgical risk assessment. *J. Clin. Monit. Comput.* **2019**, *33*, 31–38. [CrossRef] [PubMed]
26. Klinker, L.E.; Henriques, T.S.; Costa, M.D.; Davis, R.B.; Mittleman, M.A.; Mathur, P.; Subramaniam, B. Comparison of invasive and noninvasive blood pressure measurements for assessing signal complexity and surgical risk in cardiac surgical patients. *Anesth. Analg.* **2020**, *130*, 1653.
27. Li, Z.; Zhang, Y.K. Multi-scale entropy analysis of Mississippi River flow. *Stoch. Environ. Res. Risk Assess.* **2008**, *22*, 507–512. [CrossRef]

28. De Carvalho Barreto, I.D.; Stosic, T.; Filho, M.C.; Delrieux, C.; Singh, V.P.; Stosic, B. Complexity analyses of Sao Francisco river streamflow: Influence of dams and reservoirs. *J. Hydrol. Eng.* **2020**, *25*, 05020036. [CrossRef]
29. Wu, S.D.; Wu, C.W.; Lin, S.G.; Lee, K.Y.; Peng, C.K. Analysis of complex time series using refined composite multiscale entropy. *Phys. Lett. A* **2014**, *378*, 1369–1374. [CrossRef]
30. Wu, S.D.; Wu, C.W.; Lin, S.G.; Wang, C.C.; Lee, K.Y. Time series analysis using composite multiscale entropy. *Entropy* **2013**, *15*, 1069–1084. [CrossRef]
31. Huang, N.E.; Shen, Z.; Long, S.R.; Wu, M.L.C.; Shih, H.H.; Zheng, Q.N.; Yen, N.C.; Tung, C.C.; Liu, H.H. The empirical mode decomposition and the Hilbert spectrum for nonlinear and non-stationary time series analysis. *Proc. R. Soc. A-Math. Phys. Eng. Sci.* **1998**, *454*, 903–995. [CrossRef]
32. Torres, M.E.; Colominas, M.A.; Schlotthauer, G.; Flandrin, P. A complete ensemble empirical mode decomposition with adaptive noise. In Proceedings of the 2011 IEEE International Conference on Acoustics, Speech and Signal Processing (ICASSP), Prague, Czech Republic, 22–27 May 2011; pp. 4144–4147.
33. Li, T.; Qian, Z.; He, T. Short-term load forecasting with improved CEEMDAN and GWO-based multiple kernel ELM. *Complexity* **2020**, *2020*, 1209547. [CrossRef]
34. Liu, Y.; Wang, L.; Yang, L.; Liu, X.; Wang, L. Runoff prediction and analysis based on improved CEEMDAN-OS-QR-ELM. *IEEE Access* **2021**, *9*, 57311–57324. [CrossRef]
35. Chen, W.; Li, J.; Wang, Q.; Han, K. Fault feature extraction and diagnosis of rolling bearings based on wavelet thresholding denoising with CEEMDAN energy entropy and PSO-LSSVM. *Measurement* **2021**, *172*, 108901. [CrossRef]
36. Wang, C.; Zhao, H. Analysis of remote sensing time-series data to foster ecosystem sustainability: Use of temporal information entropy. *Int. J. Remote Sens.* **2019**, *40*, 2880–2894. [CrossRef]
37. Bai, J.; Zhou, Z.; Zou, Y.; Pulatov, B.; Siddique, K.H. Watershed drought and ecosystem services: Spatiotemporal characteristics and gray relational analysis. *ISPRS Int. J. Geo-Inf.* **2021**, *10*, 43. [CrossRef]
38. Li, E.H.; Mu, X.M.; Zhao, G.J.; Gao, P.; Shao, H.B. Variation of runoff and precipitation in the hekou-longmen region of the yellow river based on elasticity analysis. *Sci. World J.* **2014**, *2014*, 929858. [CrossRef]
39. Lian, Y.; Sun, M.; Wang, J.; Luan, Q.; Jiao, M.; Zhao, X.; Gao, X. Quantitative impacts of climate change and human activities on the runoff evolution process in the Yanhe River Basin. *Phys. Chem. Earth* **2021**, *122*, 102998. [CrossRef]
40. The hydrological yearbook of the Yellow River Basin. Available online: http://www.yrcc.gov.cn/ (accessed on 12 December 2021).
41. Advanced Spaceborne Thermal Emission and Reflection Radiometer Global Digital Elevation Model. Available online: http://www.gscloud.cn/ (accessed on 6 December 2021).
42. Yan'an Statistical Yearbook data for 2020. Available online: http://tjj.yanan.gov.cn/Item/5085.aspx/ (accessed on 6 December 2021).
43. Cao, J.; Li, Z.; Li, J. Financial time series forecasting model based on CEEMDAN and LSTM. *Phys. A* **2019**, *519*, 127–139. [CrossRef]
44. Wu, Z.; Huang, N.E. Ensemble empirical mode decomposition: A noise-assisted data analysis method. *Adv. Adapt. Data Anal.* **2009**, *1*, 1–41. [CrossRef]
45. Meng, F.; Su, F.; Yang, D.; Tong, K.; Hao, Z. Impacts of recent climate change on the hydrology in the source region of the Yellow River basin. *J. Hydrol.-Reg. Stud.* **2016**, *6*, 66–81. [CrossRef]
46. Cao, S.; Chen, L.; Yu, X. Impact of China's Grain for Green Project on the landscape of vulnerable arid and semi-arid agricultural regions: A case study in northern Shaanxi Province. *J. Appl. Ecol.* **2009**, *46*, 536–543. [CrossRef]
47. Xu, X.; Zhang, D.; Zhang, Y.; Yao, S.; Zhang, J. Evaluating the vegetation restoration potential achievement of ecological projects: A case study of Yan'an, China. *Land Use Policy* **2020**, *90*, 104293. [CrossRef]
48. Notice on Improving the Policy of Grain for Green Project from the State Council. Available online: http://www.gov.cn/zwgk/2007-08/14/content_716617.htm (accessed on 14 August 2007).
49. Feng, X.; Fu, B.; Piao, S.; Wang, S.; Ciais, P.; Zeng, Z.; Lü, Y.; Zeng, Y.; Li, Y.; Jiang, X.; et al. Revegetation in China's Loess Plateau is approaching sustainable water resource limits. *Nat. Clim. Chang.* **2016**, *6*, 1019–1022. [CrossRef]

entropy

MDPI

Article

Spatial-Temporal Sensitivity Analysis of Flood Control Capability in China Based on MADM-GIS Model

Weihan Zhang [1,2], Xianghe Liu [1,2], Weihua Yu [1,2,*], Chenfeng Cui [1,2] and Ailei Zheng [2]

1 Key Laboratory of Agricultural Soil and Water Engineering in Arid and Semiarid Areas of Ministry of Education, Northwest A&F University, Xianyang 712100, China; zwh2660218787@gmail.com (W.Z.); liuhe710529@gmail.com (X.L.); cuichenfeng@nwafu.edu.cn (C.C.)

2 College of Water Resources and Architecture Engineering, Northwest A&F University, Xianyang 712100, China; zal@nwafu.edu.cn

* Correspondence: yuweihua8681@163.com

Abstract: To facilitate better implementation of flood control and risk mitigation strategies, a model for evaluating the flood defense capability of China is proposed in this study. First, nine indicators such as slope and precipitation intensity are extracted from four aspects: objective inclusiveness, subjective prevention, etc. Secondly, the entropy weight method in the multi-attribute decision making (MADM) model and the improved three-dimensional technique for order preference by similarity to ideal solution (3D-TOPSIS) method were combined to construct a flood defense capacity index evaluation system. Finally, the receiver operating characteristic (ROC) curve and the Taylor plot method were innovatively used to test the model and indicators. The results show that nationwide, there is fine flood defense performance in Shandong, Jiangsu and room for improvement in Guangxi, Chongqing, Tibet and Qinghai. The good representativity of nine indicators selected by the model was verified by the Taylor plot. Simultaneously, the ROC calculated area under the curve (AUC) was 70%, which proved the good problem-solving ability of the MADM-GIS model. An accurate assessment of the sensitivity of flood control capacity in China was achieved, and it is suitable for situations where data is scarce or discontinuous. It provided scientific reference value for the planning and implementation of China's flood defense and disaster reduction projects and emergency safety strategies.

Keywords: MADM-GIS model; 3D-TOPSIS method; ROC curve; visualization of flood control capacity; entropy weight method

Citation: Zhang, W.; Liu, X.; Yu, W.; Cui, C.; Zheng, A. Spatial-Temporal Sensitivity Analysis of Flood Control Capability in China Based on MADM-GIS Model. *Entropy* **2022**, *24*, 772. https://doi.org/10.3390/e24060772

Academic Editors: Matteo Convertino and Jie Li

Received: 13 March 2022
Accepted: 25 May 2022
Published: 30 May 2022

1. Introduction

Promoted by the need for economic development, the global water cycle has accelerated substantially, causing a series of abnormal climate changes [1,2]. According to data released by the Ministry of Water Resources of China, the areas affected by flood disasters in southern China in 2020 included 198 rivers in 27 districts across the country, with a total of 30.2 million people involved, and a loss of CNY 61.79 billion as of 14:00 on 9 July 2020. The healthy and stable development of China's economy and society [3,4] has been hindered by the expansion in the scope and degree of extreme climate events. Therefore, the establishment of a flood control capability evaluation model is crucial for maintaining infrastructure such as reservoirs, reasonably managing and controlling water resources, and effectively protecting the lives and properties of the Chinese people. It is increasingly necessary to assess the flood disaster defense capability in urban hydrology.

As a common natural disaster, urban floods generate multitudes of negative impacts on people across the country. Multi-level studies on the evaluation of flood defense capacity have been carried out worldwide. The main reasons for flood disasters are unreasonable urban planning and uncoordinated flood management systems. The emergence of excess runoff is caused by short-term intense rainfall that exceeds the capacity of the drainage system [5,6], which reflects the deep relationship between flood disasters and geographical, social and other factors. Studies reveal that urban waterlogging is mainly attributable to dual aspects [7], including (1) a natural perspective: global warming indirectly increases the probability of urban rainstorms, and (2) a social standpoint: it reflects the impact on urbanization and the urban water cycle. There are many approaches that have been developed for the evaluation of pluvial flooding. The related research [8] mainly revolves around the above two aspects, including six factors, such as meteorological factors and urbanization factors. Additionally, Elmer, F. et al. [9] showed that the increase in direct flood damage observed over the last decades may have been caused by changes in the meteorological drivers of floods, or by socioeconomic development. At the same time, varieties of urban waterlogging caused by climate change have a significant influence on economic growth worldwide [10]. Economic factors are not only one of the causes of urban floods, but also passively affected by the occurrence of disasters. So, the forecasting task for urban development in flood-prone areas will be further complicated because of the interaction between the two factors [11]. Basically, the most general type of forecasting method is based on GIS and remote sensing (RS) technology to draw flood maps [12–20]. The flood disasters are monitored based on the information from different RS platforms and bands, and the possibility of applying composite information to monitor potential flooding is analyzed. The disadvantage of this method is that the investment in building and maintaining a platform is heavier than that in the theoretical research method. Another approach is a comprehensive weighting method [21–24] that can more comprehensively solve the problem by constructing an index system. However, there is no fixed system; if the indicators of different zoning systems vary, there will be extreme contradictions between the results of the risk assessment and the conclusions drawn. Research studies have reportedly evaluated the flood defense capacity of a specific area [25] by using a combination of the entropy weight method and the analytic hierarchy process (AHP) method [26–30] or the individual TOPSIS method [31]. As a hierarchical multi-objective comprehensive evaluation method, AHP plays a good evaluation role when the evaluated object's attributes are difficult to quantify. However, it is incredibly subjective to rely on the weight of expert scoring. The problem is that there are too many decision-making layers for evaluation, and the gap between the judgment matrix and the consensus matrix is too large to use AHP. This is well solved by TOPSIS. The weight is wholly derived from the data, and the objectivity is better than that of AHP. However, the degree of data dependence is deep, and some of the indicators are not always adaptable to the information, which will harm the model's accuracy.

Various research studies have been carried out by scientists in other parts of the world to approach the topic. Zhengzhou, a city in southern China, was selected as the study area by Lin, Lin et al. to build a flood-susceptibility map, which was generated by using GIS spatial analysis tools and the analytic hierarchy process method [32]. Wang, Yamei et al. used a semi-quantitative model and fuzzy analytic hierarchy process (FAHP) weighting approach to assessing flood risk in the Dongting Lake region of Hunan Province, which is in central China [33]. The results of this article were compared with the above two articles, expanding the study area. Additionally, some articles published by Chun, Xiang et al. [34–37] combined the theoretical method of analysis with actual flood control capacity evaluations, and analyzed the relevant influencing factors. They presented relevant theoretical frameworks, but many lacked accurate data support, were not experimentally validated, and could only be used as theoretical references.

This study presents a regional flood defense capacity evaluation model, called MADM-GIS, to facilitate the development of flood mitigation strategies and better flood control. It is the first attempt to treat all regions of China as the research area in the field and over a time range that is more extended than other papers for a more comprehensive study. According to a survey of the existing literature, a richer indicator evaluation system than that of Shuqi, Wu et al. [38,39] was created, and the single indicator problem in the traditional 2D-TOPSIS was avoided by using the improved 3D-TOPSIS method in this paper [40]. The entropy weight method was also creatively combined with multiple spatial analysis tools in the ArcGIS platform [41–43] to evaluate the flood control capability comprehensively and realize the visualization of data. The ROC curve [44] was firstly used to test the flood defense capability evaluation model, and the Taylor diagram [45–47] was innovatively applied to test the degree of representation of the nine indicators. The obtained verification results were satisfactory.

2. Study Area and Data Resources

2.1. Study Area

China is hit by heavy precipitation and flood disasters almost every year, so this paper targets the whole of China as the research area for the first time in the flood defense capability assessment. China's longitude ranges from 73°33′ E to 135°05′ E, and its latitude ranges from 3°51′ N to 53°33′ N. The terrain is high in the west and low in the east, forming three steps from west to east. The land area is about 9.6 million square kilometers, of which mountains, plateaus and hills account for about 67%. At present, China has 34 provincial-level administrative regions, including 23 provinces, 5 autonomous territories, 4 municipalities directly under the Central Government, and 2 particular administrative regions. Due to the lack of data on Hong Kong, Macau and Taiwan, this article will not discuss them.

China's climate is complex and diverse, spanning five climatic zones from south to north. According to the 2020 "China Climate Bulletin", a total of 37 nationwide regional rainstorms occurred in 2020, and the annual number of rainstorm days since 1961 was second only to 2016. The Yangtze River is the longest in Asia, and the Yellow River is the second longest river in China. In the summer of 2020, they received the highest precipitation since 1961, while the second-highest was recorded in the Huaihe River and Lake Taihu during the same period. In 2020, the country's total annual precipitation resources measured 6592.65 billion cubic meters, 616.33 billion more than average.

2.2. Data Resources

Among the evaluation indicators related to flood defense capability that were selected in this paper, the regional green area, the population density, the number of medical institutions, GDP, and related data on the flood control area were obtained from the annual data for each province in China's National Statistical Yearbook.

DEM, the spatial resolution of the precipitation intensity data, is 1:1,000,000 from the National Basic Geographic Information Center in China. The fragmented data were implemented in accordance with the GB/T 13989-2012 "National Basic Scale Topographic Map Framing and Numbering". The space storage unit was 6° (longitude difference) × 4° (latitude difference). The slope data was obtained by surface analysis in the spatial analysis tool of ArcGIS based on DEM data. The main river data's spatial resolution was 1:4,000,000, which came from the National Basic Geographic Information System Vector Data. It needed to be converted into raster format by using the conversion tool in ArcGIS. The raster data on river network density could be obtained through the line density analysis in the spatial analysis tool. They were converted into a scale of 1:1,000,000 through the resampling extension module in ArcGIS to ensure that the raster calculation was performed based on the same spatial resolution. Otherwise, the subsequent algebraic calculation of raster data would be meaningless. The objective flood data in the final inspection process were from the Global Disaster Data Platform. The return period for a general flood is 10 years; for a large flood, 10–20 years; and for a major flood, 20–50 years. Thus, in this paper, we selected the data for various indicators in China from 2001 to 2020. Abbreviations for every Chinese region are shown in Table 1.

Table 1. Abbreviations for every Chinese region.

Number	Area Name	Abbreviation	Number	Area Name	Abbreviation
0	Heilongjiang	HL	16	Guangxi	GX
1	Xinjiang	XJ	17	Guangdong	GD
2	Shanxi	SX	18	Hainan	HI
3	Ningxia	NX	19	Jilin	JL
4	Tibet	XZ	20	Liaoning	LN
5	Shandong	SD	21	Tianjin	TJ
6	Henan	HA	22	Qinghai	QH
7	Jiangsu	JS	23	Gansu	GS
8	Anhui	AH	24	Shaanxi	SN
9	Hubei	HB	25	Inner Mongolia	NM
10	Zhejiang	ZJ	26	Chongqing	CQ
11	Jiangxi	JX	27	Hebei	HE
12	Hunan	HN	28	Shanghai	SH
13	Yunnan	YN	29	Beijing	BJ
14	Guizhou	GZ	33	Sichuan	SC
15	Fujian	FJ			

3. Methodology

3.1. Analysis Framework

The new flood control capability analysis framework is mainly divided into four parts. The detailed content will be introduced in terms of the following points:

(1) Data collection and processing;
(2) Flood control capability evaluation index system;
(3) Flood control capability evaluation and calculation;
(4) Model validation.

There is a progressive relationship between the above four parts, forming a complete evaluation system. As long as the relevant required data are input, a systematic evaluation of the flood control capacity of any region can be realized. At the same time, the corresponding parameters can be collectively analyzed and adjusted according to the specific situation, and the comparative experiments can be carried out.

Every five years is regarded as one research period; the 20 years span was divided into a total of four periods: 2001–2005, 2006–1010, 2011–2015, and 2016–2020. Each stage generated an evaluation map, in line with the speed of the strategic policy of China's five-year plan and social development. Then, this paper compared the changes in flood sensitivity indicators in each period to profoundly and systematically explore the impact of each indicator on the flood defense capabilities in various regions. It was shown a brief overview of the article writing process by Figure 1.

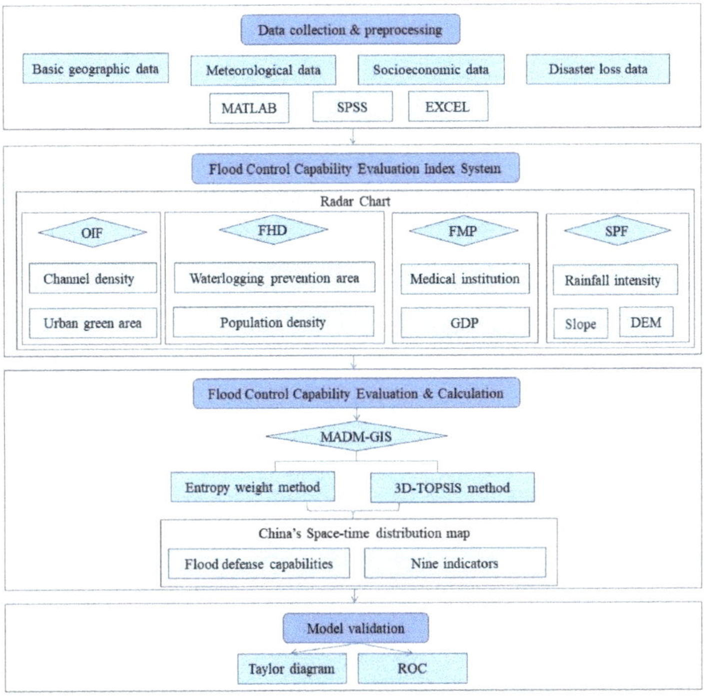

Figure 1. Flow chart for building the model for evaluation of China's flood defense capability.

3.2. Flood Control Capability Evaluation Index System

Appropriate parameter indicators play a decisive role in accurately assessing the scale of floods and implementing universal defense measures. Based on a reading of the existing literature [48,49], this paper divided the indicators into four aspects:

$$FCS = f(OIF, SPF, FMP, FHD) \tag{1}$$

where FCS denotes the evaluation results of the MADM-GIS model regarding flood defense capability. OIF refers to the region's geographical and meteorological conditions and other natural environmental aspects that can tolerate rainstorms and floods and maintain a stable state; it can reflect the macro self-regulation ability of the natural environment, thereby reducing the frequency of extreme disasters. SPF refers to the ability of humans to forecast, prevent and defend against the disasters; it reflects the nature of the ability of the Chinese government and relevant departments in various regions to respond to flood disasters promptly, protect the safety of people's lives and their properties across the country, and minimize the loss.FMP refers to the ability of human beings to participate in economic, medical, social, and other activities to reduce the negative impact of disasters under the condition that natural conditions cannot change. FHD refers to the nature of the intensity of the direct factors leading to floods. Nine indicators, including DEM, slope, precipitation intensity, gross regional product, number of medical institutions, urban population density, urban green space area, and flood control area [50,51], were extracted from four aspects to form a flood defense capacity evaluation system. Table 2 is the list of factor abbreviations and acronyms.

Table 2. List of factor abbreviations and acronyms.

Abbreviation	Parameters	Abbreviation	Parameters
OIF	The objective and inclusive factor	U C	Urban green area Channel density
SPF	The subjective preventive factor	W P	Waterlogging prevention area Population density
FMP	The flood mitigation prominence	M G	Medical institution GDP
FHD	The flood hazard degree	R D S	Rainfall intensity DEM Slope

3.3. *MADM-GIS Model*

Multi-criteria decision-making (MCDM) is a decision to choose among a finite (infinite) set of conflicting and incommensurable schemes. Its origins [52] can be traced back to the concept of Pareto optimization (1896), and it was brought into the decision-making field as a normative decision-making method in the 1960s, represented by the research on objective planning by Charnes and Cooper. MCDM methods are based on the principle of proposing the best solution among the schemes under certain criteria, so they have been used more widely recently [53]. There was a Monte Carlo Simulation used in a multicriteria decision model [54] aiming to prioritize flood risks in urban areas under climate effects. They also have been applied in many specific fields to solve some selection and ranking problems, such as information technology [55], design and development [56], civil engineering and management [57], renewable energy [58] and medical diagnosis [59]. There are more than 100 MCDM methods, and each of them has its own performance capacity and characteristics, which are often associated with the model's computational process and methodology.

MCDM is mainly divided into two categories: multi-attribute decision-making (MADM) and multiple objective decision-making (MODM), according to whether the decision-making scheme is limited or unlimited. The MCDM model has been used to evaluate flood defense capacity in the existing literature [33,60]. Compared with MADM, the scope of application of MCDM is broader, but inaccurate evaluation results are usually generated. The difficulty associated with ineffective flood control capability evaluation caused by an extensive data range processed by MCDM has been solved.

In this paper, the evaluation method for the index system was innovatively combined with the flood control capability evaluation; the two methods, including the TOPSIS method and the entropy weight method under MADM, were chosen, and MATLAB was utilized for the calculations. GIS was used to generate the distribution map of the corresponding period and location of each indicator proposed by the MADM-GIS model. As it is the first of its kind built in this field, with efficient and accurate assessment simultaneously completed, this model represents a relatively new attempt with a certain degree of practical value. Figure 2 is the MCDM logic diagram that shows the classification and content of MCDM.

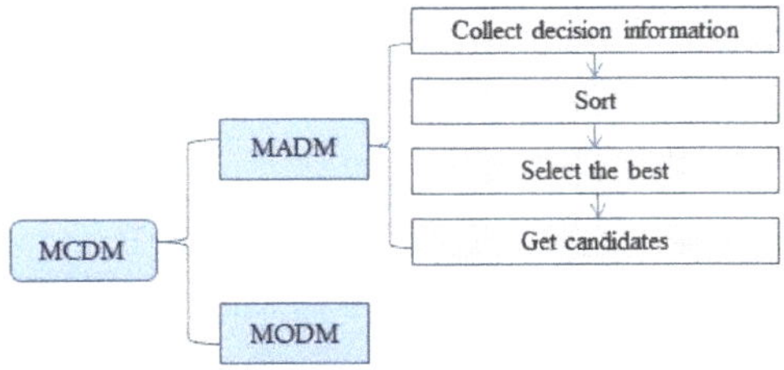

Figure 2. MCDM logic diagram.

3.3.1. The Entropy Weight Method

The entropy weight method is an objective weighting method based on the degree of variation in the index. The lower the amount of information reflected by the index, the lower the corresponding weight will be. Table 3 is the key math symbols and their meanings.

Table 3. Key math symbols and their meanings.

Symbol	Description
i	The number of evaluation objects
j	The number of evaluation indicators
P	The forward matrix
a_{ij}	The elements in a forward matrix
F	Probability matrix
F_{ij}	Corresponding probability of a_{ij}
S_{ij}	Information entropy
m_{ij}	Information utility value
W_j	Weight of each indicator

A positive matrix is established according to the data, which is convenient for subsequent data preprocessing:

$$P = \begin{bmatrix} a_{1,1} & \cdots & a_{1,9} \\ \vdots & \ddots & \vdots \\ a_{31,1} & \cdots & a_{31,1} \end{bmatrix} \tag{2}$$

The most common standardization method, known as the dispersion standardization method, was used in this paper to transform the original data linearly. The results were mapped to the [0, 1] interval, eliminating the interference of different units and incomparability between indicators.

$$\widetilde{Z_{ij}} = \frac{a_{ij} - min\{a_{1j}, a_{2j} \ldots a_{nj}\}}{max\{a_{1j}, a_{2j} \ldots a_{nj}\} - min\{a_{1j}, a_{2j} \ldots a_{nj}\}} \tag{3}$$

The proportion of the i-th sample under the j-th index is calculated, and it is regarded as the probability calculation probability matrix F used in the relative entropy calculation, where the calculation formula of F_{ij} is:

$$F_{ij} = \frac{a_{ij}}{\sum_{i=1}^{n} a_{ij}} \tag{4}$$

The greater the information entropy of the event is, the smaller the amount of existing information will be. Furthermore, the amount of information that can be supplemented will be larger.

For the *j*-th index, the calculation formula of its information entropy is:

$$s_{ij} = -\frac{1}{\ln n}\sum_{i=1}^{31} F_{ij}\log(F_{ij}) \tag{5}$$

The information utility value is defined as:

$$m_j = 1 - s_j \tag{6}$$

The weight of each indicator can be obtained by normalizing the information utility value:

$$W_j = \frac{m_j}{\sum_{j=1}^{m} m_j} \tag{7}$$

In the process of solving the MADM problem, the weight of the attribute is used to reflect the relative importance of the attribute, which plays a pivotal role. In this paper, the entropy weight method was used to calculate the respective weights of the nine indicators on the impact of flood defense capacity. Figure 3 shows that the weight of P changed abruptly from the first period to the second period and was relatively stable in the following 15 years, following a downward trend. However, the process of change in M and P was precisely the opposite. There was a sudden increase in the transition from the first five years to the second, and then there was a slight increase. Only the weight of urban green space decreased gradually and uniformly over time; the corresponding graphs of the six indicators W, D, C, S, R, and G in the radar chart all transitioned from inside to outside, which gradually increased over time.

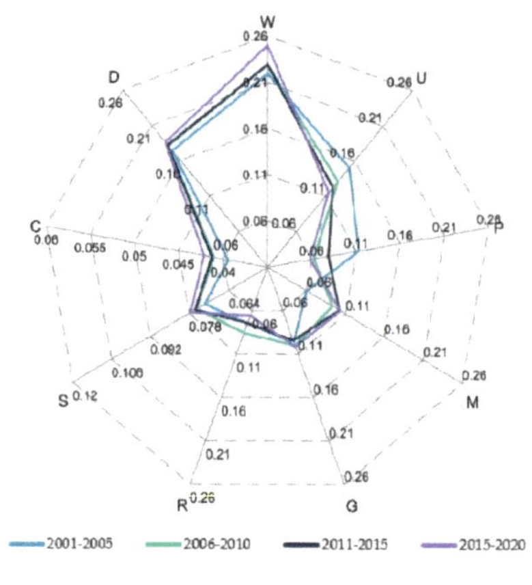

Figure 3. Nine indicator weight radar diagram.

3.3.2. 3D-TOPSIS Model

The traditional TOPSIS is based on two-dimensional indicators. Only a score for each indicator individually can be obtained, which is inconvenient for visualization. The two-dimensional model was improved to three-dimensional in this paper, and the total score for flood defense capacity in every region was obtained by superimposition. Table 4 is the key math symbols and their meanings.

Table 4. Key math symbols and their meanings.

Symbol	Description
$\dot{q}_{ij}$	Indicator raw data
q_{ij}	The index data after forwarding
T_{ij}	Normalized indicator data
Z	Weighted normalization matrix
ω_{ij}	Indicator corresponding weight
Q^+	Maximum value in q_{ij}
Q^-	Minimum value in q_{ij}
$D_i{}^+$	The distance between the evaluation object and Q^+
$D_i{}^-$	The distance between the evaluation object and Q^-
S_i	Relative proximity

The negative indicators need to be converted into positive indicators, including R, D, S, C, and the transformed data matrix is still recorded as P.

$$q_{ij} = q_{max} - \dot{q}_{ij} \tag{8}$$

The original data should be normalized to avoid large network prediction errors due to the large magnitude difference between input and output data:

$$T_{ij} = \frac{q_{ij}}{\sqrt{\sum_{i=1}^{n} {q_{ij}}^2}} \tag{9}$$

Based on the weights obtained above, a standardization matrix is constructed:

$$Z = \begin{bmatrix} q_{1\,1}\cdot\omega_1 & \cdots & q_{1\,9}\cdot\omega_1 \\ \vdots & \ddots & \vdots \\ q_{31\,1}\cdot\omega_{31} & \cdots & q_{31\,9}\cdot\omega_{31} \end{bmatrix} \tag{10}$$

The maximum and minimum values are defined as:

$$\begin{aligned} Q^+ &= (max\{a_{1\,1}, a_{2\,1} \ldots a_{n\,1}\}, max\{a_{1\,2}, a_{2\,2} \ldots a_{n\,2}\}, \ldots, max\{a_{1\,9}, a_{2\,9} \ldots a_{n\,9}\}) \\ Q^- &= (min\{a_{1\,1}, a_{2\,1} \ldots a_{n\,1}\}, min\{a_{1\,2}, a_{2\,2} \ldots a_{n\,2}\}, \ldots, min\{a_{1\,9}, a_{2\,9} \ldots a_{n\,9}\})\# \end{aligned} \tag{11}$$

The distances between the maximum and minimum values of the evaluation object i-th (i = 1, 2, ... , 31) are respectively defined as:

$$\begin{aligned} D_i{}^+ &= \sqrt{\sum_{j=1}^{m} (Q_j{}^+ - q_{ij})^2} \\ D_i{}^- &= \sqrt{\sum_{j=1}^{m} (Q_j{}^- - q_{ij})^2}\# \end{aligned} \tag{12}$$

3.4. *Taylor Diagram*

The Taylor plots in the existing literature were used to test extreme air temperature, the meteorological conditions in different models, or the correlation analysis of correlation coefficients (CC). It is a new type of ternary diagram that was firstly introduced in the studied field, which is about the evaluation of flood defense capability. In the figure, the model is represented by scattering points, the correlation coefficient (CC) is shown by the radial line, the standard deviation (SD) is represented by the horizontal and vertical axes, and the root mean square error (RMSE) is indicated by the dotted line. The CC refers to similarity between the simulation results of different indicators and the observed values. The differences between the spatial uniformity of simulation results of the models and the observations are respectively reflected by RMSE and SD. The smaller the SD is, the more stable the model is.

The results are as follows. The SD is 0.2–0.4, which is in line with the characteristics of the heterogeneous coverage of the nine indicators, and the points of data are clustered around the mean to a small degree. The correlation between the nine indicators is weak, which is shown by the CC as 0.1–0.9. The combination of the two axes shows that the nine indicators are highly representative of the research problem, and the problem of the high degree of index coincidence reducing the model's accuracy can be solved. The deviation between the observed and actual values measured by the RMSE is 0.2–0.5, which indicates that the departure of the data in this paper was slight and the applicable standard was met. The test results show that the established MADM-GIS model and the weights corresponding to the indicators are close to the research problem and have practical significance and scientific and objective persuasion. Figure 4 is the Taylor plot of nine indicators from 2001 to 2020.

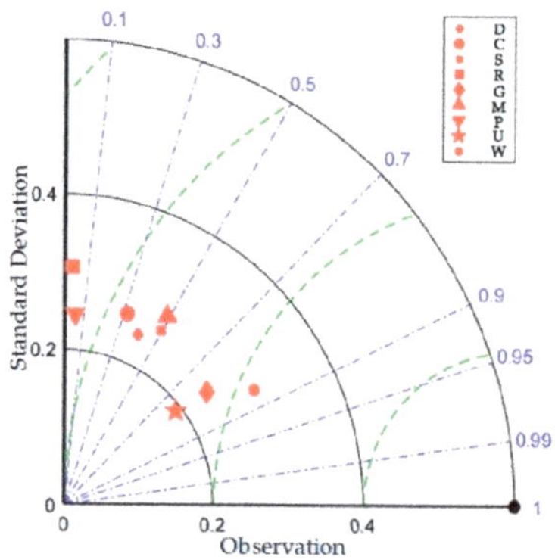

Figure 4. Taylor plot of nine indicators from 2001 to 2020.

4. Results and Discussion

4.1. Analysis and Validation

The time interval of the data is from 2001 to 2020, which is divided into four stages, showing the trend of influencing factors over time in detail. Nine indicators extracted from four aspects were shown in Table 1; from the perspective of time series, they are divided into two parts including six indicators from a timing perspective that change significantly with time U, W, P, R, G, M and D, S, C that almost don't change. The data is imported into ArcGIS, and divided into 8 categories by the natural breakpoint method. The higher the level of the region is, the better the performance of the corresponding index is, and the corresponding province shows a darker color.

4.2. 3D-TOPSIS Model Parameter Analysis

4.2.1. Urban Green Area

Figure 5 is the spatial-temporal distribution of urban green space in China. From Figure 5a, the regional differences in urban green space area from 2001 to 2005 are the most obvious. The urban green areas of the regions such as Sichuan, Guangdong, Jiangsu, Shanghai are outstanding in China. Comparison of the four maps shows that the growth rate of the value in Qinghai and Tibet was relatively slow because western China had a higher altitude, which is not conducive to vegetation growth. From Figure 5a–d, the differences of the regions were gradually decreasing. The value in Inner Mongolia and northeast China increased significantly and then tend to be stable, which indicates that the green engineering in China has been better developed with significant results under the advancement of China's five-year plan.

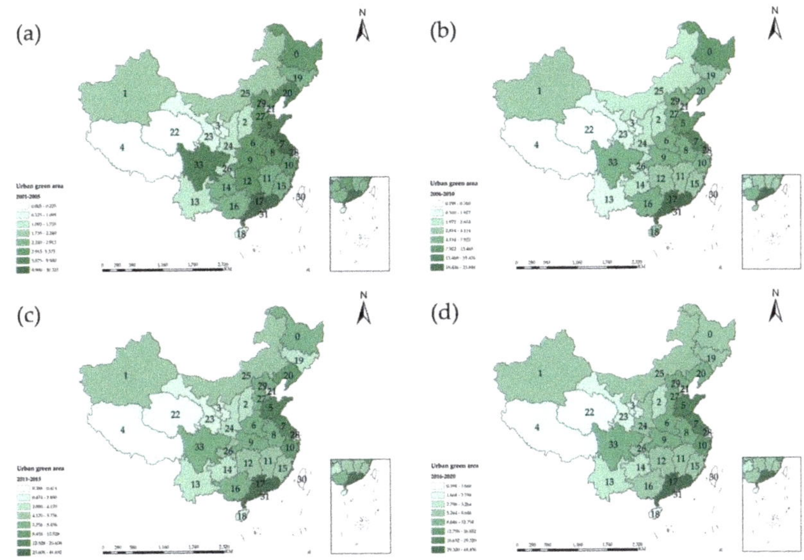

Figure 5. Spatial and temporal distribution of urban green space in China.

4.2.2. Waterlogging Prevention Area

The area of waterlogging removal refers to an area where waterlogging-prone farmland is rescued from flooding due to the construction of waterlogging control projects or other water conservancy facilities, and the standard is reached more than once in three years. It reflects the protection and utilization of water resources in China. Figure 6 is the spatial-temporal distribution of waterlogging prevention and control areas in China. From Figure 6a, the flood defense in Inner Mongolia, Tibet, Sichuan and Shandong increased significantly from 2001 to 2005. In Figure 6b, flood defense in southeastern China continued to increase within a small range. The work performed by Guangdong, Shandong, Henan, and Sichuan was excellent. To sum up, as shown in Figure 6d, the efficiency of releasing policies for flood defense and disaster reduction across the country gradually accelerated. The area of waterlogging prevention reached a balanced level, and rational water resources management was comprehensively promoted.

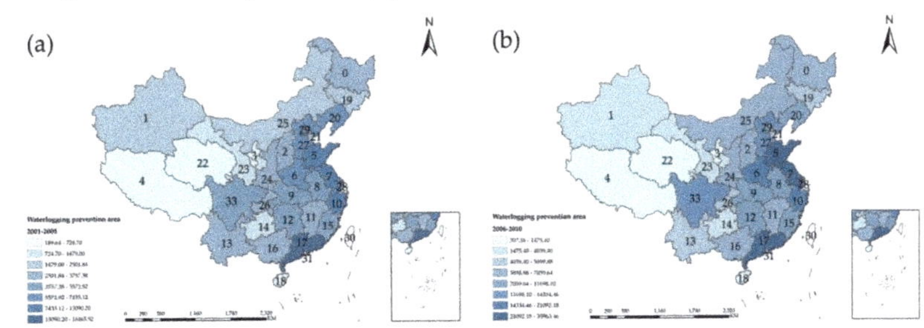

Figure 6. *Cont.*

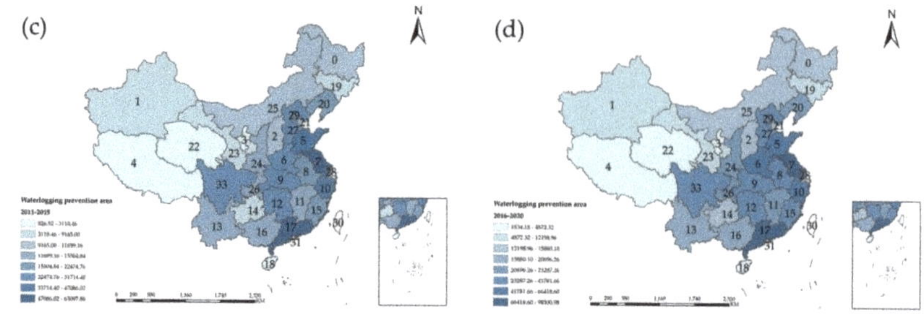

Figure 6. Spatial and temporal distribution of waterlogging prevention and control areas in China.

4.2.3. Population Density

Urban population density refers to the number of people per unit area. Figure 7 is the spatial-temporal distribution of population density in China. As vividly shown in Figure 7a, owing to the inferior backdrop of development, the population density in Xinjiang, Heilongjiang, Inner Mongolia, Sichuan, Yunnan and Guangxi is very sparse. It shows that the period of 2006–2010 was at the peak of regional differences in Figure 7b,c with the implementation of the five-year plan. In Figure 7b, population density in Xinjiang and Heilongjiang increased significantly, and also increased widely in southern China. With the continuous development of China's economy and society, the population problem had been solved, and the density of the urban population was increasing nationwide. The regional differences in terms of population density across the country were becoming increasingly slim, which is graphically depicted by Figure 7d. Due to the inception of the two-child policy, the population density across the country has basically become saturated.

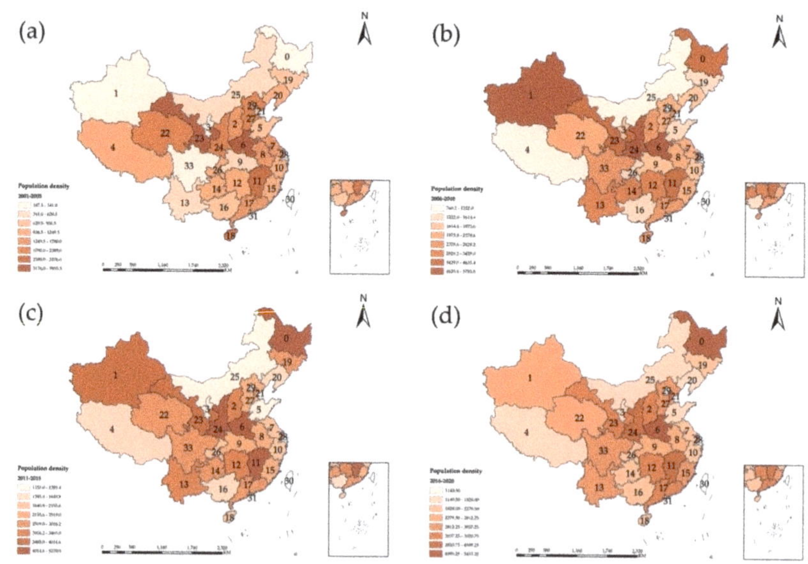

Figure 7. Spatial and temporal distribution of population density in China.

4.2.4. Medical Institutions

The number of medical institutions refers to the total number of health institutions established in accordance with legal procedures that engage in disease diagnosis and treatment activities. Figure 8 is the spatial-temporal changes in the number of medical institutions in China from 2001 to 2020. It represents the number of medical and health resources and reflects the ability of a region to provide medical services when floods come. The higher the flood control ability score climbs, the better the medical configuration is. There will be more medical resources that can be used to ensure the safety of people's lives and property. During 2001–2005, Sichuan, Hunan, and Guangdong's numbers of medical institutions were among the top three nationwide. In the second period, Beijing, Hebei, and Henan performed more outstandingly in this regard, and the number of medical institutions increased significantly.

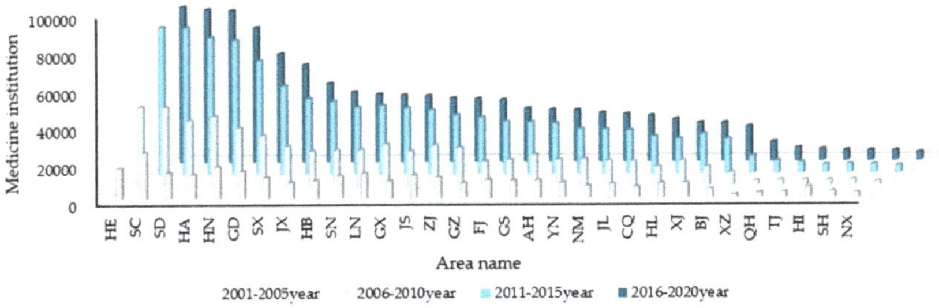

Figure 8. Spatial and temporal changes in the number of medical institutions in China from 2001 to 2020.

4.2.5. GDP

Gross domestic product (GDP) is the core indicator of national economic accounting, which can measure the financial status and development level of a country or region and represents the financial ability to deal with floods. Figure 9 is the chart of temporal and spatial changes in China's GDP. The higher GDP is, the more investment that can be allocated to flood recovery. Meanwhile, the damage caused by a flood can be quickly mitigated to a great extent. From 2001 to2005, Jiangsu, Shandong, Guangdong, Zhejiang, and Shanghai were in the leading position in China. Due to the impact of China's five-year plan on the economy, economic development improved steadily in Beijing, Hebei, Henan and Sichuan from 2006 to 2010. The northern regions also responded to the call, which made a big difference. From the graph, the southern region's GDP improved in an all-around way. People's living standards and social development levels were promoted under the economic drive.

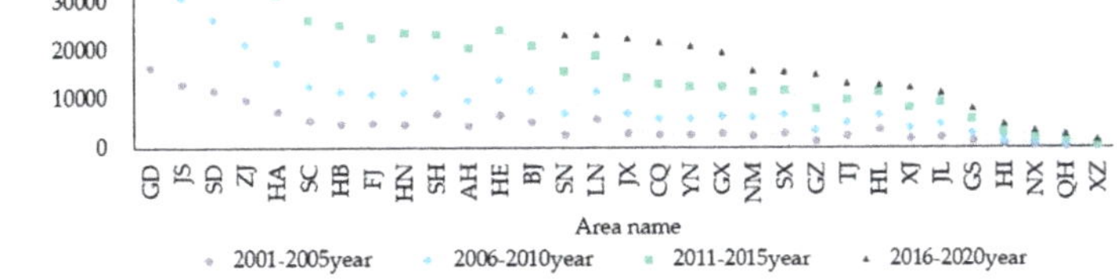

Figure 9. Chart of temporal and spatial changes in China's GDP.

4.2.6. Rainfall Intensity

Rainfall intensity refers to the amount of rainfall in a unit of time. Figure 10 is the spatial-temporal distribution of rainfall intensity in China from 2001 to 2020. It can be seen from the four figures that the spatial distribution of rainfall intensity in China is strong in the south and weak in the north, and there is an apparent planar aggregation phenomenon in the rainfall in the south. In Figure 10a, the rainfall intensity was the strongest in the four periods. The rainfall intensity was lower than the previous five years in Figure 10b. Yunnan, Sichuan, Guangxi, Hunan, Jiangxi and Chongqing are prominent gathering centers. The transition process from the four figures shows that the aggregation phenomenon was becoming more and more apparent, and a point-like aggregation can be seen in Figure 10c. Hunan, Guangxi, Jiangxi, Guangdong, and Fujian were the main centers of the point aggregation phenomenon. Close attention to each gathering center should be encouraged, and the management of extreme precipitation events should be strengthened. The efficiency of monitoring and forecasting needs to be improved. The government should provide exceptional policy support, and the surrounding provinces should also provide assistance.

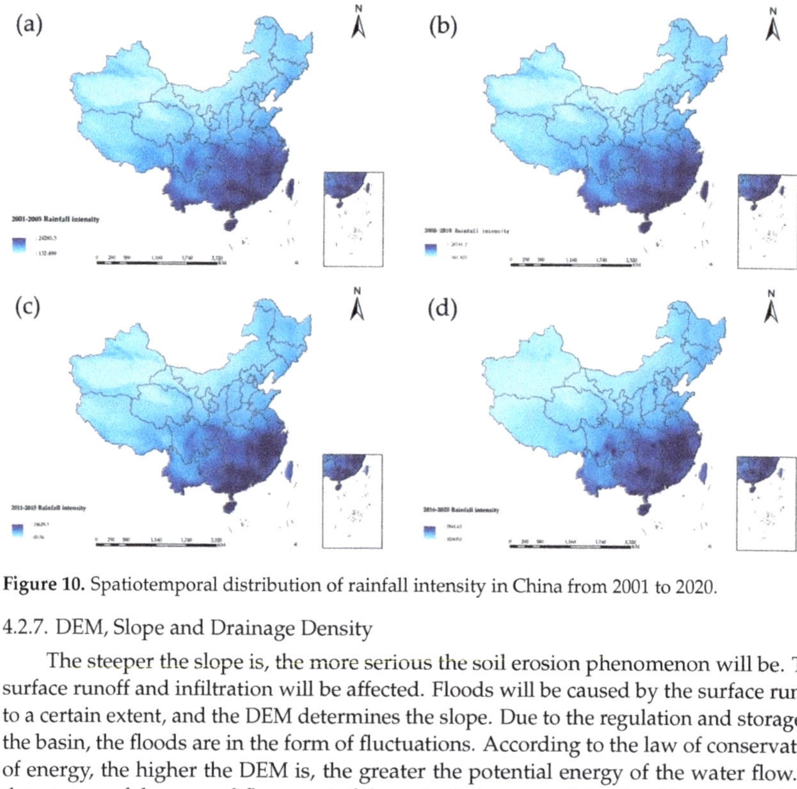

Figure 10. Spatiotemporal distribution of rainfall intensity in China from 2001 to 2020.

4.2.7. DEM, Slope and Drainage Density

The steeper the slope is, the more serious the soil erosion phenomenon will be. The surface runoff and infiltration will be affected. Floods will be caused by the surface runoff to a certain extent, and the DEM determines the slope. Due to the regulation and storage of the basin, the floods are in the form of fluctuations. According to the law of conservation of energy, the higher the DEM is, the greater the potential energy of the water flow. In the process of downward flow, part of the potential energy of the flood is converted into kinetic energy, and the flow velocity is accelerated, increasing the risk that the resulting severe shock will cause harm downstream. Figure 11 is the spatial distribution map of DEM. Figure 12 is the spatial distribution map of slope. Figure 13 is the spatial distribution map of major river networks.

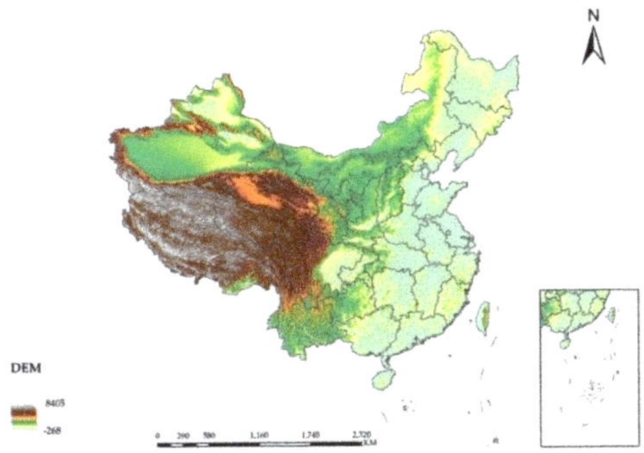

Figure 11. Spatial distribution map of DEM.

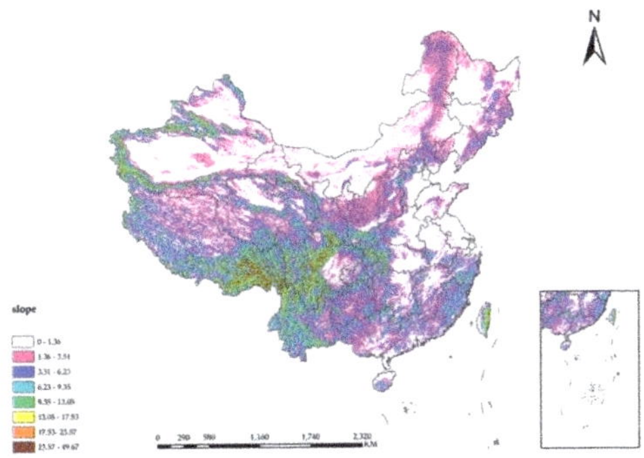

Figure 12. Spatial distribution map of slope.

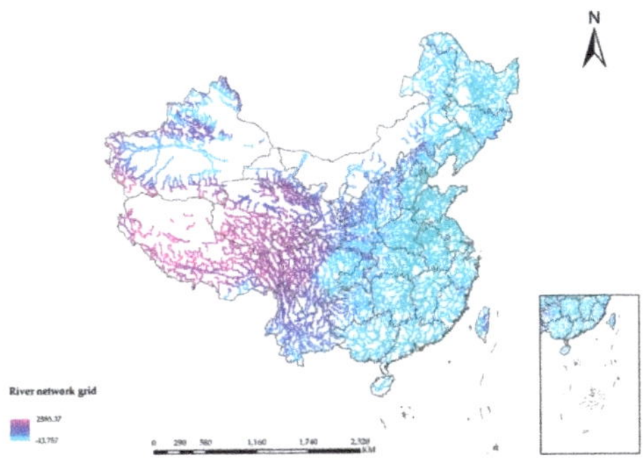

Figure 13. Spatial distribution map of major river networks.

4.3. *MADM-GIS Model*

4.3.1. Analysis of Results of Entropy Weight Method

The results show that the weight of P changed abruptly from the first period to the second period and was relatively stable in the following 15 years, showing a downward trend. However, the process of change in M and P is precisely the opposite. There is a sudden increase in the transition from the first five years to the second, and then there is a slight increase. Only the weight of urban green space decreases gradually and uniformly over time; the corresponding graphs of the six indicators W, D, C, S, R, and G in the radar chart all transition from inside to outside, which gradually increases over time.

4.3.2. Analysis of Results of D-TOPSIS Model

S_i is defined as the relative proximity, and the calculation formula is:

$$S_i = \frac{D_i^-}{D_i^+ + D_i^-} \tag{13}$$

The larger the obtained evaluation matrix S_i is, the greater the degree of closeness to the optimal solution, and the flood defense capability of the area is more robust. When $S_i = 0$, it is the lowest flood control capability; $S_i = 1$ is the highest. Finally, sort according to the size of the proximity. The larger the value is, the closer the evaluation object to the optimal solution is.

The weights in 2001–2005 are taken as an example:

$$\begin{aligned}
OIF &= \sum_{i=1}^{2} x_i\omega_i = 0.15063 \times U + 0.03942 \times C \\
FMP &= \sum_{i=1}^{2} x_i\omega_i = 0.06064 \times M + 0.09591 \times G \\
SPF &= \sum_{i=1}^{2} x_i\omega_i = 0.21922 \times W + 0.11295 \times P \\
FHD &= \sum_{i=1}^{2} x_i\omega_i = 0.07479 \times R + 0.17401 \times D + 0.17143 \times S \\
FCS &= OIF + FMP + SPF + FHD
\end{aligned} \tag{14}$$

To obtain the flood defense capacity index for each period, the linear weighted sum method (LWSM) was used to perform statistical processing on the data. The LWSM is an evaluation function method, which can solve multi-objective programming problems by assigning corresponding weight coefficients to each objective and then optimizing its linear combination. Each basic indicator was multiplied by the corresponding weight and then summed up. Multiple base metrics were converted into a single numerical index that was regarded as the final score. Through the value of the final score, it was possible to quantify and compare the flood defense capability of each city.

From Figure 14, it can be seen that the evaluation scores and absolute values of the direct economic losses in the 31 provinces show roughly the same trend. As is shown, the results of the model solution are in line with objective reality, which verifies the validity of the model. However, the evaluation score does not match the absolute value of the direct economic loss in Sichuan, Shaanxi, Jiangxi and Hunan. It was found that the occurrence of this phenomenon is strongly related with social factors outside the model that did not belong within the scope of the research through actual investigation. There was no effect on the analysis result, namely, that the higher the score of the MADM-GIS model is, the lower the value of the direct loss rate caused by the flood is. The evaluation results for regional flood defense capacity are only related to the actual data, and the government staff in the relevant regions need to formulate policies and implement various adjustments.

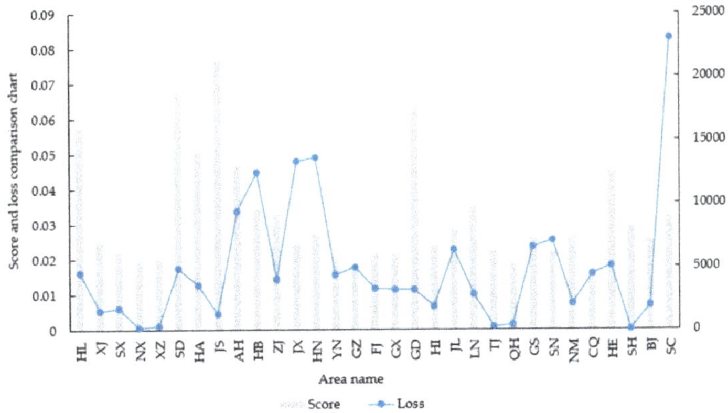

Figure 14. Combination graph of the final score and direct loss.

Figure 15a is about the flood defense capability assessment scores in China obtained with the MADM-GIS model. Figure 15b is about the direct losses that are caused by floods and is based on actual data in China. The stronger the flood defense capability of a region is, the less the direct loss is, which is in line with objective reality. The two variables of flood defense capacity and direct loss in Xinjiang, Gansu, and Inner Mongolia are all at the middle level in China. There is little difference in the comparison of two figures, whose colors are roughly the same. Tibet, Yunnan, Guizhou, Chongqing, and Guangxi are mainly located in western China, where flood defense capacity is lower, and direct losses are heavier than in other regions. The eastern regions, including Hebei, Henan, Jiangsu, Shanghai, Guangdong, and Zhejiang, played an exemplary role with higher evaluation scores in coping with severe floods.

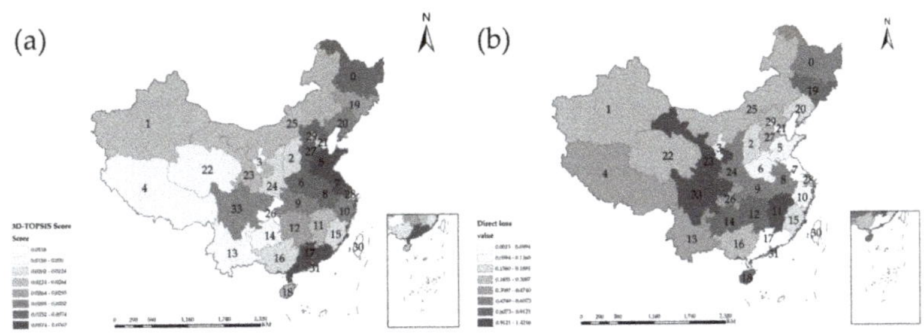

Figure 15. 3D-TOPSIS score results in spatial distribution map.

4.4. MADM-GIS Model Validation

Analysis of ROC Curve Results

Figure 16 is the ROC curve of the flood control capacity score to the MADM-GIS model. The sensitivity and specificity of variables are reflected by the ROC curve [61,62], and intuitive comparison of different test methods is provided under the same scale. It was drawn by SPSS software to test the accuracy and feasibility of the model. There are two advantages of this method. First, there is a broader application range to indirectly analyze various types of raw data. Secondly, the systematic analysis of multiple covariates can be carried out, which is more advanced than the ROC curve in analyzing univariate raw data described in the current literature [50,51,63].

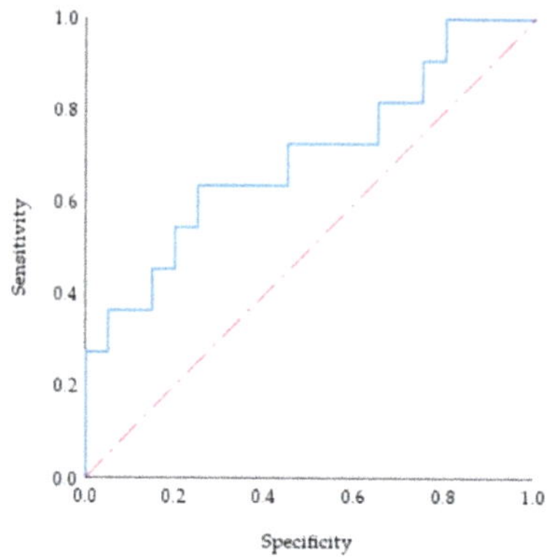

Figure 16. The ROC curve of the flood control capacity score to the MADM-GIS model.

The closer the inflection point is to the upper left corner of the figure, the greater the diagnostic value of the model is. The accuracy of test objects was indicated by the AUC of 0.7 to 0.9; the closer the AUC is to 1, the better the diagnostic effect of the model. As is shown, the AUC was 70%, and the accuracy and feasibility of the MADM-GIS model were meaningful.

5. Conclusions

MADM-GIS, a flood defense capability evaluation model with strong inclusiveness and adaptability, was highlighted in this paper. It is the first attempt to evaluate the flood defense capability in every region of China, and the problem of China's poor response to flooding disasters in the past was resolved. First, the data visualization platform based on ArcGIS was assisted by MATLAB and SPSS software. It was established to display the time-space pattern of the calculation results of the indicators and analyze the process of change in the nine indicators during the past 20 years. Combined with the entropy weight method and the improved 3D-TOPSIS method, the defense capability scores of various regions were obtained, and the corresponding spatiotemporal distribution map was generated. Then, the ROC curve was utilized as the model to test the method. As to the AUC = 70%, it was a satisfactory result, and the Taylor diagram was combined to intuitively represent enough adaptation of each index. Finally, the evaluation ability under different index modes was compared, and the usability and accuracy of the MADM-GIS model were verified.

The results of analysis show that the main influencing factors in the evaluation of flood defense capacity were W, U, and D, and that their weights hardly changed with time, which indicated they were the top three of the nine indicators. From 2001 to 2020, the differences in urban green space between regions were gradually narrowing, and there were upstanding effects in waterlogging control displayed in Guangdong, Shandong, Henan, and Sichuan. GDP in the southern region increased significantly, and the capacity of medical resources support for rainstorm and flood disasters was strengthened. From 2016 to 2020, the regional differences in overall population density across the country gradually decreased, and the population density was nearly saturated. The most obvious point-like aggregation phenomenon in regional rainfall intensity was shown from 2011 to 2015, and Hunan, Guangxi, Jiangxi, Guangdong, and Fujian were mainly at the center of the point-like aggregation. The final evaluation results show that the flood control capacity of Tibet, Yunnan, Guizhou, Chongqing and Guangxi needs to be improved significantly, and there is a plenty of room for improvement. Hebei, Henan, Jiangsu, Shanghai, Guangdong, and Zhejiang are in the leading positions in China.

The spatial situation for flood defense capabilities in China was described in this paper and will contribute to enriching the system for evaluation of urban flood control capabilities. An objective and scientific reference plan was provided for the measures taken by the Ministry of Water Resources and related departments to deal with flood control and disaster mitigation.

Since the research scope covered the whole area of China, only the main factors were selected for the evaluation indicators to prevent interference from secondary factors with the research conclusions of the model. Therefore, there were regional limitations in the promotion and application of the model. For the evaluation of flood defense capacity in smaller areas, the application of the model needs to be combined with the actual situation in each region to obtain more accurate and reliable research results with detailed evaluation indicators. This paper selected the data from 2001 to 2020, which were acquired in combination with data on China's political development and economic development. For more targeted research, the time range can be expanded and the database can be enriched to analyze more factors that contribute to the results. More indicators were selected in this paper than for ordinary evaluation models; hence, there was no detailed analysis of the sensitivity between indicators. Follow-up data and further experiments are needed to enrich the contents of the model and improve its efficiency.

Author Contributions: Conceptualization, W.Y.; Data curation, X.L.; Software, X.L.; Validation, A.Z.; Visualization, C.C.; Writing—review & editing, W.Z. All authors have read and agreed to the published version of the manuscript.

Funding: This research received no external funding.

Institutional Review Board Statement: Not applicable.

Informed Consent Statement: Not applicable.

Data Availability Statement: http://www.stats.gov.cn/tjsj/ndsj/, https://www.gddat.cn/newGlobalWeb/#/chinaDisasterDatabase, https://www.resdc.cn/ (accessed on 12 March 2022).

Conflicts of Interest: The authors declare no conflict of interest.

References

1. Peng, L.; Ma, Y.; Wei, R.; He, Y.; Mu, Z.; Li, X.; Liu, C. Using GIS Grid Dataset to Simulate and Identify Flooding Risks in Irrigation Areas in Yarkant in Xinjiang. *J. Irrig. Drain. Eng.* **2020**, *39*, 124–131.
2. Feng, K.; Fan, L.; Yifei, W.; Jian, F.; Lili, L. Spatial Relationship between Flooding Frequency and Climate Change based on Climate Change Regionalization in China from 1961 to 2010. *J. Catastr.* **2017**, *32*, 35–42. [CrossRef]
3. Lyu, H.-M.; Wang, G.-F.; Shen, J.; Lu, L.-H.; Wang, G.-Q.; Wang, G.-F.; Shen, J.; Lu, L.-H.; Wang, G.-Q. Analysis and GIS Mapping of Flooding Hazards on 10 May 2016, Guangzhou, China. *Water* **2016**, *8*, 447. [CrossRef]
4. Li, S.; Du, J.; Zhang, Z.; Xin, H. Risk regionalization for flood hazard in southern Shaanxi Province based on grey target model and GIS. *Yangtze River* **2021**, *52*, 17–21.
5. Abebe, Y.; Kabir, G.; Tesfamariam, S.; Kabir, G.; Tesfamariam, S. Assessing urban areas vulnerability to pluvial flooding using GIS applications and Bayesian Belief Network model. *J. Clean. Prod.* **2018**, *174*, 1629–1641. [CrossRef]
6. Miller, J.D.; Hutchins, M.; Hutchins, M. The impacts of urbanisation and climate change on urban flooding and urban water quality: A review of the evidence concerning the United Kingdom. *J. Hydrol. Reg. Stud.* **2017**, *12*, 345–362. [CrossRef]
7. Zhang, D.; Yan, D.; Wang, Y.; Lu, F.; Liu, S. Research Progress on Risk Assessment and Integrated Strategies for Urban Pluvial Flooding. *J. Catastr.* **2014**, *29*, 144–149. [CrossRef]
8. Ning, Y.-F.; Dong, W.-Y.; Lin, L.-S.; Zhang, Q.; Dong, W.-Y.; Lin, L.-S.; Zhang, Q. Analyzing the causes of urban waterlogging and sponge city technology in China. *IOP Conf. Ser. Earth Environ. Sci.* **2017**, *59*, 12047. [CrossRef]
9. Elmer, F.; Hoymann, J.; Düthmann, D.; Vorogushyn, S.; Kreibich, H.; Hoymann, J.; Düthmann, D.; Vorogushyn, S.; Kreibich, H. Drivers of flood risk change in residential areas. *Nat. Hazards Earth Syst. Sci.* **2012**, *12*, 1641–1657. [CrossRef]
10. Lyu, H.M.; Shen, S.L.; Sun, W.J.; Cheng, W.C. Discussion on Sponge City Construction under the Circumstance of Urban Waterlogging. In Proceedings of the International Conference on Transportation Infrastructure and Materials, Qingdao, China, 9–12 June 2017.
11. Mobilia, M.; Califano, F.; Longobardi, A.; Califano, F.; Longobardi, A. Analysis of Rainfall Events Driving MDHEs Occurred in the Solofrana River Basin, Southern Italy. *Procedia Eng.* **2015**, *119*, 1139–1146. [CrossRef]
12. Feng, B.; Wang, J.; Zhang, Y.; Hall, B.; Zeng, C.; Wang, J.; Zhang, Y.; Hall, B.; Zeng, C. Urban flood hazard mapping using a hydraulic–GIS combined model. *Nat. Hazards* **2020**, *100*, 1089–1104. [CrossRef]

13. Hong, H.; Panahi, M.; Shirzadi, A.; Ma, T.; Liu, J.; Zhu, A.-X.; Chen, W.; Kougias, I.; Kazakis, N.; Panahi, M.; et al. Flood susceptibility assessment in Hengfeng area coupling adaptive neuro-fuzzy inference system with genetic algorithm and differential evolution. *Sci. Total Environ.* **2018**, *621*, 1124–1141. [CrossRef] [PubMed]
14. Lu, C.; Meichen, F.; Li, W. Comprehensive flood risk assessment based on RS and GIS in a county arca. *South-to-North Water Transf. Water Sci. Technol.* **2019**, *17*, 37–44.
15. Malczewski, J. Local Weighted Linear Combination. *Trans. GIS* **2011**, *15*, 439–455. [CrossRef]
16. Peng, W.; Hongwei, D. Study on Flood Hazard Risk Zoning Based on GIS and Logistic Regression Model. *Adv. Earth Sci.* **2020**, *35*, 1064–1072.
17. Wang, Q.; Zeng, J.; Xin, R. Risk Identification of Storm and Flood Disaster Based on GIS Multi-criteria Evaluation and BP Neural Network: A Case Study on the Min delta. *J. Catastr.* **2021**, *36*, 192–200. [CrossRef]
18. Xu, X.; Zhuang, D. Automated extraction of drainages in china based on dem in gis environment. *Resour. Environ. Yangtze Basin* **2004**, *13*, 343–348. [CrossRef]
19. Chen, X. The application of remote sensing and geographical information system (GIS) in flood damage analysis. *J. Hydraul. Eng.* **1997**, 71–74.
20. Xu, Y.; Zhang, B.; Zhou, B.; Dong, S.; Yu, L.; Li, R. Projected Risk of Flooding Disaster in China Based on CMIP5 Mode. *Adv. Clim. Change Res.* **2014**, *10*, 268–275.
21. Liu, H.; Qing, C.; Ni, Z. Risk Analysis of Flood Disaster in Guangxi Based on Combination of Combined Weight and ArcGIS. *J. Catastr.* **2015**, *30*, 76–79.
22. Yang, W.; Xu, K.; Lian, J.; Bin, L.; Ma, C.; Xu, K.; Lian, J.; Bin, L.; Ma, C. Multiple flood vulnerability assessment approach based on fuzzy comprehensive evaluation method and coordinated development degree model. *J. Environ. Manag.* **2018**, *213*, 440–450. [CrossRef] [PubMed]
23. Wang, Y. The Risk Regionalization of Torrential Flood Disaster in Loess Plateau Region in Gansu Based on GIS and Comprehensive Weight Method. *China Rura. Water Hydropower* **2018**, 118–122. [CrossRef]
24. Xu, Y.; Xu, X.; Ma, N. Assessment and regionalization of flood disaster risk Shaanxi Province at the county scale. *Arid Land Geogr.* **2018**, *41*, 306–313.
25. Xu, N.Z.; Bai, W.R. Effects of flood and drought on development of agriculture and social economy in China. *J. Catastr.* **2002**, *97*, 57–62.
26. Das, S.; Gupta, A.; Gupta, A. Multi-criteria decision based geospatial mapping of flood susceptibility and temporal hydro-geomorphic changes in the Subarnarekha basin, India. *Geosci. Front.* **2021**, *12*, 101206. [CrossRef]
27. Jiang, W.; Cheng, C.; Zhang, Y.; Zhao, H. Risk assessment of flood disasters in Zhejiang province based on GIS/AHP integration method. *Bull. Surv. Mapp.* **2019**, *2*, 125–130.
28. Xiao, Y.; Yi, S.; Tang, Z. Integrated flood hazard assessment based on spatial ordered weighted averaging method considering spatial heterogeneity of risk preference. *Sci. Total Environ.* **2017**, *599*, 1034–1046. [CrossRef]
29. Zhan, X. Risk evaluation for flood disasters in the plain area based on gis. *Resour. Environ. Yangtze Basin* **2003**, *12*, 388–392.
30. Chen, X.; Han, R. An Analysis of the Influencing Factors of Flood Evolution in Haihe River Basin Based on Hydrologic Regionalizations. *China Rural Water Hydropower* **2021**, 69–77. [CrossRef]
31. Arabameri, A.; Rezaei, K.; Cerdà, A.; Conoscenti, C.; Kalantari, Z.; Rezaei, K.; Cerdà, A.; Conoscenti, C.; Kalantari, Z. A comparison of statistical methods and multi-criteria decision making to map flood hazard susceptibility in Northern Iran. *Sci. Total Environ.* **2019**, *660*, 443–458. [CrossRef]
32. Lin, L.; Wu, Z.; Liang, Q.; Wu, Z.; Liang, Q. Urban flood susceptibility analysis using a GIS-based multi-criteria analysis framework. *Nat. Hazards* **2019**, *97*, 455–475. [CrossRef]
33. Wang, Y.; Li, Z.; Tang, Z.; Zeng, G.; Li, Z.; Tang, Z.; Zeng, G. A GIS-Based Spatial Multi-Criteria Approach for Flood Risk Assessment in the Dongting Lake Region, Hunan, Central China. *Water Resour. Manag.* **2011**, *25*, 3465–3484. [CrossRef]
34. Xiang, C.; Wang, Y.; Shi, G. Evaluation on Flood Control and Disaster Reduction in Hangjiahua Polder Area. *J. Zhejiang Water Conserv. Hydropower Coll.* **2014**, *26*, 29–32.
35. Wu, S.; Zhao, W.; Wang, Z.; Cang, S.; Qu, X.; Wang, Y.; Li, S. Assessment and Zoning of Flood Disasters Risk Based on GIS: A Case Study of Zhejiang Province. *China Rural Water Hydropower* **2020**, 51–57.
36. Wang, X.; Bao, Z. Evaluation Method and its Application of Flood Control and Disaster Reduction Ability in Polder Areas. *J. Irrig. Drain. Eng.* **2013**, *32*, 128–131.
37. Zhao, Y.; Tang, D.T.; Haung, X. Actual Urban Flood Control Capacity Evaluation Model Based on Energy Analysis. *Yellow River* **2017**, *39*, 29–31, +35.
38. Zhang, H.; Zhang, J.Q.; Han, J.S. GIS—Based assessment and zoning of flood/waterlogging disaster risk: A case study on middle and lower reaches of Liaohe River. *J. Nat. Disasters* **2005**, *14*, 141–146.
39. Zhou, Q.; Xu, X.; Su, J. Application of GIS Geometric Analysis for Identification of Flood-prone Areas and Construction of Surface Structure. *China Water Wastewater* **2018**, *34*, 114–117.
40. Liu, C.; Shang, S.; Zhao, Q.; Xie, H. Flood Resilience Assessment in Hunan Province Based on GIS and TOPSIS-PSR. *Water Resour. Power* **2018**, *36*, 70–73.

41. Haq, M.; Akhtar, M.; Muhammad, S.; Paras, S.; Rahmatullah, J.; Akhtar, M.; Muhammad, S.; Paras, S.; Rahmatullah, J. Techniques of Remote Sensing and GIS for flood monitoring and damage assessment: A case study of Sindh province, Pakistan. *Egypt. J. Remote Sens. Space Sci.* **2012**, *15*, 135–141. [CrossRef]
42. Xu, Y.; Du, J.; Zhang, L.; Ge, X.; Peng, B.; Du, J.; Zhang, L.; Ge, X.; Peng, B. Research on system of flood disaster control and reduction supported by gis in medium and small basins. *Chin. Geogr. Sci.* **2002**, *12*, 30–34. [CrossRef]
43. Zhao, G.; Pang, B.; Xu, Z.; Yue, J.; Tu, T.; Pang, B.; Xu, Z.; Yue, J.; Tu, T. Mapping flood susceptibility in mountainous areas on a national scale in China. *Sci. Total Environ.* **2018**, *615*, 1133–1142. [CrossRef] [PubMed]
44. Khosravi, K.; Pham, B.T.; Chapi, K.; Shirzadi, A.; Shahabi, H.; Revhaug, I.; Prakash, I.; Tien Bui, D.; Pham, B.T.; Chapi, K.; et al. A comparative assessment of decision trees algorithms for flash flood susceptibility modeling at Haraz watershed, northern Iran. *Sci. Total Environ.* **2018**, *627*, 744–755. [CrossRef] [PubMed]
45. Jiang, S.; Jiang, Z.; Li, W.; Shen, Y. Evaluation of the Extreme Temperature and Its Trend in China Simulated by CMIP5 Models. *Adv. Clim. Change Res.* **2017**, *13*, 11–24.
46. Taylor, K.E. Summarizing multiple aspects of model performance in a single diagram. *J. Geophys. Res. Atmos.* **2001**, *106*, 7183–7192. [CrossRef]
47. Li, X.F.; Xu, C.C.; Li, L.; Luo, Y.; Yang, Q.; Yang, Y. Evaluation of air temperature of the typical river basin in desert area of Northwest China by the CMIP5 models: A case of the Kaidu-Kongqi River Basin. *Res. Sci.* **2019**, *41*, 1141–1153.
48. Cao, L.; Li, J. Integrated risk assessment of flood disaster in Zhejiang Province based on RS and GIS. *J. Nat. Disaster* **2015**, *24*, 111–119.
49. Li, S. Risk Assessment and Zonation of Rainstorm and Flood Disasters in the Southeast Region of Inner Mongolia: A Case Study of Tongliao. *Meteorol. J. Inn. Mong.* **2019**, 23–28. [CrossRef]
50. Choubin, B.; Moradi, E.; Golshan, M.; Adamowski, J.; Sajedi-Hosseini, F.; Mosavi, A.; Moradi, E.; Golshan, M.; Adamowski, J.; Sajedi-Hosseini, F.; et al. An ensemble prediction of flood susceptibility using multivariate discriminant analysis, classification and regression trees, and support vector machines. *Sci. Total Environ.* **2019**, *651*, 2087–2096. [CrossRef]
51. Mahmoud, S.H.; Gan, T.Y.; Gan, T.Y. Multi-criteria approach to develop flood susceptibility maps in arid regions of Middle East. *J. Clean. Prod.* **2018**, *196*, 216–229. [CrossRef]
52. Yang, M.; Li, J.; Guang, Y.; Zuo, J. Model building and accuracy analysis of domestic SW-LiDAR airborne integration system. *Sci. Surv. Mapp.* **2014**, *39*, 9–12.
53. Baydaş, M.; Pamučar, D.; Pamučar, D. Determining Objective Characteristics of MCDM Methods under Uncertainty: An Exploration Study with Financial Data. *Mathematics* **2022**, *10*, 1115. [CrossRef]
54. Da Silva, L.B.L.; Alencar, M.H.; de Almeida, A.T.; Alencar, M.H.; Almeida, A.T.d. Exploring global sensitivity analysis on a risk-based MCDM/A model to support urban adaptation policies against floods. *Int. J. Disaster Risk Reduct.* **2022**, *73*, 102898. [CrossRef]
55. Serrai, W.; Abdelli, A.; Mokdad, L.; Hammal, Y.; Abdelli, A.; Mokdad, L.; Hammal, Y. Towards an efficient and a more accurate web service selection using MCDM methods. *J. Comput. Sci.* **2017**, *22*, 253–267. [CrossRef]
56. Mousavi-Nasab, S.H.; Sotoudeh-Anvari, A.; Sotoudeh-Anvari, A. A comprehensive MCDM-based approach using TOPSIS, COPRAS and DEA as an auxiliary tool for material selection problems. *Mater. Des.* **2017**, *121*, 237–253. [CrossRef]
57. Hsieh, T.-Y.; Lu, S.-T.; Tzeng, G.-H.; Lu, S.-T.; Tzeng, G.-H. Fuzzy MCDM approach for planning and design tenders selection in public office buildings. *Int. J. Proj. Manag.* **2004**, *22*, 573–584. [CrossRef]
58. Lee, H.-C.; Chang, C.-T.; Chang, C.-T. Comparative analysis of MCDM methods for ranking renewable energy sources in Taiwan. *Renew. Sust. Energy Rev.* **2018**, *92*, 883–896. [CrossRef]
59. Molla, M.U.; Giri, B.C.; Biswas, P.; Giri, B.C.; Biswas, P. Extended PROMETHEE method with Pythagorean fuzzy sets for medical diagnosis problems. *Soft Comput.* **2021**, *25*, 4503–4512. [CrossRef]
60. Tahmasebi Birgani, Y.; Yazdandoost, F.; Yazdandoost, F. An Integrated Framework to Evaluate Resilient-Sustainable Urban Drainage Management Plans Using a Combined-adaptive MCDM Technique. *Water Resour. Manag.* **2018**, *32*, 2817–2835. [CrossRef]
61. Razavi Termeh, S.V.; Kornejady, A.; Pourghasemi, H.R.; Keesstra, S.; Kornejady, A.; Pourghasemi, H.R.; Keesstra, S. Flood susceptibility mapping using novel ensembles of adaptive neuro fuzzy inference system and metaheuristic algorithms. *Sci. Total Environ.* **2018**, *615*, 438–451. [CrossRef]
62. Liu, R. ROC curve analysis of multivariate observations using SPSS. *Chin. J. Public Health* **2003**, *19*, 1151–1152. [CrossRef]
63. Haghizadeh, A.; Siahkamari, S.; Haghiabi, A.H.; Rahmati, O. Forecasting flood-prone areas using Shannon's entropy model. *J. Earth Syst. Sci.* **2017**, *126*, 39. [CrossRef]

remote sensing

MDPI

Communication

Comparing Global Sentinel-2 Land Cover Maps for Regional Species Distribution Modeling

Zander S. Venter [1,*], Ruben E. Roos [1], Megan S. Nowell [1], Graciela M. Rusch [1], Gunnar M. Kvifte [2] and Markus A. K. Sydenham [1]

[1] Norwegian Institute for Nature Research, Sognsveien 68, N-0855 Oslo, Norway
[2] Faculty of Biosciences and Aquaculture, Nord University, N-7729 Steinkjer, Norway
* Correspondence: zander.venter@nina.no

Abstract: Mapping the spatial and temporal dynamics of species distributions is necessary for biodiversity conservation land-use planning decisions. Recent advances in remote sensing and machine learning have allowed for high-resolution species distribution modeling that can inform landscape-level decision-making. Here we compare the performance of three popular Sentinel-2 (10-m) land cover maps, including dynamic world (DW), European land cover (ELC10), and world cover (WC), in predicting wild bee species richness over southern Norway. The proportion of grassland habitat within 250 m (derived from the land cover maps), along with temperature and distance to sandy soils, were used as predictors in both Bayesian regularized neural network and random forest models. Models using grassland habitat from DW performed best ($RMSE$ = 2.8 ± 0.03; average ± standard deviation across models), followed by ELC10 ($RMSE$ = 2.85 ± 0.03) and WC ($RMSE$ = 2.87 ± 0.02). All satellite-derived maps outperformed a manually mapped Norwegian land cover dataset called AR5 ($RMSE$ = 3.02 ± 0.02). When validating the model predictions of bee species richness against citizen science data on solitary bee occurrences using generalized linear models, we found that ELC10 performed best (AIC = 2278 ± 4), followed by WC (AIC = 2367 ± 3), and DW (AIC = 2376 ± 3). While the differences in $RMSE$ we observed between models were small, they may be significant when such models are used to prioritize grassland patches within a landscape for conservation subsidies or management policies. Partial dependencies in our models showed that increasing the proportion of grassland habitat is positively associated with wild bee species richness, thereby justifying bee conservation schemes that aim to enhance semi-natural grassland habitat. Our results confirm the utility of satellite-derived land cover maps in supporting high-resolution species distribution modeling and suggest there is scope to monitor changes in species distributions over time given the dense time series provided by products such as DW.

Keywords: pollinators; grassland; wild bees; management; conservation; spatial modeling

Citation: Venter, Z.S.; Roos, R.E.; Nowell, M.S.; Rusch, G.M.; Kvifte, G.M.; Sydenham, M.A.K. Comparing Global Sentinel-2 Land Cover Maps for Regional Species Distribution Modeling. *Remote Sens.* **2023**, *15*, 1749. https://doi.org/10.3390/rs15071749

Academic Editors: Matteo Convertino and Jie Li

Received: 31 January 2023
Revised: 16 March 2023
Accepted: 22 March 2023
Published: 24 March 2023

1. Introduction

The Anthropocene has heralded an unprecedented loss of biodiversity, primarily due to changes in land use and land cover, climate change, pollution, (over-)exploitation, and biological invasions [1]. In response, governments have established frameworks that address biodiversity loss, including the United Nations' (UN) Sustainable Development Goals (SDGs, 2030 Agenda) as well as the Aichi biodiversity targets (Strategic Plan for Biodiversity 2011–2020) and the Post-2020 Global Biodiversity Framework of the Convention on Biological Diversity. Recently, the UN established the statistical standards for ecosystem accounting (EA) under the System of Environmental–Economic Accounting (SEEA), which require countries to account for changes in ecosystems over time [2]. To achieve the SDGs, meet biodiversity conservation targets, and to account for ecosystem changes, we require monitoring and evaluation tools that are both globally available and locally relevant [3].

One such tool is species distribution modeling, also known as ecological niche or habitat modeling, which has been widely used to inform decision-making in conservation planning [4].

By modeling and mapping the distribution of species over space and time, we are able to make data-driven decisions about which areas to prioritize for restoration or conservation [5]. The predictive and explanatory power of species distribution models also allows for identifying critical environmental variables (e.g., precipitation, habitat availability) that drive species communities [6]. Insect conservation strategies are a good case in point, whereby mapping and monitoring of insect species richness are prioritized both at the international [7,8] and national levels [9,10]. In Norway, field surveys of bee diversity have been combined with habitat and climate models to create prediction maps that can help determine where wild bee habitat enhancement schemes can be most efficient [11]. However, such priority maps and many species distribution models are currently based on static environmental predictor variables such as land cover maps, whereas management requires more flexible solutions that can detect temporal dynamics in ecosystem conditions and species distributions over time [12].

Recent advances in satellite remote sensing and earth observation have filled data gaps and improved the spatio-temporal transferability of species distribution models [13]. As such, satellite-derived products can capture the environmental processes that underlie the distribution of biodiversity, such as vegetation productivity, water availability, temperature, and perhaps most importantly, land use and land cover. The Sentinel satellites under the Copernicus Programme have been used to produce annually updatable regional and global land cover maps at 10 m resolution, including ELC10 [14], Dynamic World (DW) [15], and World Cover 2020 (WC) [16]. All three products have the capacity to be multi-temporal, but only DW is operationally delivering near real-time land cover maps as new Sentinel-2 scenes become available (every 2–5 days). Due to the novelty of freely available, medium-resolution, high-frequency land cover maps, their use in species distribution modeling is still in its infancy. It is also not clear whether global Sentinel-based land cover maps can replace or improve upon the contribution of regional land cover datasets [17], given that regional maps produced by national mapping agencies through manual methods such as photogrammetry are often more precise and tailored to local conditions.

The aim of this study was to compare Sentinel-based, 10-meter-resolution land cover maps in their ability to predict wild bee species richness distributions across gradients in temperature and habitat availability. To do so, we use a Norwegian land cover dataset called AR5 [18] as a benchmark to evaluate three satellite-based maps, including DW, WC, and ELC10, that are annually updatable. The proportion of semi-natural grassland habitat within 250 m derived from the land cover maps is used as a predictor variable in species distribution models that predict the richness of solitary bee species in southern Norway.

2. Methods

To compare the utility of Sentinel-based land cover maps in species distribution modeling, we chose wild bee species as a model system. We do this for three main reasons, which are elaborated on below: (1) bees are keystone species in grassy ecosystems globally that are good indicators for ecosystem condition; (2) they are experiencing significant local declines in population numbers; and (3) they have a limited home range and are dependent on local resources for survival, so we expect their distribution to respond strongly to landscape-scale land use gradients in the Sentinel-based land cover maps.

Insects in general and wild bees in particular are key components of many ecosystems and provide important contributions to people [19,20]. For example, the economic value of pollinating insects is considerable, estimated at EUR 153 billion globally [21]. Although pollination is often associated with domestic insects such as the honey bee (*Apis mellifera*), a large and diverse community of wild bee species contributes significantly to the pollination of wild plants and crops [22,23]. Insect abundance, biomass [24], and diversity [25] are declining in some regions due to urbanization, deforestation, climate change, pesticide use, and invasive species [26]. The same is true for bee species [27]. Although the diversity of wild bees is predominantly driven by temperature at global and regional scales, habitat and landscape characteristics are important drivers at local scales [28,29]. Bees are central place foragers that travel back and forth from a nesting site to collect resources [30], and therefore, the abundance and diversity of wild bees correlate to the diversity and abundance of floral resources (vascular plants) in the landscape [28,31]. Land cover types such as semi-natural grasslands, including hay and flower meadows, provide ample floral [32] and non-floral resources to wild bees [33]. Non-floral resources include nesting sites, and although some wild bee species nest in dead wood, the majority use sandy soil sediments as nest sites [34]. Consequently, wild bee diversity can be predicted by the availability of suitable land cover types for resources (i.e., semi-natural grasslands) and for nests (i.e., sandy soils), in combination with climatic variables [29].

2.1. Solitary Bee Surveys

We surveyed solitary bee communities in 72 traditionally managed (mowed) semi-natural grasslands distributed along a climatic gradient from south-eastern Norway to mid-Norway (Figure 1A). In 2019, we sampled 32 semi-natural grasslands in south-eastern Norway [11], adding another 20 study sites in 2020 [29]. To capture potential influences of climatic conditions on solitary bee diversity at regional scales, solitary bee communities were sampled in another set of 20 semi-natural grasslands in mid-Norway in 2021. All surveys were conducted using pan traps, which are an efficient method for surveying wild bee communities [35], in particular when the aim is to survey the solitary as opposed to social bee species [36]. Each pan trap consisted of three white plastic soup bowls, coated with fluorescent yellow or blue, or left white, mounted on a fence pole at the height of the surrounding vegetation [11,29]. We deployed 3 traps per site in 2019 and 2 traps per site for sites sampled in 2020 and 2021. The number of traps per site was reduced from 3 to 2 after 2019 to allow more sites to be sampled. This resulted in 176 samples (the sum of traps across all sites) distributed across 72 study sites. Within sites, traps were always placed at least 20 m apart to avoid inter-trap competition [37].

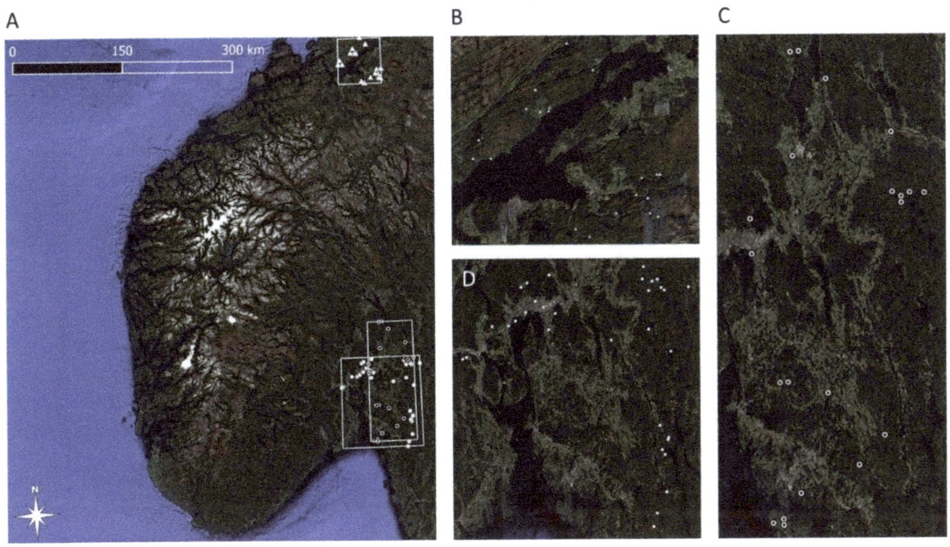

Figure 1. Extent of study area with solitary bee sampling locations and individual survey extents (**A**). Location of study sites in Trøndelag County, sampled in 2021 (**B**). Location of study sites in Oslo, Viken, and Innlandet counties, sampled in 2020 (**C**). Locations of study sites in Oslo, Viken, and Innlandet, sampled in 2019 (**D**).

In all years, sites were sampled in May, June, and July at similar times of day and on days with weather conditions that are optimal for bee activity (i.e., low wind and temperatures above 15 °C). For each sampling period, we activated the pan traps at a site by mounting the bowls and filling them with water and a drop of detergent. Sampled bee specimens were collected after 48 h. In 2019, sampling was initiated on 13 May, 21 June, 9 July, and 23 July [11]. In 2020, sampling was initiated on 13 May, 25 May, 14 June, and 16 July [29]. In 2021, sampling began on 29 May, 22 June, and 24 July. Collected bees were stored in 96% laboratory ethanol prior to pinning and identification. Voucher specimens are stored in the entomological collections at the Norwegian Institute for Nature Research. We tallied the number of solitary bee species sampled per trap across the season.

2.2. Land Cover Maps

Data pre-processing and extraction of land cover maps took place in the Google Earth Engine [38]. The AR5 map obtained from the Norwegian Institute of Bioeconomy Research [18] is provided as a vector map at 1:5000 scale, which we rasterized at 5 m resolution. AR5 is updated every 5–8 years and therefore represents a mosaic of years across the country. WC and ELC10 are available in Google Earth Engine for the years 2020 and 2018, respectively. However, DW is provided as a collection of classified Sentinel-2 images with less than 35% cloud cover. To generate an annual land cover composite for 2020 comparable with WC and ELC10, we calculated the mode predicted land cover class in the image band named "label" across all DW images during June, July, and August. For all land cover maps, we used a focal mean function to calculate the proportion of grassland habitat within 250 m of each pixel in the study area (Figure 1) at 10 m resolution. We used the 250-m radius because solitary bee diversity has been shown to respond strongly to habitat availability at this spatial scale [39]. To isolate grassland pixels, we used the "grassland" class from WC and ELC10 and the "grass" class from DW, which are both defined as areas dominated by natural herbaceous vegetation, including grasslands, prairies, steppes, savannahs, and pastures. For AR5, we used the "innmarksbeite" and "åpent fastmark" classes, which are defined as open ecosystems dominated by herbaceous vegetation and often used for extensive grazing [18]. We used a radius of 250 m as solitary bee species richness has previously been shown to respond to habitat area at this spatial scale [39].

2.3. Modeling

Data modeling and visualization were performed in R [40]. As in [29], we included predictor variables related to climatic conditions, habitat availability, and distances to high-quality nesting substrates for below ground nesting bees, which account for the vast majority of solitary bees. In [29], the spatial variation in climatic conditions was estimated using a digital elevation model (DEM) together with latitude. Here, we used the average temperature for the warmest quarter (June, July, and August), during the current 30-year climate reference period (1990–2021), calculated from daily estimates and interpolated station measurements at a 1 km resolution from the Norwegian Meteorological Institute's database [41]. We used the average temperature during the warmest quarter instead of the annual mean temperature because high winter temperatures along the coast result in annual mean temperatures not reflecting the gradient in ambient temperature experienced by bees during their main activity periods in Norway (spring to autumn). Using modeled climate data instead of DEMs to estimate the effects of temperature on bee diversity enables one to project changes in bee diversity as a function of future climate scenarios. In [29] we used a habitat suitability model to estimate the potential habitat area surrounding each pan trap at a 60-m radius. In contrast, in the current study, we use the proportion of pixels identified as "grassland" by satellite-derived land cover maps as estimates of habitat availability within a 250-m radius surrounding each trap. A benefit of using satellite-derived grassland classifications instead of estimates of habitat suitability maps that may partly rely on existing land cover products is that one reduces the number of modeling steps required to produce maps of pollinator diversity from remote sensing data. In addition, we used the geographic distance to areas with soils mapped as having a high water infiltration capacity by the Norwegian geological survey [42], as such areas are typically located on sandy soils, which is the preferred nesting substrate for the majority of Norwegian soil nesting bees.

We ran and compared eight models of solitary bee species richness that all followed the general formula:

Solitary bee species richness (survey year + average temperature during the warmest quarter + Distance to sandy soils + proportion of grassland within a 250-m radius).

The eight models differed in terms of the data source (map) used to estimate the proportion of grassland around each pan trap and the type of model used to predict solitary bee species richness. The survey year was included as a categorical variable to account for inter-annual variation in bee species richness as well as for annual differences in climatic conditions, which could influence the number of species sampled in a trap. For each land cover map, we trained a Random Forest (RF) regression model [43] and a Bayesian Regularized Neural Network (BRNN) model [44] in Caret [45] in R. BRNN models were tuned by selecting the number of neurons (1, 2, or 3) that resulted in the lowest root mean square error (*RMSE*) following 25 bootstrap resamples of the training data. RF models were tuned by selecting the mtry (number of parameters tested at each node) and split-rule variance or extra trees that resulted in the lowest *RMSE*, following 25 bootstrap resamples. We adopted leave-one-out cross-validation (LOOCV) to assess model predictive performance due to the small number of study sites. Small training datasets can result in large variances in model performance when using traditional training-testing splits (e.g., 70% training and 30% testing) before model fitting, compared to LOOCV [46]. In LOOCV, each model was iteratively trained and tuned on data from 71 study sites and then used to predict solitary bee species richness in pan traps from the one held-out site. Predictive performance is calculated across all LOOCV iterations. Following this nested cross-validation procedure ensures independence between the data used for tuning the models and the data used for final validations.

For each land cover-map model prediction, we evaluated the predictive power in terms of the Pearson correlation coefficient (Cor), the root-mean-square deviation (*RMSE*), and the mean absolute error (MAE) between observed values of solitary bee species richness and the solitary bee species richness predicted by the model. As an alternative form of model validation, we also tested the ability of model predictions of bee species richness to predict the variance in occurrence of solidary bee records obtained from citizen science data. We first downloaded all post-2015 solitary bee species observations from the Global Information Biodiversity Facility (GBIF) that intersected our study area (xmin: 9.836, xmax: 12.719, ymin: 58.909, ymax: 63.962) and had a GPS error of less than 50 m (n = 2111). To reduce spatial bias in the data—i.e., that some areas have been surveyed more frequently or intensively than others—we only included one occurrence per square kilometer. We randomly sampled 10,000 pseudo-absences from within our study area and used binomial generalized linear models to quantify how well the predicted species richness scores explained the variance in the GBIF presence-absence data. We used the Akaike information criterion (*AIC*) to compare the explanatory capacity of the different models ($\delta AIC > 2$). Finally, we used partial dependency plots through the R-package pdp [47] as a means to visualize the estimated marginal effect of the proportion of grassland, using the different land cover maps, on solitary bee species richness.

In summary, our modeling workflow included 16 unique models for combinations of reference data (survey, GBIF), model type (RF, BRNN), and land cover dataset (AR5, DW, ELC10, WC). Each model was iterated 100 times in order to estimate the mean and standard deviation in model performance metrics.

3. Results

We did not find that models using the proportion of grassland estimated from the vector-based Norwegian land cover map (AR5) outperformed models where grassland had been estimated from satellite-derived land use models with 10 m resolution. Specifically, satellite-derived maps exhibited an average *RMSE* of 2.87 ± 0.03 (±standard deviation), whereas the Norwegian AR5 map produced models with a *RMSE* of 3.02 ± 0.02 (Figure 2). Similarly, for the AR5-based BRNN models, the correlation coefficient between observed and predicted solitary bee species richness (SR) was slightly lower (Figure 2A) than for the satellite-based models (Figure 2B–D). This was also the case for the Random Forest (RF) models (Figure 2E–H).

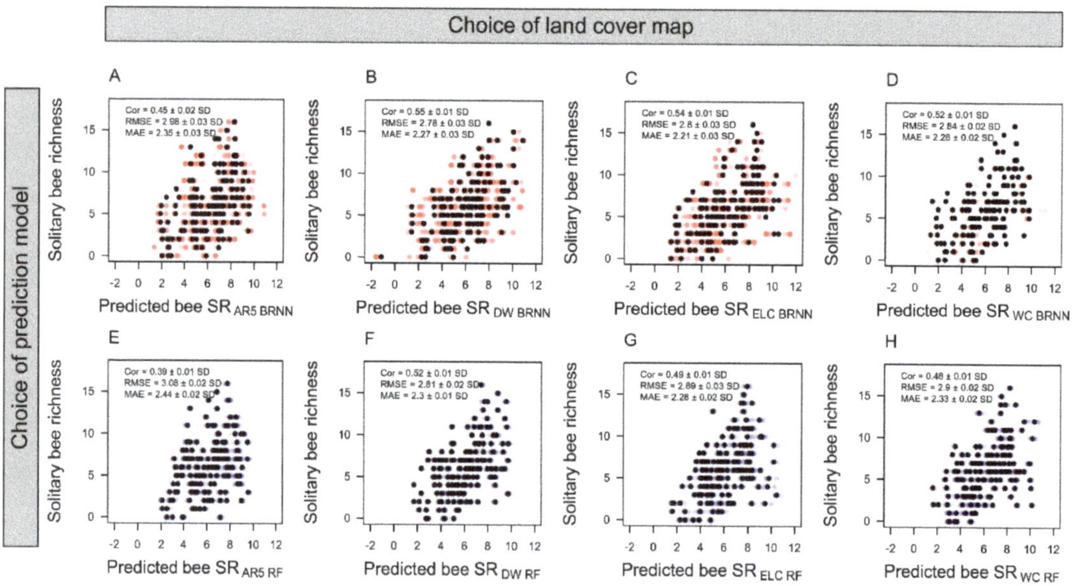

Figure 2. Performance of Bayesian regularized neural network (BRNN) and random forest (RF) models for predicting solitary bee species richness (SR). Data points represent model predictions against observed SR following a leave-one-out cross-validation procedure. The average and standard deviation of correlation coefficients (Cor), root-mean and mean-absolute error (*RMSE*, MAE) from 100 iterations of each leave-one-out cross validation are reported for models trained with land cover data from (**A**,**E**) a Norwegian map (AR5), (**B**,**F**) dynamic world (DW), (**C**,**G**) European land cover (ELC10), and (**D**,**H**) world cover (WC). Red and blue points show the predicted and observed solitary bee species richness for each pan trap across the 100 leave-one-out iterations, and black points show the average predicted bee species richness and the associated observed species richness for each specific trap.

Among the satellite-derived models (Figure 2B–D,F–H), there were only marginal differences in their performances as predictors of solitary bee SR. Models using grassland habitat from DW performed best (*RMSE* = 2.8 ± 0.03; averaged across RF and BRNN models), followed by ELC10 (*RMSE* = 2.85 ± 0.03) and WC (*RMSE* = 2.87 ± 0.02). Grassland habitat was ranked the third most important predictor variable in BRNN and RF models (Figure 3). The mean summer temperature and sampling year were the most important predictors in the models. All models, independent of data source or model type, produced partial dependence plots that corroborated a positive association between grassland habitat and solitary bee SR (Figure 4). The association was less distinct in the AR5 model (Figure 4A) compared to the satellite-based models (Figure 4B–D).

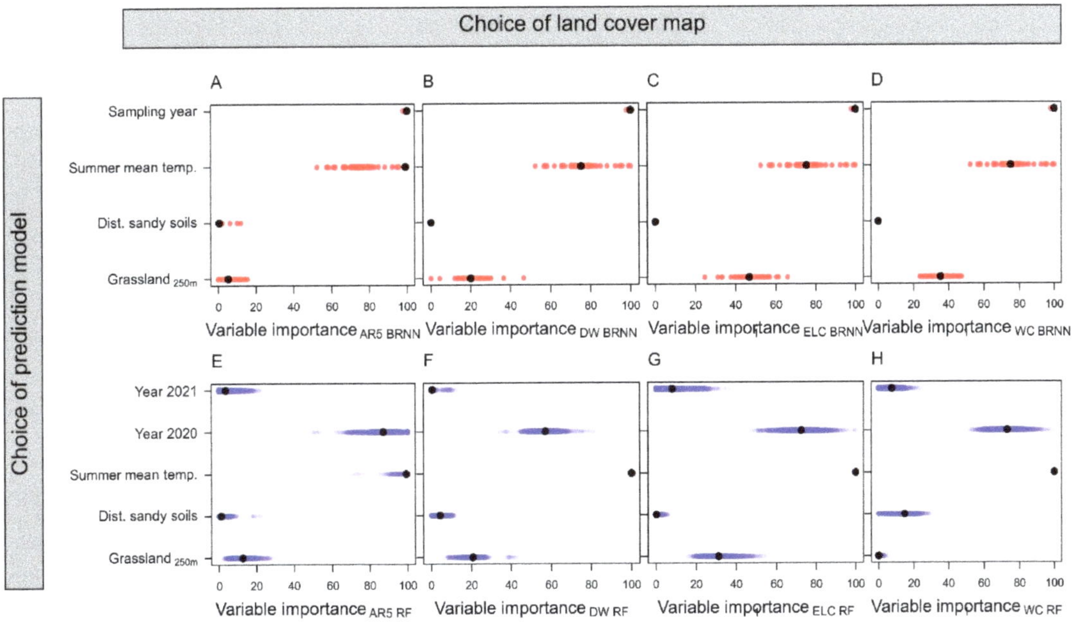

Figure 3. Variable importance plots showing the scaled relative importance of sampling year (3 levels), summer mean temperature, distance to sandy soils, and grassland proportion within 250 m for predicting solitary bee species richness. For the sampling year, the year 2019 was used as the reference year and does therefore not appear in the figures. Variable importance is derived from Bayesian regularized neural network (BRNN; **A–D**) and random forest (RF; **E–H**) models for predicting solitary bee species richness (SR). Models were trained with land cover data from (**A**,**E**) a Norwegian map (AR5), (**B**,**F**) dynamic world (DW), (**C**,**G**) European land cover (ELC10), and (**D**,**H**) world cover (WC).

When validating the model predictions of bee SR against citizen science data on solitary bee occurrences using generalized linear models (Figure 5), we found that the order of performance was changed. We found that ELC10 performed best ($AIC = 2278 \pm 4$), followed by WC ($AIC = 2367 \pm 3$), and DW ($AIC = 2376 \pm 3$). The BRNN models performed slightly better than RF models, both in terms of leave-one-out cross-validation (Figure 2) and in terms of explaining the occurrence of solitary bees from GBIF (Figure 5). Therefore, we used the BRNN models to generate wall-to-wall prediction maps of bee SR over the study region (Figure 6). A qualitative visual comparison shows that the broad-scale spatial patterns of bee SR predictions are similar between models, with bee SR increasing with north-south temperature gradients and along populated valleys where extensive grazing practices have established semi-natural grassland patches. At the landscape scale (Figure 7), AR5 predictions show less spatial variation than the satellite-derived maps. All models appear to pick up the grassland habitat adjacent to the runways at Gardermoen International Airport; however, ELC10 appears to pick up the most habitat in the agricultural landscapes southwest of the airport (Figure 7C).

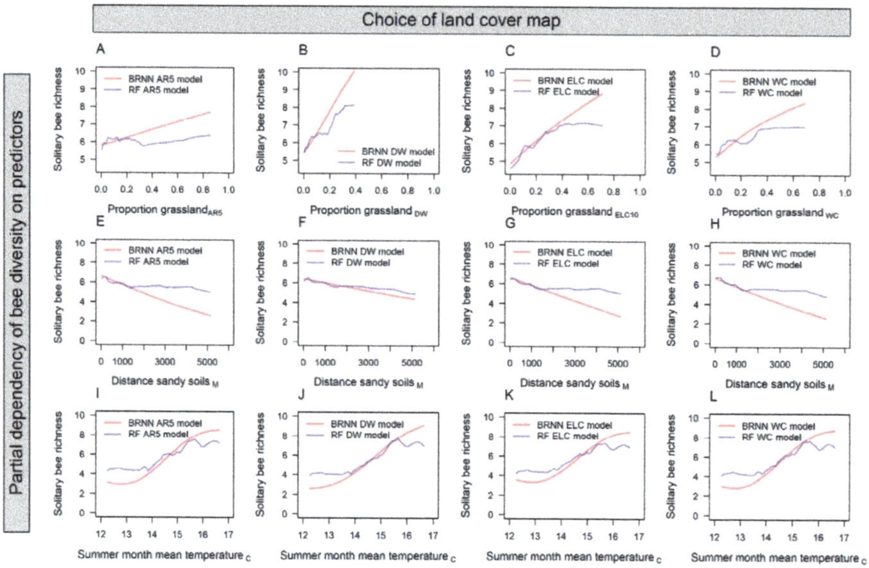

Figure 4. Partial dependence plots from Bayesian regularized neural network (BRNN) and random forest (RF) models of bee species richness. Partial dependencies are shown for the spatial predictors: proportion of grassland habitat within 250 m, derived from (**A**) a Norwegian map (AR5), (**B**) dynamic world (DW), (**C**) European land cover (ELC), and (**D**) world cover (WC); (**E–H**) distance to sandy soils; and (**I–L**) mean temperature during the summer months.

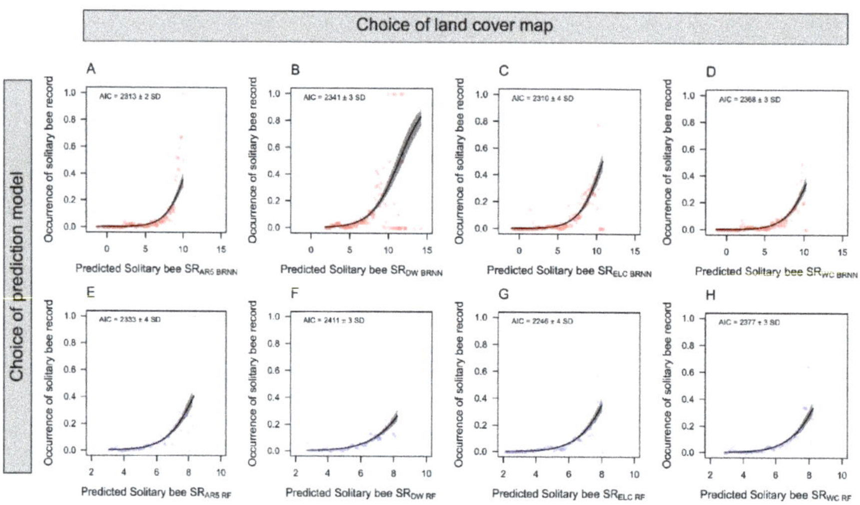

Figure 5. Explanatory power of bee species richness (SR) predictions from Bayesian regularized neural network (BRNN) and random forest (RF) models. Effects plots are shown along with *AIC* scores from binomial generalized linear models of solitary bee species occurrence data from GBIF as a function of predicted bee SR from models trained with land cover data from (**A**,**E**) a Norwegian map (AR5), (**B**,**F**) dynamic world (DW), (**C**,**G**) European land cover (ELC10), and (**D**,**H**) world cover (WC). Points in the figures show the mean occurrence of solitary bee records calculated within bins of one decimal. Individual GLMs were run for each of the 100 spatially filtered datasets of solitary bee records in order to calculate the average *AIC* and its standard deviation.

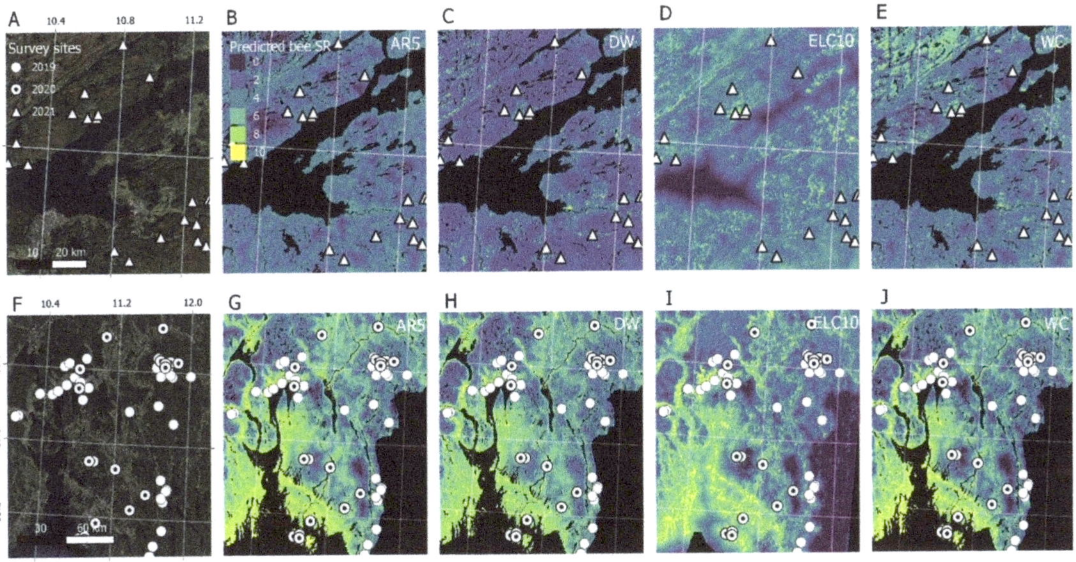

Figure 6. Solitary bee species richness (SR) prediction maps for Trøndelag county (**A–E**) and Oslo and Innlandet county (**F–J**) from Bayesian regularized neural network (BRNN) models trained on grassland habitat data from (**B**,**G**) a Norwegian map (AR5), (**C**,**H**) dynamic world (DW), (**D**,**I**) European land cover (ELC10), and (**E**,**J**) world cover (WC). Each map is overlaid with the bee survey sites.

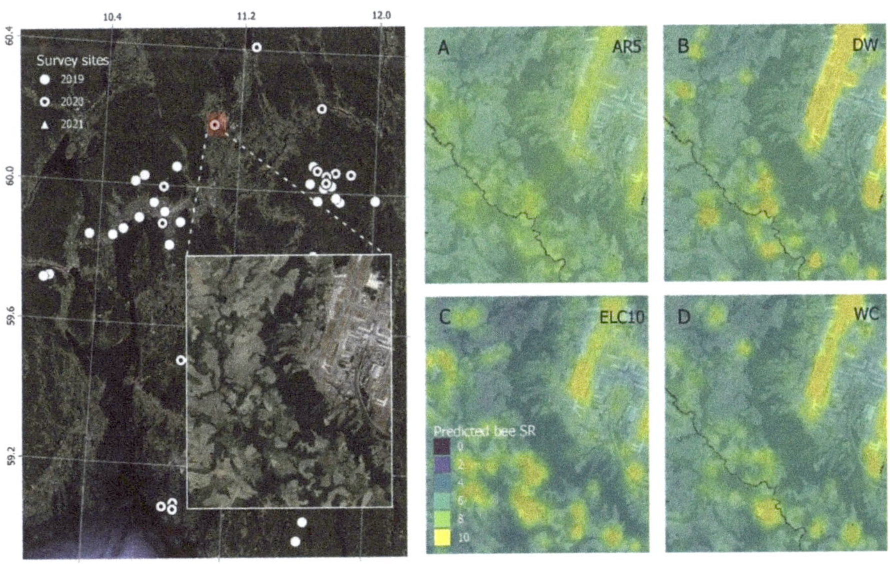

Figure 7. Solitary bee species richness (SR) prediction maps over the landscape surrounding Gardermoen international airport (left panel) from Bayesian regularized neural network (BRNN) models trained on grassland habitat data from (**A**) a Norwegian map (AR5), (**B**) dynamic world (DW), (**C**) European land cover (ELC10), and (**D**) world cover (WC).

4. Discussion

Species distribution modeling that incorporates high-resolution satellite data is still in its infancy [13], yet the application to solitary bee species richness presented in this study confirms its potential. Our results show that the Sentinel satellite-based land cover maps outperformed a regional manually digitized land cover map over southern Norway (AR5) in predicting solitary bee species richness (Figure 2). While the use of satellite imagery to map vegetation types or individual forest species is common [48], using derived products to predict habitat suitability for animal or plant species that are not directly visible in satellite images is far less common [11,49]. Here we show that the availability of grassland habitat within 250 m, measured from satellite data, is positively associated with solitary bee species richness (Figure 4) and is therefore predictive of solitary bee occurrences at regional scales (Figure 5).

We found larger differences in the predictive capacity of grassland habitat derived from manually mapped versus satellite-derived land cover maps than differences between the satellite-based maps themselves (Figure 2). The AR5 map uses a minimum mapping unit of 2000 m^2, which results in very small grassland fragments being subsumed into a broader land cover class [18]. For example, road verges or small urban parks will be classified as "built" in the AR5 maps. However, such small grassland patches can harbor significant floral resources for bees, and the fact that AR5 does not map these areas may therefore be why AR5 was less predictive of species richness than satellite-based maps. Furthermore, AR5 is not as up-to-date as the satellite-based maps and may misrepresent the conditions on the ground during 2019, 2020, and 2021, when the field surveys were conducted. In contrast, ELC10 and WC use a minimum mapping unit of 100 m^2, while DW uses 2500 m^2. At the landscape scale, it is evident that predictions of bee species richness with ELC10 were more spatially heterogeneous than AR5 or DW (Figure 6), probably due to its smaller minimum mapping unit and ability to detect smaller grassland patches. This may have contributed to its greater predictive capacity compared to AR5. This also explains why, when validating our models using citizen science data from GBIF, ELC10 outperformed DW. In contrast to the survey dataset, GBIF data are spatially clustered and biased toward urban landscapes, which are easily accessible but also have complex landscapes. Due to ELC10's small minimum mapping unit, it captures the landscape complexity more than DW does and is able to predict the GBIF bee SR better.

The accuracies of the solitary bee species richness models presented here are arguably high enough to make data-driven management decisions at the landscape scale. The average *RMSE* of 2.87 means that one can at least distinguish very species-rich areas (maximum species richness of 16 in our study) from species-poor areas (zero species). The differences in model accuracy between satellite-based grassland maps were marginal (*RMSE* difference of 0.04); however, when visualized at a landscape scale (Figure 6), small nuances may have important implications for management and decision-making. For instance, a ranking or prioritization of grassland patches for receiving conservation subsidies based on the maps in Figure 6 may yield different results depending on the data source used. Therefore, post-stratified accuracy assessment of species distribution models in specific landscapes may be necessary before they can be adopted in practice [50].

Based on several limitations identified in our study, we outline avenues for further research on the integration of high-resolution satellite data in species distribution modeling. Firstly, it is not necessary to use derived products such as land cover maps, as we have carried out here. Instead, one can use the spectral signatures themselves from a satellite image to calibrate distribution models [13], although this would not allow ecologists or policymakers to relate species-rich areas to specific land use types. Secondly, maps with a higher thematic resolution than those used in this study would produce more detailed species distribution maps [51]. For example, all four maps tested here contained a single broad category for grassland, without distinguishing between intensively managed grasslands and extensively managed grasslands, such as the mowed meadows from which we sampled solitary bees [11,29]. Therefore, measuring aspects of ecosystem condition or use, such as grassland use intensity [52], might further improve the accuracy of species distribution models. Thirdly, we did not explore how accurately satellite-based prediction models can detect real changes in species distributions over time because we did not implement a field survey design that was comprehensive enough to capture changes in species ranges over time. However, we know from earlier studies that changes in land cover that are detectable from space are direct drivers of species range shifts [53]. To this end, DW is probably the most suitable Sentinel-based land cover dataset to quantify land cover and use dynamics [54] because of its continuous updates and delivery and would be well-suited to such dynamic species distribution modeling. This also strengthens the call for investment in long-term biodiversity monitoring programs so that satellite-based distribution models can be calibrated and validated with in situ data [55].

5. Conclusions

The proliferation of high-resolution earth observation data and derived land cover products provides scope for mapping biodiversity distributions with models that are both locally relevant for decision making and scalable to the globe. Here we found that globally available Sentinel-based land cover maps can improve upon manually digitized regional land cover maps for predicting the richness of solitary bee species in southern Norway. The differences in predictive performance between DW, WC, and ELC10 were marginal; however, at the landscape scale, the smaller minimum mapping units of WC and ELC10 allow them to resolve smaller habitat patches, which are reflected in the landscape variations in predicted species richness. Furthermore, the rich time series provided by maps such as DW (from 2015 to present) offer unique opportunities to model short-term changes in species distributions in response to land use changes if paired with in-situ temporal monitoring data. We conclude that the use of satellite-derived land cover maps can facilitate high-resolution species distribution models that can guide decision-making relevant to landscape ecology. To this end, future modeling efforts should be aimed at those species that perform key roles in ecosystems, are indicators of ecosystem status, and support nature's contribution to people.

Author Contributions: Conceptualization, M.A.K.S., Z.S.V. and R.E.R.; methodology, M.A.K.S., Z.S.V. and R.E.R.; formal analysis, M.A.K.S., Z.S.V. and R.E.R.; data curation, M.A.K.S., G.M.K. and Z.S.V.; writing—original draft preparation, Z.S.V., R.E.R. and M.A.K.S.; writing—review and editing, Z.S.V., R.E.R., M.A.K.S., G.M.K., G.M.R. and M.S.N.; project administration, M.A.K.S.; funding acquisition, M.A.K.S. All authors have read and agreed to the published version of the manuscript.

Funding: Insect sampling from semi-natural grasslands in 2019 and 2020 was funded by the Norwegian Agricultural Agency (Klima-og Miljøprogrammet: POLLILAND, grant number 2018/72806). Sampling in 2021 was financed by The Research Council of Norway, project no. 160022/F40 NINA basic funding. This study was financed by the Norwegian Agricultural Agency (Klima- og Miljøprogrammet: POLLILAND-MIDT, grant number 2021/40219).

Data Availability Statement: The data and code to reproduce this analysis are archived on Zenodo: 10.5281/zenodo.7762665.

Acknowledgments: We thank Mikaela E.G.P. Olsen, Solveig Haug, Jonas Lystrup Andresen, April McKay, Stian Brønner, and Ida Elise Løvall Rastad for operating the traps installed in the semi-natural grasslands.

Conflicts of Interest: The authors declare no conflict of interest.

References

1. Johnson, C.N.; Balmford, A.; Brook, B.W.; Buettel, J.C.; Galetti, M.; Guangchun, L.; Wilmshurst, J.M. Biodiversity Losses and Conservation Responses in the Anthropocene. *Science* **2017**, *356*, 270–275. [CrossRef] [PubMed]
2. Edens, B.; Maes, J.; Hein, L.; Obst, C.; Siikamaki, J.; Schenau, S.; Javorsek, M.; Chow, J.; Chan, J.Y.; Steurer, A.; et al. Establishing the SEEA Ecosystem Accounting as a Global Standard. *Ecosyst. Serv.* **2022**, *54*, 101413. [CrossRef]
3. Schmeller, D.S.; Böhm, M.; Arvanitidis, C.; Barber-Meyer, S.; Brummitt, N.; Chandler, M.; Chatzinikolaou, E.; Costello, M.J.; Ding, H.; García-Moreno, J.; et al. Building Capacity in Biodiversity Monitoring at the Global Scale. *Biodivers. Conserv.* **2017**, *26*, 2765–2790. [CrossRef]

4. Villero, D.; Pla, M.; Camps, D.; Ruiz-Olmo, J.; Brotons, L. Integrating Species Distribution Modelling into Decision-Making to Inform Conservation Actions. *Biodivers. Conserv.* **2017**, *26*, 251–271. [CrossRef]
5. Guisan, A.; Tingley, R.; Baumgartner, J.B.; Naujokaitis-Lewis, I.; Sutcliffe, P.R.; Tulloch, A.I.T.; Regan, T.J.; Brotons, L.; McDonald-Madden, E.; Mantyka-Pringle, C.; et al. Predicting Species Distributions for Conservation Decisions. *Ecol. Lett.* **2013**, *16*, 1424–1435. [CrossRef]
6. McShea, W.J. What Are the Roles of Species Distribution Models in Conservation Planning? *Environ. Conserv.* **2014**, *41*, 93–96. [CrossRef]
7. Harvey, J.A.; Heinen, R.; Armbrecht, I.; Basset, Y.; Baxter-Gilbert, J.H.; Bezemer, T.M.; Böhm, M.; Bommarco, R.; Borges, P.A.; Cardoso, P. International Scientists Formulate a Roadmap for Insect Conservation and Recovery. *Nat. Ecol. Evol.* **2020**, *4*, 174–176. [CrossRef]
8. Senapathi, D.; Goddard, M.A.; Kunin, W.E.; Baldock, K.C. Landscape Impacts on Pollinator Communities in Temperate Systems: Evidence and Knowledge Gaps. *Funct. Ecol.* **2017**, *31*, 26–37. [CrossRef]
9. Norwegian Ministries. *National Pollinator Strategy A Strategy for Viable Populations of Wild Bees and Other Pollinating Insects*; Norwegian Government Security and Service Organisation: Oslo, Norway, 2018; p. 67.
10. IPBES. *The Assessment Report of the Intergovernmental Science-Policy Platform on Biodiversity and Ecosystem Services on Pollinators, Pollination and Food Production*; Potts, S.G., Imperatriz-Fonseca, V.L., Ngo, H.T., Eds.; Secretariat of the Intergovernmental Science-Policy Platform on Biodiversity and Ecosystem Services: Bonn, Germany, 2016; p. 552. [CrossRef]
11. Sydenham, M.A.K.; Venter, Z.S.; Eldegard, K.; Moe, S.R.; Steinert, M.; Staverløkk, A.; Dahle, S.; Skoog, D.I.J.; Hanevik, K.A.; Skrindo, A.; et al. High Resolution Prediction Maps of Solitary Bee Diversity Can Guide Conservation Measures. *Landsc. Urban Plan.* **2022**, *217*, 104267. [CrossRef]
12. Zurell, D.; Thuiller, W.; Pagel, J.; Cabral, J.S.; Münkemüller, T.; Gravel, D.; Dullinger, S.; Normand, S.; Schiffers, K.H.; Moore, K.A.; et al. Benchmarking Novel Approaches for Modelling Species Range Dynamics. *Glob. Change Biol.* **2016**, *22*, 2651–2664. [CrossRef]
13. Randin, C.F.; Ashcroft, M.B.; Bolliger, J.; Cavender-Bares, J.; Coops, N.C.; Dullinger, S.; Dirnböck, T.; Eckert, S.; Ellis, E.; Fernández, N.; et al. Monitoring Biodiversity in the Anthropocene Using Remote Sensing in Species Distribution Models. *Remote Sens. Environ.* **2020**, *239*, 111626. [CrossRef]
14. Venter, Z.S.; Sydenham, M.A.K. Continental-Scale Land Cover Mapping at 10 m Resolution Over Europe (ELC10). *Remote Sens.* **2021**, *13*, 2301. [CrossRef]
15. Brown, C.F.; Brumby, S.P.; Guzder-Williams, B.; Birch, T.; Hyde, S.B.; Mazzariello, J.; Czerwinski, W.; Pasquarella, V.J.; Haertel, R.; Ilyushchenko, S.; et al. Dynamic World, Near Real-Time Global 10 m Land Use Land Cover Mapping. *Sci. Data* **2022**, *9*, 251. [CrossRef]
16. Zanaga, D.; Van De Kerchove, R.; De Keersmaecker, W.; Souverijns, N.; Brockmann, C.; Quast, R.; Wevers, J.; Grosu, A.; Paccini, A.; Vergnaud, S.; et al. ESA WorldCover 10 m 2020 V100. *Zenodo* **2021**. [CrossRef]
17. Tulbure, M.G.; Hostert, P.; Kuemmerle, T.; Broich, M. Regional Matters: On the Usefulness of Regional Land-Cover Datasets in Times of Global Change. *Remote Sens. Ecol. Conserv.* **2022**, *8*, 272–283. [CrossRef]
18. Bjørdal, I.; Bjørkelo, K. AR5 Klassifikasjonssystem: Klassifikasjon Av Arealressurser. In *Håndbok Fra Skog Og Landskap*; The Norwegian Institute of Bioeconomy Research: Ås, Norway, 2006.
19. Noriega, J.A.; Hortal, J.; Azcárate, F.M.; Berg, M.P.; Bonada, N.; Briones, M.J.I.; Del Toro, I.; Goulson, D.; Ibanez, S.; Landis, D.A.; et al. Research Trends in Ecosystem Services Provided by Insects. *Basic Appl. Ecol.* **2018**, *26*, 8–23. [CrossRef]
20. Prather, C.M.; Pelini, S.L.; Laws, A.; Rivest, E.; Woltz, M.; Bloch, C.P.; Del Toro, I.; Ho, C.; Kominoski, J.; Newbold, T.S. Invertebrates, Ecosystem Services and Climate Change. *Biol. Rev.* **2013**, *88*, 327–348. [CrossRef]
21. Gallai, N.; Salles, J.-M.; Settele, J.; Vaissière, B.E. Economic Valuation of the Vulnerability of World Agriculture Confronted with Pollinator Decline. *Ecol. Econ.* **2009**, *68*, 810–821. [CrossRef]
22. Losey, J.E.; Vaughan, M. The Economic Value of Ecological Services Provided by Insects. *Bioscience* **2006**, *56*, 311–323. [CrossRef]
23. Smith, T.J.; Saunders, M.E. Honey Bees: The Queens of Mass Media, despite Minority Rule among Insect Pollinators. *Insect Conserv. Divers.* **2016**, *9*, 384–390. [CrossRef]
24. Hallmann, C.A.; Sorg, M.; Jongejans, E.; Siepel, H.; Hofland, N.; Schwan, H.; Stenmans, W.; Müller, A.; Sumser, H.; Hörren, T. More than 75 Percent Decline over 27 Years in Total Flying Insect Biomass in Protected Areas. *PLoS ONE* **2017**, *12*, e0185809. [CrossRef] [PubMed]
25. Seibold, S.; Gossner, M.M.; Simons, N.K.; Blüthgen, N.; Müller, J.; Ambarlı, D.; Ammer, C.; Bauhus, J.; Fischer, M.; Habel, J.C. Arthropod Decline in Grasslands and Forests Is Associated with Landscape-Level Drivers. *Nature* **2019**, *574*, 671–674. [CrossRef] [PubMed]
26. Wagner, D.L.; Grames, E.M.; Forister, M.L.; Berenbaum, M.R.; Stopak, D. Insect Decline in the Anthropocene: Death by a Thousand Cuts. *Proc. Natl. Acad. Sci. USA* **2021**, *118*, e2023989118. [CrossRef] [PubMed]
27. Zattara, E.E.; Aizen, M.A. Worldwide Occurrence Records Suggest a Global Decline in Bee Species Richness. *One Earth* **2021**, *4*, 114–123. [CrossRef]
28. Orr, M.C.; Hughes, A.C.; Chesters, D.; Pickering, J.; Zhu, C.-D.; Ascher, J.S. Global Patterns and Drivers of Bee Distribution. *Curr. Biol.* **2021**, *31*, 451–458. [CrossRef]
29. Sydenham, M.A.K.; Eldegard, K.; Venter, Z.S.; Evju, M.; Åström, J.; Rusch, G.M. Priority Maps for Pollinator Habitat Enhancement Schemes in Semi-Natural Grasslands. *Landsc. Urban Plan.* **2022**, *220*, 104354. [CrossRef]

30. Westrich, P. *Habitat Requirements of Central European Bees and the Problems of Partial Habitats*; Academic Press Limited: Cambridge, MA, USA, 1996; Volume 18, pp. 1–16.
31. Woodard, S.H.; Jha, S. Wild Bee Nutritional Ecology: Predicting Pollinator Population Dynamics, Movement, and Services from Floral Resources. *Curr. Opin. Insect Sci.* **2017**, *21*, 83–90. [CrossRef]
32. Carrié, R.; Lopes, M.; Ouin, A.; Andrieu, E. Bee Diversity in Crop Fields Is Influenced by Remotely-Sensed Nesting Resources in Surrounding Permanent Grasslands. *Ecol. Indic.* **2018**, *90*, 606–614. [CrossRef]
33. Requier, F.; Leonhardt, S.D. Beyond Flowers: Including Non-Floral Resources in Bee Conservation Schemes. *J. Insect Conserv.* **2020**, *24*, 5–16. [CrossRef]
34. Antoine, C.M.; Forrest, J.R. Nesting Habitat of Ground-nesting Bees: A Review. *Ecol. Entomol.* **2021**, *46*, 143–159. [CrossRef]
35. O'Connor, R.S.; Kunin, W.E.; Garratt, M.P.; Potts, S.G.; Roy, H.E.; Andrews, C.; Jones, C.M.; Peyton, J.M.; Savage, J.; Harvey, M.C. Monitoring Insect Pollinators and Flower Visitation: The Effectiveness and Feasibility of Different Survey Methods. *Methods Ecol. Evol.* **2019**, *10*, 2129–2140. [CrossRef]
36. Hutchinson, L.A.; Oliver, T.H.; Breeze, T.D.; O'Connor, R.S.; Potts, S.G.; Roberts, S.P.; Garratt, M.P. Inventorying and Monitoring Crop Pollinating Bees: Evaluating the Effectiveness of Common Sampling Methods. *Insect Conserv. Divers.* **2022**, *15*, 299–311. [CrossRef]
37. Droege, S.; Tepedino, V.J.; LeBuhn, G.; Link, W.; Minckley, R.L.; Chen, Q.; Conrad, C. Spatial Patterns of Bee Captures in North American Bowl Trapping Surveys. *Insect Conserv. Divers.* **2010**, *3*, 15–23. [CrossRef]
38. Gorelick, N.; Hancher, M.; Dixon, M.; Ilyushchenko, S.; Thau, D.; Moore, R. Google Earth Engine: Planetary-Scale Geospatial Analysis for Everyone. *Remote Sens. Environ.* **2017**, *202*, 18–27. [CrossRef]
39. Steffan-Dewenter, I.; Münzenberg, U.; Bürger, C.; Thies, C.; Tscharntke, T. Scale-dependent Effects of Landscape Context on Three Pollinator Guilds. *Ecology* **2002**, *83*, 1421–1432. [CrossRef]
40. R Core Team. *R: A Language and Environment for Statistical Computing 2021*; R Core Team: Vienna, Austria, 2021.
41. Lussana, C.; Tveito, O.; Uboldi, F. Three-dimensional Spatial Interpolation of 2 m Temperature over Norway. *Q. J. R. Meteorol. Soc.* **2018**, *144*, 344–364. [CrossRef]
42. Geological Survey of Norway Løsmasser WMS. Available online: https://kartkatalog.geonorge.no/metadata/norges-geologiske-undersokelse/losmasser-wms/aa780848-5de8-4562-8f35-3d5c80ea8b48/ (accessed on 24 October 2022).
43. Wright, M.N.; Wager, S.; Probst, P. *R Package*, Version 0.12; Ranger: A Fast Implementation of Random Forests; R Core Team: Vienna, Austria, 2020; Volume 1.
44. Rodriguez, P.P.; Gianola, D. *R Package*, Version 0.6; BRNN: Bayesian Regularization for Feed-Forward Neural Networks; R Core Team: Vienna, Austria, 2016.
45. Kuhn, M. Building Predictive Models in R Using the Caret Package. *J. Stat. Softw.* **2008**, *28*, 1–26. [CrossRef]
46. Singh, V.; Pencina, M.; Einstein, A.J.; Liang, J.X.; Berman, D.S.; Slomka, P. Impact of Train/Test Sample Regimen on Performance Estimate Stability of Machine Learning in Cardiovascular Imaging. *Sci. Rep.* **2021**, *11*, 14490. [CrossRef]
47. Greenwell, B.M. Pdp: An R Package for Constructing Partial Dependence Plots. *R J.* **2017**, *9*, 421. [CrossRef]
48. Nkhwanana, N.; Adam, E.; Ramoelo, A. Assessing the Utility of Sentinel-2 MSI in Mapping an Encroaching Serephium Plumosum in South African Rangeland. *Appl. Geomat.* **2022**, *14*, 435–449. [CrossRef]
49. De Simone, W.; Allegrezza, M.; Frattaroli, A.R.; Montecchiari, S.; Tesei, G.; Zuccarello, V.; Di Musciano, M. From Remote Sensing to Species Distribution Modelling: An Integrated Workflow to Monitor Spreading Species in Key Grassland Habitats. *Remote Sens.* **2021**, *13*, 1904. [CrossRef]
50. Olofsson, P.; Foody, G.M.; Herold, M.; Stehman, S.V.; Woodcock, C.E.; Wulder, M.A. Good Practices for Estimating Area and Assessing Accuracy of Land Change. *Remote Sens. Environ.* **2014**, *148*, 42–57. [CrossRef]
51. Marshall, L.; Beckers, V.; Vray, S.; Rasmont, P.; Vereecken, N.J.; Dendoncker, N. High Thematic Resolution Land Use Change Models Refine Biodiversity Scenarios: A Case Study with Belgian Bumblebees. *J. Biogeogr.* **2021**, *48*, 345–358. [CrossRef]
52. Griffiths, P.; Nendel, C.; Pickert, J.; Hostert, P. Towards National-Scale Characterization of Grassland Use Intensity from Integrated Sentinel-2 and Landsat Time Series. *Remote Sens. Environ.* **2020**, *238*, 111124. [CrossRef]
53. Coops, N.C.; Waring, R.H.; Plowright, A.; Lee, J.; Dilts, T.E. Using Remotely-Sensed Land Cover and Distribution Modeling to Estimate Tree Species Migration in the Pacific Northwest Region of North America. *Remote Sens.* **2016**, *8*, 65. [CrossRef]
54. Venter, Z.S.; Barton, D.N.; Chakraborty, T.; Simensen, T.; Singh, G. Global 10 mL and Use Land Cover Datasets: A Comparison of Dynamic World, World Cover and Esri Land Cover. *Remote Sens.* **2022**, *14*, 4101. [CrossRef]
55. White, E.R. Minimum Time Required to Detect Population Trends: The Need for Long-Term Monitoring Programs. *BioScience* **2019**, *69*, 40–46. [CrossRef]

Travis Oliphant in Provo, UT, USA; and Pandas 1.2.1 created by Wes McKinney in New York City, NY, USA. Three methods of modeling HS were used. Spatial data were obtained for LULC, species richness, and region extents. Table 1 lists the data and their sources.

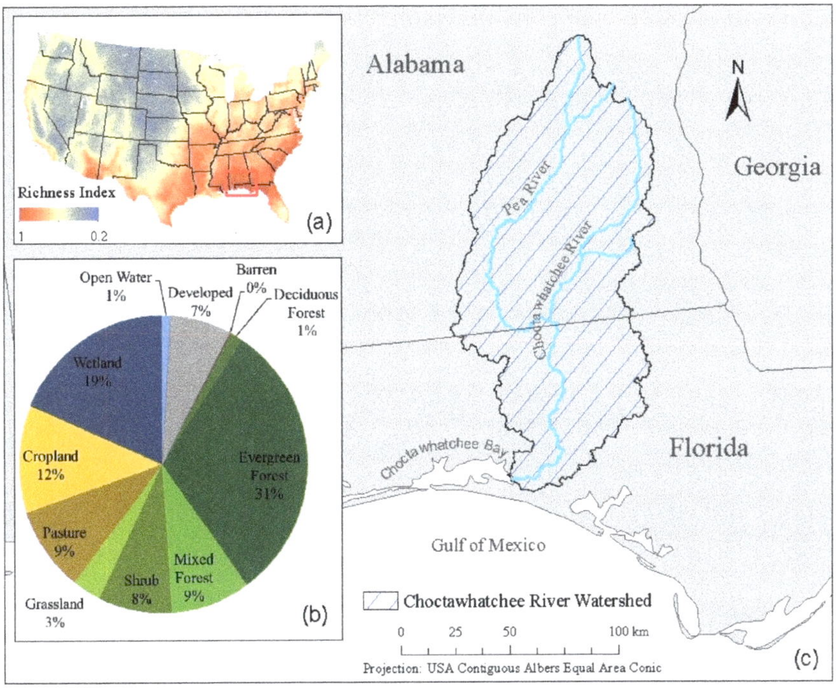

Figure 1. Location of the Choctawhatchee River Watershed. (**a**) The species richness index of the United States is shown as are where the watershed is in the United States, higher richness index values are shown as red and lower values in blue. (**b**) A pie chart of the proportion of each LULC. (**c**) The full extent of the watershed and the locations of the Choctawhatchee and Pea Rivers.

Table 1. The details of the spatial data used in the study.

Data	Year	Source	Reference
National Land Cover Dataset (NLCD)	2019	USGS	[33]
Species Richness	2018	Florida International University	[34]
Choctawhatchee River Watershed Extents	2016	USGS	[35]
Southeast Plains Ecoregion extents	2017	EPA	[36]
Eastern Temperate Forest Ecoregion Extents	2018	EPA	[37]

Method 1—Binary Method: Step 1: Download the Multi-Resolution Land Characteristics Consortium (MRLC) LULC data [38]; the land use/land cover data from the National Land Cover Dataset (NLCD) for the year 2019 and its legend were obtained [33]. Step 2: Classify the natural and unnatural groups and assign index; agriculture and developed land use classes were deemed unnatural, and the land cover classes open water, wetland, grassland, shrubland, and forests were considered as natural. Step 3: Obtain the binary

HS index; Unnatural LULC classes were assigned an HS index value of 0 and natural LULC classes were given an HS index value of 1. Step 4: Create an index map (natural and unnatural) where the LULC values were replaced with the corresponding HS index values (0/1) to create an index map. The analysis involved using the following software: ArcMap 10.4.1 (Clip tool from Raster Processing toolbox) and Python codes (Geospatial Data Abstraction Library with Jupyter notebook).

Method 2—Literature Review: Step 1 was the same as in method 2. Step 2: A document search was performed using Google Scholar (e.g., habitat suitability), which gave 500,000 plus results, and a more focused search using words in quotes, additional words (e.g., "habitat suitability", "InVEST"), and selected time-period (2010–2020) was carried out. Priority was given to articles based on the following three criteria: (1) the full articles were accessible; (2) the articles used a similar definition for habitat suitability; and (3) the articles provided the numerical habitat suitability values for LULCs comparable to those found in the study region. A total of 21 articles were retrieved. In Scholar search, InVEST, as an additional search term, was used in the search to narrow down the results in a systematic way. Step 3: Obtaining HS index (between 0 and 1). The results of the search were summarized in a table and graphs provided in the Results section. Habitat suitability values were organized by the specific LULC. The LULC were further grouped into broad classifications according to the LULC descriptions provided by the MRLC [39], for example, rivers, lakes, reservoirs, and glaciers were grouped as simply "water", while open forests, orchards, and native forests were grouped as simply "forest". When multiple values for a group were obtained from the literature, the average values for suitability were calculated and used to create the table. The box plot was created using Python. The HS vales were placed into single column arrays for each LULC group. A box plot was then created for each LULC group and displayed in the same figure. Step 4: Create an index map (natural and unnatural). The LULC values between 0 and 1 were replaced with corresponding HS index values (0/1) to create an index map. The analysis was carried out using the software described in Method 1. A table (Table 2) lists the references for each land use/cover type along with the number of data points obtained for them. The LULC values in the map clipped to the watershed's extents were replaced with the mean values obtained in the literature review.

Method 3—Species Richness Method: Step 1: Bring the datasets to uniform scales and obtain the species richness data [34]. National Land Cover Data (NLCD) land use/land cover map's resolution (30 m × 30 m) were rescaled to species richness maps with a 10 km × 10 km resolution. ArcMap 10.4.1 software with the Resample tool using the MAJORITY technique was used [40]. Step 2: Average the richness/land use. The major LULC from ~111 pixels (30 m) now represent the LULC for the 10 km map. The richness and LULC data were merged into one raster file by using the Combine ArcMap Spatial Analysist tool to observe both the number of species and LULC for each pixel. Step 3: Clip the watershed area. Shapefiles for the Choctawhatchee watershed (HUC 031402) and the Southeast Plains and Eastern Temperate Forest ecoregions were obtained [35–37]. The combined richness raster was clipped to the extent of each shapefile. Step 4: Estimate the average richness value. The average species richness for each LULC was then calculated for each region as well as the entire contiguous United States. Habitat suitability indices were normalized using the species richness results by dividing the species richness of each LULC by the highest species richness value for each region. If a 10 km grid cell for a forest within the Southeast Plains has a richness value of 300 and the highest richness value in that region is 600, the HS index would be 0.5 (300 divided by 600). Step 5: Mapping HS. The LULC values in the original 30 m map were replaced with the corresponding habitat suitability index values.

A table (Table 3) listing the total species richness, standard deviation, and count of grid cells was created by importing the attribute tables of the species richness raster images clipped to the extents of the Choctawhatchee River Watershed, Southeast Plains Ecoregion, Eastern Temperate Forest Ecoregion, and the contiguous United States.

3. Results

3.1. Method 1—Binary Method

A hypothesis map was created using the binary method where it was assumed that developed lands such as urban and agriculture have a suitability index score of 0, and every other landscape was assumed to have an index score of 1. The number of grid cells with a value of 0 were counted and compared to the number of grid cells that had a value of 1. Approximately 27.33% of the watershed had low suitability. The areas with low suitability appeared near the middle and northern parts of the watershed. Most grid cells with zero suitability occurred on the Alabama side of the watershed. Figure 2 displays the resulting HS index map of the watershed.

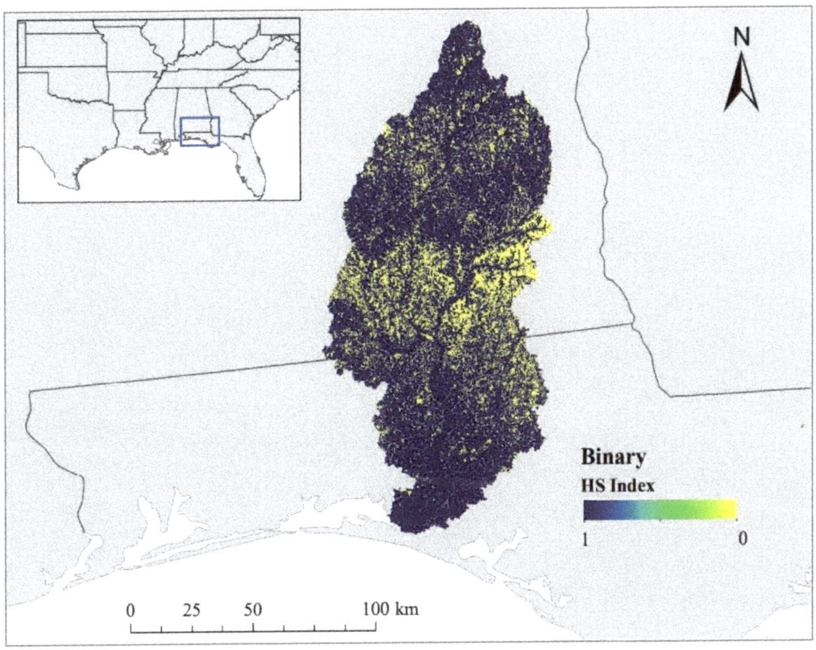

Figure 2. The habitat suitability map of the Choctawhatchee Watershed using the binary method (Method 1).

3.2. Method 2—Literature Review Method

A total of 21 studies were analyzed. Of the 21, 15 studies originated in China [9,41–54], two studies were from Ethiopia [29,55], and only one study each originated in India [56], Indonesia [57], Spain [58], and the United States [19]. A total of 36 values were used to calculate the average suitability for water, 17 were used for bare lands, 30 were used for grasslands, ten were used for shrub lands, 36 values were used for forests, 18 were used for wetlands, 24 were used for agricultural, and 15 were used for developed lands. The habitat with the highest mean value was forest land. Next was shrubland, followed by water and wetlands that had nearly the same average suitability. Developed lands predictably had the lowest mean suitability.

Table 2 breaks down the broad land use/cover classes into specific types and lists the references that site each LULC. The number of values obtained for each LULC type is listed, along with the average suitability. Different LULC within the same class sometimes had very different suitability values. For instance, raw land and beaches were both considered bare land, but had an HS of 0.05 and 0.9, respectively. The overall average HS for each broad LULC class is also listed.

Table 2. The average habitat suitability for each land use/cover type obtained from the literature review.

LULC Class	Types	References	Data Points	Average HS
Water	Water	[19,29,41,42,51,53–56]	10	0.75
	Rivers	[9,44,46–48]	6	0.88
	Lakes	[9,44–47,50,54]	6	0.98
	Pond	[9]	1	0.9
	Reservoirs	[44,45,47,50,58]	6	0.83
	Shallows	[44]	1	0.6
	Streams	[58]	4	0.73
	Channels	[45]	1	1
	Canals	[50]	1	0
		Overall		0.807
Bare Land	Bare Land	[19,29,50,56]	4	0.125
	Dry Land	[43,44,47]	7	0.243
	Desert	[49]	1	0.1
	Raw	[54]	2	0.05
	Unused Land	[41,48]	2	0.255
	Beach	[9]	1	0.9
		Overall		0.224
Grassland	Grass	[9,19,42–45,47–49,51,53–55,58]	29	0.727
	Meadow	[41]	1	1
		Overall		0.736
Shrub Land	Shrub	[9,29,43–45,50,51,54,58]	9	0.84
	Bush	[48]	1	0.8
		Overall		0.837
Forest	Forest	[9,19,29,41–43,45,47,49,51–53,57,58]	25	0.931
	Woodland	[9,29,44,47,50]	8	0.844
	Orchard	[43,52]	2	0.25
	Forestry	[48]	1	0.9
		Overall		0.873
Wetland	Wetland	[9,19,45,49,51,54,56]	9	0.844
	Marsh	[43,46,47]	5	0.74
	Mudflat	[46]	1	0.8
	Bottom Land	[44]	1	0.6
	Mangrove	[57]	1	0.8
	Swampy Bush	[57]	1	1
		Overall		0.806
Agriculture	Agricultural land	[19,41,58]	6	0.375
	Farmland	[42,53,55,57]	5	0.4
	Cropland	[29,45,48,54]	4	0.363
	Pasture	[29]	1	0.5
	Irrigable Land	[43,54]	2	0.35
	Paddy Field	[9,43,44,47,50]	6	0.267
		Overall		0.354

Table 2. *Cont.*

LULC Class	Types	References	Data Points	Average HS
Developed Land	Built-up Land	[29,42,53]	3	0
	Urban	[19,41,48,58]	5	0.03
	Suburban	[54]	1	0
	Construction	[54]	1	0
	Rural Residence	[53]	1	0
	Roads	[46,54]	2	0
	Infrastructure	[54]	1	0
	Transportation	[54]	1	0
		Overall		0.01

The landscapes with the largest range of suitability values were water habitats, which had values ranging from 0 to 1. This was followed by forest habitats with values ranging from 0.1 to 1. The land use with the lowest variability was developed land, which ranged from 0 to 0.15, with most studies reporting the suitability to be 0. The median and average values were similar for grasslands, shrub lands, wetlands, and developed lands. Median and average values for the remaining landscapes were not as close with averages falling well below the median value, except for bare lands, where the average was higher than the median. This is displayed in the box plots of Figure 3. The bold line represents the median, the diamond marker represents the mean, and the circles represent the outliers.

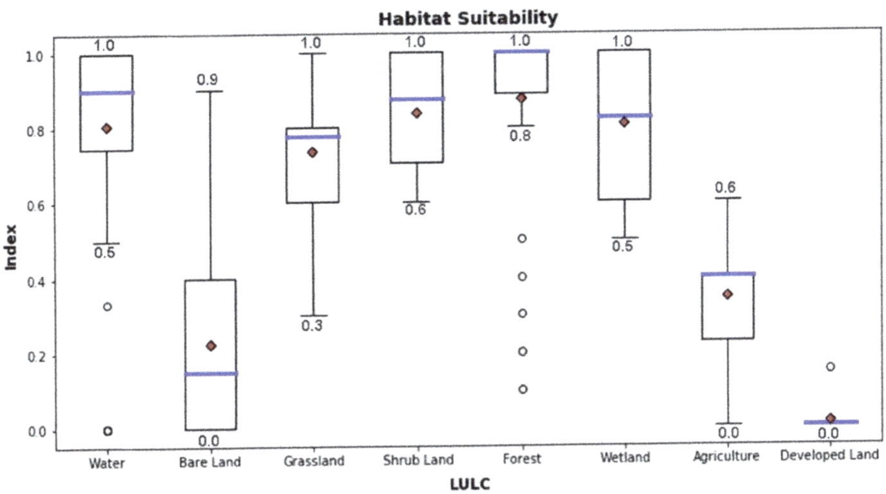

Figure 3. The suitability box plots for each LULC class. The whiskers indicate 1.5 times the interquartile range values. The bold line represents the median and the diamond shaped markers represent the mean. The circles are the outliers.

Figure 4 displays a HS map of the watershed based on the average values derived from the literature. The Alabama side of the watershed in the North generally had a lower HS when compared to the Florida side in the South. Urban areas had the lowest HS at 0.012. Urban land uses made up 6.80% of the watershed. Bare land had the second lowest HS at 0.224 and made up 0.12% of the watershed. The habitat with the third lowest HS was agriculture, having a HS near the median at 0.354. Agriculture made up 21.26% of the watershed. Overall, about 28% of the watershed had a relatively low HS.

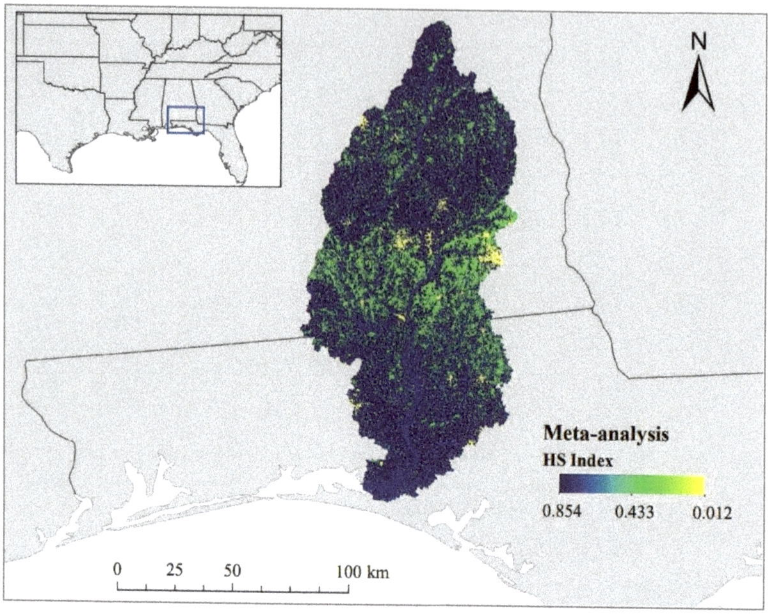

Figure 4. The habitat suitability map of the Choctawhatchee Watershed created using the results of the literature review.

3.3. Method 3—Species Richness Method

The area of interest was the Choctawhatchee River and Bay Watershed located in the Southeast United States. The watershed was within the boundaries of the Southeastern Plains ecoregion, which was in the Eastern Temperate Forest ecoregion that encompassed most of the Eastern United States. A visual representation of these regions is shown in Figure 5.

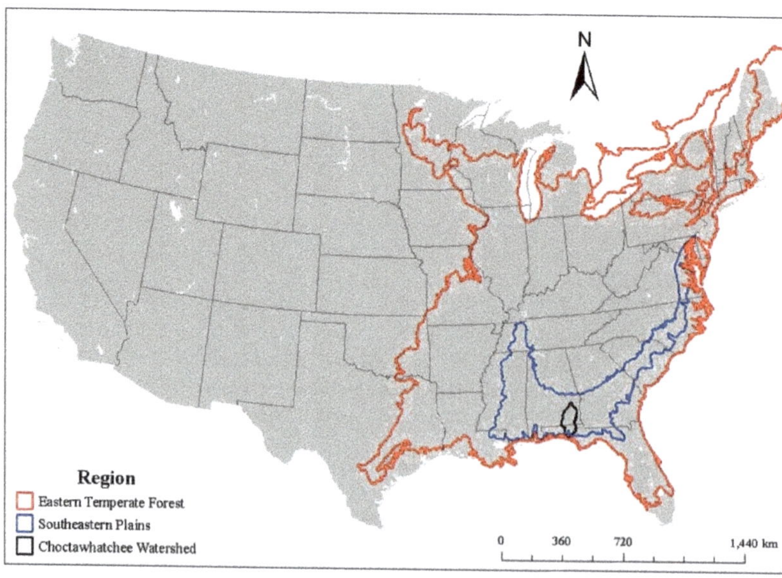

Figure 5. A location map with boundaries of the Choctawhatchee River Watershed, Southeastern Plains Ecoregion, and Eastern Temperate Forest Ecoregion overlaid in the United States.

Table 3 lists the average total richness and standard deviation values for each LULC. The number of pixels is also listed. Each pixel represents 100 square kilometers. The watershed had higher values than the averages at broader levels. Since the LULC raster was resampled from 30 m to 10,000 m, there were no pixels where high intensity developed land, barren land, or herbaceous wetland were the majority LULC. The values for these LULCs were estimated based on the most similar region (Southeast Plains). The trends also did not match across levels. Richness was high in the medium intensity developed land within the watershed, but richness generally decreased as the intensity (amount of impervious surface) increased. Broad generalizations might not be accurate when assessing a watershed. A visual representation of Table 3 can be seen in Figure S3 in the Supplementary Materials.

Table 3. The average and standard deviation of the species richness for each land use/cover type.

	Contiguous USA			Eastern Temperate Forest			Southeastern Plains			Choctawhatchee River Watershed		
LULC	**Mean**	**Std**	**Pixels**	**Mean**	**Std**	**Pixels**	**Mean**	**Std**	**Pixels**	**Mean**	**Std**	**Pixels**
Open Water	316.84	52.56	2246	345.1	40.324	1105	383.96	25.846	94	417.56	19.90	9
Developed, Open Space	335.01	46.072	2427	353.91	34.221	1671	390.16	28.627	247	414.72	18.34	19
Developed, Low Intensity	333.75	40.63	1126	348.23	31.222	753	381.41	23.211	79	415.75	14.66	4
Developed, Medium Intensity	331.57	41.125	402	349.55	34.137	212	377.25	22.493	24	432	0	1
Developed, High Intensity	339.32	40.082	95	350.31	33.576	52	384.75	22.833	8	ND	ND	ND
Barren Land	262.92	37.884	768	355.89	36.131	53	391.5	26.588	10	ND	ND	ND
Deciduous Forest	329.65	30.388	9441	340.77	24.622	6931	373.61	20.356	389	404.31	7.11	16
Evergreen Forest	311.82	54.033	10,764	384.92	34.983	2999	401.53	26.248	1069	424.55	18.76	126
Mixed Forest	328.44	40.853	1976	345.5	35.622	1358	385.3	22.14	247	403.06	3.84	16
Shrub Land	274.71	40.996	17,894	375.09	37.557	244	404.89	28.416	76	428.77	22.30	13
Grassland	251.83	38.533	10,275	368.41	44.047	146	399.43	31.225	40	450.33	3.21	3
Pasture	331.45	40.55	4099	346.19	28.541	2682	389.56	25.549	192	408.1	6.97	21
Cropland	279.5	50.227	13,052	327.76	32.584	4941	392.03	22.139	393	425.3	9.71	20
Woody Wetlands	352.32	54.173	2463	377.99	43.896	1707	397.48	24.934	441	431.61	17.34	33
Herbaceous Wetland	305.69	49.801	429	333.04	50.352	135	402.2	29.072	5	ND	ND	ND

Figure 6 shows the resulting maps. The lowest index value when using the average species index for the Southeast Plains ecoregion was close to 1, meaning that there was very little variability in the values. The variability increased as the sample size used to calculate the average increased. The lowest HS values occurred the most in the northern parts of the watershed in the Southeastern Plains and Eastern Temperate Forest maps. The map derived from using the entire contiguous United States did not seem to have a pattern aside from the highest suitability occurring in wetlands along the streams of the watershed.

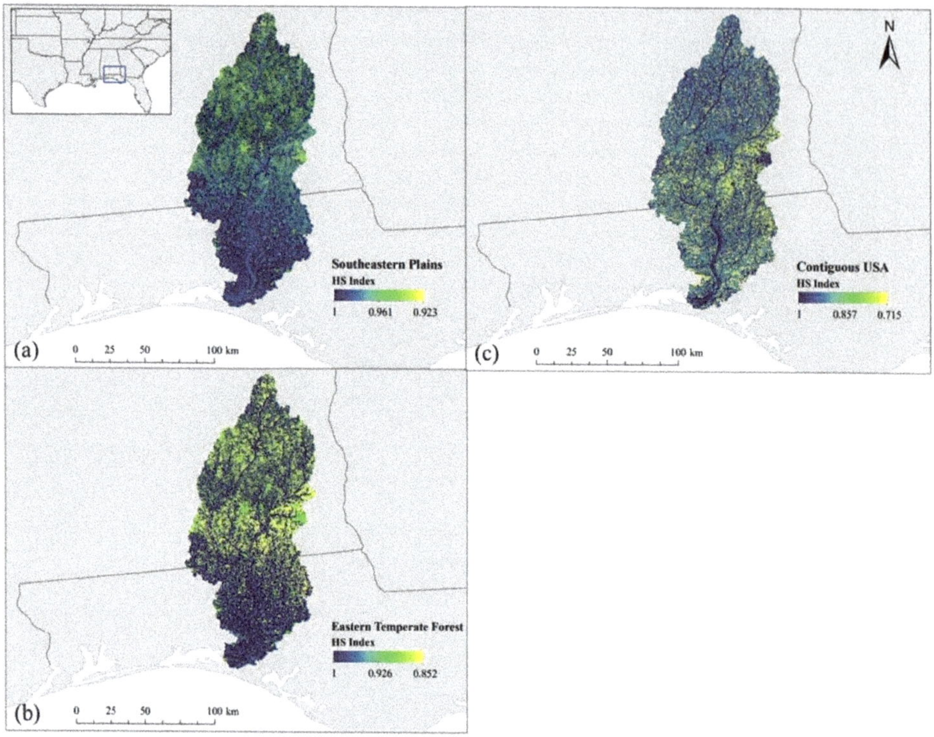

Figure 6. The habitat suitability maps of the Choctawhatchee Watershed based on species richness derived from averaging the values within regions of varying sizes. Maps created using the average values from grid cells in the (**a**) Southeastern Plains ecoregion; (**b**) Eastern Temperate Forest ecoregion; (**c**) contiguous USA.

3.4. Comparison of the Methods

The average HS values of each LULC is shown in Figure 7. The landscapes with consistently high HS values, regardless of method, are open water, forests, and wetlands. The most apparent differences in HS were seen with agriculture and urbanization. These two land uses were low when using the binary and literature review methods. However, HS was high for these land uses when using the species richness data.

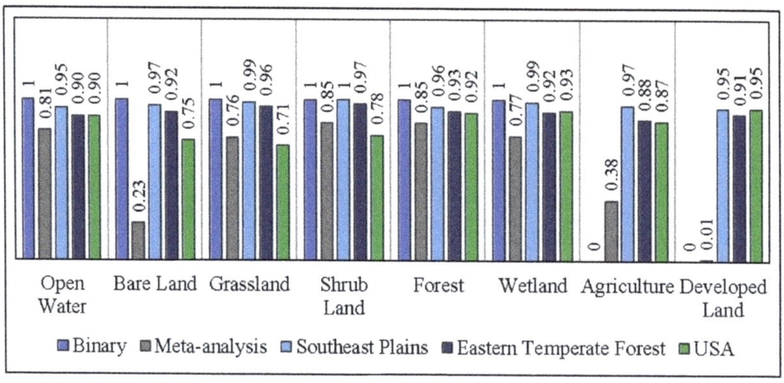

Figure 7. The habitat suitability index for the various land use/cover obtained.

The binary method is the simplest method. It only requires a list of the habitats present in a region and an understanding of which ones are natural or unnatural. The literature review method requires more research than the other methods. This method is also heavily reliant on values from previous works. The obtained values are assumed to be correct. The species richness methods require species count data. The completeness of the data has a large impact on the outcome, so the values will be inaccurate if many species are not accounted for.

3.5. Choosing a Method

Assessing what data are available is vital when deciding on a method. The natural/unnatural binary method is used when minimum data are available. If the habitats are known, it is possible to determine whether the habitat is natural or unnatural. A list of habitats is derived from the literature or datasets. Mapping HS requires spatial data. Research goals may require the HS values to be more exact. Using 0 and 1 for unnatural and natural, respectively, would be too broad. Expressing variation between the HS of different habitats requires expert knowledge of the target region. The literature review method is used when there is no access to expert knowledge. A literature review is used to synthesize the results of multiple studies [59]. Existing literature is needed to perform a literature review and gathering results from similar studies is preferred [60]. HS indicators such as species richness are used in a data driven approach. Using an indicator requires available data for the study region or a similar region. Indicators that are used for this are biophysical, socio-economic, or management attributes [61]. The characteristics of each landscape were studied to determine an indicator that could be used to model the suitability across all landscapes, which included the biotic and abiotic components of deciduous forests [62], evergreen forests [63], mixed forests [64,65], wetlands [66,67], shrub lands [68], grasslands [69,70], and bare lands [71,72]. Species richness was used as an indicator to estimate the overall health of any habitat and to identify priority conservation areas [73–75].

Figure 8 summarizes when to use each method and lists the information required to perform them. The figure indicates the specificity and complexity of the methods in relation to each other. The binary method is the simplest and least specific of the methods. Next is the literature review method, followed by the species richness method, the most complex and specific method of the three.

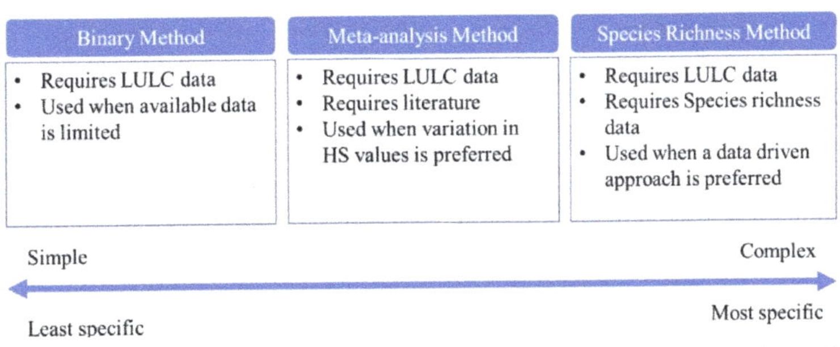

Figure 8. A comparison of the three methods. Method selection depends on the available data, complexity level, and research goals.

4. Discussion

The results revealed that there were significant differences in the habitat suitability scores when using the different methods. However, the Florida side of the watershed consistently had a higher average suitability than the Alabama side.

4.1. Assumptions and Limitations in the Case Study

The three methods to estimate habitat suitability are the natural/unnatural binary method, a literature review of published works, and the indicator method using species richness. The binary method is the simplest method since calculations are not required. This method requires the knowledge of the landscapes present in the area of interest. Some land covers such as forests and wetlands could be managed and therefore considered unnatural. It was assumed that the only unnatural LULC were urban areas, cropland, and pastures. The literature review resulted in a map that was similar to the binary map. The key difference was the absence of 0 or 1 values in the map based on the literature review. Bare land also had an index value that was less than agriculture. There was also some variability between the index values of the urban and agricultural lands instead of both being assumed to be equally unsuitable. Generally, the areas that had the lowest suitability were nearly the same.

The map of species richness (Figure 1a) showed that the index values were all close to one. This is likely due to the region being a biodiversity hotspot [31,76]. During the development of the three maps using the species richness data in conjunction with LULC, the species richness values were higher in the Choctawhatchee River Watershed than in the rest of the United States due to (1) the watershed being more diverse than average or (2) there was a smaller sample size, which resulted in higher average values. Furthermore, there were not enough data points within the Choctawhatchee River Watershed to calculate the average species richness for every LULC. Using the average richness values resulted in maps dissimilar to each other, aside from the values having low variability compared to the results of the literature review. These maps also did not resemble the hypothesis binary map. Developed land was among the land uses with the highest suitability when using the average based on the contiguous United States. However, developed land is usually thought to be 0 or very close to it [29,41,42,46,48,53,54,58]. This could either mean that the species richness is a more accurate indicator of HS than averaging the results from past studies, or that the species richness was not adequate on its own to estimate HS. It could also mean that the species richness dataset is too limited.

A study in another region would use the species richness data available in that region. If no data are available, the values from a nearby region are useable. A literature review can also be used to estimate the values. The number of species in an area could also be counted manually when working in a small area. It is also possible to use the presence of one species as an indicator of the species richness of another species based on how important the indicator species is to the diversity of a habitat [77–79].

4.2. Advantages and Limitations of the Methods Used

The advantage of the binary method (Method 1) is that it can be applied in the absence of data or expert knowledge. Its limitation is that it is broad and does not account for the differences between LULCs. Agriculture (both cropland and pastures) and all other different types of developed lands are assumed to have the same suitability. It is possible that agricultural lands are more suitable than developed lands because they are not entirely unnatural. Open space also may be more suitable than high intensity developed lands.

The advantage of obtaining HS values from literature (Method 2) is that this method does not require expert knowledge. The more variable values are more descriptive than the simple binary method. The assumption is that the values used in the literature are accurate. However, the accuracy of the values changes when using an average of the values. This method is limited by the publications available, which requires other scientists and researchers to have conducted studies beforehand. These HS values come from the literature originating in various regions due to the lack of studies conducted in the Choctawhatchee River Watershed. This may be a potential advantage as HS can be estimated in a region where no previous studies are available. Studies based on regions that are unlike the area of interest cause this method to have the same disadvantage as the binary method (i.e., LULCs are given the same suitability values despite being different). This is because the values for all types of similar habitats are used to calculate an average. Every type of developed land, agricultural land, forest, and wetland is assumed to have the same suitability, which may not be accurate. There are also habitats that have a wide range of values. Both beaches and deserts are bare land. Beaches have high HS and deserts have low HS. In this case, using the average may not be adequate to account for the range in values.

Furthermore, the HS values were the average value from 23 articles. The limitation of this method is that the average value is subjective to the literature used.

The species richness method (Method 3) presents a way to estimate the general habitat suitability, whereas other methods estimate the suitability for a single species, a species group, or a specific habitat. This method does not require knowledge of the individual species, the present, or optimum habitat conditions. The binary method where natural is suitable, and unnatural is not suitable, is currently how HS is modeled in cases where there is no specific habitat data or when the goal is to estimate HS in general [5].

The main disadvantage is that the results of these methods are sensitive to the amount of data that is present. There is currently a lack of wildlife population data in most locations. The database used in this study only presented the species richness for the vertebrates (mammals, birds, reptiles, amphibians, and fish) and trees. The available data did not cover all macro-organisms. There was no species richness data for invertebrates such as arthropods and mollusks, non-tree plant species such as grasses and shrubs, or fungi. The total population of each species group was also unavailable. The details found in Table S3 in the Supplementary Materials cannot be utilized given the obtainable data.

The resolution of the spatial data also influences the accuracy. Accuracy decreases as the grid cell size increases because it becomes harder to account for the evenness of a species. For instance, a grid cell can represent a hectare. Most of the species might live in a section that is a tenth of a hectare. However, all of the species were counted to obtain a total value for the entire hectare. Smaller grid cell sizes allowed for more precise species mapping.

Using species richness by itself is not adequate when estimating the habitat suitability.

When looking at suitability maps made from individual species group richness (Figure 9), tree richness had the highest range of index values. Despite the index values, the range of bird species was the highest, with the lowest being three species and the highest being 249 species. The distribution of values in the total richness index map (Figure 1a) was most influenced by the number of trees and birds. This means that areas in the Southeast United States and along the coasts had the highest biodiversity. The distribution of species did not seem to be driven by general land use types, but rather a combination of climate, terrain, and other factors.

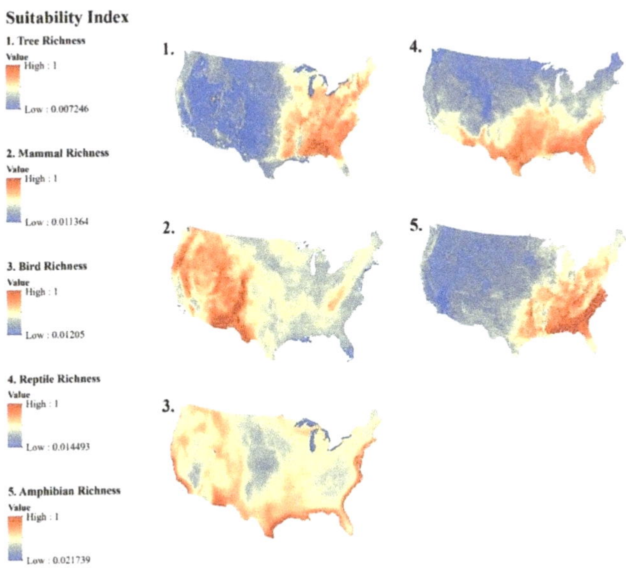

Figure 9. The national HS maps based on 1. tree richness, 2. mammal richness, 3. bird richness, 4. reptile richness, and 5. amphibian richness. Higher species richness values are shown as red and lower values in blue.

The advantage of relating the species richness to specific land uses is that it gives an extra dimension to maps in comparison to the existing methods that estimate HS using LULC alone. This method also grants the ability to determine the typical suitability of a habitat based on data. As new data become available, HS can be adjusted to reflect the changes in biodiversity [80–82]. Being based on observable data can be an advantage and a disadvantage for the method.

The HS estimation methods are based entirely on the number of species, assuming that the species are distributed independently of spatial evenness. Doing so increased the possibility of the inaccuracy in the results. In addition, functional species such as predators, raptors, or primary productivity as indicators of HS could be used as relevant factors for the study. Validating theoretical concepts is a challenge because there are not observations to validate the model [83,84]. These apply to HS, also a theoretical concept.

4.3. Implications for Conservation

The spatial representation of HS is a good tool for supplementing conservation strategies. Biodiversity maps are used to protect biodiversity in many conservation programs [85]. Studies have shown a link between habitat suitability and wildlife population viability for a variety of species [86–89]. The binary method of estimating HS may not be a good tool for biodiversity conservation as it is not a function of biodiversity or habitat conditions directly, but it does show where human land uses occur [5]. The results of the literature review provide a good idea of which habitats are the most suitable. The most suitable habitats can then be studied to determine the species viability [90]. The HS maps where suitability is an index of species richness are direct estimations of biodiversity. These maps can be used to rank habitats in an area to determine which habitats are the most viable and which habitats potentially need conservation attention.

HS are linked both directly and indirectly to almost all the 17 Sustainability Development Goals (SDGs). For example, HS is important for water and land resource conservation, which are related to SDG-14 (life below water) and SDG-15 (life on land). HS is indirectly related to SDG-6 (clean water and sanitation) because it is an integral part of water integrity, which is influenced by the physical characteristics of the waterbodies (physical integrity) and impacts the life below water (biological integrity) [91,92].

5. Conclusions

The objective was to compare the three approaches for estimating habitat suitability, summarize the advantages and disadvantages of these methods, and provide guidelines for selecting a HS method for conservation. The study focuses on the Choctawhatchee River Watershed (in Alabama and Florida, USA). The three habitat suitability estimation methods were as follows. Method 1 provides a suitability score based on the naturality of the habitat. Method 2 uses the average values from the literature with similar definitions for habitat suitability. Method 3 uses species richness. HS estimation is approachable from the perspective of a single species or species group, from a habitat-focused standpoint, or with the goal of estimating the suitability for wildlife in general. These approaches can be too specific or too broad. Estimating HS using species richness data is more specific than the existing binary method while being broad enough to use when modeling large multi-habitat areas such as a watershed. If complete species richness data are available, this method is advantageous. Using a more complete dataset may reveal that natural habitats are more suitable than developed lands. It is therefore important to gather more data before using species richness as an indicator.

In choosing a method, approaches can be chosen after determining what types of data are feasibly obtainable and based on the research goals. Things to consider are the specificity of the method, the accuracy of the data, and the assumptions made. Different methods change how conservation strategies are chosen. Broad methods assist in identifying how natural each habitat is. Specific methods assist in identifying the species or resource distribution in each habitat [74]. It is important to consider conservation goals when choosing a method. Using a method that includes one or multiple HS indicators such as species diversity, the presence of invasive species, and/or water quality makes it easier to decide on which conservation measures to take [75,93].

Steps should be taken in the future to improve HS mapping. This includes using models and techniques such as machine learning to predict species richness based on inventory data for terrestrial species [94,95] and aquatic species [96]. Modeling the change in HS in real-time is also a possibility [97]. These methods are currently not being used to produce maps for habitat conservation or the general public. Using functional species or primary productivity as indicators of HS could be used as relevant factors for study in the future. Habitat suitability modeling will become accessible and more evidence-based when accurate and complete species maps become obtainable. This will make it possible to consistently identify habitats to apply conservation actions. Currently, it is best to have expert knowledge of the region to estimate the suitability of the habitats or use the literature review carried out in this study. If this is not an option, using the natural/unnatural approach is the next best method.

Supplementary Materials: The following supporting information can be downloaded at: https://www.mdpi.com/article/10.3390/land11101754/s1, Figure S1: Venn diagram comparing deciduous forest, evergreen forest, and wetland habitats; Figure S2: Venn diagram comparing shrub land, grassland, and bare land habitats. [1] Steppe grasslands have very fertile soils. [2] Savannah grasslands have sandy/stony soil; Figure S3: Average terrestrial species richness for each LULC class.; Figure S4: Venn diagram that compares the Brillouin, Shannon–Wiener, and Hurlbert biodiversity index equations.; Table S1: Potential habitat conditions and components: A—all; B—bare land; D—deciduous forests; E—evergreen forests; G—grassland; S—shrub lands; W—wetlands; Table S2: Optimum habitat conditions; Table S3: Biodiversity Index Equations.

Author Contributions: Conceptualization, A.A., G.C. and K.M.; Methodology, K.M.; Software, K.M.; Validation, K.M.; Formal analysis, K.M.; Investigation, K.M.; Resources, A.A. and G.C.; Data curation, K.M.; Writing—original draft preparation, K.M.; Writing—review and editing, A.A., G.C. and K.M.; Visualization, A.A. and K.M.; Supervision, A.A.; Project administration, A.A. and G.C.; Funding acquisition, A.A. and G.C. All authors have read and agreed to the published version of the manuscript.

Funding: The work was supported by the National Institute of Food and Agriculture of USDA through Grant No. 2018-68002-27920 to Florida A&M University, USDA-NIFA capacity building grant 2017-38821-26405 and 2022-38821-37522; USDA-NIFA Evans-Allen Project, Grant 11979180/2016-01711; the National Science Foundation through grant no. 1735235 to Florida A&M University as part of the National Science Foundation Research Traineeship; and the U.S. Department of Education GAANN Program: GAANN: Meeting the Needs of the Nation's Infrastructure through Civil Engineering at Florida A&M University (Award #: P200A180074). The authors would like to acknowledge Mr. Teek for editing this manuscript.

Data Availability Statement: The generated data are original and specific to this study and available from the corresponding author upon request.

Acknowledgments: We especially thank the members of FAMU's NSF Research Traineeship Sustainable Food Energy and Water Systems (NRT-SFEWS) group.

Conflicts of Interest: The authors declare no conflict of interest. The funders had no role in the design of the study; in the collection, analyses, or interpretation of data; in the writing of the manuscript, or in the decision to publish the results.

References

1. FWS. *Habitat Evaluation Procedures Handbook*; U.S. Fish and Wildlife Services: Washington, DC, USA, 1980.
2. Kellner, C.; Brawn, J.; Karr, J. What is habitat suitability and how should it be measured? In *Wildlife 2001: Populations*; Springer: Dordrecht, The Netherlands, 1992; pp. 476–488.
3. Hirzel, A.H.; Le Lay, G. Habitat suitability modelling and niche theory. *J. Appl. Ecol.* **2008**, *45*, 1372–1381. [CrossRef]
4. García-Vega, A.; Fuentes-Pérez, J.F.; Fukuda, S.; Kruusmaa, M.; Sanz-Ronda, F.J.; Tuhtan, J.A. Artificial lateral line for aquatic habitat modelling: An example for Lefua echigonia. *Ecol. Inform.* **2021**, *65*, 101388. [CrossRef]

5. Nelson, E.; Ennaanay, D.; Wolny, S.; Olwero, N.; Vigerstol, K.; Pennington, D.; Mendoza, G.; Aukema, J.; Foster, J.; Forrest, J.; et al. InVEST 3.8.0. User's Guide. 2020. Available online: https://invest-userguide.readthedocs.io/_/downloads/en/3.8.3/pdf/ (accessed on 7 October 2021).
6. Ibrahim Mahmoud, M.; Duker, A.; Conrad, C.; Thiel, M.; Shaba Ahmad, H. Analysis of Settlement Expansion and Urban Growth Modelling Using Geoinformation for Assessing Potential Impacts of Urbanization on Climate in Abuja City, Nigeria. *Remote Sens.* **2016**, *8*, 220. [CrossRef]
7. Kija, H.K.; Ogutu, J.O.; Mangewa, L.J.; Bukombe, J.; Verones, F.; Graae, B.J.; Kideghesho, J.R.; Said, M.Y.; Nzunda, E.F. Spatio-Temporal Changes in Wildlife Habitat Quality in the Greater Serengeti Ecosystem. *Sustainability* **2020**, *12*, 2440. [CrossRef]
8. Di Febbraro, M.; Sallustio, L.; Vizzarri, M.; De Rosa, D.; De Lisio, L.; Loy, A.; Eichelberger, B.A.; Marchetti, M. Expert-based and correlative models to map habitat quality: Which gives better support to conservation planning? *Glob. Ecol. Conserv.* **2018**, *16*, e00513. [CrossRef]
9. Sun, X.; Jiang, Z.; Liu, F.; Zhang, D. Monitoring spatio-temporal dynamics of habitat quality in Nansihu Lake basin, eastern China, from 1980 to 2015. *Ecol. Indic.* **2019**, *102*, 716–723. [CrossRef]
10. Bentley, C.; Anandhi, A. Representing driver-response complexity in ecosystems using an improved conceptual model. *Ecol. Model.* **2020**, *437*, 109320. [CrossRef]
11. Lakoba, V.T.; Atwater, D.Z.; Thomas, V.E.; Strahm, B.D.; Barney, J.N. A global invader's niche dynamics with intercontinental introduction, novel habitats, and climate change. *Glob. Ecol. Conserv.* **2021**, *31*, e01848. [CrossRef]
12. Bailey, D.W. Identification and Creation of Optimum Habitat Conditions for Livestock. *Rangel. Ecol. Manag.* **2005**, *58*, 109–118. [CrossRef]
13. Cummings, V.; Thrush, S.; Hewitt, J.; Norkko, A.; Pickmere, S. Terrestrial deposits on intertidal sandflats: Sediment characteristics as indicators of habitat suitability for recolonising macrofauna. *Mar. Ecol. Prog. Ser.* **2002**, *253*, 39–54. [CrossRef]
14. González-Trujillo, R.; Méndez-Alonzo, R.; Arroyo-Rodríguez, V.; Vega, E.; González-Romero, A.; Reynoso, V.H. Vegetation Cover and Road Density as Indicators of Habitat Suitability for the Morelet's Crocodile. *J. Herpetol.* **2014**, *48*, 188–194. [CrossRef]
15. Bednaršek, N.; Feely, R.A.; Reum, J.C.P.; Peterson, B.; Menkel, J.; Alin, S.R.; Hales, B. *Limacina helicina* shell dissolution as an indicator of declining habitat suitability owing to ocean acidification in the California Current Ecosystem. *Proc. R. Soc. B Biol. Sci.* **2014**, *281*, 20140123. [CrossRef] [PubMed]
16. Braunisch, V.; Roder, S.; Coppes, J.; Froidevaux, J.S.P.; Arlettaz, R.; Bollmann, K. Structural complexity in managed and strictly protected mountain forests: Effects on the habitat suitability for indicator bird species. *For. Ecol. Manag.* **2019**, *448*, 139–149. [CrossRef]
17. Skidmore, A.K.; Pettorelli, N.; Coops, N.C.; Geller, G.N.; Hansen, M.; Lucas, R.; Mücher, C.A.; O'Connor, B.; Paganini, M.; Pereira, H.M.; et al. Environmental science: Agree on biodiversity metrics to track from space. *Nature* **2015**, *523*, 403–405. [CrossRef]
18. Novak, J. Evaluation of grassland quality. *Ekol. Bratisl.* **2004**, *23*, 127–143.
19. Polasky, S.; Nelson, E.; Pennington, D.; Johnson, K.A. The Impact of Land-Use Change on Ecosystem Services, Biodiversity and Returns to Landowners: A Case Study in the State of Minnesota. *Environ. Resour. Econ.* **2011**, *48*, 219–242. [CrossRef]
20. MacArthur, R.H.; Wilson, E.O. *The Theory of Island Biogeography*; Princeton University Press: Princeton, NJ, USA, 1967.
21. Wu, L.; Sun, C.; Fan, F. Estimating the Characteristic Spatiotemporal Variation in Habitat Quality Using the Invest Model—A Case Study from Guangdong–Hong Kong–Macao Greater Bay Area. *Remote Sens.* **2021**, *13*, 1008. [CrossRef]
22. Williamson, C.R.; Campa, H., III; Locher, A.B.; Winterstein, S.R.; Beyer, D.E., Jr. Applications of Integrating Wildlife Habitat Suitability and Habitat Potential Models. *Wildl. Soc. Bull.* **2021**, *45*, 70–84. [CrossRef]
23. Wang, C.; Liu, H.; Li, Y.; Dong, B.; Qiu, C.; Yang, J.; Zong, Y.; Chen, H.; Zhao, Y.; Zhang, Y. Study on habitat suitability and environmental variable thresholds of rare waterbirds. *Sci. Total Environ.* **2021**, *785*, 147316. [CrossRef]
24. Sienkiewicz, A.; Łaska, G. Application of Bayesian networks in evaluation of current status and protection of *Pulsatilla patens* (L.) Mill. *Ecosphere* **2021**, *12*, e03337. [CrossRef]
25. de Gabriel Hernando, M.; Fernández-Gil, J.; Roa, I.; Juan, J.; Ortega, F.; de la Calzada, F.; Revilla, E. Warming threatens habitat suitability and breeding occupancy of rear-edge alpine bird specialists. *Ecography* **2021**, *44*, 1191–1204. [CrossRef]
26. McNamara, S.C.; Pintar, M.R.; Resetarits, W.J., Jr. Temperature but not nutrient addition affects abundance and assemblage structure of colonizing aquatic insects. *Ecology* **2021**, *102*, e03209. [CrossRef] [PubMed]
27. Bisson, P.A.; Reeves, G.H.; Bilby, R.E.; Naiman, R.J. Watershed management and pacific salmon: Desired future conditions. In *Pacific Salmon & Their Ecosystems: Status and Future Options*; Stouder, D.J., Bisson, P.A., Naiman, R.J., Eds.; Springer US: Boston, MA, USA, 1997; pp. 447–474.
28. Zhou, C.; Wan, R.; Cao, J.; Xu, L.; Wang, X.; Zhu, J. Spatial variability of bigeye tuna habitat in the Pacific Ocean: Hindcast from a refined ecological niche model. *Fish. Oceanogr.* **2021**, *30*, 23–37. [CrossRef]
29. Berta Aneseyee, A.; Noszczyk, T.; Soromessa, T.; Elias, E. The InVEST Habitat Quality Model Associated with Land Use/Cover Changes: A Qualitative Case Study of the Winike Watershed in the Omo-Gibe Basin, Southwest Ethiopia. *Remote Sens.* **2020**, *12*, 1103. [CrossRef]
30. Nedd, R.; Light, K.; Owens, M.; James, N.; Johnson, E.; Anandhi, A. A Synthesis of Land Use/Land Cover Studies: Definitions, Classification Systems, Meta-Studies, Challenges and Knowledge Gaps on a Global Landscape. *Land* **2021**, *10*, 994. [CrossRef]
31. CPYRWMA. Choctawhatchee River. Available online: https://cpyrwma.alabama.gov/about-the-watersheds-choctawhatchee-river/ (accessed on 28 March 2020).

32. burt Fed, H.; Rosa, S.; City, P. Year One Choctawhatchee Bay Estuary Program Annual Work Plan. 2020. Available online: https://myokaloosa.com/cbec/images/docs/FINAL_YEAR_ONE_CHOCTAWHATCHEE_BAY_ESTUARY_PROGRAM_ANNUAL_WORK_PLAN_7_15_2020.pdf (accessed on 4 June 2022).
33. U.S. Geological Survey (USGS). NLCD 2019 Land Cover (CONUS). Available online: https://www.mrlc.gov/data/nlcd-2019-land-cover-conus (accessed on 15 October 2020).
34. Jenkins, C.N. GIS Layers of Biodiversity Data. Available online: https://biodiversitymapping.org/index.php/download/ (accessed on 16 November 2020).
35. USGS. Drainage Basins across the United States, Optimized for Analysis. 2016. Available online: https://www.arcgis.com/home/item.html?id=651da243132d4ed78dadbf2e5a6c8e5a (accessed on 24 May 2020).
36. EPA; U.S. Environmental Protection Agency. Level III Ecoregions of North America. Available online: https://www.arcgis.com/home/item.html?id=3a21a75ffc8244f6a1c85a19b8ff7713 (accessed on 24 May 2020).
37. EPA; U.S. Environmental Protection Agency. Level I Ecoregions of North America. Available online: https://www.arcgis.com/home/item.html?id=1c8de09728da47c387e19b17a1ffbd92 (accessed on 24 May 2020).
38. Wickham, J.; Homer, C.; Vogelmann, J.; McKerrow, A.; Mueller, R.; Herold, N.; Coulston, J. The multi-resolution land characteristics (MRLC) consortium—20 years of development and integration of USA national land cover data. *Remote Sens.* **2014**, *6*, 7424–7441. [CrossRef]
39. MRLC. National Land Cover Database Class Legend and Description. Available online: https://www.mrlc.gov/data/legends/national-land-cover-database-class-legend-and-description (accessed on 26 May 2020).
40. Sukumaran, H.; Sahoo, S.N. A Methodological framework for identification of baseline scenario and assessing the impact of DEM scenarios on SWAT model outputs. *Water Resour. Manag.* **2020**, *34*, 4795–4814. [CrossRef]
41. He, J.; Huang, J.; Li, C. The evaluation for the impact of land use change on habitat quality: A joint contribution of cellular automata scenario simulation and habitat quality assessment model. *Ecol. Model.* **2017**, *366*, 58–67. [CrossRef]
42. Li, F.X.; Wang, L.Y.; Chen, Z.J.; Clarke, K.C.; Li, M.C.; Jiang, P.H. Extending the SLEUTH model to integrate habitat quality into urban growth simulation. *J. Environ. Manag.* **2018**, *217*, 486–498. [CrossRef]
43. Lina, Z.; Jun, W. Evaluation on effect of land consolidation on habitat quality based on InVEST model. *Trans. Chin. Soc. Agric. Eng.* **2017**, *33*, 250–255. [CrossRef]
44. Xu, L.T.; Chen, S.S.; Xu, Y.; Li, G.Y.; Su, W.Z. Impacts of Land-Use Change on Habitat Quality during 1985–2015 in the Taihu Lake Basin. *Sustainability* **2019**, *11*, 513. [CrossRef]
45. Yunzhe, D.; Jiangfeng, L.; Jianxin, Y. Spatiotemporal responses of habitat quality to urban sprawl in the Changsha metropolitan area. *Prog. Geogr.* **2018**, *37*, 11. [CrossRef]
46. Zhang, H.B.; Wu, F.E.; Zhang, Y.N.; Hans, S.; Liu, Y.Q. Spatial and temporal changes of habitat quality in Jiangsu Yancheng Wetland national nature reserve—Rare birds of China. *Appl. Ecol. Environ. Res.* **2019**, *17*, 4807–4821. [CrossRef]
47. Bai, L.; Xiu, C.; Feng, X.; Liu, D. Influence of urbanization on regional habitat quality: A case study of Changchun City. *Habitat Int.* **2019**, *93*, 102042. [CrossRef]
48. Gong, J.; Xie, Y.; Cao, E.; Huang, Q.; Li, H. Integration of InVEST-habitat quality model with landscape pattern indexes to assess mountain plant biodiversity change: A case study of Bailongjiang watershed in Gansu Province. *J. Geogr. Sci.* **2019**, *29*, 1193–1210. [CrossRef]
49. Song, S.; Liu, Z.; He, C.; Lu, W. Evaluating the effects of urban expansion on natural habitat quality by coupling localized shared socioeconomic pathways and the land use scenario dynamics-urban model. *Ecol. Indic.* **2020**, *112*, 106071. [CrossRef]
50. Chu, L.; Sun, T.; Wang, T.; Li, Z.; Cai, C. Evolution and Prediction of Landscape Pattern and Habitat Quality Based on CA-Markov and InVEST Model in Hubei Section of Three Gorges Reservoir Area (TGRA). *Sustainability* **2018**, *10*, 3854. [CrossRef]
51. Zhu, J.; Ding, N.; Li, D.; Sun, W.; Xie, Y.; Wang, X. Spatiotemporal Analysis of the Nonlinear Negative Relationship between Urbanization and Habitat Quality in Metropolitan Areas. *Sustainability* **2020**, *12*, 669. [CrossRef]
52. Liu, D.; Liang, X.; Chen, H.; Zhang, H.; Mao, N. A Quantitative Assessment of Comprehensive Ecological Risk for a Loess Erosion Gully: A Case Study of Dujiashi Gully, Northern Shaanxi Province, China. *Sustainability* **2018**, *10*, 3239. [CrossRef]
53. Liu, J.; Zhang, G.; Zhuang, Z.; Cheng, Q.; Gao, Y.; Chen, T.; Huang, Q.; Xu, L.; Chen, D. A new perspective for urban development boundary delineation based on SLEUTH-InVEST model. *Habitat Int.* **2017**, *70*, 13–23. [CrossRef]
54. Chengxin, W.; Changbo, Q.; Hongdi, L.; Yan, S. Exploration of Ecological Space Identification and Ecological Impact Assessment in Planning Environmental Impact Assessment—A Case Study of Changchun New District Development Planning. *Chin. J. Environ. Manag.* **2017**, *9*, 88–94. [CrossRef]
55. Yohannes, H.; Soromessa, T.; Argaw, M.; Dewan, A. Spatio-temporal changes in habitat quality and linkage with landscape characteristics in the Beressa watershed, Blue Nile basin of Ethiopian highlands. *J. Environ. Manag.* **2021**, *281*, 111885. [CrossRef] [PubMed]
56. Choudhary, A.; Deval, K.; Joshi, P.K. Study of habitat quality assessment using geospatial techniques in Keoladeo National Park, India. *Environ. Sci. Pollut. Res.* **2021**, *28*, 14105–14114. [CrossRef] [PubMed]
57. Sharma, R.; Nehren, U.; Rahman, S.A.; Meyer, M.; Rimal, B.; Aria Seta, G.; Baral, H. Modeling Land Use and Land Cover Changes and Their Effects on Biodiversity in Central Kalimantan, Indonesia. *Land* **2018**, *7*, 57. [CrossRef]
58. Terrado, M.; Sabater, S.; Chaplin-Kramer, B.; Mandle, L.; Ziv, G.; Acuna, V. Model development for the assessment of terrestrial and aquatic habitat quality in conservation planning. *Sci. Total Environ.* **2016**, *540*, 63–70. [CrossRef]

59. Hedges, L.V. Meta-Analysis. *J. Educ. Stat.* **1992**, *17*, 279–296. [CrossRef]
60. Field, A.P.; Gillett, R. How to do a meta-analysis. *Br. J. Math. Stat. Psychol.* **2010**, *63*, 665–694. [CrossRef] [PubMed]
61. Kurtener, D.; Torbert, H.A.; Krueger, E. Evaluation of Agricultural Land Suitability: Application of Fuzzy Indicators. In Proceedings of the International Conference on Computational Science and Its Applications, Perugia, Italy, 30 June–3 July 2008; Springer: Berlin/Heidelberg, Germany, 2008; pp. 475–490.
62. Britannica. Deciduous Forest. Available online: https://www.britannica.com/science/deciduous-forest (accessed on 4 June 2022).
63. Britannica. Coniferous Forest. Available online: https://www.britannica.com/science/coniferous-forest (accessed on 4 June 2022).
64. Britannica. Mixed Forest. Available online: https://www.britannica.com/science/mixed-forest (accessed on 4 June 2022).
65. Prokhorov, A. Mixed Forest. Available online: https://encyclopedia2.thefreedictionary.com/Mixed+Forest (accessed on 4 June 2022).
66. Crandell, C.J. Wetland. Available online: https://www.britannica.com/science/wetland (accessed on 18 June 2022).
67. NSW Department of Planning and Environment. Plants and Animals in Wetlands. Available online: https://www.environment.nsw.gov.au/topics/water/wetlands/plants-and-animals-in-wetlands/ (accessed on 4 June 2022).
68. Smith, J.M.B. Scrubland. Available online: https://www.britannica.com/science/scrubland (accessed on 4 June 2022).
69. Smith, J.M.B. Grassland. Available online: https://www.britannica.com/science/grassland (accessed on 18 June 2022).
70. WWF. Grasslands. Available online: https://www.worldwildlife.org/habitats/grasslands (accessed on 18 June 2022).
71. DNR. Barren. Available online: https://secure.in.gov/dnr/fishwild/files/SWAP/SWAPHabitatSummary_Barren.pdf (accessed on 12 January 2022).
72. MRLC. National Land Cover Database 2016 (NLCD2016) Legend. Available online: https://www.mrlc.gov/data/legends/national-land-cover-database-2016-nlcd2016-legend (accessed on 18 June 2022).
73. Rondinini, C.; Di Marco, M.; Chiozza, F.; Santulli, G.; Baisero, D.; Visconti, P.; Hoffmann, M.; Schipper, J.; Stuart, S.N.; Tognelli, M.F.; et al. Global habitat suitability models of terrestrial mammals. *Philos. Trans. R. Soc. B Biol. Sci.* **2011**, *366*, 2633–2641. [CrossRef]
74. Alatawi, A.S.; Gilbert, F.; Reader, T. Modelling terrestrial reptile species richness, distributions and habitat suitability in Saudi Arabia. *J. Arid. Environ.* **2020**, *178*, 104153. [CrossRef]
75. Hernández-Quiroz, N.S.; Badano, E.I.; Barragán-Torres, F.; Flores, J.; Pinedo-Álvarez, C. Habitat suitability models to make conservation decisions based on areas of high species richness and endemism. *Biodivers. Conserv.* **2018**, *27*, 3185–3200. [CrossRef]
76. Muhammed, K.; Anandhi, A.; Chen, G.; Poole, K. Define–Investigate–Estimate–Map (DIEM) Framework for Modeling Habitat Threats. *Sustainability* **2021**, *13*, 11259. [CrossRef]
77. Chase, M.K.; Kristan, W.B.; Lynam, A.J.; Price, M.V.; Rotenberry, J.T. Single Species as Indicators of Species Richness and Composition in California Coastal Sage Scrub Birds and Small Mammals. *Conserv. Biol.* **2000**, *14*, 474–487. [CrossRef]
78. Finch, O.-D.; Löffler, J. Indicators of species richness at the local scale in an alpine region: A comparative approach between plant and invertebrate taxa. *Biodivers. Conserv.* **2010**, *19*, 1341–1352. [CrossRef]
79. Margalef, R. Information Theory in Ecology. *Gen. Syst.* **1958**, *3*, 36–71.
80. Pielou, E.C. *An Introduction to Mathematical Ecology*; Wiley-Inter-Science: New York, NY, USA, 1969; p. 286.
81. Whittaker, R.H. Evolution of species diversity in land communities. *Evol. Biol.* **1977**, *10*, 1–67.
82. Shannon, C.E. A mathematical theory of communication. *Bell Syst. Tech. J.* **1948**, *27*, 379–423. [CrossRef]
83. Anandhi, A. CISTA-A: Conceptual model using indicators selected by systems thinking for adaptation strategies in a changing climate: Case study in agro-ecosystems. *Ecol. Model.* **2017**, *345*, 41–55. [CrossRef]
84. Anandhi, A.; Kannan, N. Vulnerability assessment of water resources–translating a theoretical concept to an operational framework using systems thinking approach in a changing climate: Case study in Ogallala Aquifer. *J. Hydrol.* **2018**, *557*, 460–474. [CrossRef]
85. Malavasi, M. The map of biodiversity mapping. *Biol. Conserv.* **2020**, *252*, 108843. [CrossRef]
86. Andersen, L.H.; Sunde, P.; Pellegrino, I.; Loeschcke, V.; Pertoldi, C. Using population viability analysis, genomics, and habitat suitability to forecast future population patterns of Little Owl Athene noctua across Europe. *Ecol. Evol.* **2017**, *7*, 10987–11001. [CrossRef]
87. Larson, M.A.; Thompson, F.R.; Millspaugh, J.J.; Dijak, W.D.; Shifley, S.R. Linking population viability, habitat suitability, and landscape simulation models for conservation planning. *Ecol. Model.* **2004**, *180*, 103–118. [CrossRef]
88. Reşit Akçakaya, H.; McCarthy, M.A.; Pearce, J.L. Linking landscape data with population viability analysis: Management options for the helmeted honeyeater Lichenostomus melanops cassidix. *Biol. Conserv.* **1995**, *73*, 169–176. [CrossRef]
89. Crawford, B.A.; Maerz, J.C.; Moore, C.T. Expert-Informed Habitat Suitability Analysis for At-Risk Species Assessment and Conservation Planning. *J. Fish Wildl. Manag.* **2020**, *11*, 130–150. [CrossRef]
90. McDonald, R.I.; Güneralp, B.; Huang, C.-W.; Seto, K.C.; You, M. Conservation priorities to protect vertebrate endemics from global urban expansion. *Biol. Conserv.* **2018**, *224*, 290–299. [CrossRef]
91. Biermann, F.; Kanie, N.; Kim, R.E. Global governance by goal-setting: The novel approach of the UN Sustainable Development Goals. *Curr. Opin. Environ. Sustain.* **2017**, *26*, 26–31. [CrossRef]

92. Dhyani, S.; Singh, S.; Kadaverugu, R.; Pujari, P.; Verma, P. Habitat Suitability Modelling and Nature-Based Solutions: An Efficient Combination to Realise the Targets of Bonn Challenge and SDGs in South Asia. In *Nature-Based Solutions for Resilient Ecosystems and Societies*; Dhyani, S., Gupta, A.K., Karki, M., Eds.; Springer: Singapore, 2020; pp. 347–364.
93. Thi Nguyen, T.H.; Boets, P.; Lock, K.; Damanik Ambarita, M.N.; Forio, M.A.E.; Sasha, P.; Dominguez-Granda, L.E.; Thi Hoang, T.H.; Everaert, G.; Goethals, P.L.M. Habitat suitability of the invasive water hyacinth and its relation to water quality and macroinvertebrate diversity in a tropical reservoir. *Limnologica* **2015**, *52*, 67–74. [CrossRef]
94. Agrillo, E.; Filipponi, F.; Pezzarossa, A.; Casella, L.; Smiraglia, D.; Orasi, A.; Attorre, F.; Taramelli, A. Earth Observation and Biodiversity Big Data for Forest Habitat Types Classification and Mapping. *Remote Sens.* **2021**, *13*, 1231. [CrossRef]
95. Jetz, W.; McGeoch, M.A.; Guralnick, R.; Ferrier, S.; Beck, J.; Costello, M.J.; Fernandez, M.; Geller, G.N.; Keil, P.; Merow, C.; et al. Essential biodiversity variables for mapping and monitoring species populations. *Nat. Ecol. Evol.* **2019**, *3*, 539–551. [CrossRef]
96. Blackman, R.C.; Osathanunkul, M.; Brantschen, J.; Di Muri, C.; Harper, L.R.; Mächler, E.; Hänfling, B.; Altermatt, F. Mapping biodiversity hotspots of fish communities in subtropical streams through environmental DNA. *Sci. Rep.* **2021**, *11*, 10375. [CrossRef]
97. Eastwood, N.; Stubbings, W.A.; Abou-Elwafa Abdallah, M.A.; Durance, I.; Paavola, J.; Dallimer, M.; Pantel, J.H.; Johnson, S.; Zhou, J.; Hosking, J.S.; et al. The Time Machine framework: Monitoring and prediction of biodiversity loss. *Trends Ecol. Evol.* **2021**, *37*, 138–146. [CrossRef]

land

MDPI

Article

Ecological Environment Dynamic Monitoring and Driving Force Analysis of Karst World Heritage Sites Based on Remote-Sensing: A Case Study of Shibing Karst

Ning Zhang [1,2], Kangning Xiong [1,2,*], Hua Xiao [1,2,*], Juan Zhang [1,2] and Chuhong Shen [1,2]

1 School of Karst Science, Guizhou Normal University, Guiyang 550001, China
2 State Engineering Technology Institute for Karst Desertification Control, Guiyang 550001, China
* Correspondence: xiongkn@gznu.edu.cn (K.X.); 201407030@gznu.edu.cn (H.X.)

Abstract: The evaluation and monitoring of the ecological environment quality of heritage sites can help provide sustainable and healthy development strategies for heritage management organizations. In this study, an ecological evaluation model based on the remote sensing ecological index (RSEI) was used to measure the ecological environment of the Shibing Karst World Heritage Site and its buffer zone and the Moran index and geographic probe model were combined to quantify the ecological environment. The results show that, (1) from 2013 to 2020, the ecological environment quality of the heritage site and buffer zone was moderate to high and the mean RSEI values in the three periods studied were 0.720, 0.723 and 0.742, showing an overall upward and improving trend; (2) ecological environment quality grades of moderate and good accounted for more than 70% of the area, the distribution pattern of ecological environment quality is significantly better at the heritage site than in the buffer zone and the southwest is better than the northeast; (3) the Moran index increased from 0.600 in 2013 to 0.661 in 2020, residing in the first and third quadrants, respectively, with significantly spatial aggregation; and (4) greenness and humidity were shown to play a positive feedback role on the ecological environment quality and the spatial influence ability of humidity and dryness was greater. Overall, the RSEI is an effective method of evaluating and monitoring the ecological environment quality of heritage sites, the ecological environment quality of the Karst heritage site in Shibing is in a steady state of improvement and the relevant departments of heritage conservation need to further coordinate the relationship between conservation and development to promote the sustainable development of the heritage site and provide effective solutions for the monitoring of other Karst World Heritage sites.

Keywords: remote sensing; ecological environment; world natural neritage; Shibing Karst; South China Karst

Citation: Zhang, N.; Xiong, K.; Xiao, H.; Zhang, J.; Shen, C. Ecological Environment Dynamic Monitoring and Driving Force Analysis of Karst World Heritage Sites Based on Remote-Sensing: A Case Study of Shibing Karst. *Land* **2023**, *12*, 184. https://doi.org/10.3390/land12010184

Academic Editor: Shiliang Liu

Received: 17 November 2022
Revised: 20 December 2022
Accepted: 4 January 2023
Published: 6 January 2023

1. Introduction

To protect the common natural and cultural heritage of mankind, UNESCO adopted the Convention Concerning the Protection of the World Culture and Natural Heritage at its 17th General Conference, held at its headquarters in Paris, in November 1972 and initiated the organization of the World Heritage Organization in 1976 [1]. World Heritage refers to cultural and natural heritage with outstanding value that is a precious treasure of nature and a symbol of human history, culture and civilization, representing the most valuable cultural and natural landscapes and the common wealth of mankind [2]. In many countries around the world, World Heritage Sites have been considered representatives of national culture and even as symbols of the country [3,4]; however, in recent years, World Natural Heritage Sites (WNHs) have suffered extensive damage due to earthquakes, tsunamis, soil erosion, human activities, etc. [5,6]. To date, 17 World Natural Heritage (WNH) properties have been inscribed on the List of World Heritage in Danger.

Karst is one of the most remarkable landscapes in the world, mainly consisting of special topography and associated ecosystems developed on carbonate rocks, characterized by vadose flows, caves, dark rivers, canyons, depressions and conical and towering peaks [7]. China is one of the countries with the most extensive karst distribution in the world, with 1.25 million km^2 of carbonate rock outcrops [8] accounting for 13% of the total land area of China. The "South China Karst" is a series of heritage sites, the first phase of which consists of three constituent sites: Shilin Karst (Yunnan), Libo Karst (Guizhou) and Wulong Karst (Chongqing), which were inscribed on the World Heritage List in 2007 for meeting World Heritage criteria (vii) and (viii) [9]. The second phase of heritage sites consists of Guilin Karst (Guangxi), Shibing Karst (Guizhou) and Jinfo Mountain Karst (Chongqing), which were inscribed on the World Heritage List in 2014 for meeting World Heritage criteria (vii) and (viii). As karst areas are characterized by soil vulnerability, hydrological vulnerability, vegetation vulnerability and human vulnerability [10], it is necessary to monitor them regularly and to protect the sustainability of karst WNHs.

To protect World Heritage properties, the World Heritage Centre developed a program of periodic monitoring reports, conservation status reports, etc., in the 1970s to investigate and track the health of natural World Heritage properties [11]. Monitoring is the process of collecting information, analyzing data [12] and subsequently using the information to assess the status, threat factors and severity of resources [13]. Current research applications to monitor ecological quality through integrated survey data by remote sensing [14], PSR models [15], urbanization factors [16] and natural succession of the landscape [17] are also common. However, in the process of monitoring heritage sites, it is important to consider human and financial constraints and to choose an appropriate strategy. With the development of remote sensing science, multi-source remote sensing technology is frequently applied in the monitoring of WNHs [18], especially the combination of GIS and remote sensing technology, which enables researchers and managers of heritage sites to effectively and reliably monitor the ecological environment quality [19,20]. Currently, one method for measuring ecological quality is single-factor change analysis, including analysis of land use change [21,22], NPP [23] and eco-efficiency change [24], as well as other factor changes closely related to the ecological environment. Another method is integrated multi-factor change analysis, which is more comprehensive and accurate compared with single-factor analysis, and scholars have proposed various evaluation index systems for this purpose [25,26]. Xu Hanqiu et al. [27] proposed a pure remote sensing-driven Remote Sensing based Ecological Index (RSEI) to reflect the ecological status comprehensively, which was normalized and subjected to principal component analysis from four aspects: greenness NDVI, heat LST, humidity WET and dryness NDISI, to achieve remote sensing ecological status evaluation. Subsequently, several scholars have conducted practical studies in areas such as Bayinbrook WNHs and Karajun-Kurdening WNHs in Xinjiang, China [28]. Although this method has been successfully applied in different regions, no remote sensing ecological evaluation study on karst WNHs have been conducted.

As a typical representative of dolomite karst landscapes, this study uses Landsat remote sensing images as the data source, explores the characteristics of ecological environment quality changes in the heritage site by constructing a RSEI model and using principal component analysis, Moran index [29] and geographic probe, further analyzes the driving factors of its ecological environment changes and provides a reference basis for the ecological environmental protection and sustainable development of karst heritage sites.

2. Materials and Methods

2.1. Study Area

Shibing Karst WNHs is located at 108°01′36″–108°10′52″ E, 27°13′56″–27°04′51″ N in Shibing County, Guizhou Province. The average elevation is 912 m (ranging from 600 to 1250 m). It is located on the slope of the overlap between the mountains of Qianzhong and the hills of western Hunan (Figure 1). It has a typical and complete dolomite karst landscape, which is deeply cut by rivers and has a surface form of crested canyons and crested valleys. The heritage site represents continental tropical–subtropical dolomite karst geological evolution and bio-ecological processes and is an ideal place and natural test site for the study of dolomite karst forest vegetation.

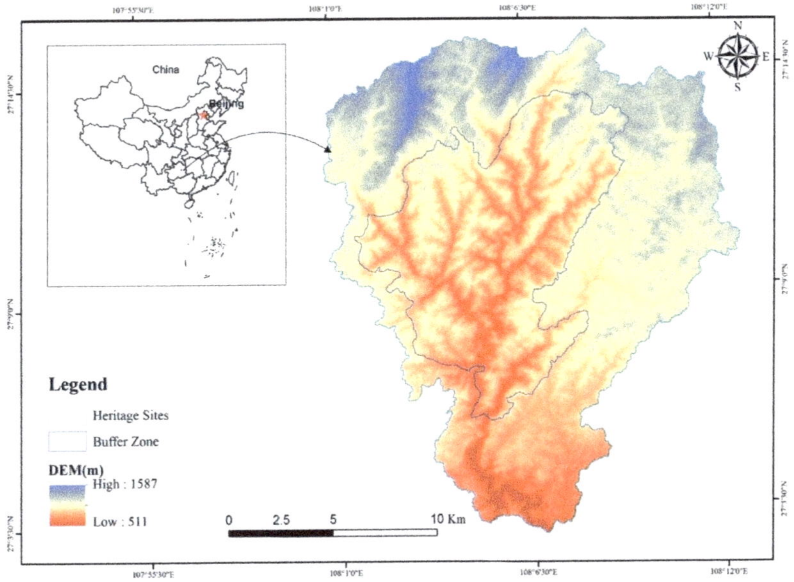

Figure 1. Location of the study area.

2.2. *Data Source and Pre-Processing*

The image data used for this study were Landsat 8 OLI remote sensing images of the Shibing Karst WNHs from 2013 to 2020, which are available through the USGS (https://earthexplorer.usgs.gov/; last accessed on 12 October 2022) and Geospatial Data Cloud (http://www.gscloud.cn/; last accessed on 12 October 2022). The image data are available for free. The acquired images were all in the third quarter of the year and cloud-free in the study area and the image quality was good. The data are preprocessed with radiometric calibration, atmospheric correction, geometric correction, image mosaic, image alignment, etc.

2.3. *Methodology*

2.3.1. Ecological Indicators Extracted

RSEI is a new ecological index that uses multi-source remote sensing data and integrates several natural factors as the main driving factors to monitor and evaluate the ecological environment quality of a certain area, which has the characteristics of short cycle time and wide application when compared with the code EI index and complements the EI index well. The index couples four quantifiable indexes, namely, greenness, humidity, dryness and heat, and is constructed using a principal component analysis (PCA), which can quickly evaluate the ecological environment quality for a certain study area.

Greenness. The Normalized Difference Vegetation Index (NDVI) is one of the best indicators of healthy vegetation growth, vegetation distribution and vegetation density distribution and has the following formula [30,31]:

$$NDVI = \frac{\rho_4 - \rho_3}{\rho_4 + \rho_3} \quad (1)$$

Humidity. The wetness component of the tasseled cap transformation (WET) component reflects the moisture content of water bodies, soil and vegetation and is mainly obtained from the remote sensing tassel cap transformation.

$$WET_{OLT} = 0.1511\rho_1 + 0.1973\rho_2 + 0.3283\rho_3 + 0.3407\rho_4 - 0.7117\rho_5 - 0.4559\rho_6 \quad (2)$$

Dryness. The degree of soil drying is an important factor that affects the ecological environment and is positively related to it [32]. However, in practice, the presence of certain building sites in the area is also an important factor affecting the ecological environment. Therefore, the Normalized Difference Imperviousness and Soil Index (NDISI) is synthesized using both the bare soil index (SI) and the Impervious Built-up Index (IBI).

$$NDISI = \frac{SI + IBI}{2} \quad (3)$$

$$SI = \frac{(\rho_3 + \rho_5) - (\rho_1 + \rho_4)}{(\rho_3 + \rho_5) + (\rho_1 + \rho_4)} \quad (4)$$

$$IBI = \frac{2\rho_5/(\rho_5 + \rho_4) - [\rho_4/(\rho_4 + \rho_3) + \rho_2/(\rho_2 + \rho_5)]}{2\rho_5/(\rho_5 + \rho_4) + [\rho_4/(\rho_4 + \rho_3) + \rho_2/(\rho_2 + \rho_5)]} \quad (5)$$

Heat. Land surface temperatures (LST) were calculated using the Landsat user manual model [33] and the surface temperature was chosen to represent the heat index.

$$L_\rho = [\varepsilon P(T_s) + (1 - \varepsilon)L_\downarrow]\tau + L_\uparrow \quad (6)$$

$$P(T_s) = \frac{L_\rho - L_\uparrow - \tau(1 - \varepsilon)L_\downarrow}{\tau\varepsilon} \quad (7)$$

$$T_s = \frac{K_2}{Ln[K_1/P(T_S) + 1]} \quad (8)$$

In the equation, L_ρ is the thermal infrared radiance brightness value; ε is the surface-specific emissivity; T_s is the real surface temperature; and k is the calibration parameter.

Construction of RSEI. The above four indicators were standardized in order to facilitate comparisons under the same system and the standardization formula was as follows:

$$NI_i = \frac{(I_i - I_{\min})}{(I_{\max} - I_{\min})} \quad (9)$$

where NI_i is the index value after the normalization of an image element; I_i is the DN value of the index at an image element i; $I_{\max}$ is the maximum value of the index; and $I_{\min}$ is the minimum value of the index.

The four standardized images were then synthesized and the four indicators were coupled using principal component analysis (PCA), a multidimensional data compression technique that selects a few important variables via the orthogonal linear transformation of multiple variables that has the advantages of integrating the weights of each indicator, avoiding human determination and automatically and objectively determining each indicator based on the nature of the data themselves and the contribution of each indicator to each principal component. The following equation was used for the initial RSEI calculation:

$$RSEI = 1 - PCA[f(NDVI, WET, NDBSI, LST)] \quad (10)$$

Similarly, the calculated RSEI values were standardized to obtain the final RSEI.

2.3.2. Exploratory Spatial Data Analysis

The first law of geography states that the correlation between features is related to distance and, in general, the closer the distance, the greater the correlation between features; the farther the distance, the greater the dissimilarity between features [34]; therefore, the law of spatial correlation is often used in spatiotemporal evolution studies [35] and the global Moran's I index is used to express the global spatial autocorrelation.

$$I = \frac{\sum_{i=1}^{n}\sum_{i=1}^{n} \omega_{ij}(x_i - \overline{x})(x_j - \overline{x})}{S^2 \sum_{i=1}^{n}\sum_{j=1}^{n} \omega_{ij}} \tag{11}$$

$$S^2 = \frac{1}{n}\sum(x_i - \overline{x})^2 \tag{12}$$

where I is the global Moran's I index with a value range of $(-1, 1)$, $I < 0$ indicates a negative correlation and $I > 0$ indicates a positive correlation; ω_{ij} is the weighting coefficient; and X_i and X_j are the remotely sensed ecological indices at i and j in the study area, respectively.

Local spatial autocorrelation in terms of local Moran's I index:

$$I_i = \frac{(x_i - \overline{x})}{s^2}\sum_j \omega_{ij}(x_j - \overline{x}) \tag{13}$$

2.3.3. Geographical Detector

A geographic detector is a statistical method that detects spatial differentiation and reveal the driving forces behind it. It is mainly used to analyze the interaction between multiple factors [36] and the analysis of spatial differences in regional variables, such as changes in spatial patterns. A model consists of four main detectors: the divergence and factor detector, interaction, risk and ecological detector.

Divergence and factor detection. This is mainly used to detect the spatial heterogeneity of the attributes and the ability of the drivers to explain the RSEI attributes.

$$q = 1 - \frac{1}{N\sigma^2}\sum_{h=1}^{L} N_h\sigma_h^2 \tag{14}$$

where q is the explanatory power of an influencing factor on RSEI, which takes values in the range [0, 1]—the larger the value, the stronger the explanatory power; h is the sub-region of the image factor; L is the number of grades and classifications of RSEI and the influence factor; N_h and N are the numbers of units in different grades of the region and the whole region, respectively; σ_h^2 and σ^2 is the variance of RSEI in different grades of the region and the whole region.

Risk Area Detection. This is mainly used to determine whether there is a significant difference in the mean values of attributes between two sub-regions using the t-statistic test.

$$t_{\overline{y}_{h=1}-\overline{y}_{h=2}} = \frac{\overline{Y}_{h=1} - \overline{Y}_{h=2}}{\left[\frac{Var(\overline{Y}_{h=1})}{n_{h=1}} + \frac{Var(\overline{Y}_{h=2})}{n_{h=2}}\right]^{1/2}} \tag{15}$$

where $\overline{Y}_{h=\mathrm{i}}$ denotes the mean value of the attributes in subregion h, n_h denotes the number of samples in subregion h and Var indicates the variance. If H_0 is rejected at the confidence level, there is a significant difference in the mean values of the attributes between the two sub-regions.

2.3.4. Processing Flow

A flow chart of the ecological environment dynamic monitoring and driver analysis is shown in Figure 2.

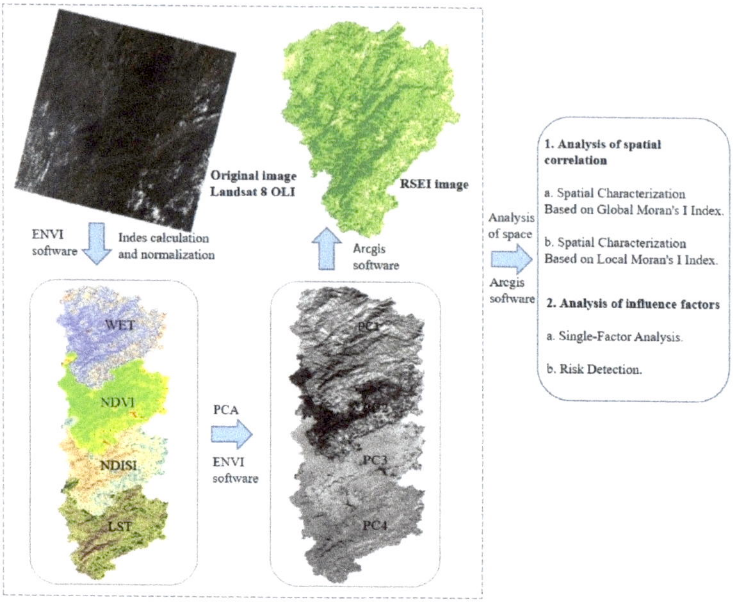

Figure 2. Methodological framework for the ecological quality analysis of the Shibing Karst.

3. Results

3.1. Characteristics of the Ecological Environment

In a principal component analysis, the principal component is a linear combination of individual indicators and the weights of the indicators are the eigenvectors. These indicate the contribution of each indicator to the principal component and determine the actual significance of the principal component. The four normalized indicators were analyzed using the principal component analysis module in ENVI. The eigenvalues and contribution rates of the principal components were obtained to demonstrate the applicability of RSEI in karst heritage sites (Table 1). The results showed that PC1 had the largest eigenvalues among the four PCs in the study years, with a proportion of 80–87%, indicating that PC1 collected the most information on the variability of the four indicators compared to PC2, PC3 and PC4. Therefore, all four indicator variables are represented by PC1.

Table 1. Results of principal component analysis.

Year		PC1	PC2	PC3	PC4
2013	Eigenvalues/λ	0.012	0.002	0.001	0.000
	Contribution Ratio/%	80.21%	11.51%	7.81%	0.47%
2016	Eigenvalues/λ	0.660	0.001	0.001	0.0001
	Contribution Ratio/%	87.42%	7.9%	3.46%	1.22%
2020	Eigenvalues/λ	0.008	0.002	0.001	0.000
	Contribution Ratio/%	80.56%	10.94%	6.17%	2.33%

Table 2 provides the mean values of RSEI, which range from 0.720 to 0.742 (corresponding to level 4). This indicates that the overall ecological quality of Shibing karst improved during the study period. The mean values of the four indicators during the study period are also provided. Over all four study years, WET, which contributed the most to PC1, increased by 18.8%, from 0.753 to 0.895 and NDVI increased by 0.056 (7.2%). Of the other two indicators, NDISI decreased by 60.7% and LST increased by 7.5%. The increases in NDVI and WET and the decreases in NDISI and LST can be offset by an increase in RSEI by

3.1% over the study period. Together, the four indicators of RSEI produce a quantitative signal of response to ecological stressors. The strength of RSEI lies not only in its ability to provide a specific area final score but also in its interpretation of the scores of four indicators representing specific spatial characteristics of ecological states. Thus, as an ecological quality evaluation indicator, RSEI is more comprehensive than other individual indicators.

Table 2. RSEI and mean value of each index.

Year	Index				RSEI
	NDVI	WET	NDISI	LST	
2013	0.773	0.753	0.880	0.469	0.720
2016	0.782	0.776	0.509	0.570	0.723
2020	0.829	0.895	0.345	0.436	0.742

3.2. Analysis of Spatial and Temporal Variation in RSEI

In order to more accurately reflect the characteristics of ecological environment quality changes at the Shibing Karst WNHs, the RSEI was divided into five levels of parity, with Level 1 0–0.2 (very poor), Level 2 0.2–0.4 (poor), Level 3 0.4–0.6 (moderate), Level 4 0.6–0.8 (good) and Level 5 0.8–1 (excellent). In the RSEI grade images of the study years (Figure 3), the changes in the ecological environment quality of the Shibing Karst WNHs from 2013 before the inscription to 2020 after the inscription are comprehensively illustrated. The red patches representing areas with poor to very poor ecological conditions are mainly concentrated in the southern part of the buffer zone, near the urban area of Shibing County, and have been developed intensively. The green patches, ranging from good to excellent, are widely distributed over the site, which is dominated by dolomite karst with good forest and shrub cover. The overall greening trend indicates that the ecological condition within the site is very good, while the eastern and southern areas of the buffer zone near the urban area need to be improved.

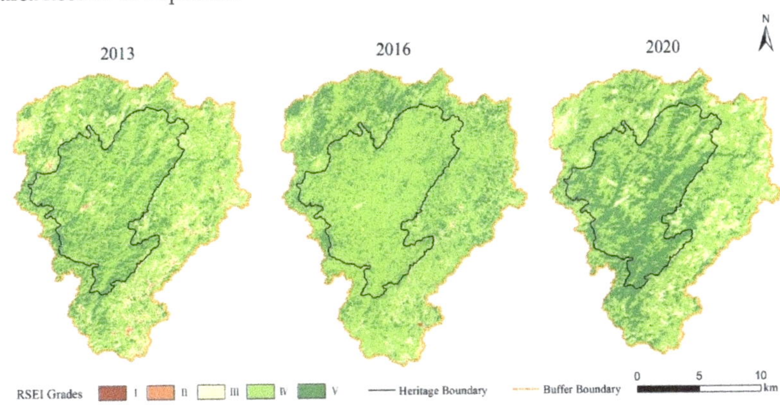

Figure 3. Spatial and temporal evolution of the RSEI grades at the Shibing Karst.

The area of each ecological status class was calculated and the data are shown in Figure 4. The area of the five areas in 2013 was 0.167 km^2, 3.124 km^2, 30.012 km^2, 189.774 km^2 and 60.222 km^2, respectively. From 2013 to 2020, the fifth level increased to 84.861 km^2 and the first, second, third and fourth levels decreased to 0.002 km^2, 0.0981 km^2, 23.076 km^2 and 175.262 km^2, respectively. The area of the region with poor and worse RSEI levels decreases period by period and the regional transition mainly occurs between adjacent layers, which is manifested as a transition from lower to higher levels. The main transition types are level 1 to level 2, level 2 to level 3, level 3 to level 4 and level 4 to level 5.

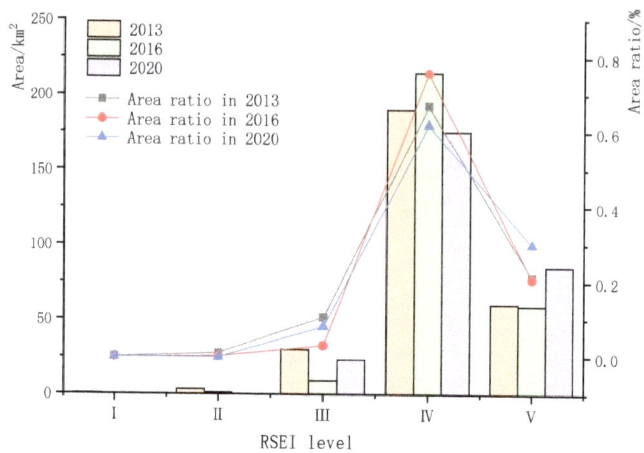

Figure 4. Area and proportion of each ecological class from 2013 to 2020.

We counted the area of the three RSEI grades. Combined with Table 3, the area of good and excellent regions in the third RSEI classification of the Shibing Karst WNHs increased from 88.24% in 2013 to 91.82% in 2020, indicating that the overall ecological environment quality within the site is good. The area of the medium region decreased from 10.6% in 2013 to 8% in 2020, indicating that the medium region is more stable. The areas with poor and worse RSEI grades decreased period by period, from 1.15% in 2013 to 0.04% in 2020 and the degree of decrease was more obvious.

Table 3. RSEI and mean value of each index.

Area (km^2)	2020					
2013	**I**	**II**	**III**	**IV**	**V**	**Total**
I	0	0.019	0.095	0.051	0	0.165
II	0	0.027	1.980	1.109	0.005	3.121
III	0.001	0.041	13.095	16.797	0.077	30.011
IV	0.001	0.006	7.869	144.407	37.489	189.772
V	0	0.002	0.035	12.895	47.288	60.220
Total	0.002	0.095	23.074	175.259	84.859	283.289

As can be seen from Table 4, from 2013 to 2020, 37 km^2 of the Shibing Karst WNHs were converted from good to excellent, accounting for 13.23% of the total area. The area of excellent turned to poor is 0% and the area of good and medium turned to poor is less than 0.1%. In general, the ecological environment quality of Shibing Karst WNHs is in a better state after its inscription.

Table 4. Single-factor detection q-value.

Factors	2013	2016	2020
NDVI	0.418	0.611	0.364
NDISI	0.823	0.900	0.798
WET	0.907	0.750	0.834
LST	0.279	0.271	0.531
DEM	0.011	0.033	0.129
LUCC	0.612	0.341	0.655

To further indicate the spatial and temporal changes in the ecological environmental quality of the heritage site, this paper analyzes the changes in RSEI values between 2013 and 2020 (Figure 5) and divides the results into five conversion levels: significantly lower, slightly lower, basically unchanged, slightly higher and significantly higher, for a comprehensive analysis of the data of the three phases of the Shibing Karst WNHs. It can be seen that, from 2013 to 2016, the ecological environment quality within the heritage site basically remained unchanged. Between 2016 and 2020, the ecological environment quality of the heritage site significantly improved and the parts that showed a decreased index were almost all within the buffer zone. Through on-site investigations and by understanding the relevant local areas, we found that both the buffer zone and the heritage site were developing tourism during this period and human activities were more obvious, which caused damage to the ecological environment in the buffer zone. In general, the ecological environment quality of the heritage site and buffer zone is improving and only in the buffer zone are there scattered areas with a decreasing trend.

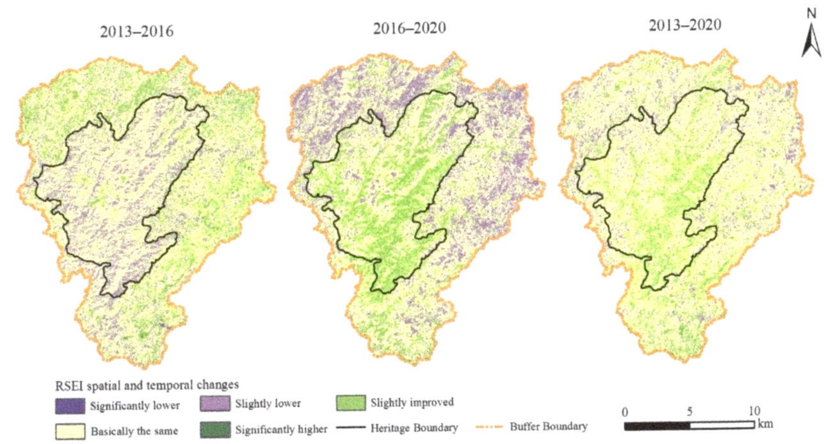

Figure 5. Analysis of RSEI changes from 2013 to 2020.

3.3. Characteristics of Ecological and Environmental Quality Changes

3.3.1. Spatial Characterization Based on Global Moran's I Index

The RSEI maps for 2013–2020 were used to examine the global Moran's I index, which can describe the overall correlation (Figure 6). The global Moran's I indices all passed the significance test (0.02 significance level), indicating that the ecological and environmental quality of Shibing Karst WNHs has significant spatial autocorrelation characteristics. In terms of evolutionary trends, the global Moran's I index showed an increasing characteristic from 2007 to 2018, the degree of spatial clustering of the ecological environment in the study area was continuously strengthened and the level of clustering was improved.

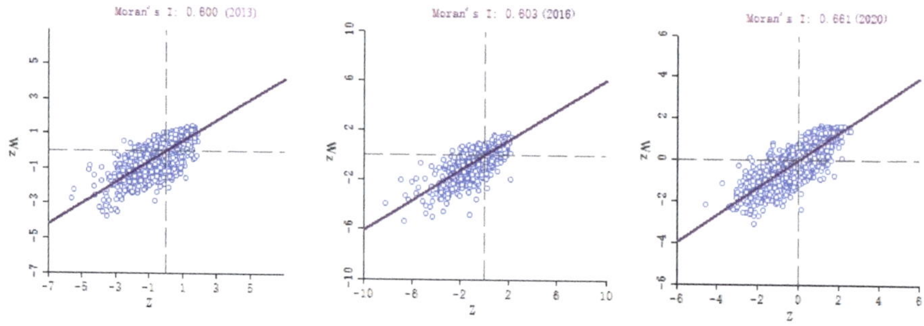

Figure 6. Moran's I scatter plot.

3.3.2. Spatial Characterization Based on Local Moran's I Index

The local Moran index was used to assess the spatial dependence between samples. The following four types of spatial association were proposed. High–high cluster value type, high–low outlier type, low–high outlier type and low–low cluster type. High–high and low–low clusters correspond to positive spatial autocorrelation, while high–low and low–high outliers correspond to negative spatial autocorrelation. The characteristics of local spatial clustering are summarized in Figure 7. 1. High–high clustering type. The spatial differences of the high–high clustering type are small. The values of neighboring samples are highly sampled and show significant positive correlations. Most of the high-clustering samples are distributed in the middle, i.e., within the heritage site. 2. High–low outlier type. The sample itself has high ecological quality, while its neighboring samples have low ecological quality, showing a negative correlation of "high itself, low surrounding". The high–low outlier samples were scattered at the edge of the study area, showing a point-like distribution structure. 3. Low–high outlier type. The value of this sample is very low, while the value of the neighboring samples is high. The low–high outlier type shows a negative correlation of "low itself, high surrounding". They are mainly distributed in parts of the buffer zone. 4. The spatial variation of the low–low clustering type is small. These samples and their neighboring samples are of low ecological quality and show a significant positive correlation. The number of low–low clustering type samples gradually decreased and showed a blocky distribution, mainly in the buffer area.

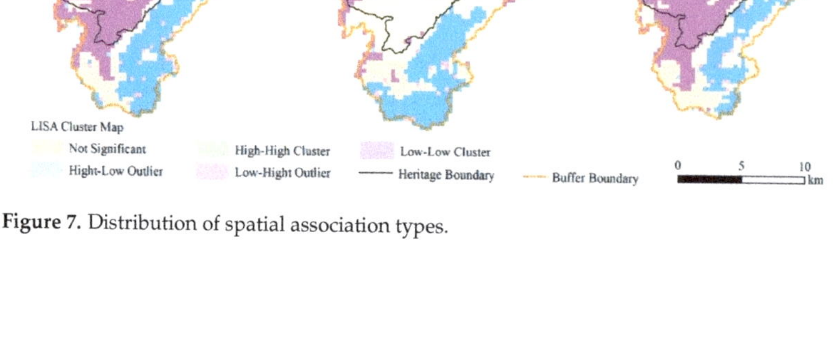

Figure 7. Distribution of spatial association types.

3.4. Detection and Analysis of Factors Influencing Ecological Environment Quality

3.4.1. Single-Factor Analysis

In this study, the remote sensing ecological index was used as the dependent variable in geo-detector and land use type, elevation, NDVI, NDISI, WET and LST in the study area were selected as independent variables. Data integration of dependent and independent variables was performed before model construction. Geographic detectors generally use a grid to represent spatial statistical units. Therefore, in this study, the natural interruption method was used to classify the independent variables into five classes and a 300*300 image element grid was created in the study area to obtain the grid center points as sample points, of which there were 3135 in total. Additionally, the values of the dependent and independent variables in the three periods were factor-probed. The q value represents the degree of influence of the independent variable on the dependent variable, i.e., the explanatory power of the spatially divergent characteristics of the RSEI (Table 4). The results showed that the q-statistics of the factors detected in 2007 were ranked as follows: WET (0.970) > NDISI (0.823) > LUCC (0.612) > NDVI (0.418) > LST (0.279) > DEM (0.011).The q-statistics in 2016 were ranked as follows: NDISI (0.900) > WET (0.750) > NDVI (0.611) > LUCC (0.341) > LST (0.271) > DEM (0.033). The q-statistic ranking in 2020 was as follows: WET (0.834) > NDISI (0.798) > LUCC (0.655) > LST (0.531) > NDVI (0.364) > DEM (0.129). WET and NDISI have a greater ability to spatially influence the quality of the ecological environment, while elevation has a smaller ability to influence the quality of the ecological environment.

3.4.2. Risk Detection

Risk area detection refers to the degree of influence of each factor on RSEI at different levels and the results indicate that NDVI and WET positively affect ecological environment quality, indicating that areas with higher vegetation cover and higher humidity have a greater influence on the ecological environment quality. Additionally, NDISI and LST have a negative feedback effect on the ecological environment quality, indicating that the higher the dryness value and the higher the surface temperature, the higher the ecological pressure and the poorer the area's quality. The risk area detection of land use shows that shrubs, grasslands, water bodies and woodlands play a positive feedback role in ecological quality and building land plays a negative feedback role. Altitude-specific risk zone detection shows that areas at sufficiently high or low elevations negatively affect ecological quality. Conversely, areas at intermediate elevations have a positive feedback effect.

4. Discussion

There are numerous methods for monitoring and evaluating the ecological environment quality. Li Hailong et al. established an urban evaluation index system from five target layers: resource conservation, environmental friendliness, economic sustainability, social harmony, economic sustainability and social harmony [37]; Peng Tao et al. established an ecological evaluation system for coastal wetlands including 17 specific indicators such as population density, degree of eutrophication of water bodies, biodiversity and awareness of wetland protection [38]. In recent years, national and regional government agencies have also established some ecological evaluation index systems, such as the ecological environment status index (EI) proposed by the Chinese Ministry of Environmental Protection in 2006. Remote sensing images are used to select indicators as well as to evaluate the ecological environment quality of heritage sites so as to visualize and analyze them spatially. With the development of remote sensing data, the development of large-scale remote sensing satellites, medium-scale UAV monitoring and small-scale ground monitoring stations provides a multi-scale, multi-data source approach for the monitoring of the ecological environment quality of heritage sites. Many scholars also obtained spatial data of ecological environment in cities, economic circles and mining areas by remote sensing technology, selected indicators of natural environmental conditions, environmental quality, natural landscape pattern and urbanization impact and assimilated environmental pollution monitoring data and socioeconomic statistics by using GIS spatial analysis technology to make a comprehensive evaluation of the ecological environment status [39–41]. Currently, the RSEI is mainly applied to the evaluation of ecological environment quality in urban and mining areas. There are fewer applications in karst areas. It is challenging to apply RSEI to monitor and evaluate the ecological quality of karst areas with a fragile ecological background, unique hydrogeological dichotomy development and particularly contradictory human-land relationship.

In this study, the monitoring and evaluation of the ecological environment quality of Shibing Karst WNHs was carried out using the RSEI index model, the indicators of heat, humidity, dryness and greenness were selected as evaluation indicators and the weight was determined by the contribution rate of each indicator to the first principal component, which was found to be above 80% by measurement. This indicates that PC1 concentrates most of the information of the four indicators. The selection of indicators and the determination of weights are more reasonable. An analysis of Table 4 reveals that the greenness index and humidity index are the preferred factors affecting the ecological environment quality of the heritage site, which is strongly related to the vegetation cover and precipitation of the Shibing Karst WNHs and increases in temperature and precipitation are favorable to the growth of vegetation. This finding is also consistent with the relevant studies of Wangguo-Qing et al. [42] and Chen-Juan et al. [43]. In this paper, satellite images were used to invert surface temperature instead of air temperature and humidity to represent precipitation, which were used to analyze the effects of changes in air temperature and precipitation on ecological environment quality.

The Shibing Karst has the properties of a WNHs and a Scenic Area for conservation and heritage display as one of the important functions [44] and is mainly concentrated in the heritage showcase area, including the Sugimu River and Yuntai Mountain scenic areas in the southern part of the nominated site. The ecological environment in this area is well maintained, the geological and geomorphological features are outstanding, the Science and Research development is in its early stages and the conditions for carrying out science and education tourism are relatively mature. As Shibing belongs to a karst region, its fragile ecological environment is vulnerable to the natural environment and human activities. However, this study reveals that since the listing of Shibing Karst as a World Heritage Site, the heritage management departments at all levels of government have protected and managed the site in accordance with the law and the local aboriginal village rules and regulations and the ecological quality of the site has been well preserved, in line with the sustainable heritage tourism advocated by UNESCO. Karst is mainly composed of a special topography developed in carbonate rocks and related ecosystems [45]. In this study, lithological data were not used as the main evaluation index due to the high vegetation cover in the Shibing Karst WNHs and the surface morphology of the peak canyon and peak valley in the Shibing Karst WNHs and thus elevation was used as an important index to study its spatial differentiation. In this paper, we considered that there are only a few human activities in the heritage site, all of them are at the junction of the heritage site and the buffer zone and most of them involve tourists and tourism employees, Because the study year was 2020, the development of the tourism industry was suspended due to the impact of the COVID-19 epidemic and thus the population and economic data were not used as evaluation indexes [46]. Unlike other regions, there is not only surface loss but also subsurface leakage of soil and water in karst areas [47]. Therefore, in the future environmental monitoring of the Karst World Natural Heritage, we should focus on the above-ground and below-ground monitoring in the field, in order to protect the ecological environment of the heritage sites more comprehensively. Remote sensing images were used to select indicators as well as to evaluate the ecological and environmental quality of heritage sites so as to visualize and analyze them spatially. With the development of remote sensing data, large-scale remote sensing satellites, medium-scale UAV monitoring and small-scale ground monitoring stations have been developed to provide a multi-scale, multi-data source approach to monitoring the ecological and environmental quality of heritage sites. In future ecological environment quality monitoring, full consideration should be given to UAV monitoring as well as ground monitoring stations to give full play to the synergy of sky–ground integration [48,49] and provide effective solutions for the conservation and sustainable development of WNHs.

5. Conclusions

The proposed method for karst-like heritage sites using RSEI and the geo-detector model, combined with GIS spatial analysis and statistical methods, provides a quantifiable and visualized method with large temporal and spatial scales for assessing and monitoring the ecological environment quality, which can be used for monitoring the ecological environment quality of WNHs, especially karst-like WNHs with a more fragile ecology. The evaluation of ecological environment quality is important for the management, conservation and sustainable development of heritage sites. This study uses remote sensing data and geospatial analysis methods to evaluate the ecological environment quality of karst heritage sites, taking the pre-application, post-application and current conditions of the heritage sites as the temporal research scale. The study found that:

(1) In the evaluation of the ecological environment quality of the three phases of the Karst Heritage Site in Shibing, the contribution of the PC1 principal component eigenvalues reached more than 80% and the RSEI was applicable to the ecological environment quality assessment of the karst WNHs.

(2) The proportion of the areas with good and excellent ecological quality rose from 88.24% in 2013 to 91.82% in 2020, while the proportion of areas with poor RSEI grades declined period by period, from 1.15% in 2013 to 0.04% in 2020, showing a significant decline.

(3) The ecological environment quality shows a positive spatial correlation and the Moran index value is steadily increasing. The ecological environment quality has significant spatial aggregation characteristics, generally showing that the west side of the heritage site is more aggregated than the east side of the heritage site. The high–high aggregation is mainly distributed in the heritage site and low–low aggregation is mainly distributed in the buffer zone.

(4) The results based on the geo-detectors model show that WET and NDISI have greater explanatory power for the spatially divergent features of ecological environmental quality in single-factor detection and are the dominant factors of environmental quality in Shibing WNHs. NDVI and WET were found to play a positive feedback role in ecological environmental quality in terms of risk area detection and NDISI and LST play a negative feedback role in terms of ecological environmental quality.

Author Contributions: Writing—original draft, N.Z.; Conceptualization, N.Z.; Writing—review & editing, N.Z.; Data curation, N.Z.; Methodology, N.Z.; Funding acquisition, K.X.; Project administration, K.X.; Resources, H.X.; Validation, H.X.; Visualization, H.X.; Supervision, J.Z.; Investigation, C.S. All authors have read and agreed to the published version of the manuscript.

Funding: This study was supported by the Philosophy and Social Science Planning Key Project of Guizhou Province (Grant No. 21GZZB43), the Key Project of Science and Technology Program of Guizhou Province (No. 5411 2017 Qiankehe Pingtai Rencai) and the China Overseas Expertise Introduction Program for Discipline Innovation (No. D17016).

Institutional Review Board Statement: Not applicable.

Informed Consent Statement: Not applicable.

Data Availability Statement: Not applicable.

Conflicts of Interest: The authors declare no conflict of interest.

References

1. Wu, B.H.; Li, M.M.; Huang, G.P. A study on relationship of conservation and tourism demand of World Heritage Sites in China. *Geogr. Res.* **2002**, *21*, 617–626.
2. Sun, K.Q. *Heritage Conservation and Development*; Tourism Education Press: Beijin, China, 2008.
3. Jaafar, M.; Noor, S.M.; Rasoolimanesh, S.M. Perception of young local residents toward sustainable conservation programmes: A case study of the Lenggong World Cultural Heritage Site. *Tour. Manag.* **2015**, *48*, 154–163. [CrossRef]
4. Lollino, G.; Giordan, D.; Marunteanu, C.; Christaras, B.; Yoshinori, I.; Margottini, C. Engineering geology for society and territory. In *Volume 8 The Monviso Ophiolite Geopark, a Symbol of the Alpine Chain and Geological Heritage in Piemonte, Italy*; Springer: Cham, Switzerland, 2015; Volume 40, pp. 239–243.
5. Chen, X.Q.; Chen, J.G.; Cui, P.; You, Y.; Hu, K.H.; Yang, Z.J.; Zhang, W.F.; Li, X.P.; Wu, Y. Assessment of prospective hazards resulting from the 2017 earthquake at the world heritage site Jiuzhaigou Valley, Sichuan, China. *J. Mt. Sci.* **2018**, *15*, 779–792. [CrossRef]
6. Pavlova, I. Global overview of the geological hazard exposure and disaster risk awareness at world heritage sites. *J. Cult. Herit.* **2017**, *28*, 151–157. [CrossRef]
7. Li, G.C.; Xiong, K.N.; Xiao, S.Z. Comparison study of World Heritage geomorphologic value of Dolomite Karst in Shibing and Huanjiang. *Acta Sci. Nat. Univ. Sunyatseni* **2014**, *53*, 142–148.
8. Yuan, D.X. Challenges and opportunities for karst research of our country under the new situation. *Carsologica Sin.* **2009**, *28*, 329–331.
9. Xiong, K.N.; Xiao, S.Z.; Liu, Z.Q.; Chen, P.D. Comparative analysis on World Natural Heritage value of South China Karst. *Strateg. Study CAE* **2008**, *10*, 17–28.
10. Xiong, K.N.; Chi, Y.K. The problems in Southern China Karst ecosystem in Southern of China and its countermeasures. *Ecol. Econ.* **2015**, *31*, 23–30.
11. Titchen, S.M. On the construction of 'outstanding universal value': Some comments on the implementation of the 1972 UNESCO World Heritage Convention. *Conserv. Manag. Archaeol. Sites* **1996**, *1*, 235–242. [CrossRef]
12. Alexander, M. Survey, surveillance, monitoring and recording. *Manag. Plan. Nat. Conserv. A Theor. Basis Pract. Guide* **2008**, *1*, 49–62.
13. Bruner, A.G. Effectiveness of parks in protecting tropical biodiversity. *Science* **2001**, *291*, 125–128. [CrossRef] [PubMed]
14. Irfan, Z.B.; Venkatachalam, L.; Jayakumar, S. Ecological health assessment of the Ousteri wetland in India through synthesizing remote sensing and inventory data. *Lakes Reserv. Res. Manag.* **2020**, *25*, 84–92. [CrossRef]
15. Das, S. Assessment of wetland ecosystem health using the pressure–state–response (PSR) model: A case study of mursidabad district of West Bengal (India). *Sustainability* **2020**, *12*, 5923. [CrossRef]
16. Cui, N. Impact of urbanization on ecosystem health: A case study in Zhuhai, China. *Int. J. Environ. Res. Public Health* **2019**, *16*, 4717. [CrossRef]
17. Wu, N. An Assessment framework for grassland ecosystem health with consideration of natural succession: A case study in Bayinxile, China. *Sustainability* **2019**, *11*, 1096. [CrossRef]
18. Smith, A.M. Remote sensing the vulnerability of vegetation in natural terrestrial ecosystems. *Remote Sens. Environ.* **2014**, *154*, 322–337. [CrossRef]
19. Groom, G. Remote sensing in landscape ecology: Experiences and perspectives in a European context. *Landsc. Ecol.* **2006**, *21*, 391–408. [CrossRef]
20. Hu, X.S.; Xu, H.Q. A new remote sensing index for assessing the spatial heterogeneity in urban ecological quality: A case from Fuzhou City, China. *Ecol. Indic.* **2018**, *89*, 11–21. [CrossRef]
21. Ren, W.; Zhang, X.S.; Shi, Y.B. Evaluation of ecological environment effect of villages land use and cover change: A Case study of some Villages in Yudian Town, Guangshui city, Hubei Province. *Land* **2021**, *10*, 251. [CrossRef]
22. Yu, Z.Y.; Deng, X.Z. Assessment of land degradation in the North China Plain driven by food security goals. *Ecol. Eng.* **2022**, *183*, 106766. [CrossRef]
23. Li, Z.H. Tradeoffs between agricultural production and ecosystem services: A case study in Zhangye, Northwest China. *Sci. Total Environ.* **2020**, *707*, 136032. [CrossRef] [PubMed]
24. He, D.W. Ecological efficiency of grass-based livestock husbandry under the background of rural revitalization: An empirical study of Agro-Pastoral Ecotone. *Front. Environ. Sci.* **2022**, *10*, 100. [CrossRef]
25. Li, Z.H. Multilevel modelling of impacts of human and natural factors on ecosystem services change in an oasis, Northwest China. Resources. *Conserv. Recycl.* **2021**, *169*, 105474. [CrossRef]
26. Wang, W.X. Impacts of infrastructure construction on ecosystem services in new-type urbanization area of North China Plain. Resources. *Conserv. Recycl.* **2022**, *185*, 106376. [CrossRef]
27. Xu, H.Q. A remote sensing index for assessment of regional ecological changes. *China Environ. Sci.* **2013**, *33*, 889–897.
28. Liu, Q. Ecological environment assessment in World Natural Heritage site based on Remote-Sensing Data. A case study from the Bayinbuluke. *Sustainability* **2019**, *11*, 6385. [CrossRef]
29. Balducci, F.; Ferrara, A. Using urban environmental policy data to understand the domains of smartness: An analysis of spatial autocorrelation for all the Italian chief towns. *Ecol. Indic.* **2018**, *89*, 386–396. [CrossRef]

30. Rouse, J.W.; Haas, R.H.; Schell, J.A.; Deering, D.W. Monitoring vegetation systems in the great plains with ERTS. In Proceedings of the 3rd ERTS Symposium, Washington, DC, USA, 10–14 December 1974; Volume 12, pp. 309–317.
31. Freden, S.C.; Mercanti, E.P.; Becker, M.A. *Third Earth Resources Technology Satellite-1 Symposium-Volume I: Technical Presentations*; NASA SP-351; NASA: Washinton, DC, USA, 1974; p. 351.
32. Wang, H.S. Discussion on feedback effect of soil desiccation by vegetation and related issues. *Prog. Geogr.* **2007**, *26*, 33–39.
33. Huang, C.Q. Derivation of a tasselled cap transformation based on Landsat 7 at-satellite reflectance. *Int. J. Remote Sens.* **2002**, *23*, 1741–1748. [CrossRef]
34. Tobler, W.R. A computer movie simulating urban growth in the Detroit region. *Econ. Geogr.* **1970**, *46*, 234–240. [CrossRef]
35. Jiang, T.Y. Study on spatial and temporal evolution and factors of regional innovation in China. *Econ. Geogr.* **2013**, *33*, 22–29.
36. Wang, J.F.; Xu, C.D. Geodetector: Principle and prospective. *Acta Geogr. Sin.* **2017**, *72*, 116–134.
37. Li, H.L.; Yu, L. Chinese Eco-city indicator construction. *Urban Stud.* **2011**, *18*, 81–86.
38. Peng, T.; Chen, X.H.; Wang, G.X.; Li, Y.H.; Li, J. Assessment of coastal wetland ecosystem health based on set pair analysis and triangular fuzzy numbers. *Ecol. Environ. Sci.* **2014**, *23*, 917–922.
39. Xu, P.W.; Zhao, D. Ecological environmental quality assessment of Hangzhou urban area based on RS and GIS. *Chin. J. Appl. Ecol.* **2006**, *17*, 1034–1038.
40. Wang, W.; Wang, X.L.; Feng, Z.K.; Dong, S.Y. Dynamic evaluation of eco-environmental quality in the capital economic circle based on RS and GIS. *J. Anhui Agric. Univ.* **2015**, *42*, 257–262.
41. Huang, H.; Chen, W.; Zhang, Y.; Qiao, L.; Du, Y. Analysis of ecological quality in Lhasa Metropolitan Area during 1990–2017 based on remote sensing and Google Earth Engine platform. *J. Geogr. Sci.* **2021**, *31*, 265–280. [CrossRef]
42. Wang, G.Q.; Guan, X.X.; Wang, L.Y.; Wang, J. Impacts of climate change and human activities on stream flow of the key runoff generation areas of the Yellow River Basin. *Yellow River* **2019**, *41*, 26–30.
43. Chen, J.; Song, N.P.; Chen, L.; Wang, X.; Wang, Q.X. Soil moisture dynamics and its response to precipitation in different cover types of desert steppe. *J. Soil Water Conserv.* **2021**, *35*, 198–206.
44. Fang, R.N.; Zhang, J.; Xiong, K.N. Influencing factors of residents perception of responsibilities for heritage conservation in world heritage buffer zone: A case study of libo karst. *Sustainability* **2021**, *13*, 10233. [CrossRef]
45. Xiong, K.N.; Li, G.C.; Wang, L.Y. Study on the protection and sustainable development of South China Karst Libo World Natural Heritage Site. *Chin. Landsc. Archit.* **2012**, *28*, 66–71.
46. Wang, Z.J.; Zhou, X.M.; Fang, Z.Q. Evaluation on the resilience of urban tourism flow network structure under the impact of COVID-19 pandemic: A case of Chongqing. *J. Arid. Land Resour. Environ.* **2022**, *11*, 148–157.
47. Chen, H.S.; Feng, T.; Li, Z.C.; Fu, Z.Y.; Lian, J.J.; Wang, K.L. Characteristics of soil erosion in karst regions of Southwest China: Research Advance and Prospective. *J. Soil Water Conserv.* **2018**, *32*, 10–16.
48. Li, Q.; Yang, Y.G.; Ti, B.; Xu, K.; Zhao, X.Q. Integrated aerial-space-ground ecological monitoring technology system of Baima Snow Mountain National Nature Reserve. *J. West China For. Sci.* **2021**, *50*, 49–54.
49. Wu, L.X.; Li, J.; Miao, Z.L.; Wang, W.; Chen, B.Y.; Li, Z.W.; Dai, W.J.; Xu, W.B. Pattern and directions of spaceborne-airborne-ground collaborated intelligent monitoring on the geo-hazards developing environment and disastersin glacial Basin. *Acta Geod. Cartogr. Sin.* **2021**, *50*, 1109–1121.

land

MDPI

Article

Pathways towards the Sustainable Management of Woody Invasive Species: Understanding What Drives Land Users' Decisions to Adopt and Use Land Management Practices

Beatrice Adoyo [1,2,*], Urs Schaffner [3], Stellah Mukhovi [1], Boniface Kiteme [2], Purity Rima Mbaabu [4], Sandra Eckert [5,6], Simon Choge [7] and Albrecht Ehrensperger [5]

1 Department of Geography, Population and Environmental Studies, University of Nairobi, Nairobi P.O. Box 30197-00100, Kenya; smmukhovi@uonbi.ac.ke
2 Centre for Training and Integrated Research in ASAL Development (CETRAD), Nanyuki P.O. Box 144-10400, Kenya; b.kiteme@africaonline.co.ke
3 Centre for Agriculture and Bioscience International (CABI), Rue des Grillons 1, 2800 Delemont, Switzerland; u.schaffner@cabi.org
4 Department of Social Sciences, Chuka University, Chuka P.O. Box 109-60400, Kenya; prima@chuka.ac.ke
5 Centre for Development and Environment (CDE), University of Bern, 3012 Bern, Switzerland; sandra.eckert@unibe.ch (S.E.); albrecht.ehrensperger@unibe.ch (A.E.)
6 Institute of Geography, University of Bern, 3012 Bern, Switzerland
7 Kenya Forestry Research Institute, Marigat P.O. Box 20412-30403, Kenya; schoge@kefri.org
* Correspondence: beatriceadoyo01@students.uonbi.ac.ke

Citation: Adoyo, B.; Schaffner, U.; Mukhovi, S.; Kiteme, B.; Mbaabu, P.R.; Eckert, S.; Choge, S.; Ehrensperger, A. Pathways towards the Sustainable Management of Woody Invasive Species: Understanding What Drives Land Users' Decisions to Adopt and Use Land Management Practices. *Land* **2022**, *11*, 550. https://doi.org/10.3390/land11040550

Academic Editors: Matteo Convertino and Jie Li

Received: 14 March 2022
Accepted: 4 April 2022
Published: 8 April 2022

Abstract: Sustainable land management (SLM) practices are key for achieving land degradation neutrality, but their continued implementation lag behind the progression of various forms of land degradation. While many scholars have assessed the drivers of SLM uptake for restoring land affected by desertification, drought, and floods (SDG 15.3 and partly SDG 2.4), little is known about the implication of SLM implementation on invasive alien species (IAS) management. This study aimed at understanding the challenges and proposing solutions for the uptake of SLMs with respect to the management of the invasive tree, *Prosopis juliflora*, in Baringo County, Kenya. Data were collected with semi-structured questionnaires, the responses were coded into themes, and c-coefficient tables were used to determine code linkages. Our results show that the availability of incentives is the main motivation for invasion management. Thus, management efforts have often focused on private parcels, while communally shared lands tended to be neglected despite their vulnerability to invasion. We conclude that sustainable IAS management lies at a landscape scale, and thus the national IAS management strategies should adopt a collective approach by empowering local actors to engage in SLM implementation.

Keywords: drivers; sustainable land management; invasive alien species; *Prosopis juliflora*

1. Introduction

Land degradation is a global challenge that is caused by climate change-induced disasters and anthropogenic activities such as urbanization, unsustainable agricultural activities, deforestation as well as loss of native biodiversity [1]. Sustainable Land Management (SLM) practices are seen as a solution for mitigating land degradation [2]. SLM is the rational utilization of land and its related resources to produce goods and services that enhance human life without jeopardizing the land's long-term productive potential and environmental functions [3]. However, no matter how effective SLM practices are, land users' decisions to ultimately adopt and continuously implement them or not will determine their success in combating land degradation. In this regard, it is evident that the slow uptake of SLM practices as compared to the fast progression of land degradation has so far hindered the achievement of degradation neutrality in many places [4].

Invasion by alien plant species such as *Prosopis juliflora* (Sw.) DC. (hereafter referred to as *Prosopis*) is a particular form of land degradation—one with severe adverse impacts, including the loss of biodiversity [5], encroachment of agricultural and pasture lands [6], and depletion of ground and surface water [7,8]. Recent studies have stressed the urgency to sustainably manage *Prosopis* invasion following its steady increase in cover and its adverse impacts on ecosystems and human livelihoods [8,9]. Even though the threat of invasive alien species (IAS) is explicitly addressed in the 2030 Agenda (Target 15.8), it does not seem to get the same attention as other forms of land degradation such as soil erosion, deforestation, or desertification [10]. Studies on the adoption and continued use of SLM practices [11,12] focus on the management of these other forms of degradation by single land users on isolated or individualized parcels of land. This makes them ineffective for managing IAS, which have complex invasion processes, spreading over large tracks of land within short periods of time.

Owing to the complexity, speed, and extent of invasion processes, the management of IAS often demands a coordinated strategy beyond an individual land user's plot and reach. Such a strategy must combine various approaches in a spatially differentiated manner: prevention, early detection and rapid response, control of invasive species as well as the rehabilitation of cleared land. This requires coordination at the community or landscape level rather than at the farm or household level [13]. For example, communities in the invaded area and neighboring communities in areas that have not yet been invaded must consensually agree on a management approach that addresses the spatial differences of invasion levels among or between the communities [14]. This is attributed to variations in the spread and density of *Prosopis,* necessitating different management objectives and different investments in management. Strategies for managing IAS such as *Prosopis* also require collective decisions on whether to invest or not in the adoption and continued use of IAS management practices [1,15].

The need to understand drivers that influence land users' management decisions and their mutual linkages is widely acknowledged as vital for informing sustainable invasion management strategies. Thus, previous studies [14,16,17] have identified governance decisions, environmental conditions, and social and economic dynamics as the key drivers of invasion management. These drivers are multi-dimensional and have complex systemic interactions, making it necessary to adopt a holistic approach to addressing them [11,16]. However, to inform spatially explicit management approaches at a landscape level, there is a need to contextualize these drivers based on their spatiotemporal extent.

To address the above-mentioned challenges, this paper aimed to assess the drivers of land users' management decisions with respect to IAS management. Based on the spatiotemporal trajectories of *Prosopis* cover in a case study area—Marigat Sub-County, Baringo County in Kenya—we assess the drivers of land users' uptake and continued use of SLM practices aimed at managing *Prosopis* invasion. We also identify promising entry points for the uptake and continued use of SLMs in the context of IAS management. Finally, we derive key lessons from a successful implementation of national and sub-national IAS management strategies such as the recently launched Kenya National *Prosopis* Strategy. Owing to the associated high costs of managing invasive species, this paper presumes that land users are likely to engage in invasion management only if the benefits derived substantially exceed the cost of management. Likewise, considering the complex invasion pattern of IAS, we predict that sustainable management may be attained through a joint communal action at the landscape level rather than at the individual farm level.

2. Materials and Methods

2.1. Study Area

The study was conducted in Marigat Sub-County (Figure 1), in the eastern lowlands of Baringo County, Kenya (11,075 km^2). Marigat was selected because it is representative of a densely invaded agro-pastoral area that has experienced a substantial impact on local rural livelihoods. Since the 1980s, the native vegetation—trees such as *Vachellia tortillis, Boscia* spp., and *Balanites aegyptiae* as well as native grass species and shrubs such as *Salvadora* spp.—has been severely invaded and displaced by *Prosopis* [9,18]. *Prosopis* stands in Baringo County are mainly confined to Marigat sub-county [19].

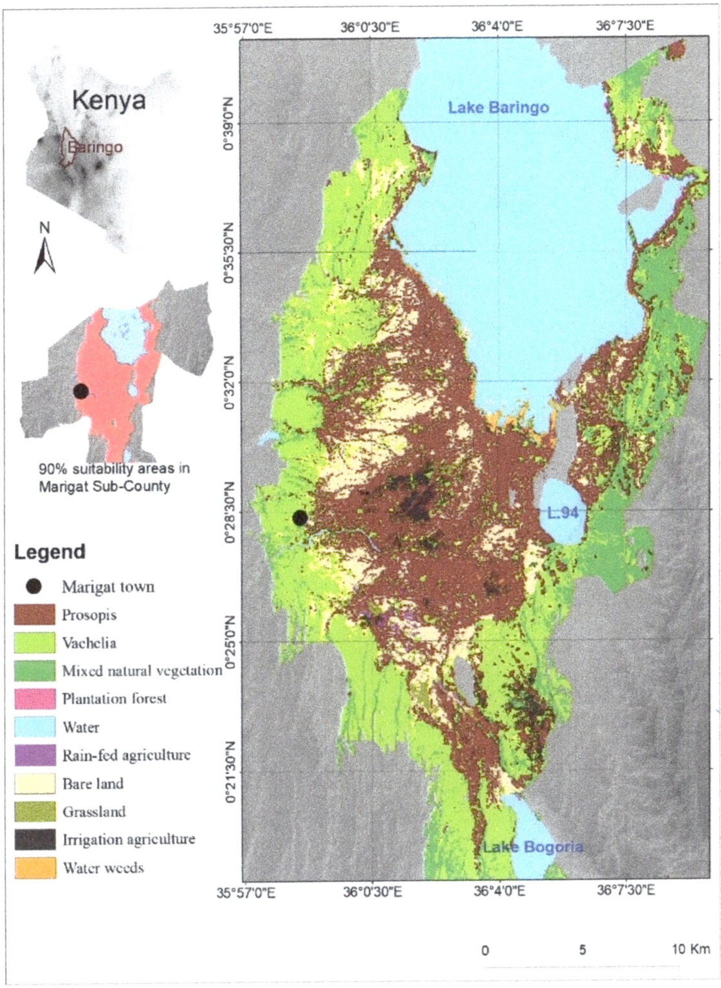

Figure 1. Map of the study area showing different land-use categories in 2016. Adapted from [9].

The annual rainfall in Baringo ranges between 600 mm and 650 mm, with annual average temperatures of 30–35 °C; the area is dominated by andosols (young volcanic soils) [20–22]. The population of Marigat Sub-County is just above 100,000, with nine out of ten inhabitants being rural dwellers [23]. While pastoralism is the primary source of livelihood, furrow irrigation agriculture is prominent and is practiced in areas adjacent to water bodies [19]. The area is dominated by rangelands, which cover 47.25%, while farming, a major form of livelihood diversification, constitutes 21.7% of the land cover. Water bodies, settlements, forests, and bare land cover 9.7%, 9.6%, 8.4%, and 3.3% of the area, respectively [21]. De jure, most of the land in Marigat is strictly communal, but de facto, a large percentage is privately claimed [24].

2.2. Conceptual Framework

We have adopted a systemic conceptual approach to identify and evaluate the interaction among drivers of land users' decisions on whether to adopt and continually use SLM practices or not in the context of *Prosopis* management. We consider drivers as positive when they promote the implementation of SLMs, while negative drivers limit SLM implementation by land users. This approach is informed by the human actors' model [25], which emphasizes the crucial role of actors in development outcomes. The conceptual framework is founded on the acknowledgment of actors such as land users, who are the key agents of transformation in any form of development. Thus, the human actors' model is structured based on three main ideologies—that actors' decisions to create change are an interplay of three factors: (1) their prioritized activities, (2) their understanding of the effects of those activities on their desired outcomes, and (3) the availability of means needed for them to implement their prioritized activities. This actors' strategy of actions is non-linear and dynamic. For instance, an undesired outcome of actors' activities may change the meanings of such activities, leading to modifications of the activities or the re-assignment of their available means to other activities.

The model also considers external and dynamic conditions (environmental, political, economic, and social) as being important in determining actors' actions and in influencing their activities and stock of means—finances, technology, knowledge, and skills. Similar to the strategy of actions explained above, the effect of these dynamic conditions depends on actors' perceptions, beliefs, and values. Their outcome leads to either the adaptation or modification of activities to fit into, improve, or stabilize the dynamic conditions.

The actors' model, therefore, illustrates that actors' strategies of actions are strongly embedded in their social and cultural institutions as defined by their beliefs, rules, and regulations as well as by the enforcement of such institutions. This is because such institutions define their perceptions, valuations, and meaning of actions. We adapted the model to assist in identifying relevant entry points to address the effect of negative drivers and to evaluate possible pathways for attaining sustainable management strategies.

2.3. Data Collection and Analysis

2.3.1. Data Collection

We spatially structured data collection based on the occurrence of three different temporal patterns of *Prosopis* invasion:

1. Areas that were cleared of *Prosopis* and never re-invaded, indicating land users' willingness and efforts to continuously implement SLM practices.
2. Areas that were cleared but later re-invaded, denoting the uptake and subsequent abandonment of SLM practices by land users.
3. Areas that were never cleared since they were first invaded, suggesting that land users never implemented or tried but failed to implement SLM practices that would help to manage invasion.

Based on the data from [9], five spatial "hotspot areas" were identified with visual evidence of all three invasion patterns above. The five hotspot areas were distributed over eight sub-locations, which are the lowest of the administrative units in Kenya. In total, eighty land users from these eight sub-locations were interviewed, with the number of interviewees proportional to the population size in each of the sub-locations (Figure 2). Details of the distribution of sample size proportionate to the population of respective sub-locations are found in Appendix A.

Three large-scale maps, each illustrating one of the three *Prosopis* invasion patterns, were produced for each of the five hotspot areas (see Figures S1–S15, Supplementary Material) and used during the interviews. The three patterns of *Prosopis* presence/absence were explained, and each interviewee was asked to respond to open-ended questions relating to drivers contributing to these patterns in the hotspot areas falling within his/her sub-location. This was to assist in assessing differences in responses from different sub-locations, which were associated with different invasion levels. The level of invasion per sub-location was derived from [9]'s fractional cover dataset. A 40% fractional cover, beyond which it is often perceived impossible to sustainably control invasion, was used as a cut-off point for densely and sparsely invaded areas. Advanced invasion is used in this paper to refer to a perceived invasion level beyond which respondents considered SLM implementation to be impossible or meaningless as the associated management costs would exceed the benefits to be derived from management efforts. The main guiding questions for the interviews were as follows:

1. Do you agree that the general pattern of *Prosopis* cover in your sub-location is accurately represented on the map? This question was to confirm the occurrence of the three spatial-temporal patterns that reflect different scenarios of adoption and continued use of SLM practices. Any contradicting opinion was noted and clarification/revision of patterns was sought.
2. What are the factors that could have contributed to the occurrence of the respective patterns? Respondents were expected to reflect on all factors they could think of while answering this open-ended question.

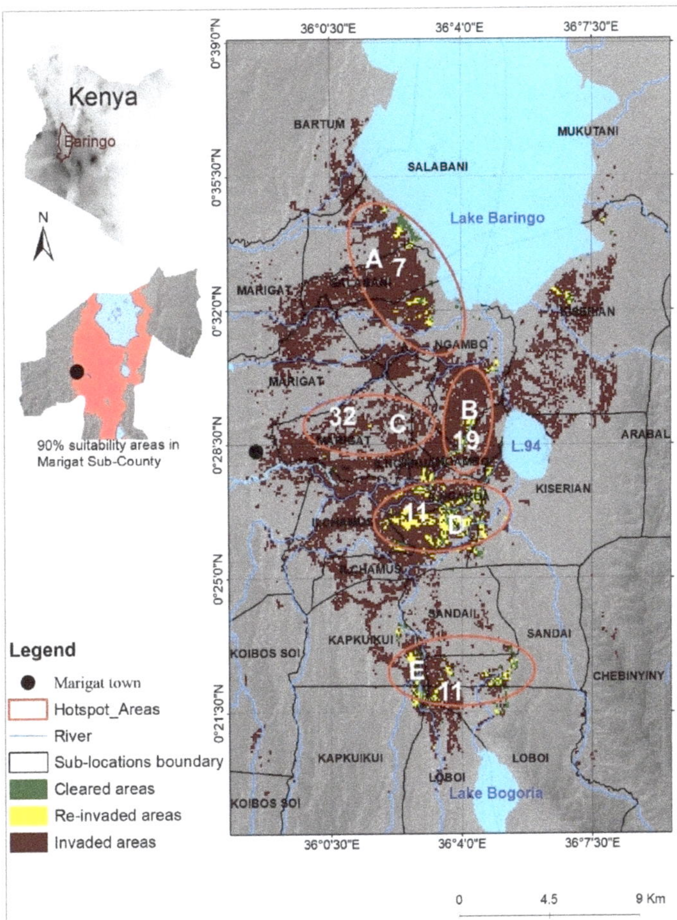

Figure 2. A map showing hotspot areas (red ellipses) that were selected for the study due to a visual representation of invasion patterns. The numbers of respondents per hotspot area (A–E) are indicated within the hotspots.

In addition to the questions on drivers, we obtained details regarding the parcels that the land users owned, such as the previous and current land cover/use and the approximate distance of the farms from their homesteads. We later validated the findings by interviewing all the area chiefs using the same questionnaires in order to clarify aspects that seemed ambiguous.

2.3.2. Data Analysis

Responses to the open-ended questions and the corresponding socio-demographic data from the survey were entered into MS Excel, combined to form a single string ID, and converted into a text format before being exported to ATLAS.ti for analysis (Appendix B). Responses were then coded and categorized into related thematic codes falling within four distinct categories of drivers that influence the uptake and continued use of SLM practices: socio-cultural, environmental, economic, and political drivers. The process of assigning a code to each response was limited to the interviewer's understanding of the subject matter and the context under which each response was given. To avoid a misinterpretation of the responses and thus the assignment of the wrong codes, the researcher was present at each interview to seek clarification for the responses that might have been unclear.

The dataset was then entered into c-coefficient tables to analyze the association between codes, depending on whether the codes would co-occur whenever they were assigned to a quotation or response. The c-coefficients range from 0, for codes that do not co-occur, to 1, for codes that are always assigned to the same quotation or response. The c-coefficient (c), also known as the c-index, is calculated as follows:

$$c = n_{1,2}/(n_1 + n_2 - n_{1,2})$$

where n_1 and n_2 are the occurrence frequency of two co-occurring codes—c_1 and c_2—while $n_{1,2}$ is the co-occurrence frequency of the two codes.

The output was displayed in network views that visualized the relationship between and among the codes. The codes in the network view were organized depending on their groundedness—the number of quotations/responses assigned to them—to illustrate their hierarchical significance. Some quotations/responses had multiple meanings and were hence assigned to more than one code. For instance, a decision to seek an alternative source of income due to drought was assigned two codes: livelihood diversification and natural disasters. In such a case, the two codes were linked to the same quotation/response. The number of linkages to a code is denoted as its density (D) and is displayed in the network view. According to [26], only the values of a code's groundedness are relevant in determining the significance of a code; the higher the value, the more number of times that code was mentioned.

The application of c-indices is instrumental in depicting the association between codes in that it performs a quantitative analysis of qualitative data [26]. However, their values may not reflect the actual strength of association, especially if the codes are distorted by unequal frequencies [27]. Therefore, while c-index values close to one indicate a strong association, two codes varying greatly in frequency (usually if the ratio between the two frequencies is greater than five) may also reveal a strong association at a considerably lower value. This is indicated by c-indices highlighted in the co-occurrence coefficient tables (Appendix B). Our data were distorted by variations in frequencies since each respondent was free to mention as many drivers of a particular invasion trend as he or she wished. To enable a comparison between and among the codes, the c-indices were multiplied by a system-generated factor to normalize them. The ATLAS.ti software automatically highlights significant c-indices if their low values are associated with such distortions [27]. This simplified the selection of codes, whose co-occurrence was significant for our analyses.

3. Results

3.1. Categories of Drivers Influencing Land Management Decisions

The results revealed a broad range of drivers of land users' decisions to adopt and continuously implement SLM practices for the control of the spread of *Prosopis*. These were grouped into four broad categories, including socio-cultural, environmental, economic, and political drivers, in the order of their importance (Figure 3). Socio-cultural drivers were the most often mentioned, while political drivers were the least often mentioned. Below, we present the findings on how drivers assigned to these four categories influenced land users' management decisions to adopt and continuously implement SLM practices.

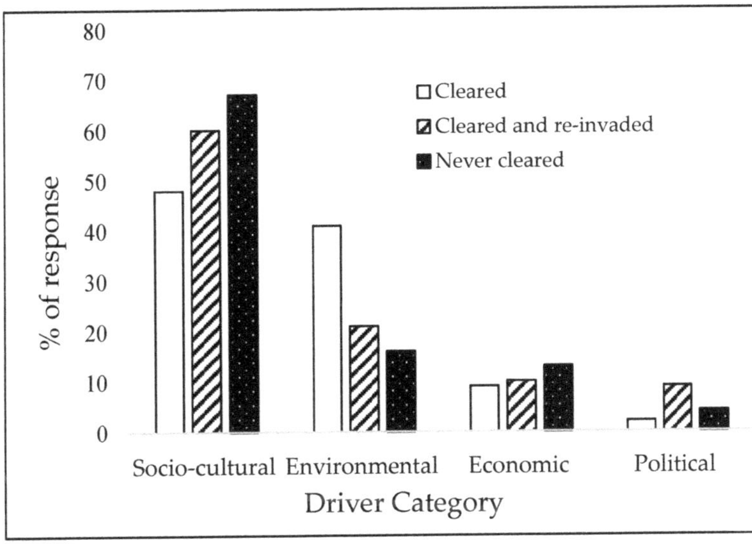

Figure 3. Influence of socio-cultural, environmental, economic, and political drivers on the adoption, abandonment, and continued use of SLMs.

3.2. Drivers for Adoption and Continuous Implementation of SLMs

3.2.1. Intensive Land Management Practices

Intensive land management practices such as irrigation, cultivation, the planting of perennial plants such as fruit trees, and crop rotation (Figure 4), all of which ensure that land is constantly managed and not left idle, are perceived to be the key drivers in promoting the continuous implementation of SLM practices.

Continuous cultivation was reported to lower the risk of (re-)invasion as communities were keen on retaining their farms, from which they derived their livelihoods. Despite this, the implementation of these practices is only possible with the availability of means such as labor, financial resources, water for irrigation as well as knowledge of the potential impacts of *Prosopis* and the skills to manage it.

Proximity to settlements was perceived to contribute to continuous *Prosopis* clearance. This is because land users are committed to maintaining the safety of their residential areas by constantly clearing *Prosopis* thickets, which often harbor thieves and wild animals, or by cultivating parcels close to their homesteads. Although it was difficult to map settlement areas due to the pastoralist nature of the study area, physical visits to areas and responses from interviewees confirmed that settlements are a factor that drives constant clearance within homesteads. This is supported by the fact that the proximity to settlements and continuous cultivation are co-occurring in our findings (c-index = 0.05).

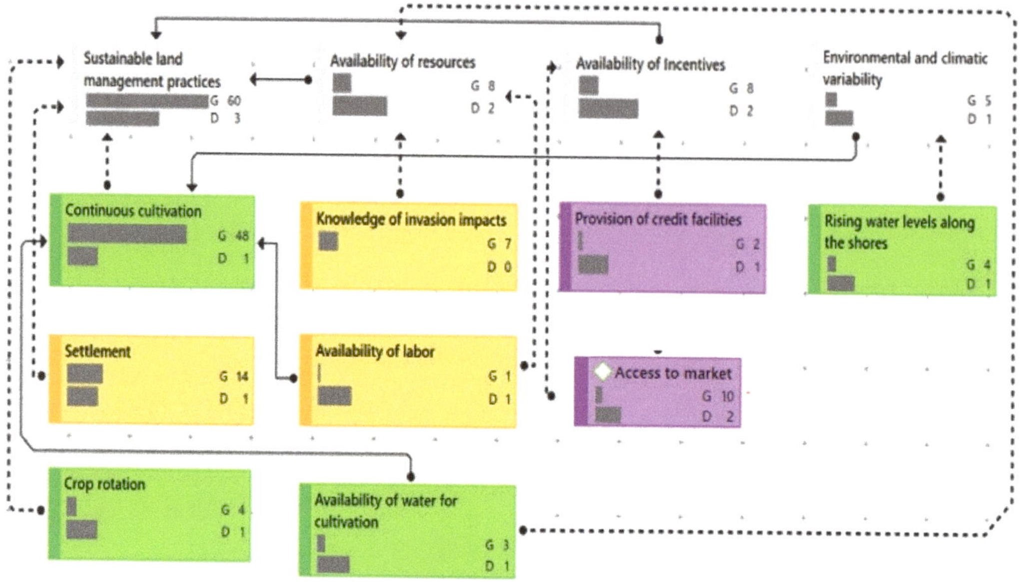

Figure 4. Drivers of continuous clearance (yellow = socio-cultural; green = environmental; purple = economic). Arrows indicate the nature and direction of relations (dotted = "indicates"; solid = "promotes"). G is the groundedness, referring to the number of linked quotations. D is the density, referring to the number of linked codes. The length of bars within each box indicates the frequency of G and D. The significance of a code is determined by its groundedness (G): The G-value increases with an increasing number of respondents who mentioned it.

3.2.2. Availability of Resources and Incentives

Respondents reported that incentives are significant in enabling the sustainable management of IAS. Land users are unlikely to invest in SLM practices on parcels for which management is not incentivized or where benefits are not guaranteed. Credit facilities, access to market, and better pricing of farm produce (Figure 3) help farmers in availing the means needed for the implementation of practices on cropland. For example, the National Irrigation Board (NIB) engages community members through the contractual farming of maize seeds (personal interviews with area chiefs). This arrangement helps land users to access a ready market for their produce. The NIB also extends credit facilities and extension services to farmers through irrigation schemes—factors that promote the clearance of land for cultivation purposes. This is perceived to be the reason why land users consistently clear land within established croplands, especially along the shorelines of lakes and rivers, where constant water supply sustains year-round irrigation. This finding is supported by the fact that 51.5 hectares (62%) of parcels that were cleared of *Prosopis* were converted into irrigated farms, while 106 ha (88%) of irrigated farms have never been converted to other uses (Figure 5).

3.2.3. Knowledge of Impacts and Associated High Cost of Clearance

Land users' knowledge of the impacts of invasion, benefits of management, and effective implementation of SLM practices was perceived as a key driver of successful IAS management. The understanding of the implications of adopting and implementing (or not) a land management practice is shaped by land users' past experiences. Land users who bear the costs of clearing *Prosopis* continue to clear it to prevent re-invasion for fear of incurring similar expenses. Similarly, respondents from densely invaded areas who understand the implication of delaying management were keen to engage in managing *Prosopis*, the perceived possible benefits from direct utilization notwithstanding. Knowledge about high clearance costs, which is gained from first-hand experience, was also reported to contribute to the continuous implementation of SLM practices. However, only 9% of the interviewees responded that knowledge of the impacts had motivated land users to clear their land from *Prosopis;* in contrast, continuous cultivation was mentioned by more than half (56%) of the respondents.

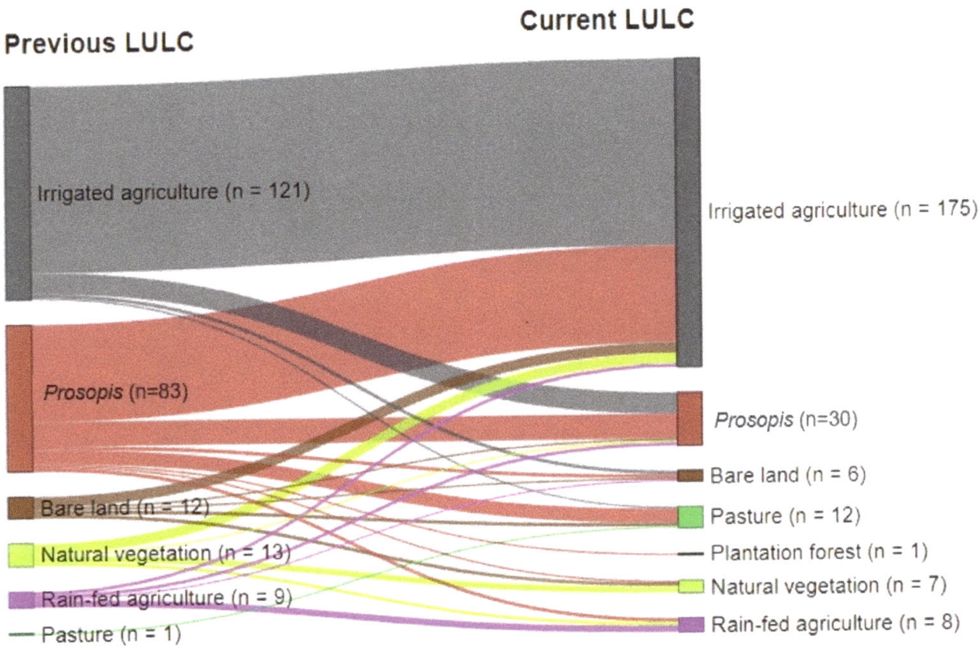

Figure 5. Changes in land use and land cover within respondents' parcels of land (*n* values indicate the size of land in hectares). Source: Survey data.

3.3. Drivers of Non-Adoption or Abandonment of Adopted SLM Practices

Failure to adopt and the abandonment of adopted SLM practices were mainly attributed to socio-cultural factors (Figure 3) and were perceived to dominate on *communally* owned lands by 87% of the respondents. The diversification of local livelihoods, the prevailing land tenure system, and unreliable knowledge about invasive alien species are perceived to be the main contributors to the non-adoption or abandonment of land management practices (Figure 6).

3.3.1. Diversification of Livelihoods

The transformation of the traditional way of life from pastoralism—due to the loss of grazing land—to cultivation and charcoal production is perceived to reduce land users' dependence on land. This, in turn, leads to less engagement in management and maintenance, thus facilitating the invasion by *Prosopis*. Respondents also argued that apart from the emergence of new opportunities, the occurrence of natural disasters such as droughts and floods and the associated degradation of grasslands were some of the main causes of this livelihood transformation. Indeed, data show the co-occurrence of livelihood diversification with the above-the-ground cutting of *Prosopis* for charcoal production (c-index = 0.36) and the natural disasters caused by drought (c-index = 0.29).

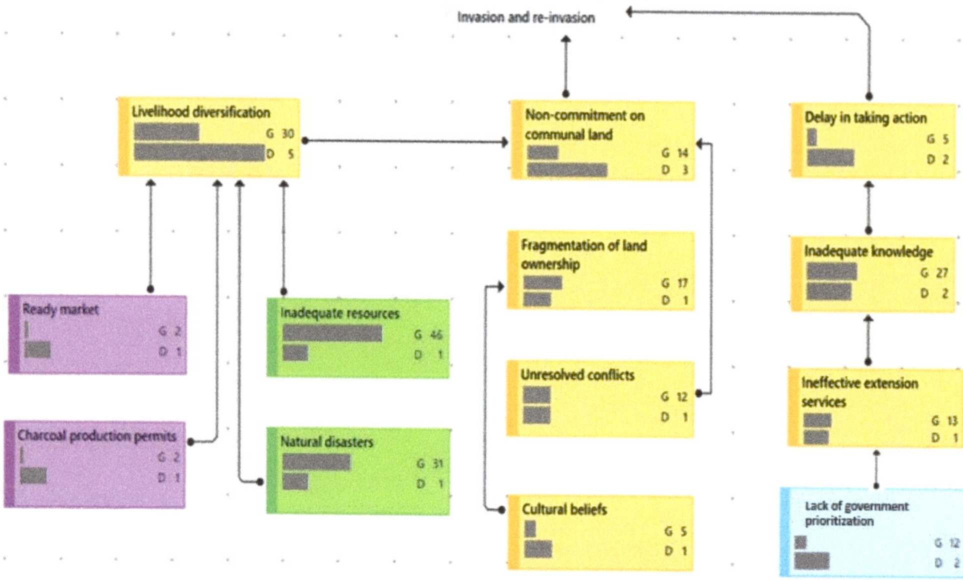

Figure 6. Drivers of continuous invasion and re-invasion (yellow = socio-cultural; green = environmental; purple = economic; blue = political). Arrows indicate which drivers promote another. G (Groundedness) refers to the number of linked quotations. D (Density) refers to the number of linked codes. The length of bars in each box indicates the frequency of G and D.

Respondents perceived that severe drought, especially in 1995, prompted communities to find alternative income sources by engaging in charcoal production, an activity that later facilitated the coppicing of *Prosopis*. The degradation of pasturelands due to either drought or heavy invasion was believed to have accelerated the loss of livestock that pastoral communities depended on, thus forcing them to seek alternative means of survival. Charcoal licensing by the government and the high market demand (c-index = 0.25), together with the widespread availability of *Prosopis* trees, were considered to be a strong incentive that promoted charcoal production as a source of livelihood. This vicious circle further reduced their dependence on and motivation to manage communal grazing lands, thus exposing them to invasion (personal interviews, area chiefs, October 2019).

3.3.2. Ineffective Land Governance

Ineffective land governance was perceived to hinder the uptake and continued use of SLM practices. According to a total of 48 respondents, ineffective land governance was associated with the fragmentation of parcels, unresolved land disputes, and the neglected management of communally shared lands. Under the customary land tenure system, the traditional land acquisition process guided access to land for private use such as farmlands and homesteads. According to the respondents, this approach bestowed upon men absolute freedom to own unlimited parcels of land as long as they had a history of utilizing such parcels in the past. Consequently, the aged community members own many fragmented parcels, most of which they are not capable of managing due to limited labor and financial resources. Fully invaded parcels of cropland were found to be the furthest from their owners' homesteads (Figure 7). This supports the respondents' views that land users with many fragmented parcels prioritize managing *Prosopis* on the farms closest to their homesteads while leaving the distant parcels to be invaded. Therefore, the two explanatory factors of 'ownership of many parcels' and 'distance from home' co-occurred (c-index = 0.09).

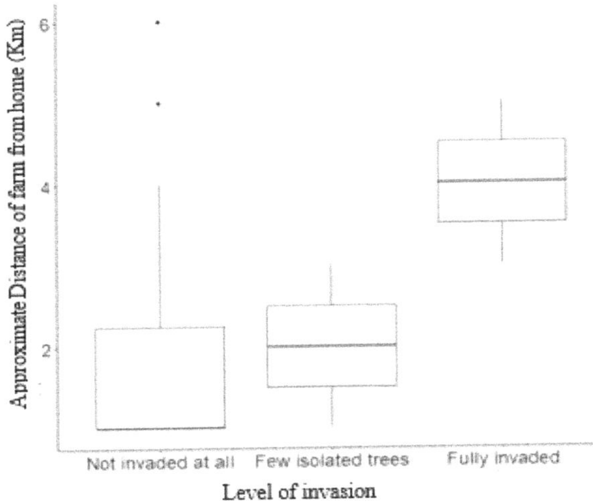

Figure 7. Relationship between the level of *Prosopis* invasion and the distance of farms from the homesteads. The boxes indicate the median (the thick line separating the box into two) and the lower and upper quartiles (lower and upper border of the boxes); the whiskers extending from the lower and upper quartile marks indicate the scores falling outside 50% of the total scores, while the dots represent outliers.

In addition to distance from homesteads, unresolved land disputes associated with the weak enforcement of tenure rights by responsible institutions were perceived to be a reason for land abandonment and subsequent *Prosopis* invasion. Such disputes can remain unresolved for a long time, owing to the failure of concerned parties to respect decisions by local authorities. Land disputes that had dragged on for years led to a stop of all maintenance activities—a free pass for *Prosopis* invasion. Conflicts over communally owned resources are most prominent in areas of low invasion levels (c-index = 0.02). Based on interviews with the area chiefs, the migration of livestock from densely invaded to sparsely invaded communal grazing areas contributed to the invasion of the latter, leading to conflicts over the diminishing grazing lands.

3.3.3. Unreliable Knowledge about the Impacts of IAS and the Benefits of Timely Management

Late and unreliable information about the nature and potential impacts of *Prosopis* was considered by 34% of the respondents as a key driver of the non-adoption of SLM practices as it delays the onset of management efforts. Delayed management was reported to cause invasion to progress to a level where taking action would require more labor and finances. This has discouraged land users from taking any action. The situation was perceived to be worsened by inappropriate information about suitable approaches for *Prosopis* control and removal; initial efforts to cut the trees above the ground were believed to have favored rapid spread as land users were initially not aware of the ability of *Prosopis* to coppice.

Delays in knowing the impacts of *Prosopis* had a stronger co-occurrence with respondents from densely invaded areas (c-index = 0.14) as compared to respondents from sparsely invaded areas (c-index = 0.01). It also co-occurred with perceived advanced invasion (c-index = 0.22), an indication that it contributed to the advanced invasion in densely invaded areas. Inversely, respondents in sparsely invaded areas did not consider late and unreliable knowledge to be a hindrance to the management of IAS. However, they considered the perceived benefits of *Prosopis* (c-index = 0.05) to have a negative impact on the willingness of land users to adopt SLM practices as they feared the loss of these benefits. Some of the benefits include a source of income from charcoal production and the provision of shade. Respondents from heavily invaded areas have a similar opinion but with a lower co-occurrence (c-index = 0.02).

As already mentioned, inadequate knowledge of effective land management practices was seen to limit their implementation. This was evident in the Perkerra area, the main irrigation scheme in Marigat, where many agricultural lands have been invaded by *Prosopis* following the loss of their initial productive potential. Respondents attributed this to poor farming methods that resulted in waterlogging and salinization. This perception by interviewees was confirmed by the National Irrigation Board's officer during an interview. Respondents reported that such parcels were abandoned by their owners as they were unwilling to invest in them due to their low production potential. Findings from interviews with community members show that 10% of the irrigated farms owned by the respondents have been lost to the invasion (Figure 4).

4. Discussion

This study has shown that land users' decisions to implement SLM practices to manage *Prosopis* in Baringo County are reflected in patterns of invasion, and that such decisions are influenced by socio-cultural, economic, environmental, and political drivers. Continuous cultivation was found to be the main factor contributing to invasion management. Even so, this depends on land users' motivation and ability to access resources for SLM implementation [17]. Thus, the key drivers of continuous SLM implementation are timely access to appropriate information on the need to manage invasion as well as the provision of incentives, financial resources, and favorable market dynamics [28,29].

Unlike the positive drivers, the negative drivers are numerous, diverse, and limit the performance of the positive drivers. Addressing the negative drivers should therefore be prioritized to achieve sustainability in the management of invasive species. Hence, we will first discuss the causal linkages among negative drivers and their association with critical socio-ecological factors that limit SLM adoption and continued implementation (Figure 8). The study applied the local actors' model as an explanatory framework for relating how linkages among different drivers provide a rationale for land users' management decisions.

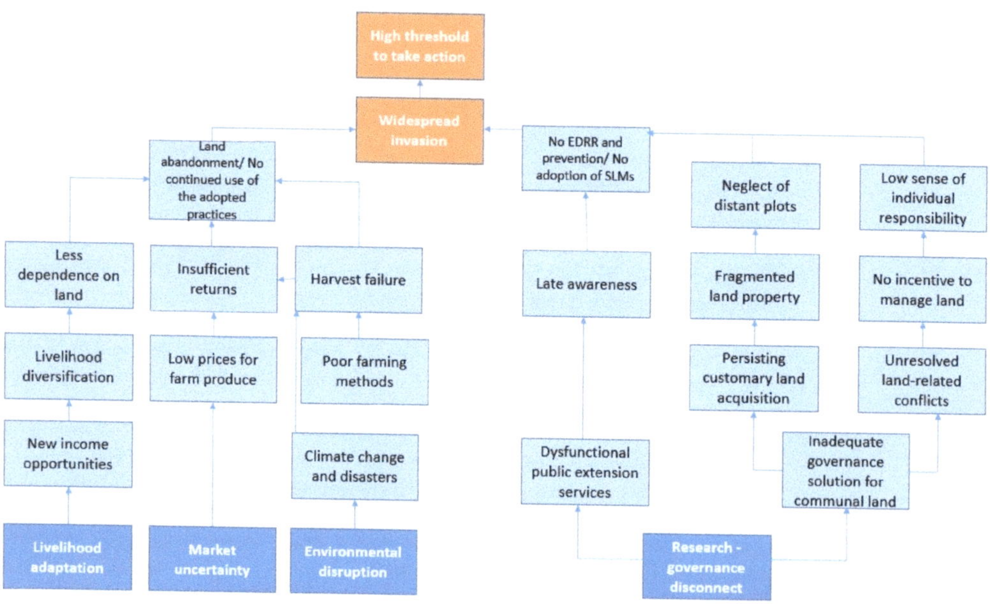

Figure 8. Socio-ecological factors and their associated causal links with drivers influencing the adoption and abandonment of SLM practices.

4.1. Critical Socio-Ecological Factors Associated with Non-Adoption and Abandonment of SLM Practices

The four main factors associated with non-adoption and abandonment of SLMs are: livelihood adaptation, market uncertainties, environmental disruptions, and the disconnect between research and land governance. These factors are important in informing management strategies as the failure to address them may limit SLM implementation and hinder the successful management of invasive species. We hereby describe each of them in relation to their causal drivers.

4.1.1. Environmental Disruptions

These include natural disasters such as drought and floods, which result in low productivity or crop failures. Such events tend to increase in the wake of global climate change [30]. They result in low returns compared to the cost of SLM implementation [31], thus leading to the abandonment of affected parcels as land users are unwilling to continue implementing SLM practices on them. This exposes initially managed parcels to invasion or re-invasion. Based on the land users' experience, they believe that environmental shocks adversely impact the outcome of their SLM implementation. These beliefs and perceptions of the impact of external drivers on land users' expected returns inform their land management decisions, as indicated in the human actors' model [25]. In this case, the land users' perceptions discourage the implementation of SLMs, hence limiting invasion management.

4.1.2. Market Uncertainty

Under contractual farming, which is dominant in the irrigation zone of the study area, land users sell their produce to the contracting companies, the NIB, who were reported to dictate the prices and sometimes purchase below market price. A decrease in demand for agricultural products and its effect on prices are also a disincentive to SLM implementation.

As explained in the actors' model, the land users' rationale for assigning their stock of means to alternative activities is exposed to dynamic factors, including the economic drivers [25]. Thus, land users perceive that market dynamics result in low returns despite their implementation of SLMs. Consequently, they adapt by re-assigning their available stock of resources to other promising economic activities [32]. This is illustrated by the fact that land users opt to discontinue farming activities and channel their resources to alternative means of survival when they perceive that their returns will be lower than the costs of farm inputs. This also exposes the abandoned parcels to re-invasion.

4.1.3. Livelihood Adaptation

Contrary to the initial purpose of its introduction as, among other things, a source of fodder for livestock [33], *Prosopis* has led to the loss of herbaceous biomass, resulting in the decline of livestock numbers [34]. Coupled with high invasion management costs and the impacts of prevailing environmental disruptions such as drought and floods, land users gradually adapt and transition to alternative sources of livelihood such as trade, charcoal production, or employment. Land users' motivation to implement SLM practices on their lands therefore declines [31], leaving affected parcels exposed to invasion. This indicates that the exposure of land users' actions to external conditions such as environmental disturbances influences them to adapt their activities to the prevailing conditions [25,35]. According to [6], the income from alternative livelihoods such as charcoal production may not replace the cultural value bestowed upon pastoralism. Thus, the main reason for community members' diversification to other livelihood sources may be due to the degradation of their initial grazing field upon which their livestock depended. This implies that the restoration of grazing fields presents the possibility of promoting the engagement of land users in communal land management.

4.1.4. The Disconnect between Research and Land Governance

In the absence of an effective science-policy interface, evidence from science does not or only slowly finds its way into land governance decisions. This results in two main factors that limit land management, specifically within communal tenure systems: dysfunctional extension services and ineffective land governance practices. Reliance on extension services that are not guided by credible research findings misleads land users' actions. Likewise, the delay of extension services limits timely response actions [32], which then enhances the progression of invasion to advanced levels, during which it will be impossible to manage invasion [13].

The land governance systems in place were perceived to be ineffective in addressing the prevailing tenure issues. The customary land acquisition process was associated with land fragmentation, which has been associated with technical inefficiencies in croplands [36]. In addition, weak land governance institutions have led to unresolved land disputes. This, in turn, has discouraged community members from holding themselves accountable for managing land due to a lack of incentives. As demonstrated in the human actors' model, community perceptions and practices are strongly embedded within their social and cultural institutions. This requires that interventions aimed at addressing land management issues should focus on reinforcing these institutional frameworks or modifying them in order to support sustainable environmental management [25].

Based on the four factors above that hinder SLM implementation, we compare our findings with a South African study, which is among studies that explicitly focus on barriers to the management of *Prosopis* [16]. Both studies found that land users' decisions are shaped by drivers of multiple dimensions ranging from economic, social, political, and environmental, with the social drivers being the most dominant. The main difference is that the current findings are contextualized on specific patterns of invasion as determined by successive presence /absence of *Prosopis*. Therefore, ours advances the previous study by offering context-based drivers of invasion management by land users. In addition to the institutional and knowledge gaps identified by [16], our findings focused on ineffective land governance, environmental disruptions, livelihood adaptations, and the science-policy gap as key challenges to IAS management.

In contrast to [16]'s study, our respondents did not mention a lack of technical knowledge about the effectiveness and cost implications of alternative control options such as biological and chemical control methods. This shows that respondents in South Africa were better acquainted with different options for IAS management as compared to those in our Kenyan study. According to [17], successful management strategies involve a combination of practices. However, the land users' rationale for applying multiple alternatives in managing invasion is determined by their knowledge and experience with such diverse practices [28]. This is because land users tend to adopt practices whose success rates they are well-acquainted with. However, to date, measures for managing *Prosopis* by chemical and biological means have not yet been made available in Baringo County. This is despite reports from previous studies explaining that biological control has the benefit that it is a "natural" management approach and does not require continuous implementation by individual land users [13,37]. Therefore, there is a need for land users to benchmark their own practice against that of implementers of alternative management practices as this may widen their knowledge and encourage them to take initiatives in their implementation.

Both studies also observed that benefits from *Prosopis* have been associated with the reluctance of land users to manage it. In this study, this challenge was majorly associated with the sparsely invaded areas, where respondents had limited experience with the high clearance costs associated with advanced invasion [17]. The failure of these land users to learn from the experiences of those coming from heavily invaded areas was mentioned as a challenge that hindered timely invasion control in the South African study.

4.2. *Potential Entry Points for Sustainable IAS Management*

The management of IAS such as *Prosopis* cannot be accomplished at the plot level. To be successful, it must be jointly implemented by multiple stakeholders on both privately and communally owned land and at the landscape scale [13,17]. Therefore, to promote the implementation of SLM practices and hence sustainably manage IAS, there is a need to address the main barriers captured in the four underlying factors mentioned above. This will allow for a better performance of the positive drivers towards the higher adoption and continuous implementation of SLM practices for sustainable IAS management. We therefore propose the following entry points for a successful IAS management strategy:

4.2.1. Enhancing Community Resilience to Environmental Disruptions

Based on our findings, the most viable entry point to address environmental disruptions should aim at convincing land users that the outcome of environmental disruptions does not necessarily have to render their management efforts futile. This calls for enhanced community resilience to environmental disturbances, which are associated with far-reaching impacts, often beyond the capacity of the local community to respond effectively. Building local capacity and strengthening community resilience need to depart from the traditional practice of targeting external support and should instead explore the options of adapting activities in order to minimize uncertainty or modify the dynamic conditions within which the actors operate. The latter is likely to yield more effective outcomes [25]. Thus, external support from governments and institutions in establishing infrastructures and building capacity to enhance resilience to natural disasters is a key aspect of early preparedness. However, modifying the conditions within which land users operate is likely to enhance their resilience. For instance, since resilience to natural disasters can be better attained through collective rather than individual actions [29], well-established social networks at the community level are likely to reinforce community capacity to organize themselves, learn from experiences with disturbances, and adopt the appropriate mechanisms to recover and adapt to shocks [38].

In the response to extreme events, especially in croplands where losses may be inevitable despite improved infrastructure, robust insurance schemes and effective programs for the compensation of losses are also necessary. Such measures enhance resilience against shocks, which foster land users' continued engagement in the implementation of SLM practices [11,38]. Apart from enhancing resilience, the establishment and enforcement of appropriate land use plans are necessary for restricting human activities in disaster-prone zones. This is particularly relevant to areas along rivers and in other flood-prone areas, which are often selected for farming due to water availability. Therefore, strengthening community-based institutions may be pivotal in enforcing land-use plans [14]. This is because land users' actions and behaviors are embedded within their socially-defined institutions [17].

4.2.2. Price Control and Regulation of Market Actors

Price control and the regulation of market actors are required to create a fair environment and to guarantee sufficient returns to smallholders, which is a key condition for them to sustainably implement SLM practices. The enforcement of an adequate regulatory framework for community members that strives to minimize the risk of their exploitation is central. The formulation of policies that uniformly regulate market prices and enhance equitable access to available markets may minimize monopolization by contracting institutions, hence checking the exploitation of land users [36]. To achieve this, it is important to integrate land users—through farmers' cooperatives which advocate for enhanced farmers' rights in terms of price control—and their access to markets. Such cooperatives should focus on creating an environment that favors a competitive market for farmers' produce rather than restricting them to a single market outlet. This way, land users may be sure of benefiting from their efforts and thus continue with SLM implementation.

4.2.3. Integrating Environmental Management with Poverty Alleviation

The findings of our study confirm the need to integrate environmental management with poverty alleviation [39] as communities tend to modernize their livelihoods with new and attractive income opportunities. A promising entry point in this regard would be to prioritize livelihood activities that are dependent on SLM implementation; for example, by taking farming as a primary rather than a supplementary source of livelihood, as indicated in the County Development Plan [19]. However, the promotion of such land-dependent economic activities would require access to land and water resources. Consequently, there would be a need to establish flexible land markets that promote the re-distribution of parcels to people who would convert their purpose to meaningful uses and prevent (re-)invasion. This approach, however, does not address the management of communally shared parcels such as roadsides and rangelands, which are prone to abandonment and invasion.

Land subdivision, although useful in managing individual parcels in the face of increasing population, adversely affects IAS management at a landscape level. This was illustrated in California, where only some landowners managed the invasive yellow starthistle (*Centaurea solstitialis)* on their parcels; the parcels whose owners opted not to manage the weed acted as a source of invader propagules [14]. This shows that the management of communally shared resources such as rangelands requires collective action that incorporates all users. Indeed, it has been reported that land users tend to adopt practices in solidarity with community collective action within which their actions and beliefs are embedded [17] as long as social relationships and trust are built among the concerned land managers [28]. As such, a probable solution may be to replant communally shared lands with native tree and grass species under the stewardship of assigned community members. Their engagement may be secured if their efforts are rewarded through a compensation scheme and village development programs such as the establishment of infrastructure as well as by securing and enforcing land use rights. This ensures that environmental management is integrated with social and economic development and with benefits that contribute to poverty reduction among the affected communities. To enhance collective participation, social sanctions on non-participating members have been reported to yield effective cooperation among community members [17].

Likewise, the harmonization of policies across sectors (from agricultural and environmental to trade) and scales (from local to national levels) is needed. For instance, the simultaneous occurrence of a market opportunity (for charcoal), policy provisions for the easier licensing of charcoal trade, and the need to diversify livelihoods away from farming activities all lead to a positive perception of *Prosopis*, thus promoting its utilization rather than control. The harmonization of such diverse policy instruments is necessary for designing effective management strategies. This is consistent with the logic that meaningful outcomes of actors' actions should holistically integrate their activities, their means for achieving such activities, and the meanings or interpretations of the outcomes of their actions [25]. Therefore, enhancing access to land (means), coupled with the perceived outcome of alleviating poverty (meaning of actions) through farming activities (actors' activity) provides a sound entry point for sustainable environmental management.

4.2.4. Community Mobilization and Awareness Creation

The findings of this study indicate that scientific evidence is fundamental for informing land governance concerning IAS management as it helps in identifying factors that influence land users' adoption of SLM practices as well as the areas where management should be prioritized. For instance, we found that communally owned land and irrigation schemes are vulnerable to invasion, thus necessitating robust management measures in order to prevent further spread in these areas.

Even though efforts have been made to disseminate research findings to community members, we propose that innovative extension service approaches be used to help promote communal actions—through the engagement of the community members in reflecting on the benefits and nuisances of ecosystem services generated by different land management practices and how to integrate stakeholder needs with sustainable environmental management at the landscape scale. Likewise, the use of demonstration plots, the introduction of attractive community field days, and the appreciation of SLM champions may not only motivate community participation in SLM practices but also provide an interactive platform for the widespread dissemination and exchange of research innovations [40]. These concur with a study in Mongolia that found that integrating diverse sources of information enhances knowledge exchange, which then promotes rangeland management [41].

4.2.5. Effective Land Governance Solutions

Apart from awareness creation, the enforcement of tenure rights may provide land governance solutions and support invasion management. According to [12], a consolidated land ownership approach is indispensable for improving access to land and simplifying land management. This may be achieved if flexible land markets are established to encourage transactions in parcels that are vulnerable to abandonment by their owners. However, this approach is only applicable to individual lands or parcels under customary rights of occupancy, as provided for in the Kenyan Community Land Act, 2016. Nevertheless, managing the often neglected communally shared parcels, which are prone to disputes and neglect, is complex [42]. Therefore, IAS management at the landscape level is hindered by land subdivisions as land users disregard the management of parcels jointly owned due to a lack of incentives [14]. Addressing the challenge of joint registration of communally shared resources, as provided for in Kenya's new Community Land Act, 2016, may yield a positive outcome. Such an approach is likely to promote the collective utilization and management of communal resources by legitimate community members [43]. This is because it helps in recognizing the legitimate beneficiaries of communally owned resources, thus holding them accountable for their management. A participatory formulation and enforcement of by-laws guiding the use and management of the Simanjiro rangeland in Tanzania is a practical example of successful collective action in integrating IAS control within rangeland management [44]. As identified in our research, the benefits achieved thereof are a great incentive that determines whether land users will manage invasion or not. This means that for communally shared lands to be effectively managed, coordinated management has to be integrated with shared benefits for land users [13,14].

The above-mentioned interventions may significantly contribute to invasion management if jointly implemented and enforced by community members, who are the agents of change. This requires an effective institutional framework to promote the timely resolution of land-related disputes as well as communal engagement in management measures. To achieve this, efforts should be made to promote coordinated and community-based tenure rights enforcement procedures in order to enhance the legitimacy of enforcement officers in the community where change is needed and for invasion management to be sustainable. The establishment of empowering institutional frameworks and powers at communal levels is likely to enhance the solidarity and collective involvement of community members [45] in managing the invasion. In this regard, the establishment of communally managed conservancies has proved to be effective in attaining collective communal action in the management of invasive species [44].

4.3. Lessons Learned and Recommendations to the NPS

The continuous implementation of SLMs is central to the successful management of *Prosopis*, and this may be sustainable if it is centered around effective communal participation. Despite this, land users' continued participation depends on whether their efforts will be compensated by benefits derived from management efforts. To enhance anticipated benefits and promote land users' contribution to managing invasion through the implementation of SLMs, the NPS should incorporate programs aimed at enhancing the resilience of community members to adverse external drivers. Integrating management strategies with the enhancement of land users' livelihood such as through the increase of access to the market for farm produce is likely to promote land users' participation in the associated SLM implementation.

This study observed that land management efforts primarily focus on privately owned lands, while the management of communally shared resources is often neglected despite their vulnerability to invasion. The implementation of SLMs to manage invasion therefore depends on the enforcement of tenure rights, whose solutions vary with the nature of the tenure system. Access to the land market is likely to enhance land re-distribution and promote the implementation of SLMs on individual parcels. However, the same challenge on communally shared resources may be best resolved through the collective registration, utilization, and management of land. In both cases, strengthening communal land management institutions plays a pivotal role in effective land governance and invasion management. Nonetheless, such institutions have to be informed by credible research output for them to offer viable solutions to challenges that face the prevailing tenure systems. Thus, the NPS should adopt means of ensuring that land management institutions operate within the guidelines of scientific research output.

5. Conclusions

We conclude that while benefits derived from management efforts are key determinants of land users' decisions to engage in invasion management at any spatial scale, sustainable management strategies lie at a landscape scale. Communally shared parcels where benefits of management efforts are not guaranteed are therefore vulnerable to neglect and subsequent invasion. Therefore, strengthening institutional capacity at a communal level may promote effective community participation in providing sustainable land governance solutions. Furthermore, drivers of land users' decisions in managing IAS are multi-dimensional with complex systemic interactions. This implies that management strategies should adopt a holistic approach that integrates all potential drivers in designing management interventions for invasive species.

Supplementary Materials: The following supporting information can be downloaded at: https://www.mdpi.com/article/10.3390/land11040550/s1.

Author Contributions: Conceptualization, B.A., U.S., S.M., B.K., S.E. and A.E.; data curation, S.M., B.K., P.R.M., S.C. and A.E.; formal analysis, B.A., U.S., S.M., B.K., P.R.M., S.E. and A.E.; funding acquisition, U.S., B.K. and A.E.; investigation, B.A., U.S., S.M., B.K., P.R.M., S.E., S.C. and A.E.; methodology, B.A., U.S., S.M., B.K., P.R.M., S.E. and A.E.; project administration, B.A., U.S., B.K., S.C. and A.E.; resources, U.S., B.K., S.C. and A.E.; software, B.A., S.E. and A.E.; supervision, U.S., S.M., B.K. and A.E.; validation, B.A., U.S., B.K., P.R.M., S.E. and A.E.; visualization, B.A., S.E. and A.E.; writing—original draft, B.A.; writing—review and editing, B.A., U.S., S.M., B.K., P.R.M., S.E., S.C. and A.E. All authors have read and agreed to the published version of the manuscript.

Funding: This research was funded by the Swiss National Science Foundation (SNSF) and the Swiss Agency for Development and Cooperation (SDC) as part of the Swiss Programme for Research on Global Issues for Development (r4d), for the project "Woody invasive alien species in East Africa: Assessing and mitigating their negative impact on ecosystem services and rural livelihood" and by the Centre for Training and Integrated Research in ASAL (Grant Number: 400440_152085).

Data Availability Statement: The data presented in this study are available on request from the corresponding author.

Conflicts of Interest: The authors declare no conflict of interest.

Appendix A. Distribution of Respondents per Hotspot Area

Hotspot Area	Sub-Location Name/s	No. of Households	Population	No. of Respondents
1	Salabani	277	1316	7
2	Ngambo and Sintaan	777	4060	19
3	Perkerra	1276	4911	32
4	Sandai/Loboi	458	2055	11
5	Eldume and Kailer	440	2155	11
	Total	**3228**	**14497**	**80**

Appendix B. Text Format of the First Five Respondents on the Drivers of Constant invasion. Responses Are in Black Font, While Respondent Characteristics Are in Red Font. Information about One Respondent Is Separated from the Next Respondent by the Spaces between Them

Reasons_for_constant_invasion
ID_String

1. Communal land-no one is committed as there is fewer benefits derived
2. Inadequate funds to control Prosopis
3. Lack of timely knowledge on potential impacts

1-Ngambo-40-High-M-Married

1. Lack of knowledge on impacts
2. Inadequate funds for management

2-Salabani-37.3-Low-F-Married

3-Salabani-37.3-Low-F-Married

1. No man's land hence no one held responsible for managing them
2. Pasture land-Restricted to grazing and not cultivation yet people can only clear areas they to cultivate
3. Suitable soils-clay and sandy

4-Salabani-37.3-Low-M-Married

References

1. UNCCD. *Land Degradation Neutrality Transformative Projects and Programmes:Operational Guidance for Country Support*; United Nations Convention to Combat Desertification: Bonn, Germany, 2019; ISBN 9789295117655.
2. Schwilch, G.; Bachmann, F.; de Graff, J. Decision Support for Selecting SLM Technologies with Stakeholders. *Appl. Geogr.* **2012**, *34*, 86–98. [CrossRef]
3. Alemu, M.M. Sustainable Land Management. *Environ. Prot.* **2016**, *7*, 502–506. [CrossRef]
4. De Graaff, J.; Amsalu, A.; Bodnár, F.; Kessler, A.; Posthumus, H.; Tenge, A. Factors Influencing Adoption and Continued Use of Long-Term Soil and Water Conservation Measures in Five Developing Countries. *Appl. Geogr.* **2008**, *28*, 271–280. [CrossRef]
5. Wise, R.M.; van Wilgen, B.W.; Le Maitre, D.C. Costs, Benefits and Management Options for an Invasive Alien Tree Species: The Case of Mesquite in the Northern Cape, South Africa. *J. Arid Environ.* **2012**, *84*, 80–90. [CrossRef]
6. Linders, T.; Bekele, K.; Schaffner, U.; Allan, E.; Alamirew, T.; Choge, S.K.; Eckert, S.; Haji, J.; Muturi, G.; Mbaabu, P.R.; et al. The Impact of Invasive Species on Social-Ecological Systems: Relating Supply and Use of Selected Provisioning Ecosystem Services. *Ecosyst. Serv.* **2020**, *41*, 101055. [CrossRef]
7. Le Maitre, D.; Blignaut, J.; Clulow, A.; Dzikiti, S.; Everson, C.; Görgens, A.; Hush, M.B. Impacts of Plant Invasions on Terrestrial Water Flows in South Africa. *Biol. Invasions* **2020**, *14*, 431–457. [CrossRef]
8. Shiferaw, H.; Bewket, W.; Alamirew, T.; Zeleke, G.; Teketay, D.; Bekele, K.; Schaffner, U.; Eckert, S. Implications of Land Use/Land Cover Dynamics and Prosopis Invasion on Ecosystem Service Values in Afar Region, Ethiopia. *Sci. Total Environ.* **2019**, *675*, 354–366. [CrossRef]
9. Mbaabu, P.R.; Ng, W.T.; Schaffner, U.; Gichaba, M.; Olago, D.; Choge, S.; Oriaso, S.; Eckert, S. Spatial Evolution of Prosopis Invasion and Its Effects on LULC and Livelihoods in Baringo, Kenya. *Remote Sens.* **2019**, *11*, 1217. [CrossRef]

10. Johnston, R.B. Arsenic and the 2030 agenda for sustainable development. In Proceedings of the Arsenic Research and Global Sustainability—6th International Congress on Arsenic in the Environment (AS 2016), Stockholm, Sweden, 19–23 June 2016; pp. 12–14. [CrossRef]
11. Banadda, N. Gaps, Barriers and Bottlenecks to Sustainable Land Management (SLM) Adoption in Uganda. *Afr. J. Agric. Res.* **2010**, *5*, 3571–3580.
12. Van Song, N.; Cuonga, H.; Huyena, V.; Rañola, R. The Determinants of Sustainable Land Management Adoption under Risks in an Upland Area of Vietnam. *J. Sustain. Futures* **2020**, *2*, 100015. [CrossRef]
13. Bagavathiannan, M.V.; Graham, S.; Ma, Z.; Barney, J.N.; Coutts, S.R.; Caicedo, A.L.; De Clerck-Floate, R.; West, N.M.; Blank, L.; Metcalf, A.L.; et al. Considering Weed Management as a Social Dilemma Bridges Individual and Collective Interests. *Nature Plants* **2019**, *5*, 343–351. [CrossRef]
14. Epanchin-Niell, R.S.; Hufford, M.B.; Asian, C.E.; Sexton, J.P.; Port, J.D.; Waring, T.M. Controlling Invasive Species in Complex Social Landscapes. *Front. Ecol. Environ.* **2010**, *8*, 210–216. [CrossRef]
15. UNCCD. *Science—Policy Brief. Sustainable Land Management for Climate and People Further Reading: Scaling up Implementation of Sustainable Land Management How Does SLM Contribute to Sustainable Development at the Sustainable Land Management Solutions*; United Nations Convention to Combat Desertification: Bonn, Germany, 2017; Volume 49.
16. Shackleton, R.T.; Le Maitre, D.C.; van Wilgen, B.W.; Richardson, D.M. Identifying Barriers to Effective Management of Widespread Invasive Alien Trees: Prosopis Species (Mesquite) in South Africa as a Case Study. *Glob. Environ. Chang.* **2016**, *38*, 183–194. [CrossRef]
17. Kropf, B.; Schmid, E.; Schönhart, M.; Mitter, H. Exploring Farmers' Behavior toward Individual and Collective Measures of Western Corn Rootworm Control—A Case Study in South-East Austria. *J. Environ. Manag.* **2020**, 264. [CrossRef]
18. Mwangi, E.; Swallow, B. Prosopis Juliflora Invasion and Rural Livelihoods in the Lake Baringo Area of Kenya. *Conserv. Soc.* **2008**, *6*, 130–140. [CrossRef]
19. County Government of Baringo. *Baringo County Integrated Development Plan. for 2013–2017—Finance and Economic Planning, Baringo*; County Government of Baringo: Kabarnet, Kenya, 2018.
20. Cherop, H.K.; Muchiri, L.; Githiri, G.J.; Ambusso, W.J. Groundwater Investigation and Characterisation in Marigat Area, Baringo County, Using Vertical Electrical Sounding Resistivity Surveys. *IOSR J. Appl. Geol. Geophys.* **2016**, *4*, 103. [CrossRef]
21. Ochuka, M.; Ikporukpo, C.; Mijinyawa, Y.; Ogendi, G. Land Use/Land Cover Dynamics and Anthropogenic Driving Factors in Lake Baringo Catchment, Rift Valley, Kenya. *Nat. Resour.* **2019**, *10*, 367–389. [CrossRef]
22. ISRIC. Soil Geographic Databases. Available online: https://www.isric.org/explore/soil-geographic-databases (accessed on 4 April 2022).
23. KNBS. *Kenya Population and Housing Census: Counting Our People for Sustainable Development and Devolution of Services*; Kenya National Bureau of Statistics: Nairobi, Kenya, 2019.
24. Greiner, C. Pastoralism and Land-Tenure Change in Kenya: The Failure of Customary Institutions. *Dev. Chang.* **2017**, *48*, 78–97. [CrossRef]
25. Wiesmann, U.; Ott, C.; Ifejika Speranza, C.; Kiteme, B.P.; Müller-Böker, U.; Messerli, P.; Zinsstag, J. A Human Actor Model in Interdisciplinary Research for Sustainable Development. In *Research for Sustainable Development: Foundations, Experiences, and Perspectives*; Swiss National Centre of Competence in Research (NCCR): Bern, Switzerland, 2011; pp. 231–256. [CrossRef]
26. Lewis, J.K. Using ATLAS.Ti to Facilitate Data Analysis for a Systematic Review of Leadership Competencies in the Completion of a Doctoral Dissertation. *SSRN Electron. J.* **2017**, 1–14. [CrossRef]
27. Friese, S. *Qualitative Data Analysis with ATLAS.Ti*; Sage: London, UK, 2019.
28. Gill, N.J.; Graham, S.; Cross, R.; Taylor, E.M. Weed hygiene practices in rural industries and public land management: Variable Knowledge, Patchy Implementation, Inconsistent Coordination. *J. Environ. Manag.* **2018**, *223*, 140–149. [CrossRef]
29. Niemiec, R.M.; McCaffrey, S.; Jones, M.S. Clarifying the Degree and Type of Public Good Collective Action Problem Posed by Natural Resource Management Challenges. *Ecol. Soc.* **2020**, 25. [CrossRef]
30. Manzoor, S.A.; Griffiths, G.; Lukac, M. Land Use and Climate Change Interaction Triggers Contrasting Trajectories of Biological Invasion. *Ecol. Indic.* **2021**, *120*, 106936. [CrossRef]
31. Cowie, A.; Orrb, B.J.; Sanchezc, V.; Chasekd, P.; Crossmane, N.D.; Erleweinf, A.; Louwagieg, G. Land in Balance: The Scientific Conceptual Framework for Land Degradation Neutrality. *Environ. Sci. Policy* **2018**, *79*, 25–35. [CrossRef]
32. Mena, M.M. Community Adoption of Watershed Management Practices at Kindo Didaye District, Southern Ethiopia. *Int. J. Environ. Sci. Nat. Resour.* **2018**, *14*, 32–39. [CrossRef]
33. Mwangi, E.; Swallow, B. *Invasion of Prosopis Juliflora and Local Livelihoods*; World of Agroforestry Centre (ICRAF): Nairobi, Kenya, 2005; 68p.
34. Ndhlovu, T.; Milton-Dean, S.J.; Esler, K.J. Impact of Prosopis (Mesquite) Invasion and Clearing on the Grazing Capacity of Semiarid Nama Karoo Rangeland, South Africa. *Afr. J. Range Forage Sci.* **2011**, *28*, 129–137. [CrossRef]
35. Trang, N.; HuuLo, H. Livelihood Sustainability of Rural Households in Adapting to Environmental Changes: An Empirical Analysis of Ecological Shrimp Aquaculture Model in the Vietnamese Mekong Delta. *Environ. Dev.* **2021**, *39*, 100653. [CrossRef]
36. Danquah, N.; Twumsi, M.A.; Korankye, B.A. Impact of Land Fragmentation on Technical Efficiency: The Case of Maize Farmers in the Transitional Zone of Ghana. *Int. J. Environ. Agric. Res.* **2019**, *5*, 14–26.

37. Zimmermann, A.; Maennling, C. Multi-Stakeholder Management. Tools for Stakeholder Analysis: 10 Building Blocks for Designing Participatory Systems of Cooperation. *Promot. Particip. Dev. German Dev. Cooperat.* **2007**, *5180*, 65726.
38. Jacobi, J.; Mukhovi, S.; Llanque, A.; Toledo, D.; Speranza, C.I.; Käser, F.; Augstburger, H.; Delgado, J.M.F.; Kiteme, B.P.; Rist, S. Actor-Specific Risk Perceptions and Strategies for Resilience Building in Different Food Systems in Kenya and Bolivia. *Reg. Environ. Chang.* **2019**, *19*, 879–892. [CrossRef]
39. Zhang, K.; Putzel, L. Institutional Innovation and Forest Landscape Restoration in China: Multi-Scale Cross-Sector Networking, Household Fiscal Modernization and Tenure Reform. *World Dev. Perspect.* **2016**, *3*, 18–21. [CrossRef]
40. Kondylis, F.; Mueller, V. *Seeing Is Believing? Evidence from a Demonstration Plot Experiment in Mozambique: Strategy Support Program Working Paper*; International Food Policy Research Institute: Washington, DC, USA, 2013.
41. Ulambayar, T.; Fernández-Giménez, M.E. How Community-Based Rangeland Management Achieves Positive Social Outcomes In Mongolia: A Moderated Mediation Analysis. *Land Use Policy* **2019**, *82*, 93–104. [CrossRef]
42. Tebbоth, M.G.L.; Few, R.; Assen, M.; Degefu, M.A. Valuing Local Perspectives on Invasive Species Management: Moving beyond the Ecosystem Service-Disservice Dichotomy. *Ecosyst. Serv.* **2020**, *42*, 101068. [CrossRef]
43. Wily, L.A. The Community Land Act in Kenya Opportunities and Challenges for Communities. *Land* **2018**, *7*, 12. [CrossRef]
44. Adoyo, B.; Choge, S.; Eckert, S.; Ehrensperger, A.; Eschen, R.; Kilawe, C.J.; Kiteme, B.; Schaffner, U. *Integrating the Control of Invasive Trees in Land Use and Management Plans. Woody Weeds Practice Brief*; Nairobi, Kenya. 2021. Available online: https://woodyweeds.org/wp-content/uploads/2021/10/WoodyWeeds-Practice-Brief-Integration.pdf (accessed on 1 March 2022).
45. Wily, L.A. Customary tenure: Remaking property for the 21st century. In *Comparative Property Law: Global Perspectives*; Edward Elgar Publishing: Northampton, MA, USA, 2017; pp. 458–477. [CrossRef]

land

MDPI

Article

Spatial Pattern and Key Environmental Determinants of Vegetation in Sand Mining and Non-Mining Sites along the Panjkora River Basin

Kishwar Ali [1,*], Nasrullah Khan [2], Rafi Ullah [2], Muzammil Shah [2], Muhammad Ezaz Hasan Khan [1], David Aaron Jones [3] and Maha Dewidar [1]

1 College of General Education, University of Doha for Science and Technology, Arab League Street, Doha P.O. Box 24449, Qatar
2 Department of Botany, University of Malakand, Khyber Pakhtunkhwa, Chakdara P.O. Box 18800, Pakistan
3 College of Health Sciences, University of Doha for Science and Technology, Arab League Street, Doha P.O. Box 24449, Qatar
* Correspondence: kishwar.ali@udst.edu.qa

Abstract: A specific set of environmental conditions characterizes plant species patterns and distribution on Earth. Similarly, riparian vegetation can be impacted by anthropogenic activities like mining practices involving the removal of vegetation cover, which destroys the structure and diversity of the habitat, adversely affecting the ecosystem services. In this study, we explored the role of environmental variables and biotic intervention in deriving spatial patterns and distribution of riparian vegetation at mining and non-mining sites along the most depleted Panjkora River basin in NW Pakistan. Vegetation data and its determining factors at 28 mining and non-mining sites (14 each) were sampled using 10 m × 10 m (100 m^2) systematic plots at 50 m intervals along transects in a downstream direction from the upper catchments to the bottom junction with the Swat River. We recorded 186 species in both mining and non-mining sites, belonging to 70 families comprising 174 angiosperms, 3 gymnosperms, and 9 Pteridophytes. Results show that annual or perennial therophytic life forms predominated in the Panjkora River system, indicating anthropogenic disturbances. At the same time, the aggressively invasive species, such as *Xanthium strumarium* and *Cannabis sativa*, further heightened plant community disturbances. Generally, the species diversity was higher in non-mining sites and may be attributed to habitat fragmentation. Likewise, the Canonical Correspondence Analysis (CCA-ordination) revealed that geographic coordinate (i.e., latitude $r = 0.80$; longitude $r = 0.75$) and elevation ($r = 0.95$) were more meaningful predictors than soil texture (i.e., silt%, $r = -0.30$), nutrients (i.e., potassium, $r = -0.35$; phosphorus, $r = 0.38$) and soil pH ($r = -0.50$) in shaping the spatial pattern and vegetation structure. Our result implies that the present vegetation composition and spatial assemblages are due to heavy anthropogenic interventions, especially mining activities. Therefore, the heavily degraded fragile riparian system of the Panjkora River and its tributaries needed to be conserved and restored by predicting the composition of communities in response to changing climatic conditions.

Keywords: Panjkora River basin; Canonical Correspondence Analysis; mining activities; socioeconomic aspects

Citation: Ali, K.; Khan, N.; Ullah, R.; Shah, M.; Khan, M.E.H.; Jones, D.A.; Dewidar, M. Spatial Pattern and Key Environmental Determinants of Vegetation in Sand Mining and Non-Mining Sites along the Panjkora River Basin. *Land* **2022**, *11*, 1801. https://doi.org/10.3390/land11101801

Academic Editor: Eva Papastergiadou

Received: 4 September 2022
Accepted: 12 October 2022
Published: 14 October 2022

1. Introduction

Plant species are not evenly or randomly distributed throughout the world; instead, they are distributed along distinct geographical units due to specific climatic and environmental factors [1,2]. Plant species distribution is influenced by biotic and abiotic variables, including environmental changes, soil types, species migration, habitat characteristics, terrain, hydrology, geology, tectonic plate movement, etc. Due to all of these dynamics, ancient habitat topography and climate are drastically altered, mountains are raised, and as a result, ecological factors (competition, hybridization, and pollination) significantly impact the patterns of phytogeographical distribution [3]. Its phytogeographic analysis traces specific plant species allocation, diversification, evolution, and origin [4].

Because riparian vegetation is regarded as one of the indicators to be used in management practices, ecologists and land managers are aware of the variety of ecosystem services it provides [5]. Variations in the riparian environment substantially influence the composition and diversity of plant species [6]. Plant species variety has increased significantly due to variations in the riparian ecosystem's soil physicochemical characteristics and microclimate [7]. Natural and artificial calamities drastically destroy an area's vegetation [8,9]. Over-exploitation is the primary factor causing the observed reduction in an area's vegetation, which often progresses more rapidly than the pace of species regeneration and natural restoration. Numerous reasons, such as the removal of trees for use as lumber and wood, the expansion of agricultural land, the rapid spread of invasive species, and grazing pressure, contribute to the ongoing loss of species diversity in these environments [7].

The health of the local ecosystems and the protection of the streams and accompanying fauna depend on the riparian vegetation [10]. Despite their significance, anthropoid activities have fragmented these forest corridors lowering biodiversity and several ecological processes, making riparian forests one of the most degraded ecosystems in the world [10,11]. Floristic and structural studies are necessary to understand a given community better and track the profile of habitats with unique environmental features in response to human stresses [11]. When addressing abiotic elements such as edaphic attributes, these studies allow the comprehension of how these abiotic factors, in addition to floristic composition, environmental quality, vegetation structure, and history, might impact the vegetation of these sites [12]. Moreover, the formation of subgroups in vegetative forms with specific affinities and edaphic factors influences vegetation significantly at the local scale [13].

Similarly, local elements such as soil, topography, and vegetation in the region significantly influence riparian vegetation, resulting in a varied floristic composition with species characteristic of river banks and distinct phytophysiognomies [14]. Additionally, due to regular changes, mostly brought on by variations in the inundating regime and natural and human instabilities, riparian vegetation often displays substantial variety in its species structure and geographic distribution [15]. Because of the abundance of forest species, these forests are often better off in terms of vegetation than non-riparian habitats [16].

Sand mining has become a significant source of employment for a substantial number of people worldwide and has brought about harmful anthropogenic alterations in the river basins where this activity takes place [17]. River sand and gravel have long been utilized as aggregate for building and road construction, and demand for these resources is still growing. Sand is primarily obtained in Indonesia via river or stream mining [18]. The natural balance of a stream channel may be drastically altered by excessive sand removal [19]. By disrupting the subterranean bed, the procedure impacts the land, causing water imbalances and harmful environmental issues in the affected regions [20,21]. A drop in the groundwater level reduces the amount of water accessible for home and agricultural use. People relying on agricultural land for living also lose their jobs due to mining operations destroying their farmland [22]. Sand mining negatively impacts the vegetation's structure and content and causes societal unrest [23].

Management and successful restoration efforts in susceptible areas affected by mining depend on an understanding of species composition, vegetation structure, and interactions between vegetation and edaphic components [24]. Past sand mining studies have focused on the advantages and disadvantages of sand mining methods on the environment and local people's means of subsistence in various places. For example, the studies conducted by [20–24] have focused on the mining activities and their socioeconomic importance. However, these studies lack vegetation characteristics and their relationship with the corresponding environmental factors. Therefore, a detailed study was conducted to compare the riparian vegetation diversity between mining and non-mining communities along the Panjkora River basin, Khyber Pakhtunkhwa, Pakistan.

The significant environmental factors that operate in riparian vegetation were also evaluated using multivariate analysis. We hypothesized that mining had resulted in a deliberate change in habitat structure that is well evident from the floristic composition and vegetation structure of communities and its diversity and can be assessed by comparing it with control sites (non-mining sites). This research will add to our understanding of the region's mining communities and will provide insight into how we can better manage these communities for socioeconomic benefit and conservation.

2. Materials and Methods

2.1. Study Area

The Panjkora River basin in Khyber Pakhtunkhwa, Pakistan, originates from tributaries of the eastern Hindu Kush Mountains and was the site of the current research (Figure 1) [25]. The Panjkora River begins as a torrent from the Hindukush Mountain's enormous icecaps. It regenerates at the intersection of Barawal and Jandool streams to the east and Ushirai and Niag streams to the west. It originates in District Dir Lower, its flow travels 113 kilometres, has a catchment area of 5905 km^2, and enters the Chakdara Valley before joining the Swat River near Qalangi [26]. Alternatively, it is separated from Kabul, and the Swat River basins by western and eastern mountain ranges and the extension of these mountain ranges then forms the Panjkora River basin. Geographically, the region stretches from 71°13′8″ to 72°22′13″ E longitude and 34°39′30″ to 35°47′17″ N latitude. These mountains are covered with snow and have numerous glaciers in valleys that reach above 4000 m above sea level (a.s.l.) climatically; the area under the present work has a cool to warm summer with temperatures ranging from 16 °C to 32 °C. There is a significant drop in the temperature between December and February, and is characterized by snowfall. Rainfall is reported annually and varies from 823 to 2149 mm [27].

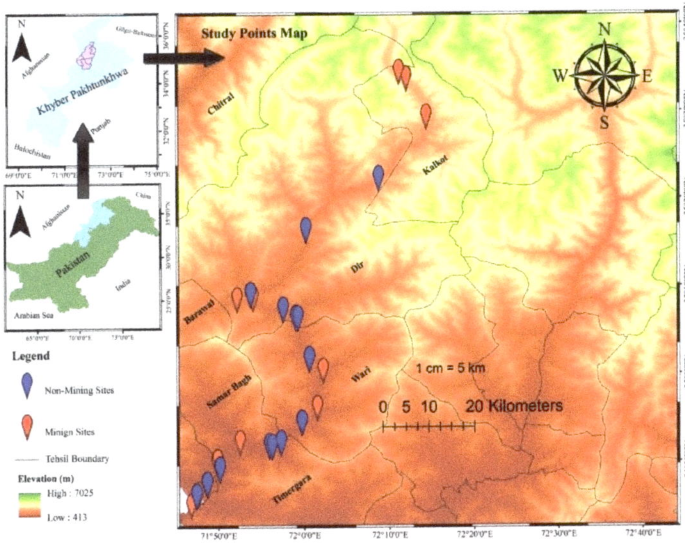

Figure 1. Study area map showing sampling points across the River Panjkora in Khyber Pakhtunkhwa, Pakistan.

2.2. Vegetation Sampling and Diversity Indices

The phytosociological vegetation attributes were assessed in mining (Experimental) and non-mining (Control) sites during the field survey for two years starting in 2020 to

2021, along the Panjkora river basin by using the quadrat method following [24]. Vegetation was recorded using a 10 × 10 m quadrat for trees, 5 × 5 m for shrubs, and 1 × 1 m for herbs in the selected stands. A total of 60 stands and 600 quadrates, which include 30 mining and 30 non-mining sites, were selected for data collection and assessment. Various quantitative variables were used including density, frequency, height, cover, importance value (IV), and importance value index (IVI), employing techniques as described by [28], were recorded in each quadrate. Herb and shrub height and cover were recorded using a forestry tap, while tree diameter was recorded at breast height (dbh). The IV was calculated as the sum of the relative frequency, relative density, and relative cover [28]. A 10-m buffer zone was removed from the stand borders, as recommended by [29], to lessen the impact of edge effects. The plants recorded were identified using *Flora of Pakistan* [30,31] plant taxonomic nomenclature.

Floristic characteristics were evaluated following the *Flora of Pakistan*, and an inventory was compiled to record species details [30]. Dried plant specimens were stored on Herbarium vouchers at the Department of Botany at Malakand Khyber Pakhtunkhwa University in Pakistan. Following [32], species density was used to compute three main diversity indices: species richness (S), evenness index (E), and Shannon-Wiener diversity index (H′). The number of species present determines the species richness of a stand. The H′ and E indices were calculated using the following formulas:

$$E = \frac{H\prime}{\ln S} \quad (1)$$

$$H\prime = -\sum_{i=1}^{S} pi \ln pi \quad (2)$$

where *pi* = proportion of the species (*i*) to a total number of species, ln = natural logarithm of pi, and lnS = natural logarithm of species richness.

2.3. Environmental Variables and Soil Analysis

A 3 kg sample was taken from two opposing corners and the middle of each site using auger borings to define the soil parameters in each location. In general, the top layer of soil is nutrient-rich. As a result, samples were collected between 0 and 30 cm deep [33], bagged, and carefully mixed to minimize variability [28,34]. A digital pH meter and an EC meter were used in the field to measure electrical conductivity and soil pH in a soil-water solution (1:5). By air-drying the samples and running them through a 2 mm filter following USDA guidelines, the physiochemical makeup and textural characteristics of the soil were ascertained [35]. The Walkley–Black technique was employed to estimate organic matter, and wet combustion with chromic acid digestion followed by dry combustion was utilized to quantify Total and Organic Carbon [36]. Micro-Kjeldahl was used to measure total nitrogen, while Yadav et al. [37] methods were used to assess accessible phosphorus (P^{2+}) and exchangeable potassium (K^+). The geometric approach was used to determine lime (calcium carbonate; percent), and geometric measurement of CO_2 evolution was done following [38]. Using an online calculator (https://www.nrcs.usda.gov (accessed on 20 November 2021)), we evaluated other soil properties, including field capacity (FC), available water (AW), conductivity (s/cm), wilting and saturation point (WSP), bulk density (BD), following [39]. The soil analysis was performed at the Swat Agriculture Research Institute (SARI).

2.4. Data Analysis

From the 28 stands, floral and phytosociological data and relevant environmental factors, i.e., geographic attributes, soil texture, and nutrients, were data-banked for analysis and interpretation. Following the flowchart in Figure 2, various floristic attributes, i.e., life form, leaf form, status in Pakistan, and Rainkiaer life form, were assigned to species following already published literature conducted in areas that resemble the study area [40,41].

The relative phytosociological qualities were converted into an importance value index (IVI). After numerically classifying mining and non-mining sites, each stand's species was allocated to a phytosociological category, either mining or non-mining sites. Recent literature was consulted to determine whether a particular species may indicate a group [31,42]. The means of the environmental and topographical factors were determined using a paired *t*-test to identify variances.

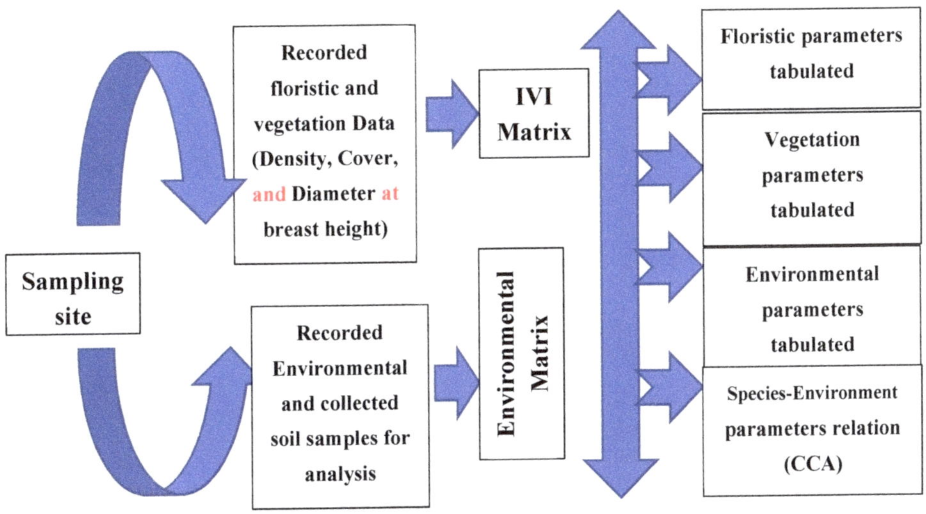

Figure 2. Flow chart of the methodology.

Using canonical correspondence analysis (CCA) of species IVI, the link between mining and non-mining vegetation and environmental factors was investigated. In the ordination graph, the vectors of the significant variables were also shown. All quantitative vegetation and environmental data analyses were conducted using Graph Pad Version 7.0 and Microsoft Excel [43].

3. Results

3.1. Floristic and Phytosociological Diversity

A total of 186 plants, representing 70 families of angiosperms, gymnosperms, and pteridophytes, were identified in mining and non-mining sites. The families with the highest species were Asteraceae (20), Poaceae (19), Lamiaceae (9), Amaranthaceae (8), Rosaceae (6), Brassicaceae (5 each of Cyperaceae, Cucurbitaceae, Polygonaceae, and Solanaceae), and Moraceae (4 species) (Table S1). One hundred seventy-four of the total plant species reported were angiosperms, compared to 9 pteridophytes and three gymnosperms. Based on life form, the flora was dominated by herbs 134 (72%), followed by shrubs 32 (17%), while the rest were tree species, and one was a climber (Figure 3). Based on life span (LS), perennials (P) made up the majority of the flora, accounting for 116 (62%), followed by annuals (A), 63 (34%), biennials (B), 1 (>1%), annual/perennials (A/P), 4 (2%), and annual/biennial/perennial (A/B/P), 1 (>1%).

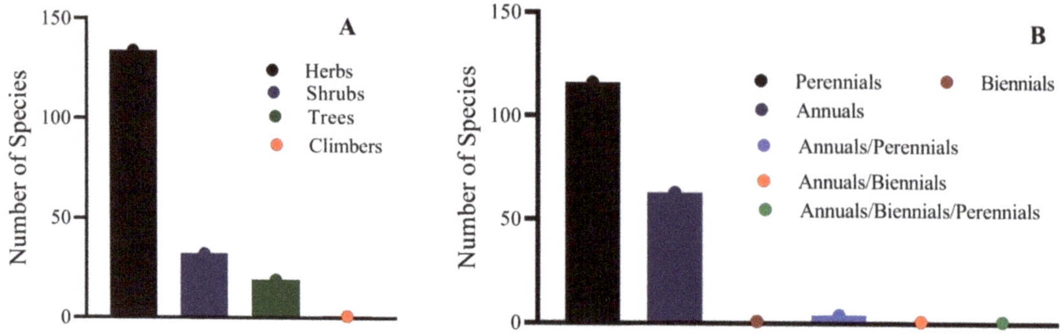

Figure 3. Life form (**A**) and Habit based spectrum (**B**) of the flora recorded in the mining and non-mining sites along the River Panjkora basin.

Results revealed that in the study area, dicots dominated with 140 (75.26%), followed by monocots 34 (18.27%), while the rest were either gymnosperms or Pteridophytes. The spectrum of flora was diverse based on Status in Pakistan (SP) ranging from native to exotic: Casual (CA) 1 (>1%), Cosmopolitan (CO) 1 (>1%), Cultivated (CU) 18 (10%), Exotic (EX) 1 (>1%), Invasive (I) are 14 (8%), Indigenous (ID) 2 (1%), Introduced (IN) 2 (1%), Naturalized (N) are 21 (11%), Native (NV) 109 (59%), and Weed (W) 17 (8%) as indicated in Figure 4.

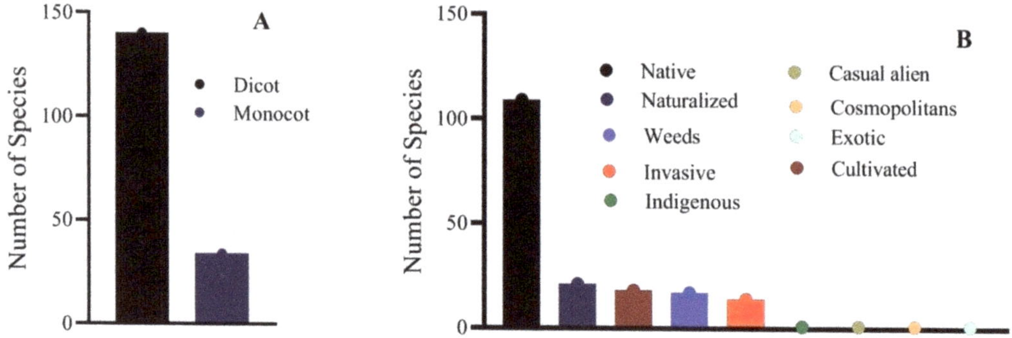

Figure 4. Spectrum based (**A**) on seed and status in Pakistan (**B**) of the flora recorded in the mining and non-mining sites along the River Panjkora basin.

Based on Raunkiaer's life form (Figure 5), 87 (47%) of the species were therophytes (Th), 31 (17%) were hemicryptophytes (H), 18 (10%) were megaphanerophytes (Megp), 17 (9%) were nanophanerophytes (Np), 14 (7%) were geophytes (G), 8 (4%) were mesophanerophytes (Mesp), 6 (3%) were chamaephytes (Ch), 4 (2%) were microphanerophyte (Micp) and 1 was parasite (P). The leaf size spectrum was similarly diverse, ranging from microphyllous to aphyllous, including microphyll 67 (36.02%), nanophyll 50 (26.88%), mesophyll 47 (25.26%), leptophyll 10 (5.37%), macrophyll 3 (4.30%), megaphyll 3 (1.61%), and aphyllous 1 (0.53%).

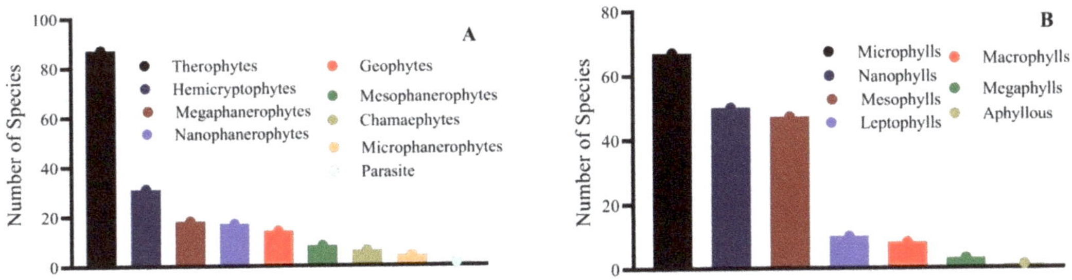

Figure 5. Raunkiaer life form (**A**) and leaf size class spectrum (**B**) of the flora recorded in the mining and non-mining sites along the River Panjkora basin.

In the mining sites (Group I), the IVI means value ranges from 4.52 ± 0.70–1.02 ± 0.48 representing *Cynodon dactylon* and *Maytenus royleanus*, respectively. In mining sites, the dominant plants were *Cynodon dactylon* (4.52 ± 0.70), having co-dominance of invasive *Xanthium strumarium* (4.23 ± 1.12) and *Cannabis sativa* (2.7 ± 0.87), exotic *Populus nigra* with IVI of 3.41 ± 0.98), and native *Mentha longifolia* having IVI *of* 3.12 ± 0.91. In addition, some of the species have IVI between 3–1, like *Calotropis procera, Centaurea cyanus, Chrozophora tinctoria, Citrus limon, Cirsium arvense, Corynephorus canescens, Punica granatum, Ricinus communis, Solanum surattense, Marsilea quadrifolia, Ranunculus sceleratus, Maytenus royleanus, Tagetes minuta, Withania somnifera*. The IVI mean values of most plant species were less than 1, as presented in Table S2.

The non-mining sites (Group II) were diverse, with more plant species than the mining sites in Group I. The mean IVI of species ranges from 4.65 ± 1.28–0.09 ± 0.09 representing *Cannabis sativa* and *Zea mays*. At the same time, co-dominated by *Pinus roxburghii* with IVI of 2.86 ± 1.23, *Cynodon dactylon* having 2.8 ± 0.94 IVI, *Dodonaea viscosa* with IVI of 2.1 ± 1.43 and *Olea ferruginea* with IVI of 2.21 ± 1.12, belonging to native species (Table 1). Similarly, the mean IVI of most plant species was less than 1.

Table 1. Mean importance value index of dominant plant species in Group I (mining sites) and (Group II) non-mining sites.

Group I (Mining Sites)		**Group II (Non-mining Sites)**	
Plant Binomial	**Mean ± SE**	**Plant Binomial**	**Mean ± SE**
Cynodon dactylon (L.)	4.52 ± 0.70	*Cannabis sativa* L.	4.65 ± 1.28
Cannabis sativa L.	2.7 ± 0.87	*Cirsium arvense* (L.) Scop.	1.88 ± 0.67
Cirsium arvense (L.) Scop.	2.4 ± 0.73	*Corynephorus canescens* (L.) P Beauv.	2.13 ± 1.18
Corynephorus canescens (L.) P. Beauv.	2.52 ± 1.22	*Cynodon dactylon* (L.)	2.8 ± 0.94
Datura metal L.	2.07 ± 0.86	*Dodonaea viscosa* (L.) Jacq.	2.1 ± 1.43
Helianthus annus L.	2.22 ± 1.01	*Duchesnea indica* (Andr.) Focke	1.79 ± 1.01
Mentha longifolia (L.) Huds	3.12 ± 0.91	*Ficus carica* L.	1.62 ± 0.71
Persicaria hydropiper (L.) Delarbre.	2.06 ± 0.74	*Monotheca buxifolia* (Falc.) A. DC.	1.86 ± 1.15
Populus nigra L.	3.41 ± 0.98	*Olea ferruginea* Royle.	2.21 ± 1.12
Ricinus communis L.	2.24 ± 0.94	*Pinus roxburghii* Sargent.	2.86 ± 1.23
Tagetes minuta L.	2.61 ± 0.72	*Rumex dentatus* L.	1.68 ± 0.49
Xanthium strumarium L.	4.23 ± 1.12	*Ficus carica* L.	1.62 ± 0.71

Note: Only twelve major dominant species having an importance value (IVI) higher than 1 are selected, and their IVI is presented as Mean ± SE.

3.2. Environmental Variables and Its Influence

Table 2 shows the various environmental factors impacting non-mining and mining areas. Non-mining sites have flat to moderately sloping topography, reasonably deep soil (up to 90 cm in depth), and typically well-drained soil. The usual soil type was sandy-loam, with minor variations. All vegetation had somewhat acidic soil, likely since their parental material is comparable (igneous rocks). From species-poor (mining) to species-rich (non-mining) plant types, the value of nitrogen and phosphorus rises while that of lime, potassium, and organic matter declines. According to [1], for *Dodonaea viscosa* in the same location, the available phosphorus was quite high. It was never over the 2.5 mg/kg detection level in other plant natural communities. Numerous soil characteristics, such as wilting point, electric conductivity, bulk density, field capacity, and saturation point, decreased with species-rich to species-poor vegetation types and showed statistically insignificant differences (Table 2). Still, they did not affect these or other vegetation physiognomies significantly.

Table 2. Descriptive value (Mean ± SD) of environmental variables affecting the vegetation sites of mining and non-mining sites.

Factor	Code	Mining Sites	Non-Mining Sites	*t*-Value	*p*-Value
Aspect Angle (°)	AA	257.5 ± 24.13	237.21 ± 21.21	0.68	0.25
Elevation (m)	ELE	1189.4 ± 179.68	1081.5 ± 111.02	1.13	0.13
Latitude (°)	Lat.	35.04 ± 0.08	35.00 ± 0.05	1.1	0.14
Longitude (°)	Long	71.95 ± 0.04	71.94 ± 0.02	0.17	0.43
Sand (%age)	Sand	47.41 ± 2.10	43.17 ± 2.27	1.2	0.12
Silt (%age)	Silt	29.82 ± 2.58	32.51 ± 2.87	0.79	0.22
Clay (%age)	Clay	22.75 ± 1.90	23.93 ± 2.19	0.66	0.33
Nitrogen (%age)	N	0.15 ± 0.05	0.34 ± 0.07	2.66	0.009
Phosphorus (mg/Kg)	P	5.12 ± 0.51	6.34 ± 0.54	1.76	0.05
Potassium (mg/Kg)	K	137.35 ± 18.06	110.92 ± 7.17	1.63	0.06
Lime (%age)	L	13.09 ± 1.02	11.17 ± 0.70	1.9	0.033
Organic matter (%age)	OM	0.86 ± 0.11	0.84 ± 0.09	0.14	0.44
pH (1:5)	pH	6.84 ± 0.14	6.47 ± 0.21	1.39	0.09
Electrical conductivity	EC	36.42 ± 2.61	37.5 ± 2.5	0.55	0.29
Wilting point	WP	0.13 ± 0.00	0.14 ± 0.01	0.65	0.22
Bulk density (g/cm)	BD	1.41 ± 0.01	1.42 ± 0.01	1.09	0.14
Available water	AW	0.12 ± 0.00	0.13 ± 0.00	−1.04	0.16
Saturation point	SP	0.46 ± 0.00	0.47 ± 0.00	−1.07	0.15

Elevation and latitude have a strong, significant association ($p < 0.01$; $r = 0.93$), elevation and longitude ($p < 0.01$; $r = 0.84$), whereas potassium has a weak, statistically significant relationship ($p < 0.05$; $r = -0.41$) with elevation as depicted from Figure 6. However, a weak negative correlation was shown between latitude and potassium ($p < 0.05$; $r = -0.43$), sand and silt ($p < 0.01$; $r = -0.68$), and latitude and longitude ($p < 0.01$; $r = 0.87$). In contrast, a positive association between saturation point and clay particles ($p < 0.01$; $r = 0.94$), while a significant negative correlation between clay particles and bulk density ($p < 0.01$; $r = -0.94$). The soil nutrients like potassium and nitrogen significantly correlated with electrical conductivity having $p < 0.05$; $r = -0.36$ and $p < 0.05$; $r = -0.42$, respectively. Likewise, potassium was found correlated with pH and lime, having $p < 0.05$; $r = 0.41$, and $p < 0.05$; $r = 0.36$, respectively. In addition, organic matter and electrical conductivity negatively correlated with $r = -0.47$; $p > 0.01$. Moreover, an important soil hydraulic property, i.e., wilting point, was significantly correlated with electrical conductivity ($p < 0.05$; $r = 0.36$). A negative correlation between available water and bulk density, electrical conductivity, and wilting point ($p < 0.01$, $r = -0.69$; $p < 0.01$, $r = -0.90$).

	AA	Elev.	Lat.	Long.	Sand	Silt	Clay	N	P	K	L	OM	pH	EC	WP	BD	AW	SP
AA	1.0	-0.2	-0.3	-0.2	-0.1	0.3	-0.2	0.0	-0.2	0.2	0.2	0.0	0.1	-0.1	-0.2	0.2	0.2	-0.2
Elev.	-0.2	1.0	0.9	0.8	0.0	-0.2	0.2	0.1	-0.2	-0.4	-0.2	-0.1	0.1	0.0	0.2	-0.2	-0.1	0.2
Lat.	-0.3	0.9	1.0	0.9	0.1	-0.3	0.3	0.1	-0.1	-0.4	-0.3	-0.1	-0.2	0.0	0.2	-0.2	-0.2	0.2
Long.	-0.2	0.8	0.9	1.0	0.1	-0.3	0.2	0.1	0.0	-0.3	-0.2	-0.2	-0.1	0.1	0.2	-0.2	-0.3	0.2
Sand	-0.1	0.0	0.1	0.1	1.0	-0.7	-0.2	-0.3	0.2	0.2	-0.1	0.1	-0.3	0.5	-0.2	0.5	-0.9	-0.5
Silt	0.3	-0.2	-0.3	-0.3	-0.7	1.0	-0.6	0.3	-0.1	-0.1	0.3	0.2	0.3	-0.7	-0.6	0.3	1.0	-0.3
Clay	-0.2	0.2	0.3	0.2	-0.2	-0.6	1.0	-0.2	-0.1	0.0	-0.3	-0.3	-0.1	0.3	1.0	-0.9	-0.3	0.9
N	0.0	0.1	0.1	0.1	-0.3	0.3	-0.2	1.0	-0.2	-0.4	0.0	0.3	0.0	-0.4	-0.2	0.1	0.3	-0.1
P	-0.2	-0.2	-0.1	0.0	0.2	-0.1	-0.1	-0.2	1.0	0.3	0.1	-0.1	-0.3	0.3	-0.1	0.1	-0.2	-0.1
K	0.2	-0.4	-0.4	-0.3	0.2	-0.1	0.0	-0.4	0.3	1.0	0.4	-0.3	-0.1	0.3	0.0	0.1	-0.1	-0.1
L	0.2	-0.2	-0.3	-0.2	-0.1	0.3	-0.3	0.0	0.1	0.4	1.0	-0.1	0.4	-0.1	-0.3	0.3	0.3	-0.3
OM	0.0	-0.1	-0.1	-0.2	0.1	0.2	-0.3	0.3	-0.1	-0.3	-0.1	1.0	0.1	-0.5	-0.3	0.3	0.1	-0.3
pH	0.1	0.1	-0.2	-0.1	-0.3	0.3	-0.1	0.0	-0.3	-0.1	0.4	0.1	1.0	-0.3	-0.1	0.0	0.3	0.0
EC	-0.1	0.0	0.0	0.1	0.5	-0.7	0.3	-0.4	0.3	0.3	-0.1	-0.5	-0.3	1.0	0.4	-0.2	-0.7	0.2
WP	-0.2	0.2	0.2	0.2	-0.2	-0.6	1.0	-0.2	-0.1	0.0	-0.3	-0.3	-0.1	0.4	1.0	-0.9	-0.3	0.9
BD	0.2	-0.2	-0.2	-0.2	0.5	0.3	-0.9	0.1	0.1	0.1	0.3	0.3	0.0	-0.2	-0.9	1.0	0.1	-1.0
AW	0.2	-0.1	-0.2	-0.3	-0.9	1.0	-0.3	0.3	-0.2	-0.1	0.3	0.1	0.3	-0.7	-0.3	0.1	1.0	-0.1
SP	-0.2	0.2	0.2	0.2	-0.5	-0.3	0.9	-0.1	-0.1	-0.1	-0.3	-0.3	0.0	0.2	0.9	-1.0	-0.1	1.0

Figure 6. Correlation heat map showing the relationship among the environmental variables of mining and non-mining sites. Note: Legends are same as that of Table 2.

3.3. Species-Environment Correlation

Canonical correspondence analysis explains the total variance of 20.9% on the ordination axes having the highest variance on axis 1 (8.9%). Similarly, the Pearson correlation on axis 1 (0.99) is higher than on axis 2 and 3. In contrast, the Kendall correlation is greater on axis three than on axis 2 and 1 (Table 3). In the axis's correlation, geographic variable i.e., elevation, latitude and longitude on axis 1 shows significant negative correlation ($r = -0.95$, $r = -0.80$, $r = -0.75$ respectively, at $p < 0.01$). Similarly, aspect angle, and pH shows significant negative correlation on axis 3 with $p < 0.05$; $r = -0.33$, and $p < 0.01$; $r = -0.54$, respectively. Similarly, soil nutrient i.e., Potassium ($p < 0.05$; $r = -0.35$) on axis 1 and Lime on axis 3 ($p < 0.05$; $r = -0.25$) shows significant correlation indicating that these factors plays important role in shaping the communities.

Table 3. Canonical correspondence analysis axes summary, correlation and biplot scores of the environmental variables.

Axes					**Axis 1**	**Axis 2**	**Axis 3**
Eigenvalue					0.701	0.504	0.438
Species data variance							
Variance explained (%)					8.9	6.4	5.6
Cumulative variance (%)					8.9	15.3	20.9
Correlation (Pearson). Spp-Envt					0.995	0.969	0.989
Correlation (Kendall). Spp-Envt					0.825	0.847	0.899
Variable		**Correlation**			**Biplot scores**		
		Axis 1	**Axis 2**	**Axis 3**	**Axis 1**	**Axis 2**	**Axis 3**
1	AA	−0.188	0.02	−0.331	−0.157	0.014	−0.224
2	ELE	0.955	−0.192	−0.034	0.795	−0.133	−0.023
3	Lat.	0.803	−0.436	0.066	0.669	−0.302	0.045
4	Long	0.752	−0.417	0.199	0.626	−0.289	0.134
5	Sand	−0.003	0.03	0.07	−0.003	0.021	0.047
6	Silt	−0.111	−0.024	−0.302	−0.092	−0.017	−0.204

Table 3. *Cont.*

Variable		Correlation Axis 1	Axis 2	Axis 3	Biplot scores Axis 1	Axis 2	Axis 3
7	Clay	0.145	−0.016	0.291	0.121	−0.011	0.196
8	N	0.146	0.125	0.268	0.122	0.087	0.181
9	P	−0.188	0.076	0.383	−0.156	0.053	0.258
10	K	−0.351	0.262	−0.03	−0.292	0.181	−0.02
11	L	−0.188	0.032	−0.255	−0.156	0.022	−0.172
12	OM	−0.003	−0.063	−0.11	−0.002	−0.043	−0.074
13	pH	0.147	0.207	−0.538	0.122	0.143	−0.363
14	EC	−0.04	−0.036	0.352	−0.034	−0.025	0.238
15	WP	0.115	−0.025	0.265	0.095	−0.017	0.179
16	BD	−0.184	−0.001	−0.316	−0.153	0	−0.213
17	AW	−0.091	−0.029	−0.252	−0.076	−0.02	−0.17
18	SP	0.176	0	0.313	0.147	0	0.211

Note: Legends are the same as that in Table 2.

Elevation, latitude, and longitude in axis 1 have negative biplot scores of −0.521, −0.464, and −0.433, respectively. The biplot shows that critical factors influencing the community structure and composition are potassium, electrical conductivity, aspect angle, pH, and silt (Figure 7).

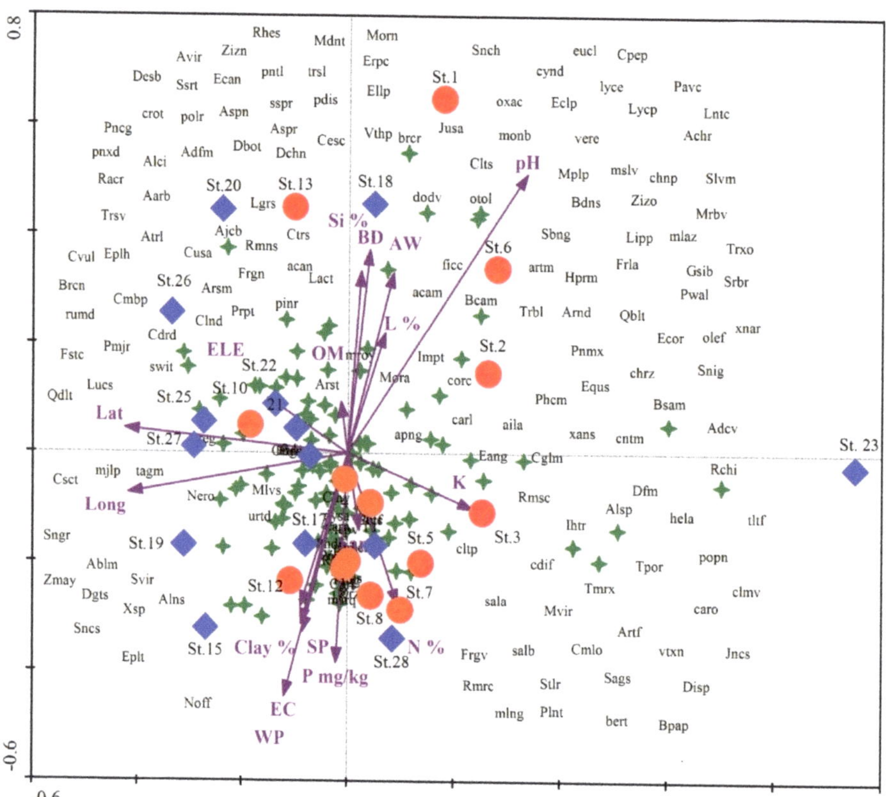

Figure 7. CCA-ordination and biplot showing vegetation distribution in 30 stands along the Panjkora river basin. Note: St. (Stand number); Environmental variables legends are same as that in Table 2; Species codes are same as that in Table S2.

4. Discussion

The present research analyzed the vegetation of riparian locations throughout the Panjkora River basin and compared mining and non-mining vegetation to determine the impacts of mining on the vegetation. The qualitative analysis of flora and its composition provides an important source of information on the dynamics of ecological changes on a geographical and temporal scale [44]. The biological spectrums of flora depict the climatic conditions and possible environmental influences that might alter the structure and composition of the community [45]. The vegetation was diversified with 186 plant species and a broad floristic spectrum. Similarly, 215 species in the riparian vegetation throughout the Hindu Kush Mountains Range was recorded by [4].

In contrast, 105 species from mining sites in four districts of Khyber Pakhtunkhwa, Pakistan was reported by [46]. The overall sum of species included in the study was greater than the sum of species listed by other studies; including [47] recorded 92 species from a mid-elevation forest in the Central Himalayas of India, while [48] reported 123 species from a tropical forest in Manipur, north India. Similarly, 85 species from Dulahazara Garjan, Bangladesh by [49], and investigators in tropical forests in various regions of the globe, for example, 89 species were reported by [50], 52 species were recorded by [51], and 94 species was reported by [52]. Most plant species in this research demonstrate the transmissible distribution of vegetation, which is thought to be the fundamental characteristic of natural vegetation [53].

Native species such *C. dactylon*, *Mentha longifolia* (L.) Huds, and *Dodonaea viscosa* (L.) Jacq predominates in the communities, whereas invasive species with greater IVI include *Xanthium strumarium* L. and *Cannabis sativa* L. The species that dominated were mostly therophytes, which make up 37.5 percent of the flora, and may damage the native plants, as revealed by [54]. The communities were also dominated by perennial and annual plant species, which is in contrast to opposing views expressed in the literature that (1) annuals, in particular, have strong effects on community homogenization [55] and (2) perennials promote community disturbance and favour alien [56,57] species, which are found in riparian communities. Smaller herb species like *X. strumarium*, *M. longifolia*, *D. viscosa*, and *C. sativa* may be at a disadvantage versus tree plant species like *Populus nigra* L. and *Pinus roxburghii* Sargent, which will eventually displace those smaller herbs [58]. This displacement eventually results in disturbed communities and will pose a risk to vegetation diversity in the future.

Understanding the ecological basis of vegetation is aided by the spectrum of diverse life forms, which depicts the physiognomy of the flora and vegetation, which is the final product of all living activities coupled with the environment [59]. According to Figure 1, therophytes, with 87 species (37.5%), predominated in the region, followed by hemicryptophytes, with 16 species (10%). The prevalence of therophytes suggests that this landform experiences frequent, minor flood disturbances and terrain modification, which has limited the occurrence of woody vegetation (phanerophytes) and forced them to relocate to riparian slopes and uplands with minimal flood disturbance. Most therophytes are annual plants that reproduce in the floodplain and marshes in the next season after surviving through the adverse season via seeds or spores. The relatively high proportion of therophytes in the riparian vegetation demonstrated that the landscape was often changed by human and natural disturbances, which encourages therophytic types of vegetation [60].

Similarly, a phytosociological study of the vegetation of Hayatabad, Peshawar conducted by [61], stated that therophytes were the leading class over others, supporting our findings. We reported therophytes followed by hemicryptophytes from the area. Because they are intimately tied to the succession pattern and life forms to the geomorphology of the river, riparian plant life forms and life history strategies are crucial to preserving these ecosystems [40]. Documenting functional vegetation's life cycle methods is also important for the ecosystem's non-native vegetation.

The area's leaf size spectrum was dominated by microphyll (36.02%), followed by manophylls, and progressed gradually with a declining percentage, with only one aphyllous species. A key indicator of climate change is leaf physiology, which is strongly affected by temperature and precipitation [62]. Nanophyll was listed as the area's major leaf size class by [63], which virtually supports our conclusion. Following our results, [64] reported that nanophylls were the dominant leaf size class, followed by Microphyll, leptophylls, and mesophyll at the University of Peshawar Botanical Garden in Azhakhel, Nowshehra. Leaf size spectra of different altitudinal zones indicated that microphylls steadily expand from lower to higher zones [65]. Similar to findings when microphylls dominate the zone at greater altitudes, nanophylls were somewhat higher in the lower zone. Microphylls were the predominant leaf size, followed by nanophylls in springs and monsoon, according to [40], who studied the leaf size spectra of Ganga Chotti and Bedori Hills. This information strengthens our current analysis since it considers the same altitudinal fluctuation. Our current results are supported by [41] study of the biological properties of plants in Tehsil Takht-e-Nasrati, which noted that microphylls were the predominant leaf size with 52.8 percent of species, followed by nanophylls (19.88 percent). The floral variety of the Frontier Region was recently explained by [66] and agreed with our results.

Species richness, which is identified as the total of various species in a particular region, is considered an important characteristic of a plant community [67]. In the mining site community in this study, the Shannon–Wiener index (H′) ranged from 2.14 to 2.87, while in the non-mining community, its value ranged from 2.72 to 3.39. In mining site communities, the equitability or evenness index (J) varied from 0.86 to 0.95, but in non-mining site communities, it ranged from 0.83 to 0.96. If the value of (H′) is high, the community should be more diverse. The H value of a community containing only one species should be zero (0) because the value of Pi should be equal to 1 and multiplied by Pi, which should equal zero (0). When the evenness is high, the H value will also be high. So, the H value not only indicates the number of species but also shows us the abundance of the species distributions in the total number of species among the whole community. Hence, the diversity index value of this study falls within the range reported for tropical forests [48,68]. Species richness along the river basin is influenced by edaphic, biotic and climatic factors [69]. Moreover, riparian forests connect different types of vegetation and adjacent flora, and soil and hydromorphic conditions create a spatial heterogeneity that favours species richness in these environments [11,70].

The link between the different vegetation indicators and the measured environmental variables was then ascertained using CCA. A permutation approach was used to assess the correlations between the canonical axes and the explanatory matrix and the importance of each species. The proposed associations between the response and explanatory factors were evaluated by normalizing the axis scores, focusing on the unit variance, and using axes scaled to best reflect each species. The findings indicate that the elements that support these communities include soil texture, potassium, aspect angle, and electrical conductivity. In both sites, the content and organization of the community were greatly influenced by the soils. The site soils were rich in organic matter, which is associated with high soil fertility. These soils also enable water and air to pass through them, allowing roots to penetrate more easily and supplying plants with nutrients and clay aggregation stability [71] revealing that the properties of the soil might explain the distribution of vegetation at local scales. The communities were found to be dominated by certain invasive generalist species, including *C. sativa*, *X. strumarium*, *P. nigra*, and *C. dactylon*, pioneer species of disturbed vegetation [72].

5. Conclusions

Riparian vegetation is considered the health indicator of aquatic vegetation along riparian corridors. The floristic diversity was dominated by perennials and therophytes that indicated the anthropogenic pressure, which mainly contributed to mining activities in the area. These floristic indicators provide baseline information for designing strategies for conserving and managing the riparian vegetation affected by the mining activities. Additionally, identifying plant communities' particular to the various mineral zones may serve as a foundation for phytoremediation strategies for mine waste rehabilitation. The mechanisms underlying these vegetation variations would benefit from further investigation. These mechanisms may be related to the preferential uptake or tolerance of specific soil minerals, such as heavy metals, as well as variations in other soil properties such as pH, water-holding capacity, and the availability of macronutrient elements. The current research results also show that soil characteristics affect species richness, structure, dominance, and establishment patterns and that ecological character can also affect the composition. In these study regions, soil fertility and physical attributes may be limiting element that serves as an environmental filter for establishing species. To fully comprehend the relationship between riparian vegetation and its socioeconomic effect, we recommend comparing riparian vegetation's socioeconomic impact at mining and non-mining sites with livelihood assessments of residents.

6. Limitations

The current study focuses on floristic compositions, vegetation characteristics, and their relationships with the environmental factors that play a vital role in shaping plant communities and provide baseline information for conservation and preservation. Despite the importance of the study, there are still some limitations to the current research, which can be beneficial for the development of future research in the study area. The present study lacks information about the socioeconomic impacts of the mining operations on the livelihoods of the local people, which is important aspect and, most of the time, creates hurdles in the restoration of mining vegetation. In addition, the possible restoration process and mechanism need further comprehensive evaluation for better utilization of the information in rehabilitating vegetation on the mining sites. Moreover, riparian vegetation can promote the growth of stress-resistant plants. Identifying such plants and their possible stress-resistance potential can better provide information about the utilization of mining flora for the possible remediation and accumulation of toxic pollutants.

Supplementary Materials: The following supporting information can be downloaded at: https://www.mdpi.com/article/10.3390/land11101801/s1, Table S1: Check List of plant species along with taxonomic attributes (Status in Pakistan, family, Raunkier life form and Leaf size classes); Table S2: Showing the IVI and Mean Standard error of 186 plant species found in both mining (Group I) and non-mining sites (Group II).

Author Contributions: Conceptualization, by K.A. and N.K.; methodology, by R.U.; software, by M.S.; validation and Data curation, by M.E.H.K. and M.D.; formal revision and validation, by D.A.J. All authors have read and agreed to the published version of the manuscript.

Funding: No funding was received for this research.

Institutional Review Board Statement: Not applicable.

Informed Consent Statement: Not applicable.

Data Availability Statement: All the research data are presented in the manuscript and are available.

Conflicts of Interest: The authors declare no conflict of interest.

References

1. Khan, S.A.; Khan, S.M.; Ullah, Z.; Ahmad, Z.; Alam, N.; Shah, S.N.; Khan, R.; Zada, M. Phytogeographic classification using multivariate approach; a case study from the Jambil Valley Swat, Pakistan. *Pak. J. Bot.* **2020**, *52*, 279–290. [CrossRef]
2. Teixeira, A.M.C.; Pinto, J.R.R.; Amaral, A.G.; Munhoz, C.B.R. Angiosperm species of "Cerrado" sensu stricto in Terra Ronca State Park, Brazil: Floristics, phytogeography and conservation. *Braz. J. Bot.* **2017**, *40*, 225–234. [CrossRef]
3. Mota, G.S.; Luz, G.R.; Mota, N.M.; Coutinho, E.S.; Veloso, M.d.D.M.; Fernandes, G.W.; Nunes, Y.R.F. Changes in species composition, vegetation structure, and life forms along an altitudinal gradient of rupestrian grasslands in south-eastern Brazil. *Flora* **2018**, *238*, 32–42. [CrossRef]
4. Zeb, S.A.; Khan, S.M.; Ahmad, Z. Phytogeographic elements and vegetation along the river Panjkora-Classification and ordination studies from the Hindu Kush Mountains range. *Bot. Rev.* **2021**, *87*, 518–542. [CrossRef]
5. Gashaw, T.; Tulu, T.; Argaw, M.; Worqlul, A.W. Modeling the hydrological impacts of land use/land cover changes in the Andassa watershed, Blue Nile Basin, Ethiopia. *Sci. Total Environ.* **2018**, *619*, 1394–1408. [CrossRef]

6. Bejarano, M.D.; Jansson, R.; Nilsson, C. The effects of hydropeaking on riverine plants: A review. *Biol. Rev.* **2018**, *93*, 658–673. [CrossRef]
7. Meragiaw, M.; Woldu, Z.; Martinsen, V.; Singh, B.R. Woody species composition and diversity of riparian vegetation along the Walga River, Southwestern Ethiopia. *PLoS ONE* **2018**, *13*, e0204733. [CrossRef]
8. Sunil, C.; Somashekar, R.K.; Nagaraja, B.C. Diversity and composition of riparian vegetation across forest and agro-ecosystem landscapes of river Cauvery, southern India. *Trop. Ecol.* **2016**, *57*, 343–354.
9. Njue, N.; Koech, E.; Hitimana, J.; Sirmah, P. Influence of land use activities on riparian vegetation, soil and water quality: An indicator of biodiversity loss, South West Mau Forest, Kenya. *Open J. For.* **2016**, *6*, 373–385. [CrossRef]
10. Dietrich, A.L.; Nilsson, C.; Jansson, R. Restoration effects on germination and survival of plants in the riparian zone: A phytometer study. *Plant Ecol.* **2015**, *216*, 465–477. [CrossRef]
11. Corenblit, D.; Davies, N.S.; Steiger, J.; Gibling, M.R.; Bornette, G. Considering river structure and stability in the light of evolution: Feedbacks between riparian vegetation and hydrogeomorphology. *Earth Surf. Process. Landf.* **2015**, *40*, 189–207. [CrossRef]
12. Caiafa, A.N.; Martins, F.R. Taxonomic identification, sampling methods, and minimum size of the tree sampled: Implications and perspectives for studies in the Brazilian Atlantic Rainforest. *Funct. Ecosyst. Communities* **2007**, *1*, 95–104.
13. Arruda, D.M.; Ferreira-Junior, W.G.; Duque-Brasil, R.; Schaefer, C.E. Phytogeographical patterns of dry forests sensu stricto in northern Minas Gerais State, Brazil. *An. Acad. Bras. Ciências* **2013**, *85*, 623–634. [CrossRef]
14. De Azevedo, I.F.P.; Nunes, Y.R.F.; de Avila, M.A.; da Silva, D.L.; Fernandes, G.W.; Veloso, R.B. Phenology of riparian tree species in a transitional region in southeastern Brazil. *Braz. J. Bot.* **2014**, *37*, 47–59. [CrossRef]
15. Rodrigues, P.M.S.; Azevedo, I.; Veloso, M.; Santos, R.; Menino, G.; Nunes, Y.; Fernandes, G. Riqueza Florística da Vegetação ciliar do rio Pandeiros, norte de Minas Gerais. *MG. Biota* **2009**, *2*, 18–35.
16. Bennett, A.F.; Nimmo, D.G.; Radford, J.Q. Riparian vegetation has disproportionate benefits for landscape-scale conservation of woodland birds in highly modified environments. *J. Appl. Ecol.* **2014**, *51*, 514–523. [CrossRef]
17. Purnomo, M.; Utomo, M.R.; Pertiwi, V.i.A.; Laili, F.; Pariasa, I.I.; Riyanto, S.; Andriatmoko, N.D.; Handono, S.Y. Resistance to mining and adaptation of Indonesia farmer's household to economic vulnerability of small scale sand mining activities. *Local Environ.* **2021**, *26*, 1498–1511. [CrossRef]
18. Peduzzi, P. Sand, rarer than one thinks. *Environ. Dev.* **2014**, *11*, 208–218.
19. Ikhsan, J.; Rezanaldy, A.; Rozainy, M.Z.M.R. Analysis of Sand Mining Impacts on Riverbed in the Downstream of the Progo River, Indonesia. *IOP Conf. Ser. Mater. Sci. Eng.* **2021**, *1144*, 012065. [CrossRef]
20. Saviour, M.N. Environmental impact of soil and sand mining: A review. *Int. J. Sci. Environ. Technol.* **2012**, *1*, 125–134.
21. Ahmad, M.S.A.; Ashraf, M.; Ali, Q. Soil salinity as a selection pressure is a key determinant for the evolution of salt tolerance in Blue Panicgrass (*Panicum antidotale* Retz.). *Flora-Morphol. Distrib. Funct. Ecol. Plants* **2010**, *205*, 37–45. [CrossRef]
22. Mtero, F. Rural livelihoods, large-scale mining and agrarian change in Mapela, Limpopo, South Africa. *Resour. Policy* **2017**, *53*, 190–200. [CrossRef]
23. Bhatawdekar, R.M.; Singh, T.N.; Tonnizam Mohamad, E.; Armaghani, D.J.; Binti Abang Hasbollah, D.Z. River sand mining vis a vis manufactured sand for sustainability. In Proceedings of the International Conference on Innovations for Sustainable and Responsible Mining, Hanoi, Vietnam, 15–17 October 2020; pp. 143–169.
24. Madeira, B.G.; Espírito-Santo, M.M.; Neto, S.D.Â.; Nunes, Y.R.; Arturo Sánchez Azofeifa, G.; Wilson Fernandes, G.; Quesada, M. Changes in tree and liana communities along a successional gradient in a tropical dry forest in south-eastern Brazil. In *Forest Ecology*; Springer: Berlin/Heidelberg, Germany, 2009; pp. 291–304.
25. Ullah, K.; Zhang, J. GIS-based flood hazard mapping using relative frequency ratio method: A case study of Panjkora River Basin, eastern Hindu Kush, Pakistan. *PLoS ONE* **2020**, *15*, e0229153. [CrossRef]
26. Ahmad, D. *Khyber Pakhtunkhwa State of the Environment*; Environmental Protection Agency, Government of Khyber Pakhtunkhwa: Peshawar, Pakistan, 2012.
27. Ali, A.; Khan, T.A.; Ahmad, S. Analysis of climate data of Khyber Pakhtunkhwa, Pakistan. *Int. Res. J. Eng. Technol.* **2018**, *5*, 4266–4282.
28. Curtis, J.T.; Mcintosh, R.P. The interrelations of certain analytic and synthetic phytosociological characters. *Ecology* **1950**, *31*, 434–455. [CrossRef]
29. Martínez-Falcón, A.P.; Zurita, G.A.; Ortega-Martínez, I.J.; Moreno, C.E. Populations and assemblages living on the edge: Dung beetles responses to forests-pasture ecotones. *PeerJ* **2018**, *6*, e6148. [CrossRef]
30. Ali, S.I.; Qaiser, M. A phytogeographical analysis of the phanerogams of Pakistan and Kashmir. *Proc. R. Soc. Edinb. Sect. B Biol. Sci.* **1986**, *89*, 89–101. [CrossRef]
31. Aeschimann, D.; Rasolofo, N.; Theurillat, J.-P. Analyse de la flore des Alpes. 1: Historique et biodiversité. *Candollea* **2011**, *66*, 27–55. [CrossRef]
32. Maan, I.; Kaur, A.; Singh, H.P.; Batish, D.R.; Kohli, R.K. Exotic avenue plantations turning foe: Invasive potential, distribution and impact of Broussonetia papyrifera in Chandigarh, India. *Urban For. Urban Green.* **2021**, *59*, 127010. [CrossRef]
33. Castillón, E.E.; Arévalo, J.R.; Quintanilla, J.Á.V.; Rodríguez, M.M.S.; Encina-Domínguez, J.A.; Rodríguez, H.G.; Ayala, C.M.C. Classification and ordination of main plant communities along an altitudinal gradient in the arid and temperate climates of northeastern Mexico. *Sci. Nat.* **2015**, *102*, 59. [CrossRef]

34. Kamrani, A.; Jalili, A.; Naqinezhad, A.; Attar, F.; Maassoumi, A.A.; Shaw, S.C. Relationships between environmental variables and vegetation across mountain wetland sites, N. Iran. *Biologia* **2011**, *66*, 76–87. [CrossRef]
35. Reis, A.T.; Coelho, J.P.; Rucandio, I.; Davidson, C.M.; Duarte, A.C.; Pereira, E. Thermo-desorption: A valid tool for mercury speciation in soils and sediments? *Geoderma* **2015**, *237*, 98–104. [CrossRef]
36. Nelson, D.W.; Sommers, L.E. Total carbon, organic carbon, and organic matter. *Methods Soil Anal. Part 3 Chem. Methods* **1996**, *5*, 961–1010.
37. Yadav, R.; Tomar, S.; Sharma, U. Output: Input ratios and apparent balances of N, P and K inputs in a rice-wheat system in North-West India. *Exp. Agric.* **2002**, *38*, 457–468. [CrossRef]
38. Robin, V.; Cuisinier, O.; Masrouri, F.; Javadi, A. Chemo-mechanical modelling of lime treated soils. *Appl. Clay Sci.* **2014**, *95*, 211–219. [CrossRef]
39. Saxton, K.; Rawls, W.J.; Romberger, J.S.; Papendick, R. Estimating generalized soil-water characteristics from texture. *Soil Sci. Soc. Am. J.* **1986**, *50*, 1031–1036. [CrossRef]
40. Malik, Z.; Hussain, F.; Malik, N. Life form and leaf size spectra of plant communities Harbouring Ganga Chotti and Bedori Hills during 1999–2000. *Int. J. Agric. Biol.* **2007**, *9*, 833–838.
41. Khan, M.; Hussain, F.; Musharaf, S. Biological characteristics of plant species in Tehsil Takht-e-Nasrati Pakistan. *J. Biodivers. Environ. Sci.* **2012**, *2*, 42–47.
42. Biondi, E.; Allegrezza, M.; Casavecchia, S.; Galdenzi, D.; Gasparri, R.; Pesaresi, S.; Soriano, P.; Tesei, G.; Blasi, C. New insight on Mediterranean and sub-Mediterranean syntaxa included in the Vegetation Prodrome of Italy. *Flora Mediterr.* **2015**, *25*, 77–102.
43. Barua, N.; Aziz, M.A.I.; Tareq, A.M.; Sayeed, M.A.; Alam, N.; ul Alam, N.; Uddin, M.A.; Lyzu, C.; Emran, T.B. In vivo and in vitro evaluation of pharmacological activities of Adenia trilobata (Roxb.). *Biochem. Biophys. Rep.* **2020**, *23*, 100772. [CrossRef]
44. Devi, K.; Singh, H.; Gupta, A. Life form in hill forest of Manipur, North-East, India. *J. Agroecol. Nat. Resour. Manag.* **2014**, *1*, 7–13.
45. Díaz, S.; Kattge, J.; Cornelissen, J.H.; Wright, I.J.; Lavorel, S.; Dray, S.; Reu, B.; Kleyer, M.; Wirth, C.; Colin Prentice, I. The global spectrum of plant form and function. *Nature* **2016**, *529*, 167–171. [CrossRef]
46. Khwaja, M.A.; Saeed, S.; Urooj, M. *Preliminary Environmental Impact Assessment (EIA) Study of China-Pakistan Economic Corridor (CPEC) Northern Route Road Construction Activities in Khyber Pakhtunkhwa (KPK), Pakistan*; Sustainable Development Policy Institute: Islamabad, Pakistan, 2018.
47. Khera, N.; Kumar, A.; Ram, J.; Tewari, A. Plant biodiversity assessment in relation to disturbances in mid-elevational forest of Central Himalaya, India. *Trop. Ecol.* **2001**, *42*, 83–95.
48. Devi, L.S.; Yadava, P. Floristic diversity assessment and vegetation analysis of tropical semievergreen forest of Manipur, north east India. *Trop. Ecol.* **2006**, *47*, 89–98.
49. Chowdhury, M.; Huda, M.; Islam, A. Phytodiversity of Dipterocarpus turbinatus Gaertn. F. (garjan) undergrowths at Dulahazara Garjan Forest, Cox's Bazar, Bangladesh. *Indian For.* **2000**, *126*, 674–684.
50. Kadavul, K.; Parthasarathy, N. Plant biodiversity and conservation of tropical semi-evergreen forest in the Shervarayan hills of Eastern Ghats, India. *Biodivers. Conserv.* **1999**, *8*, 419–437. [CrossRef]
51. Pande, P. Comparative vegetation analysis and sal (*Shorea robusta*) regeneration in relation to their disturbance magnitude in some sal forests. *Trop. Ecol.* **1999**, *40*, 51–61.
52. Fox, B.J.; Taylor, J.E.; Fox, M.D.; Williams, C. Vegetation changes across edges of rainforest remnants. *Biol. Conserv.* **1997**, *82*, 1–13. [CrossRef]
53. Ohlson, M.; Söderström, L.; Hörnberg, G.; Zackrisson, O.; Hermansson, J. Habitat qualities versus long-term continuity as determinants of biodiversity in boreal old-growth swamp forests. *Biol. Conserv.* **1997**, *81*, 221–231. [CrossRef]
54. Nautiyal, M.; Nautiyal, B.; Prakash, V. Phenology and growth form distribution in an alpine pasture at Tungnath, Garhwal, Himalaya. *Mt. Res. Dev.* **2001**, *21*, 168–174. [CrossRef]
55. Qian, H.; Guo, Q. Linking biotic homogenization to habitat type, invasiveness and growth form of naturalized alien plants in North America. *Divers. Distrib.* **2010**, *16*, 119–125. [CrossRef]
56. Tappeiner, J.; Zasada, J.; Ryan, P.; Newton, M. Salmonberry clonal and population structure: The basis for a persistent cover. *Ecology* **1991**, *72*, 609–618. [CrossRef]
57. Milchunas, D.G.; Sala, O.E.; Lauenroth, W.K. A generalized model of the effects of grazing by large herbivores on grassland community structure. *Am. Nat.* **1988**, *132*, 87–106. [CrossRef]
58. Timsina, B.; Shrestha, B.B.; Rokaya, M.B.; Münzbergová, Z. Impact of Parthenium hysterophorus L. invasion on plant species composition and soil properties of grassland communities in Nepal. *Flora-Morphol. Distrib. Funct. Ecol. Plants* **2011**, *206*, 233–240. [CrossRef]
59. Amjad, M.S.; Arshad, M.; Sadaf, H.M.; Akrim, F.; Arshad, A. Floristic composition, biological spectrum and conservation status of the vegetation in Nikyal valley, Azad Jammu and Kashmir. *Asian Pac. J. Trop. Dis.* **2016**, *6*, 63–69. [CrossRef]
60. Nazir, A.; Malik, Z. Life-form and index of similarity of plant communities recorded at Sarsawa Hills, District Kotli. *J. Res. Sci* **2006**, *17*, 27–33.
61. Shah, M.; Hussain, F. Phytosociological Study of the Vegetation of Hayat Abadpeshawar, Pakistan. *Pak. J. Plant Sci.* **2009**, *15*.
62. Traiser, C.; Klotz, S.; Uhl, D.; Mosbrugger, V. Environmental signals from leaves–a physiognomic analysis of European vegetation. *New Phytol.* **2005**, *166*, 465–484. [CrossRef]

63. Nasir, Z.A.; Sultan, S. Floristic, biological and leaf size spectra of weeds in gram, lentil, mustard and wheat fields of district Chakwal, Pakistan. *Pak. J. Biol. Sci. (Pak.)* **2002**, *5*, 758–762.
64. Hussain, T. A floristic description of flora and ethnobotany of Samahni Valley (AK), Pakistan. *Ethnobot. Leafl.* **2009**, *2009*, 6.
65. Samad, M.; Badshah, L.; Khan, S.M. Biological spectra of Lala Kalay area distrist Peshawar Khyber Pakhtunkhwa province Pakistan. *Pak. J. Weed Sci. Res.* **2018**, *24*, 353–362. [CrossRef]
66. Samreen, U.; Ibrar, M.; Badshah, L. Diversity and Ecological Characteristics of Weed Flora at Darazinda, Frontier Region Dera Ismail Khan Pakistan. *Pak. J. Weed Sci. Res.* **2018**, *24*, 223–229. [CrossRef]
67. McIntosh, R.P. An index of diversity and the relation of certain concepts to diversity. *Ecology* **1967**, *48*, 392–404. [CrossRef]
68. Singh, K.; Kumar, A.; Lal, B.; Todaria, N. Species diversity and population status of threatened plants in different landscape elements of the Rohtang Pass, Western Himalaya. *J. Mt. Sci.* **2008**, *5*, 73–83. [CrossRef]
69. Ayyappan, N.; Parthasarathy, N. Biodiversity inventory of trees in a large-scale permanent plot of tropical evergreen forest at Varagalaiar, Anamalais, Western Ghats, India. *Biodivers. Conserv.* **1999**, *8*, 1533–1554. [CrossRef]
70. Veloso, M.D.D.M.; Nunes, Y.R.F.; Azevedo, I.F.P.; Rodrigues, P.M.S.; Fernandes, L.A.; Santos, R.M.D.; Fernandes, G.W.; Pereira, J.A.A. Floristic and structural variations of the arboreal community in relation to soil properties in the Pandeiros river riparian forest, Minas Gerais, Brazil. *Interciencia* **2014**, *39*, 628–636.
71. De Lima Dantas, V.; Batalha, M.A. Vegetation structure: Fine scale relationships with soil in a cerrado site. *Flora-Morphol. Distrib. Funct. Ecol. Plants* **2011**, *206*, 341–346. [CrossRef]
72. Fagundes, N.C.A.; de Ávila, M.A.; de Souza, S.R.; de Azevedo, I.F.P.; Nunes, Y.R.F.; Fernandes, G.W.; Fernandes, L.A.; dos Santos, R.M.; Veloso, M.d.D.M. Riparian vegetation structure and soil variables in Pandeiros river, Brazil. *Rodriguésia* **2019**, *70*, e01822017. [CrossRef]

MDPI
St. Alban-Anlage 66
4052 Basel
Switzerland
www.mdpi.com

MDPI Books Editorial Office
E-mail: books@mdpi.com
www.mdpi.com/books

www.ingramcontent.com/pod-product-compliance
Lightning Source LLC
Chambersburg PA
CBHW061615220926
51850CB00005B/513
9783036592602